AF782411

BUILDING INTEGRATED PHOTOVOLTAIC THERMAL SYSTEMS

To Blue Engine LLC

Let us all work together towards

NET ZERO ENERGY

[illegible]

BUILDING INTEGRATED PHOTOVOLTAIC THERMAL SYSTEMS

Fundamentals, Designs and Applications

Huiming Yin

Department of Civil Engineering and Engineering Mechanics, Columbia University, NY, United States

Mehdi Zadshir

Department of Civil Engineering and Engineering Mechanics, Columbia University, NY, United States

Frank Pao

Sustainable Product Development Lab, NY, United States

ELSEVIER

ACADEMIC PRESS

An imprint of Elsevier

elsevier.com/books-and-journals

Academic Press is an imprint of Elsevier
125 London Wall, London EC2Y 5AS, United Kingdom
525 B Street, Suite 1650, San Diego, CA 92101, United States
50 Hampshire Street, 5th Floor, Cambridge, MA 02139, United States
The Boulevard, Langford Lane, Kidlington, Oxford OX5 1GB, United Kingdom

British Library Cataloguing-in-Publication Data
A catalogue record for this book is available from the British Library

Library of Congress Cataloging-in-Publication Data
A catalog record for this book is available from the Library of Congress

ISBN: 978-0-12-821064-2

For Information on all Academic Press publications visit our website at https://www.elsevier.com/books-and-journals

Publisher: Brian Romer
Acquisitions Editor: Lisa Reading
Editorial Project Manager: Hilary Carr
Production Project Manager: Sojan P. Pazhayattil
Cover Designer: Victoria Pearson

Typeset by Aptara, New Delhi, India

Contents

Preface

Explanation of the cover: the lighthouse serves as a guide to pilots the energy harvested by building integrated photovoltaic thermal (BIPVT) systems powers lighting, and communication devices as well as ocean farming. It bathes in sunlight, wind, and ocean wave, which exhibits great potentials to change it into an energy house through novel future designs. This book will be a guide to BIPVT designers, users, researchers, and learners.

When author Huiming Yin firstly met author Frank Pao at Columbia University, it was a cloudy day in September 2014. At that time, although the solar industry was still growing rapidly in the United States, it was significantly impacted by the reduction of oil prices and the price competition from the Chinese manufacturers. As the margin of solar panel manufacture was getting thinner, some manufacturers closed their businesses, and we could foresee the difficult time ahead for both researchers and entrepreneurs. Huiming was a researcher and the inventor of a novel building-integrated photovoltaic thermal (BIPVT) system, while Frank was an industry pioneer and veteran in building-integrated photovoltaic (BIPV) industry. When we introduced our work to each other, we were surprised that some innovations that we had were rooted in the same ideas and principles, even though we had never met before. Moreover, we have shared the same concerns on pollution and global warming, the same passion for clean energy and energy-efficient or independent buildings, and the same vision for holistic integration of solar energy harvesting into the infrastructure systems, such as buildings, transportation, and solar farms. Despite all short-term challenges and difficulties, we believe in the bright future of novel BIPVT technologies to fight for climate change and energy crisis by oil depletion.

Since Columbia has some of the first-class research facilities, and Frank owns the complete manufacturing line of solar panels, we can work and support each other in a complementary way. In 2017, we formed the National Science Foundation (NSF) Industry-University Cooperative Research Center—Center for Energy Harvesting Materials and Systems (CEHMS) at Columbia Site along with Virginia Tech as the original site. Since then, we have formed a list of projects on different aspects of BIPVT technologies. At the same time, we developed a new course of Energy Harvesting among CEHMS sites as another new site was formed at Pennsylvania State University in 2019. We found that it would be nice if we could have a book to comprehensively introduce the BIPVT technologies for training, education, and reference services of students, early stage professionals, and researchers. Although there are many books about solar energy on the market, we could not find a satisfactory one that covers our unique work and visions on BIPVT thus far. Therefore, we discussed the idea of writing a new book to cover both fundamentals and applications in this realm.

Once we truly started the preparation for writing this book, we found that it is a formidable task to reach a concrete writing plan to cover the depth and breadth of the dynamic research and technology areas. Very often, after we started a section and shifted to our other work when we came back, we found new ways to write the previous part. It is challenging to reach a convergent version. We needed help to manage and organize the writing process. Mehdi Zadshir is an eager learner and young researcher in BIPVT with the ambition of being a rising start in the BIPVT industry. Since he started his graduate studies at Columbia University in spring 2018, he has taken most courses on photovoltaic materials and technologies, participated in the CEHMS research projects on BIPVT, and exhibited solid understanding and great passion for BIPVT. Once we chatted about the challenges of this book-writing, Mehdi showed an enthusiastic desire to join this effort. Therefore, we formed this team of authors with three generations.

Frank Pao has been one of the pioneers to introduce and manufacture BIPVT in the United States and also a veteran semiconductor engineer and entrepreneur with more than 30 years of experience in semiconductors and photovoltaics. He has led numerous projects in solar module design and manufacturing and patented several BIPVT technologies. Huiming Yin has been an active researcher, a well-known expert, and a passionate educator in BIPVT materials and systems in the past decade. He serves as the Director of NSF Industry-University Cooperative Research Center at Columbia Site and principal investigators for a list of research and development projects in BIPVT sponsored by the US Department of Energy, NSF, and the industry. Mehdi Zadshir is an early stage researcher and entrepreneur in BIPVT. He has been the Entrepreneur Lead of the NSF I-Corps project and formed a small business to commercialize the BIPVT and other clean energy technologies that were developed in our research. His contribution to this book will provide a solid foundation for his career development.

The topics covered in this book are useful for scientists and engineers in the fields of mechanical, electrical, civil, material science, etc., as well as researchers and specialists in the areas of sustainable energy harvesting, solar, and renewable energy production. Also, it can serve the following audience:

1. senior undergraduate and graduate students.
2. businesses and industries related to innovative building components, and solar panel production, and

3. scholars in interdisciplinary areas such as energy-efficient building materials.

This book could be a useful reference to the wide spectrum of readers covering the fundamentals of BIPV/BIPVT to various applications, from the design to manufacture and installation, from quality control and certificate to long-term performance tests, from research to commercialization. With the rapid progress of BIPVT research and applications, many information resources can be found in the literature and the Internet at different levels of credibility and coverage. Because the BIPVT applications cover too many aspects, with a comprehensive understanding of the components to systems, we may find problems at the late stage of the applications, which lead to high costs of corrections. Although it is impossible for authors to cover every aspect and new technology of BIPVT, they introduce the fundamentals and design philosophy of BIPVT as well as reference resources, so that readers can easily find their answers and solutions starting with this book.

One of the main reasons for writing this book is our concern for the sustainable development of human civilization. Considering the scarcity of resources and the global population growth, it is vital to design and construct buildings that are enduring, well-insulated, and energy-efficient while also being sturdy to withstand high winds, snowstorms, and fire. Since the initial installation of the Sunslate by Frank Pao's company, which is one of the first BIPV roofing products in the country, it has hardly had any issues with adverse weather conditions even after 20 years. This level of credibility applies to other solar installations of his company as well, in Montana, Aspen, Co., Jones Beach, NY, Orlando, FL, and Jamaica. Despite going through heavy snowstorms and hurricanes, currently, they are all in great condition. We will use those projects for case studies to bring audiences to real-world applications. As BIPVT is a dynamic industry and research area, the book cannot cover all aspects the technologies, but aims to provide a solid foundation, building blocks, or systemic framework for future development.

Dr. Huiming Yin in January 2021

It has been a great honor and pleasure to have the opportunity to work with Prof. Huiming Yin and Mehdi Zadshir. I am writing this book in honor and memory of my parents. As an infant in war-torn Shanghai, malnutrition was a serious issue that resulted in affecting my health throughout my life. Growing up, I had problems with school work. Though my father was demanding, he constantly encouraged me to try harder and never quit. It is how that I managed to get into the top schools in Hong Kong and the United States. Luckily, my parents have been successful which has given me the opportunity to do the same. What happened to the unlucky ones born at the same time? The oil war started back in World II and to this day, it has not ended yet. The detriment of war to human suffering is unfathomable.

Silicon is in abundance and every nation large and small could produce silicon. No nations have to depend on others for energy. Why do not we start to it and became more friendly and work together to solve climate change? We all worked together to produce the COVID-19 vaccine and we should all work together to solve climate change. Finally, I would like to thank Dr. Patrick Hofer, CEO of 3S Solar Plus, who has been given me advice throughout the years, Mr. Joe Morrissey, Director of Aesthetic Green Power Corporation, who has provided a number of Atlantis project information, Prof. Ronald Lasser at Tufts University for supervising the interns at the New Hampshire BIPVT experiment, Mr. Jonathan Philips, Managing Director of Anka Funds for providing information on the Cherokee Home, and Senator Marc R. Pacheco of Massachusetts, who has been instrumental in passing the 2008 Global Warming Solution Act and 2021 Net Zero Energy Act which recommends applying cutting edge technologies.

Frank Pao in March 2021

Acknowledgments

During the preparation of this book, we have received help and support from our colleagues, students, collaborators, and customers. First, we thank Dr. Pablo Prieto-Muñoz, Mr. Michael Lackey, Dr. Po-Hua Lee, Dr. Zifeng Yuan, Dr. Dajiang Yang, Dr. Fangliang Chen, Dr. Liang Wang, Dr. Yingjie Liu, Dr. Gan Song, Dr. Qiliang Lin, Dr. Liangliang Zhang, Dr. Zhenhua Wei, Dr. Chunlin Wu, Dr. Xiaokong Yu, and Dr. Junhui Lou Lou from Dr. Yin's research team for their contribution to BIPVT research at Columbia University, Dr. Qiliang Liu, Dr. Liming Li, Mr. Freddie Wheeler, Dr. William Hunnicutt, and Dr. Adrian Brugger for their assistance to the laboratory work, as well as industry collaborators Greg Kelly, Raymond Daddazio, Jacques Garant, Joe Morrissey, Bill James, and Sunil Shah for their support to industry university cooperative research, and academic collaborators Prof. Yuan Yang, Dr. Vasilis Fthenakis, Dr. Charlie Gay, Prof. Michael Bell, Prof. Anton Falkeis, Prof. Rene Testa, Prof. Irving Herman, and Prof. Lei Zuo for the inspiring discussion with them. Some parts of this book are based on the publications and collaborative work with them. We are also grateful to the following students for their help with some of the figures: Mr. Tengxiang Wang, Mr. Arnaud Edmond Van Mieghem, Mr. Yanchu Zhang, Ms. Jessie Ruixuan Yan, Ms. Guillermina Maria Peragallo Ramonde, Mr. Roland Yan, and Mr. Henry Xu, as well as Ms. Linda Getachew Teka for proofreading the chapters.

We also would like to thank Dr. Todd J. Duellman for his contribution to sharing images and information on the net-zero house case study, Mr. William James, CEO of Jpods Inc. for providing valuable information on the personal rapid transit systems, Mr. Sylvain Cote, the architect and builder who creatively turned the 83 year-old house into "The Beauty and the Beach" with the latest green technologies and shared the information about this project with us, Mr. Jonathan Phillips, Managing Director of Anka Funds for directing Atlantis to install the BIPVT roofing system on Cherokee Home and later on providing information on the project, Mr. George Bialecki, President of Alternative Living Foundation, Dr. Tao, Prof. of Florida International University, and the late Louis Demitri who all worked tirelessly in making Future House USA a great success, and Mr. Henry Vandermark, President/Founder of Solar Wave who provided guidance on BIPVT monitoring. Finally, we would like to thank the staff of Atlantis, 3S and Golden Solar working tirelessly in contributing to the success of the solar system at the Water & Life Museum in Hemet, California, which got the Iconic Award, Public Architecture with Distinction by German Design Council 2016.

Specially, we thank Ms. Grace Yin for her brilliant artwork of illustrations, which make the book much interactive and more appealing to read. We highly appreciate the support of our families that made this time-consuming work possible. Finally, we acknowledge the financial support from the National Science Foundation (NSF) Divisions of Civil, Mechanical and Manufacturing Innovation and Industrial Innovation and Partnerships, Department of Energy Small Business Technology Transfer, Department of Agriculture National Institute of Food and Agriculture. Particularly, the authors thank the researchers and industry members of NSF IUCRC Center for Energy Harvesting Materials and Systems. Many ideas were inspired by the discussion and collaboration through the research project meetings and workshops.

Dedication

Lovingly dedicated to my wife Yali Liu, whose support, consideration, and encouragement made this book possible.

—Huiming Yin

Though plastics serve many purposes, single-use plastics have taken over our daily lives and our environment. Half of all plastics are purposed for single-use, and about 300 million tons of plastic pollution is generated every year. Plastics take thousands of years to decompose and at the current usage rate, plastic pollution has become a problem that is growing exponentially. Through the plastic trash breaking out of the crest of the wave as the surfer surfs on a water bottle, this piece attempts to show the complicit acceptance toward our pollution. This is a call for change that we shall use, manage, and maintain the resources appropriately–rely on renewables, reuse, and recycle raw materials, extend the lifetime of infrastructure, and reduce our carbon footprint–with the healthy, sustainable, and affordable development for next generations. Renewable clean energy harvesting can be the first step to power this huge change that Mother Nature is forcing us to do.

CHAPTER

1

Introduction

1.1 Background

One can think of many ills that beset our environment and infrastructure as consequences of over/ill-utilization of resources in the sense that usage surpasses what mother nature yields. Climate change, environmental deterioration, and energy depletion have called significant attention to renovate the existing civil engineering practice and promote the sustainability of infrastructures. This is not just a societal challenge; it is also a technological challenge of monumental difficulty. It involves groundbreaking research from the component level to the mega-city system level. As is well known, buildings and civil infrastructure, as the carriers for most human activities, consume the majority of all energy and natural resources, and produce significant waste and pollutions on the earth. New construction standards should evaluate the input and output of mass and energy flow in material production, building construction, and system operation for life cycle sustainability. Therefore, the life cycle cost and performance have to be predicted in the design and construction phase.

Fig. 1.1 illustrates the life cycle of an infrastructure system with the boundary limited to its service life. The sustainability of the system can

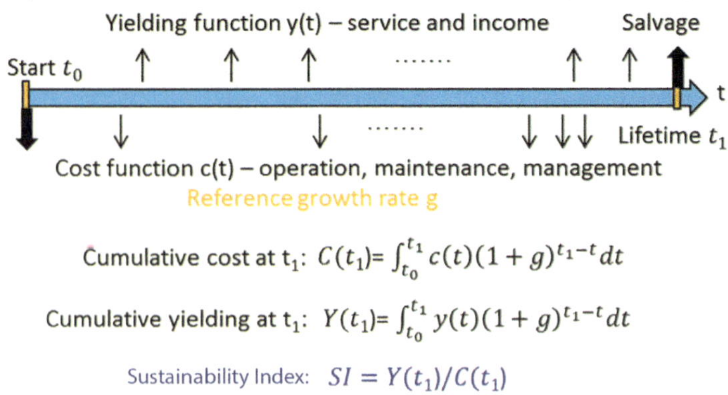

FIGURE 1.1 Quantitative sustainability of an infrastructure system in its lifecycle.

Building Integrated Photovoltaic Thermal Systems: Fundamentals, Designs and Applications
DOI: https://doi.org/10.1016/B978-0-12-821064-2.00011-0

be indicated by a sustainability index (SI), i.e., the ratio of the cumulative yielding and the cost at the end of its lifetime. The cumulative yielding may include the service value and income of the system, its salvage value, and the reusability of the building materials components for future infrastructure development; whereas the cumulative cost includes the initial investment of system construction and installation, the operation, maintenance, and management cost. A reference growth rate ($g \geq 1$) can be applied to reflect the time factor. The measurements, i.e., integral variables c(t) and y(t) can be counted through the economic models considering energy, materials, or environmental impacts, among others. The ratio of the cumulative yielding to cost shows sustainability.

Because energy has become one of the most vital factors in industrial economics that greatly affect others, we may use it as a measurement. From its definition, we can see five ways to maximize the SI of a system: (1) increase energy efficiency for lower energy consumption; (2) increase energy harvesting; (3) recycle or reuse the materials for higher salvage value; (4) extend the lifetime for larger integral limits, and; (5) maximize usage of renewable natural resources and environmental benign reagents.

However, our existing infrastructure has been mainly designed and built with contemporary requirements and costs. Due to the high speed of urbanization and population expansion in certain areas, the addition of new construction in a built environment is not trivial while the existing infrastructure is constantly subjected to increasing demand. Any major disruption, caused by either natural or man-made actions, could have a strong impact on a large part of our nation. Therefore, protection of existing civil infrastructure and enhancement of their performance including buildings, roads, highways, bridges, pipelines, and others become of extreme importance from a national security point of view.

The durability of civil infrastructure is most devastatingly affected by solar radiation, temperature, moisture, and other environmental conditions that lead to aging and degradation observed in corrosion, cracking, and spalling of surface materials in these infrastructure elements. New technologies to renew infrastructure by surface engineering with solar energy harvesting can address the above problems in a complementary way: harvesting solar energy by a durable surface provides self-supplied clean energy to the infrastructure and protects the infrastructure from aging and degradation. The main approach is to develop durable, multifunctional, smart skins, or envelope systems, which can harvest energy and exhibit high energy efficiency.

For a new technology to be technically effective and economically feasible, a holistic approach is essential to assure that the interactions and potential synergies between the various physical and chemical processes are properly understood, modeled, and where possible, exploited, while eliminating or minimizing any potentially detrimental consequences or interactions. To be specific, the overall approach toward building, protection, and retrofit of sustainable infrastructure has to start with recognizing and specifying the various performance requirements, such as structural functions, architectural requirements, aesthetic appearance, mechanical strength and durability, thermal efficiency, chemical and physical compatibility, sound absorption, moisture migration, and material recyclability. To integrate sustainable development into the building industry, the technologies should be carefully evaluated regarding their impacts on the environment, economy, social well-being, technology advancement, and performance management.

Although this book focuses on building skins or envelope systems, the design and installation methods can be easily extended to other civil infrastructure systems, such as roads, bridges, parking lots, greenhouses, etc. Buildings nationwide consume >70% of electricity, and ~40% of all energy, and a significant percentage of nonrenewable natural resources and nonrecyclable building materials, and also generated ~40% of greenhouse gas emissions, and 30% of waste output, which contributes to global warming and environmental pollutions. To make dramatic improvements with regard to conserving energy and natural resources, and improving the energy efficiency of buildings, it is necessary to revisit the way building envelopes are designed and manufactured.

The future advanced building, protection, and retrofit technologies should address the critical issues in safety, resilience, energy, and environment through infrastructure, and meet the following needs:

1. The advanced protection technologies should protect the infrastructure from the increasing environmental and man-made invasion, provide self-monitoring mechanisms to sense the performance and safety of the infrastructure, and prolong the lifetime of the infrastructure under the devastating load conditions. In emergent conditions, they should be able to provide prompt warnings to avoid further loss and malfunctions.
2. The new retrofit technologies should maximize the reuse of existing materials, extend the functions of structures for multidimensional benefits, and reshape the infrastructure to be adaptive to climate change and extreme load conditions. Particularly, energy efficiency and yielding of building envelope should be prioritized as there is great potential to significantly improve the energy performance in existing buildings.
3. The new building technologies should minimize energy and other resource consumption as well as environmental footprint, consider net-zero or near-zero energy technologies, provide smart control and management mechanisms while satisfying all performance specifications,

and still being economically feasible and sustainable in all stages of its life-cycle.

Novel multifunctional building envelope systems could provide one comprehensive solution to cover the needs for infrastructure design, construction, protection, retrofit. It can be useful for both building retrofitting and new construction.

In past decades, solar energy harvesting has evolved as a promising solution for energy and environmental challenges and global warming threats (Crabtree and Lewis, 2007; Lewis and Crabtree, 2005; U.S.P.I.R.S. Committee, 2001). However, some constraints, such as large land use, energy storage and transmission, and high initial investment, have imposed bottlenecks for future applications of these technologies (Hegedus, 2006; Hasnain, 1998). Considering most energy will be used in civil infrastructure, we can directly use the infrastructure surface to harvest energy, immediately provide energy supply to the infrastructure reducing the large expense in energy storage and transmission, and leverage the first investment of the solar system by the synergistic benefits to the infrastructure.

Taking a holistic view of the building envelope and incorporating solar energy technologies within the envelope systems will overhaul the way we are looking at the building technologies today and ultimately lead to a strategy to achieve zero net energy and sustainable buildings. Envelopes of buildings, whether commercial or residential, must fulfill several functions simultaneously (Niachou et al., 2001; Dunnett and Kingsbury, 2008). By assigning multiple tasks to materials and components to be developed, new economies and sustainability goals will be achieved, thereby transforming existing design and construction methodologies for higher sustainability. First and foremost, the envelope serves as a space enclosure to protect the building interior from the elements. A separate but related requirement is that it can safely carry all external loads applied to it, such as wind and self-weight, and transfer these to the building frame. To minimize energy usage, expenditures for heating in winter and cooling in summer should be minimized, which calls for effective thermal insulation. In urban areas, it is also necessary to protect the building interior from outside noise, especially in the vicinity of major traffic arteries and airports. Finally, the envelope should lend itself to esthetic treatment so that the architect can use it as an expressive tool. In the past, these criteria could be satisfied simultaneously by using massive walls.

In modern engineered structures, a premium is placed on minimizing the required amount of materials and their cost and maximizing the synergistic benefits for high sustainability. In this context, most systems used at present satisfy these diverse requirements by relying on sandwich construction, in which different layers of the construction are assigned different tasks. The costs of such a specially designed envelope system can be as high as the cost of the structural building frame. The ultimate goal is therefore to develop a new type of construction that satisfies all design criteria, yet at a considerably lower cost than that of currently available alternatives. By developing new materials that perform several such functions simultaneously, economic benefits can be derived. In addition, with the ever-increasing demands of sustainable development, which are getting more and more important in the construction industry, the reduction of required natural resources, whether material or energy, has become a fundamental requirement as well.

The multifunctional envelope systems will bring innovations in materials science and the multiphysical properties of new materials to bear on the overall assembly of a newly conceived layered and coordinated panel. The basic needs of the building envelope will be met with some other synergistic benefits in workability, aesthetic appearance, recyclability, and climate adaptation. Overall, an advanced building, protection, and retrofit technology with multifunctional envelope systems can integrate solar energy harvesting onto building skin. This holistic design concept for the infrastructure envelope has the potential of achieving the sustainable goal of zero net energy for infrastructure while protecting infrastructure itself and human activities inside from climate change and environmental devastation. The improvement of energy efficiency follows a two-fold strategy: (1) the holistically designed building envelope with thermal management will drastically reduce the energy consumption by minimizing the need for cooling in summer and heating in winter; and (2) the conversion of solar energy to electricity and heat reduces or eliminates the need for power/heat supplies toward energy independent buildings.

1.2 Solar energy harvesting methods

Solar energy, as a renewable and clean energy resource, has been considered as a promising solution for the energy and environmental challenges and global warming threat. While we are suffering from energy crisis and environmental pollution related to fossil fuel combustion, the sun delivers to the earth 1.2×10^5 terawatts, which is about 10^4 times of the rate at which human civilization currently produces and uses energy. In summer, the unused solar energy unfavorably heats the living area, which requests more energy for air conditioning. In some metropolitan areas, the outage of electric service has caused huge economical losses and fatal accidents due to heatstroke. Solar roofing panel technologies exhibit significant advantages, such as:

- Long-term free and renewable electricity without consumption of any fuels.
- Clean energy resources without harmfulness to the environment, such as emissions or noise.

- Stable and available energy resources for most houses and buildings in a remote or urban area.
- Improvement of thermal comfort and material life by reducing solar radiation exposure.

Therefore, the utilization of solar energy with a roofing panel is a sustainable strategy for roof construction and energy transmission and security. The standard *Extraterrestrial Solar Spectrum* is shown in Fig. 1.2 with the overall irradiance density as 1366 W/m^2. Solar energy is transmitted to the earth through photons, which exhibit a higher energy level at a lower wavelength. Solar energy is typically harvested with three methods. Photovoltaic (PV) utilization directly converts sunlight to electricity by the PV effect; thermoelectric (TE) utilization can transfer solar heat to electricity by the TE effect; whereas solar heat utilization can directly harvest heat for different applications. The first two are based on solid-state physics and yield electricity directly at a lower conversion efficiency. The last one is based on the heat exchange and the conversion efficiency can be very high.

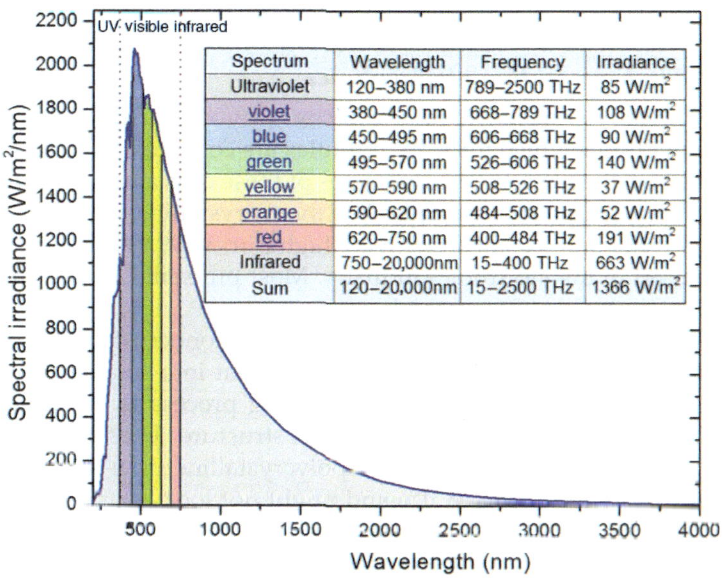

FIGURE 1.2 Solar spectral irradiance in the wavelength and colors.

1.2.1 Photovoltaic utilization

Although the PV phenomenon has been well understood since the 1950s, it was not applied in solar energy conversion and storage until the 1970s. Since then, more than 30 years of research on terrestrial PV has resulted in significant improvement in utilization efficiency and reduction of cost. In the past decade, PV electricity was becoming a promising industry, and PV modules have been used in roof construction for solar energy harvesting. However, the application of the PV panel is seriously impeded by the high cost and low efficiency. The majority of PV module production is based on crystalline silicon (c-Si) wafer technologies.

The thermodynamic analysis showed the efficiency limit for a single p-n junction is about 34%, and thus a large portion of solar energy is wasted through heat dissipation. Although some emerging technologies can considerably improve energy utilization efficiency, such as multijunction cells, optical frequency shifting, multiple exciton generation cells, multiple energy level cells, hot carrier cells, and concentration PV system, due to the higher cost and complex service conditions, they have not been used on roof construction yet. Besides, because the energy utilization efficiency greatly depends on the temperature, most current solar roofing panels only serve at the temperature range from −40 to 85°C. Because silicon production is energy-consuming and expensive, low-cost solar cells are of special interest for roofing panel manufacturing. Especially, emerging polymer solar cells embracing nanotechnology provide very promising solutions if the durability and efficiency can be significantly improved.

1.2.1.1 *First-generation photovoltaics*

The crystalline silicon (c-Si) PV technology comprising of interconnected small cells which form PV modules are considered the first generation of PV in the market. The two types of these cells are monocrystalline and multicrystalline silicone cells. The monocrystalline silicon (mono-Si) solar cells are made of silicon with N7 high purity (99.99999%), similar to what is used in the electronics industry. Most pure silicons are produced using the *Czochralski (CZ)* method.

In multicrystalline silicon wafers, similar to monocrystalline materials, the pure molten silicon is cast in blocks and cut into smaller blocks and eventually thin wafers, however, the casting process is different in the sense that it produces a multigrain crystal structure. Thus, the multicrystalline silicone cells, also known as polycrystalline or p-Si, results in a slight efficiency reduction of ˜1% and might not look as appealing as the monocrystalline cells to the end-user, however, the downside is offset by a simpler manufacturing process and a lower cost. Here, instead of the cells being cut in a pseudo-square shape, they are cut in a square or rectangular shape, thus, making them easier to be packed closely in modules. In terms of performance, there is not much difference between monocrystalline and multicrystalline PV modules. The principle for the silicon solar cells is the single p-n junction as the building block of the semiconductors. Similar to diodes, the electrons in n-type material and the holes in p-type are the major players here. However, with irradiation of photons on the solar cell, the electron-hole pairs in the crystal lattice are created at the junction, resulting in the generation of a current (Fthenakis and Lynn, 2018).

1.2.1.2 Second-generation photovoltaics

Unlike the first generation, the second generation of the solar cells involves more complicated processes including deposition of thin films on substrates and then dividing them to form the cells, thus, making them with a thickness at a hundredth level of the silicon wafers. The three types of thin-film technologies are amorphous thin-film silicone (a-Si), Cadmium Telluride (CdTe), and copper indium gallium diselenide (CIGS). The amorphous silicon cells have low efficiency in the range of 6%–9% about half that of the crystalline silicon, and are used in consumer products such as watches and calculators. The advantages of this technology are their low cost and the fact that the products where they are used require low power, thus, making their low efficiency relatively unimportant. In recent years, however, the a-Si technology has been further developed and scaled up for higher power generation, where they have been used in buildings facades.

Another thin-film technology is Cadmium telluride (CdTe). The 1.5 μm thickness of this material and its direct bandgap of 1.45 eV allows it to optimally capture the sun's spectrum using a single-junction device. While initially, there were some health and environmental concerns because of Cadmium's toxicity, further lifecycle analysis by experts has allayed this concern and prove it to be safe. The other materials used to create thin-film solar cells are Copper Indium Gallium Diselenide (CIGS). The compound copper indium selenide is an excellent absorber of light in an extremely thin layer, and with further modification with gallium, the material can produce an effective heterojunction, passing the 20% efficiency for the cell in the lab, while for commercial modules, this number is much lower at 10% to 12%. The tradeoff of second-generation solar cells being so thin and flexible, therefore, making them good candidates to be laminated onto windows, roof tiles, and other substrates is their low efficiency.

1.2.1.3 Third-generation photovoltaics

Finally, the third and the latest solar cell technologies are organic cells, dye-sensitized cells, and multijunction cells. While both the first- and second-generation cells can only absorb a photon that matches a narrow wavelength, the third generation technologies have the goal of surpassing the Shockley-Quessier limit (34%) and bringing the best features of both first and second-generation cells, having a higher efficiency while being cheaper and more versatile. Some of these technologies include gallium arsenide (GaAs) multi junctions, dye-sensitized cells, organic solar cells, and Perovskite solar cells. In Fig. 1.3, a roadmap and comparison of all the solar cell technologies have been demonstrated.

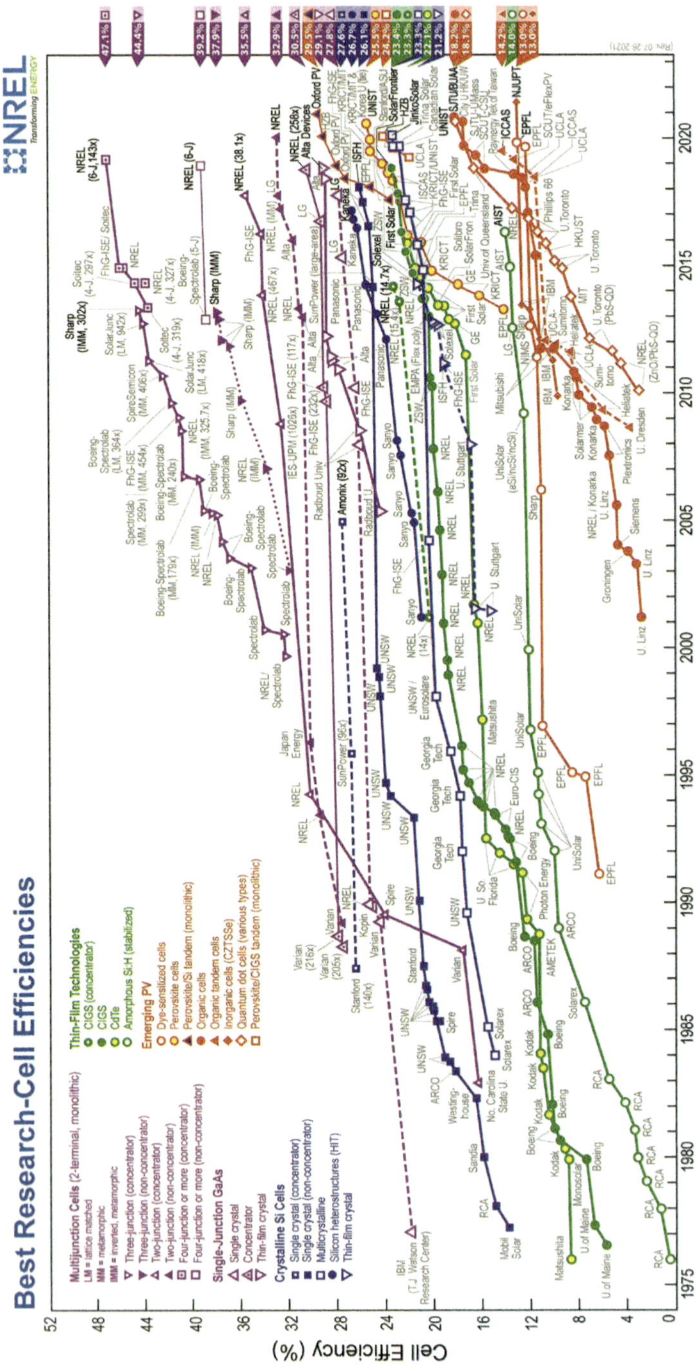

FIGURE 1.3 Best research-cell efficiency chart (Source: Courtesy of National Renewable Energy Laboratory; NREL, (2020)).

1.2.2 Thermoelectric utilization

Because the PV utilization focuses on photons with energy higher than the PV bandgap, which covers about half of the total solar energy, the unused portion of solar energy, mainly in the infrared range (Fig. 1.2), will heat the solar panel, and thus further reduce the PV utilization efficiency. In 1821, Seebeck discovered that a voltage is created in the presence of a temperature difference between two metals so that a TE generator can be assembled with multiple p-n junctions connected in parallel or series (Fig. 1.4). Peltier found that the reverse was also true: an electric current could produce a temperature difference. Since then, TE devices have been used for cooling and electricity generation. The main challenge of TE utilization is to find optimal TE materials with high electric conductivity and low thermal conductivity. With the rise of nanotechnology, higher thermodynamic efficiencies can be obtained by fabricating the TE device at the nanoscale, which will enable a wide range of applications. Currently, most TE generators are used at a higher temperature for higher efficiency. However, solar roofing panels generally need to work at ambient temperatures below 40°C. It is difficult to make a large temperature difference between two sides of a TE module, and thus only relatively low utilization efficiency can be achieved. Although several teams have used TE modules in solar roofing panels, the economic viability is an issue for large-scale applications. Moreover, the long-term performance of roofing panels under multiple weathering conditions needs to be addressed. However, TE utilization may still have great potential under two cases: (1) new low-cost and high efficient TE modules emerge and (2) concentrated solar power applications at high temperature and energy density.

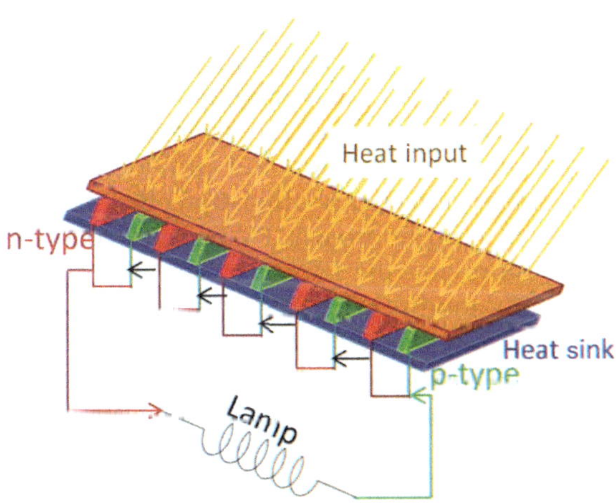

FIGURE 1.4 Schematic illustration of a Seebeck TE cell. *TE*, thermoelectric.

1.2.3 Heat harvesting

Solar thermal energy can be collected and stored for later use or directly be harvested by heating water for domestic or industrial use. Integral collector storage solar water heating systems have been developed in the past decades. A solar collector essentially includes an absorber, water tubes, water tanks, and a control system. The absorber receives solar radiation and transfers heat to circulating water. It is typically made of metals such as copper, aluminum, and steel, but polymer materials are becoming a promising substitution for their excellent corrosion resistance and multiphysical properties. In recent years, solar heat harvesting has been integrated into roof construction not only for providing heat supply but also for improving thermal comfort in the building. However, the cost, durability, and efficiency of these technologies are the main issues to be addressed in the future. In addition, internal condensation is commonly observed due to the temperature difference between cold panel surfaces and warm indoor air. However, the needs of heat are limited, and only a small roof area can meet the needs of most residential homes. Although the energy harvesting efficiency can be very high, say commonly >85%, the temperature is relatively low. Therefore the low-quality heat may not be as favorable as electricity by PV or TE utilization because it requires a high volume of storage but cannot be converted to other forms of energy.

1.2.4 Hybrid solar panels

The next-generation solar roofing panel pursues high efficiency of energy utilization, low cost of manufacturing and construction, and excellent durability in long-term service. Although emerging nano-technologies and novel polymeric materials make it possible to significantly improve the performance of the above three solar energy utilization approaches, each of them has its limitations that may seriously hinder their applications. The combination of two or three of the approaches is not a simple superposition of the materials and cost but provides a viable solution to significantly increase overall energy utilization efficiency while alleviating the disadvantages of each approach. PV-thermal collector enables heat harvesting while improving the PV utilization efficiency by controlling the temperature of PV modules. PV-TE hybrid systems were proposed to utilize the full spectrum of solar radiation for higher efficiency. Thermo-PV conversion cells were originally made for nuclear electric generation and used with a solar concentrator to transfer infrared photons into electricity.

The roof of a building typically serves multifunctional purposes, including insulation of thermal, water, and airflows, structural integrity under the wind, rain, snow, and hail loading, mechanical strength for temperature and moisture-induced stresses, material integrity under aging, weathering, and deterioration environments, and its aesthetical outlook. Because the demand for heat and electricity in a building is supplementary, a thermal and electrical hybrid solar utilization system may reach a remarkable efficiency while fulfilling multifunctions.

1.3 Challenges and opportunities of solar panels

Conventionally, solar panels are made of a glass surface, solar cell, and back sheet packaged with encapsulant and aluminum frame. They are used in building applications but, to date, have achieved only modest penetration into the vast commercial and residential roof markets because of relatively high cost-benefit ratios and long payback periods, which are very sensitive to the price of other energy resources. Although the cost of solar panels has been reduced significantly, due to the relatively low price of oil in the past years, the cost/benefit ratio of solar systems is still unattractive compared with conventional energy supply, which results from both high upfront costs, in terms of both equipment and installation, and low in-service efficiencies. However, the gap has been reducing. With governmental solar subsidies, now solar energy becomes a very competitive energy source. Currently, the majority of PV module production is based on crystalline silicon (c-Si) wafer technologies. Their energy utilization efficiencies greatly depend on the temperature; most current solar roofing panels only operate within the temperature range from −40°C to 85°C, with a temperature coefficient of reduced power efficiency at about −0.4% per°C.

Without a means for cooling the panel, in-service surface temperatures are commonly 40–50°C higher than the ambient temperature. Therefore, when the panels are subject to the highest solar energy levels, the efficiency of conventional PV panels is greatly reduced. This is often the case in the southern regions of the United States where ambient temperatures in the summer often exceed 35°C, resulting in roof surface temperatures exceeding 60°C and panel surface temperatures of 80°C.

Conventional solar panels are configured as mounted equipment attached to the structural elements of the building skin, which is less than optimal, as shown in Fig. 1.5. The power generating elements of such panels are typically adhered to a structural substrate, which is supported by a structural framework that is fastened through the building's waterproofing system, to the structural elements of the wall or roof.

Because of the turbulent nature of the wind fields surrounding buildings, traditional solar arrays do not shield the building's skin from wind loads. This conventional configuration necessitates several redundancies, as the panel substrate and frame must be designed to resist the same wind and snow loads as the building skin. Building-integrated solar panels can

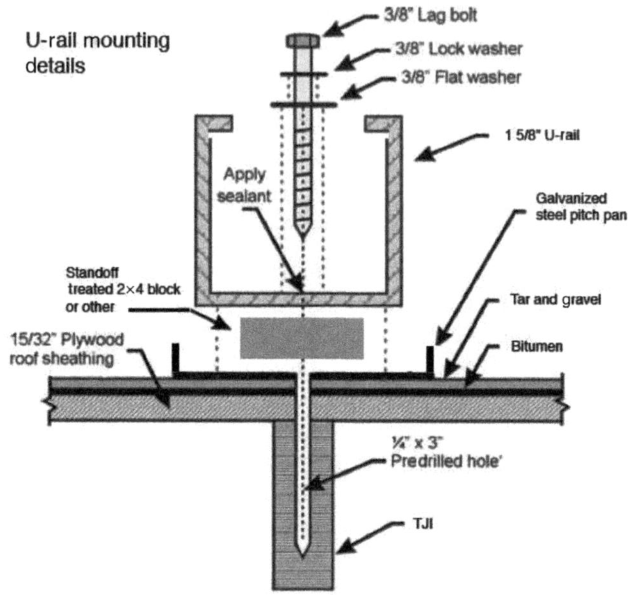

FIGURE 1.5 Schematic illustration for penetration waterproofing system (Daddazio et al., 2010).

eliminate these redundant elements and reduce the installed cost of solar energy systems. The benefits, however, extend beyond the financial realm; the embodied carbon of these redundant elements will also be reduced and the reduction of embodied carbon is a significant strategy in our battle against climate change.

The conventional mounting of solar arrays also produces waterproofing problems because the connection of the arrays vastly increases the number of penetrations in the waterproofing system. Each penetration must be waterproofed and, at each location, the potential exists for a leak. From a statistical standpoint, increasing the number of penetrations increases the risk of leaks. The additional roof layer of solar panels will increase the maintenance and operations cost.

In summary, conventional roofing panels commonly exhibit the following challenges:

- When a traditional solar roofing panel is subjected to a large temperature change, delamination between layers caused by thermal stress will severely reduce the life and efficiency of the panels.
- The efficiency of PV modules significantly reduces at high temperatures. When solar irradiation reaches the maximum, the panels may exhibit the worst efficiency.

- Exposure to weathering environments, solar roofing material properties, and structural strength decay with the service time. The long-term performance needs to be quantitatively studied.
- When the panel's efficiency reduces to a certain level, the replacement, and recycling of the materials will create more challenges, whereas some components of solar panels are still usable.

Through appropriate design, the above-mentioned problems can be resolved or alleviated. Particularly, building integrated photovoltaic thermal (BIPVT) systems may create great opportunities to extend the lifetime and optimize the life-cycle performance.

1.4 Building integrated photovoltaic and building integrated photovoltaic thermal systems

Building integrated photovoltaic (BIPV) systems integrate solar cells into the building envelope as a part of the roof structure for energy harvesting, which can be in the form of roof/wall panels or roof shingles. By cutting the material redundancy of the conventional solar roofing system, the initial installation cost can be offset by reducing the amount spent on building materials and labor that would normally be used to construct the part of the building that the BIPV modules replace. These advantages make BIPV one of the fastest-growing segments of the PV industry. BIPV systems will become the backbone of zero-energy buildings. However, the unused solar energy will heat the building's skin, so the temperature-induced efficiency loss is still a problem.

BIPVT systems introduce thermal management components into the BIPV system, which create dual benefits: heat energy harvesting and thermal management of the building skin. The heat collection is conducted by a thermal fluid or air circulation system. BIPVT overcomes the difficulties of conventional solar panels and provides a cornerstone for future solar panel manufacturing. Its main benefits are summarized as follows:

1. Due to the temperature control, the PV module can work at lower temperatures in the summer and thus stably obtain a higher efficiency for PV utilization.
2. The thermal fluid, whose temperature is partially controlled by the flow rate, can be directly utilized by water or floor heating systems for domestic usage.
3. Due to the temperature control on the roof, the room temperature can be significantly reduced and thermal comfort in the building can be much improved.

4. In winter, the thermal fluid system can be used to remove ice and snow on the roof, clean solar panels, and thus restore and enhance solar energy utilization.
5. Through the modular design, BIPVT can be designed in such a manner that the panel can be easily replaced and delaminated for reuse of the useful components.

The most common BIPVT systems are realized through a heat transfer fluid in an open-loop (usually air) or closed-loop (usually liquid) configuration, which are shown in Fig. 1.6A and B, respectively.

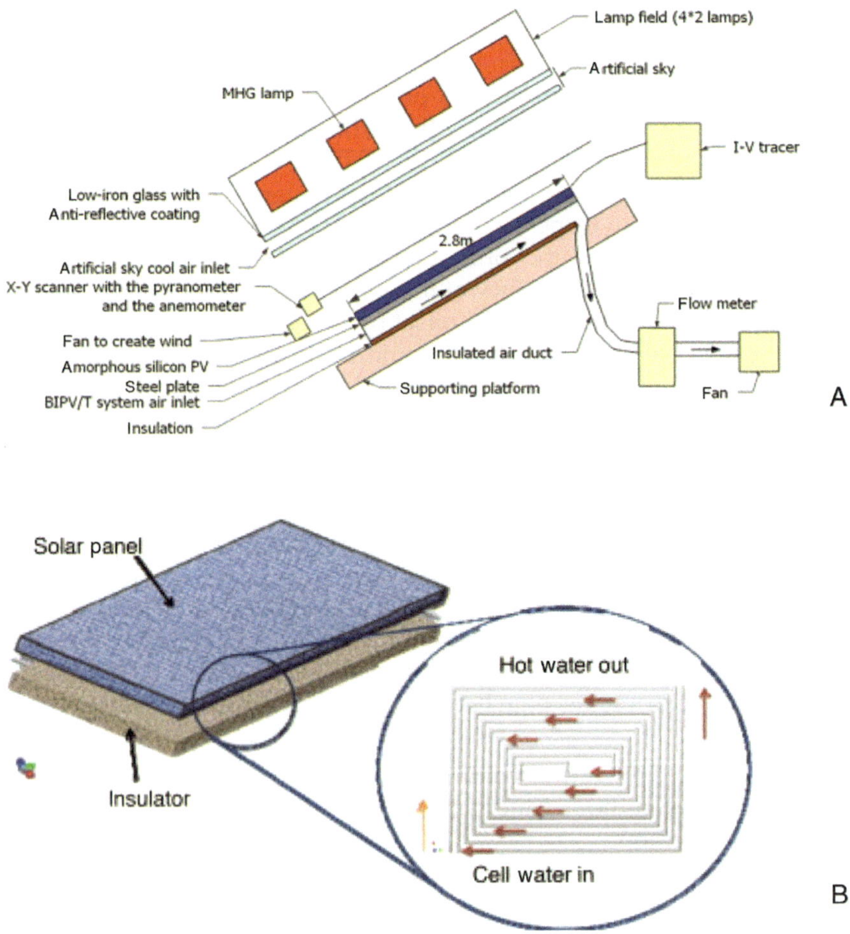

FIGURE 1.6 Schematic illustration of the BIPVT system cooled by (A) open-loop (Yang and Athienitis, 2014); and (B) closed-loop (Ibrahim et al., 2011). *BIPVT*, building integrated photovoltaic thermal.

Generally, the closed-loop configuration with liquid is more efficient than the open-loop with air as heat transfer media due to the high thermophysical properties of liquid compared to air. Most closed-loop BIPVT systems employ water tubes for cooling and thermal energy collection. Normally, the water tubes are embedded in insulation materials and covered by absorber materials in contact with the PV elements above. However, such designs commonly exhibit poor heat conduction due to the small contact area between the absorber and water pipes, and thus not able to efficiently cool the PV cells to increase the PV efficiency. Moreover, the system should address the potential problems of leakage, condensation, and freezing. To avoid winterizing procedures in some geographic areas, glycol has been used to replace water as the thermal fluid.

The idea of combining solar and thermal energies—which attributes to the use of the sun's infrared spectrum to collect heat, in addition to PVs, which is the generation of electricity using the sun's visible spectrum—has been a motivation for many researchers. Yet, the limiting factor is to find a method or material to benefit from the wide range of the solar spectrum. However, considering the fact that PVs absorb the infrared wavelength of the sunshine, a less complex way to combine both solar PV and thermal to indirectly harness the heat that is generated in the PVs. Yang and Yin introduced a novel idea by using a TE module for thermal energy harvesting (Yang and Yin, 2011). In this multilayer design, a semiconductor material, being bismuth telluride (Bi2TE3) is used as a TE generator layer underneath the PV cells, and then a back layer of functionally graded material (FGM) with cast water tubes made out of copper is implemented. The novelty of the design was the application of the FGM layer, in which the thermal resistance gradually changes with depth. The heat from the solar panels first transfers to the TE layer to generate electricity and then is dissipated to the water through the FGM material. In a similar work from the same group, experiments on the hybrid solar panel with PV cells and FGM layer with water tubes showed that the system can lower the temperature of the panel from 55°C for 1100 W/m^2 irradiation to 32°C (Yang et al., 2012). The authors used a finite element model simulation for the thermal performance of their panel. However, the temperature distribution of the working fluid was simplified as a linear function of the path and the forced convection as a Nusselt number-related quantity obtained by Bejan (Bejan, 2013). The improved version of the panel included the highly thermal conductive aluminum layer on the top with the thermally insulated high-density polyethylene at the bottom of the FGM (Yin et al., 2013). The designed panel can be used as an integrated part of the roof in the building envelope and thus, bear structural load due to its rigidity and strength. Finally, Chen et al. further matured this BIPVT design by performing more tests and showing that at a water flow rate of 150 ml/min and under 1000 W/m2 solar irradiation, the efficiency is enhanced by 24% compared with the panel with no cooling system (Chen and Yin, 2016).

The three general types of thermal management methods for BIPVT are active, passive, and hybrid which is a combination of the two. In terms of the hybrid methods, one way is to incorporate the fluid-driven (or air-driven) cooling system with phase change materials (PCMs). The PCM can store the excess heat generated during the operation of the panel and release it at night when there is no solar energy intake (Fig. 1.7). Thus, aside from the electricity generation feature of the panel, this will result in bridging the gap between supply and demand of the thermal energy that is collected from the panels. The mechanism is as follows: during the day, when the panel is exposed to sunlight and becomes hot, the heat is transferred from the back of the panel via a thermally conductive layer to the PCM, which will result in reducing the temperature of the PV. The PCM stores the heat isothermally until it reaches its melting point and completely melts, and then the excessive heat is transferred to the thermal management fluid which could be air, water, glycol, etc., and be used to partially heating the building's inside air, or domestic hot water (Preet, 2018; Chow, 2010). Studies have shown that this innovative system can lead to a reduction of 56.8%–104.4% in the final energy consumptions of the building, approaching a nearly net-zero or even net-positive energy building (Kant et al., 2019).

The PVT collector of BIPVT can be integrated with building components such as rooftops, façades, windows, and shading devices. An overall classification of the BIPVT system in buildings' façade has been shown in Fig. 1.8.

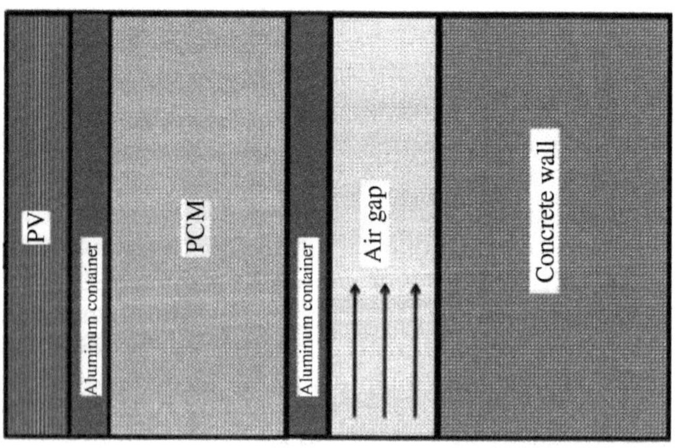

FIGURE 1.7 Schematic illustration of a hybrid air-driven BIPVT with PCM (Kant et al., 2019). *BIPVT*, building integrated photovoltaic thermal; *PCM*, phase change materials.

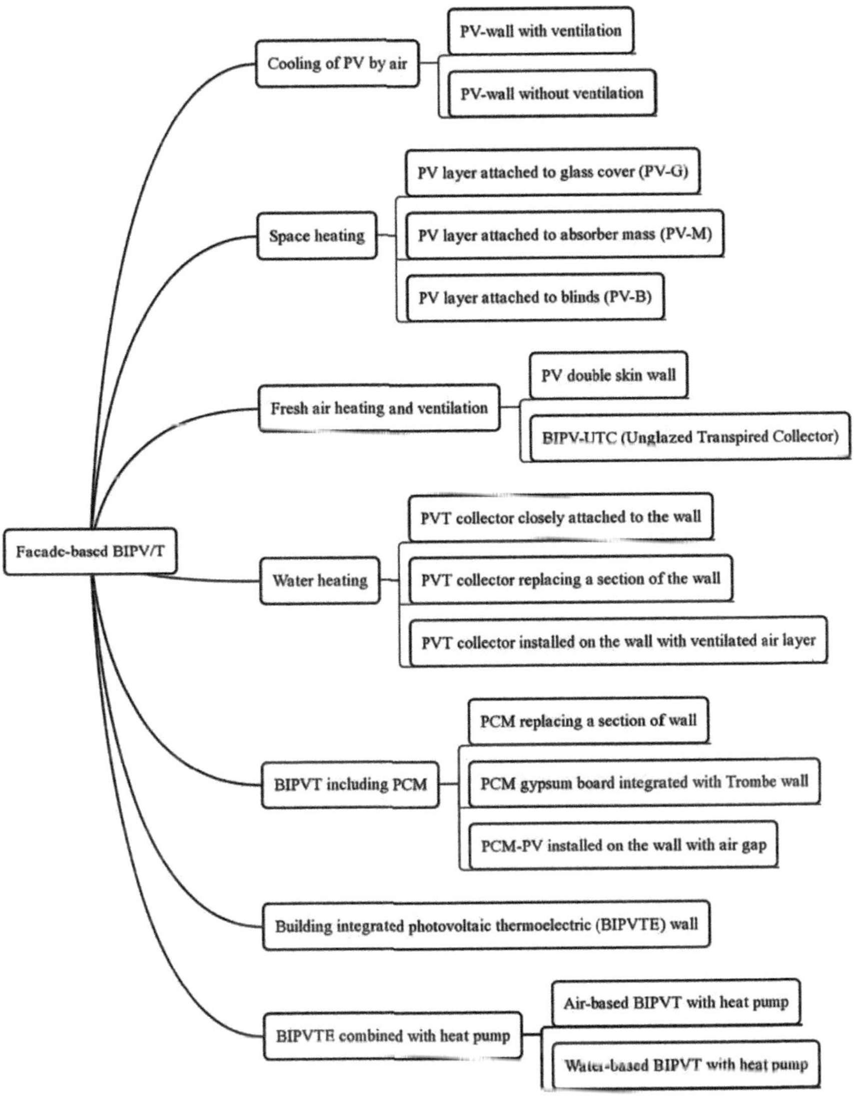

FIGURE 1.8 Classification of façade-based BIPVT systems (Yu et al., 2020). *BIPVT*, building integrated photovoltaic thermal.

Because the solar panel configuration can be tailored to a most modern building with curtain wall, air-based BIPVT are commonly used for the integration with the HVAC system for higher building efficiency as well. The BIPVT applications in other building parts follow similar classifications but fluid-based BIPVT is more common for roof applications as a higher heat exchange rate is required. The BIPVT applications for windows and other shading devices may require customized design and control of the thermal management parts. Because the BIPVT in those applications covers much smaller coverage areas on the moveable parts, solid-state physics-based methods exhibit advantages as the nimble integration does not require other gas/fluid circulation connections and can avoid leakage problems.

The case study at end of this Chapter will introduce two BIPVT systems for case studies: building integrated thermal electric roofing system (BITERS) by Atlantis Energy Systems, Inc, and a novel BIPVT solar roof panel developed by Columbia University.

1.5 Solar energy industry in the United States and the world

Solar energy is one of the fastest-growing US industries. In 2019, more than 13 GW solar capacity was installed and the annual residential solar PV installations reached 2800 MW, more than a 16% increase in the installations from 2018 (SEIA, 2019). In 2019, the US solar market grew to an $18.7 billion industry and more than 250,000 Americans worked in the solar—more than double the number in 2012—at roughly 10,000 businesses, across 50 states. In the United States, the solar industry had a 49% average annual growth just in the last decade, due to the federal and state policies and lower module prices. Today, there are more than 81 Gigawatts (GW) solar capacity installed nationwide, enough to power 15.7 million homes. Solar power is becoming more affordable with technology innovation, scaled-up manufacturing, faster installation techniques, and new financing options. The installation cost of solar had a decline of more than 70% since 2010, and the prices as of the first quarter of 2020 are at their lowest levels in history across all market segments. The solar industry is segmented by application areas of residential, commercial, and industrial. The price for an average-sized residential system without any incentive is $18,000 today, compared with $40,000 in 2010. The utility-scale prices ranging from $16/MWh to $35/MWh are also competitive with other forms of electricity generation.

It is expected that ample rooftop and land availability, and unrivaled electricity consumption will allow the United States to continue to be considered among the most valuable PV growth markets. According to analysts at National Renewable Energy Laboratory who modeled the fraction of customers likely to invest in PV for a range of payback times, the percent of customers likely to invest in a system with a 7–8 years payback is 12%–10%, respectively (Drury et al., 2010).

With costs continuing to fall and solar expanding into more state markets, the residential solar market experienced a record year in 2019. There is also a rapid rise in the nonresidential segment as more companies such as Walmart, Apple, Target, and Amazon are increasing the number of their rooftop installations. Additionally, with a capacity of over 1000 Megawatts (MW) of thermal energy generation, more than 1.5 million homes and businesses use solar water heating in the United States. According to the US Department of Energy, another 400 MW is likely to be installed in the next 3–5 years (EESI, 2006). In 2018, with 85 GWh, the United States was the world's second-largest producer of electricity by solar energy, after China having 178 GWh of electricity production from solar (IRENA, 2018).

While the cost of solar energy has declined rapidly in the recent past, it remains much higher than the cost of conventional energy technologies. Such is the case with many renewable energy technologies, solar energy benefits from fiscal and regulatory incentives and mandates, including tax credits and exemptions, feed-in-tariff, preferential interest rates, renewable portfolio standards, and voluntary green power programs in many countries. Additionally, the potential expansion of carbon credit markets also would provide additional incentives to solar energy deployment. However, despite the huge technical potential, development, and large-scale, market-driven deployment of solar energy technologies worldwide still has to overcome several technical and financial barriers. If these barriers are not overcome, maintaining and increasing electricity supplies from solar energy will require the continuation of potentially costly policy supports (Timilsina et al., 2011).

The daunting challenge of energy policy decision-makers across the globe has been to make the cost of renewable energy competitive with heavily subsidized conventional energy. Historically, householders or energy companies who wanted to install solar panels have been faced with lengthy pay-back times. Such consumers have been forced to make a choice based on ethics rather than economics. Without increased consumer demand and political measures to facilitate access to the market, manufacturers of solar PV panels cannot produce the unit volumes needed to bring prices down and drive technological innovation. The Feed-In Tariff has proven to be the most effective policy instrument in overcoming these barriers, and close to 75% of the world's residential solar PV installations have occurred with the support of the national, state, or provincial FIT policies (Gregg, 2011).

1.6 Scope of this book

This book first introduces the fundamentals of the physics and mechanics in BIPVT including PV phenomenon, heat transfer, energy analysis, and solar panel design in Chapter 2, and then describes the silicon solar cell production and solar module packaging in Chapter 3, in which the mainstream manufacturing methods were introduced. Chapter 4 introduces the production and acceptance of BIPV panels. The standard quality control, certification, and inspection are described. Chapter 5 introduces the design and installation of BIPV and BIPVT systems. Some actual projects were introduced as examples for future houses. One problem existing in BIPVT applications is the unbalanced heat between the harvesting and demanding, i.e., in summer, much more heat is collected than the indoor needs, and the opposite in winter. To address this problem, Chapter 6 introduces the BIPVT coupling with geothermal systems for seasonal heat storage and reuse. Because solar energy harvesting requires large land surface coverage, to avoid land competition, Chapter 7 shows BIPVT's application in seashore and offshore. Coupling with wind and wave energy can significantly increase the energy harvesting capacity without improved efficiency and reduce the initial investment for unit energy yield, which may reduce the barrier for the BIPVT to the seashore or offshore applications. Chapter 8 presents the application of BIPVT covering transportation infrastructure surfaces. It can create an additional transportation mode and provide an electric power supply to future vehicles. It will change the lifestyle of future generations. Chapter 9 introduces the BIPVT applications to farms, which can transform marginal land to arable farms for food, energy, and water supplies. In the end, some cutting edging BIPVT research and development, as well as emerging technologies and applications, are summarized in Chapter 10. Each chapter starts with an iconic image for the significance or content of the chapter and finishes with one or two case studies of BIPVT projects or designs.

1.7 Two building integrated photovoltaic thermal systems

This section presents the state-of-the-art BIPVT. A commercial BITERS developed by Atlantis Energy Systems, Inc. and a novel BIPVT solar roof panel developed by Columbia University are introduced in detail. Both BITERS and BIPVT are the most advanced solar roof panels to harvest solar energy in the form of PV electricity as well as heat. The performance of the BIPVT roof panel has been evaluated in terms of its thermal efficiency via warm water collection and PV efficiency via the output electricity.

1.7.1 Building an integrated thermal electric roofing system

Founded in 2002, Atlantis Energy Systems produces solar BIPV glass and solar PV **Sunslates and TallSlates Roof Tiles** (Fig. 1.9) The Sunslates and TallSlates are both BIPV solar roofing products. Despite all the recent developments over the past decade, the SunSlates system is still one of the most advanced BIPV Class A roofing systems in the market. It is the only

FIGURE 1.9 Solar roof system made by AES: (A) Sunslates, and (B) TallSlates.

product of its type that offers a roof warranty and has a UL Class A fire certification. The TallSlates is a solar module that generates electricity and acts as a roof slate as well, which is ideal for straight slope roof applications. The gutters on each side of the module act as drain channels as well as support to the mounting system.

It was reported from field experience that the Sunslate installation could be time-consuming and prone to mistakes. To prevent installation errors and lower the labor and maintenance costs while providing the aesthetic appearance and functionality, Atlantis Energy Systems, Inc. (AES) developed a roof covering that combines the best features of legacy products with the utility and economy that customers desire. The new hybrid product, BITERS, the modular system could be hoisted preassembled onto an existing roof or a new construction roof with excellent workability for installation (Pao, 2021). Each part of the module can be replaced, recycled, or reused in the life cycle of the system, which leads to lower life cycle costs.

The BITERS is a solar hybrid system that can harvest solar energy in the form of both thermal energy and electricity. The thermal component is located directly beneath the Sunslates or the TallSlates. Therefore, solar modules convert solar energy into DC power at a much higher efficiency. All internal systems can be preinstalled in the home before the units' arrival, which produces dramatic labor and insurance savings for the roofing contractor. The units are then lowered into place by crane onto preset cleats attached to the roof and securely locked down. They are then plumbed to the in-house connecting lines. Once installed a roof can either be finished out to its surrounding edge with slates or left as is on an asphalt shingle roof.

The energy harvesting mechanism of the BITERS is illustrated in Fig. 1.10, the liquid going through the thermal system is being pumped and circulated throughout the slope roof. The thermal energy is being captured and transferred to the heat exchanger that can provide domestic hot water. Meanwhile, DC electricity is being generated by the Sunslates and fed

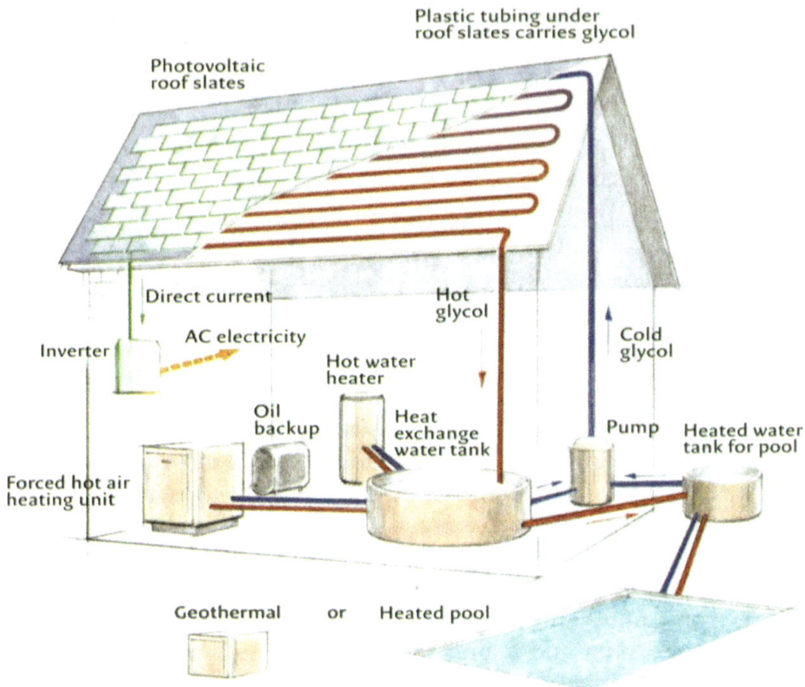

FIGURE 1.10 Operation of the BITERS in a building. *BITERS*, Building Integrated Thermal Electric Roofing System.

into the power grid through the inverter or the batteries if the customer prefers a standalone system. In addition, one gets money back on the slate investment.

For a Sunslates system, once all strings are checked and if they reach the acceptable power, they will be connected to the inverter. Then the PV system would be turned on and start supplying AC power to the grid. In the case of a standalone system, the strings could be connected to the battery via a charge controller and start storing electricity that could be used to power the DC appliances and lighting applications.

One of the pilot projects of the BITERS is shown in Fig. 1.11 The BITERS thermal system uses a pump to circulate the thermal liquid throughout the system. As the intensity of the solar power increases so would be the throughput of the liquid in the thermal tubing resulting in generating more thermal energy. In addition, while heat is being extracted through the thermal tubing from the Sunslates that is being cooled thereby prevents the degradation of efficiency in converting sunlight to DC and ultimately AC electricity. Since much thermal energy had been taken out from the attic, the house would be cooler which results in a reduction in the air

FIGURE 1.11 A pilot project of the BITERS. *BITERS*, Building Integrated Thermal Electric Roofing System.

conditioning energy consumption. In case, there is an excess of energy available on standard domestic hot water application, it could be stored for later use or it could be used for hot tub or heater pools or geothermal, which would also be subject to locations.

Because of its high performance coupled with the aesthetic design of the BITERS and quality of all the BIPV installations, AES received the Frost & Sullivan "2009 World Building-Integrated PV Niche Player of the Year Award." The award was in recognition of AES's successful initiatives in entering a niche market with innovative and user-friendly solutions, understanding the needs of the customers and providing solutions, blending innovative developments, years of experience, superior quality, and service.

1.7.2 Building integrated photovoltaic thermal solar roof

By integrating FGMs as a cooling substrate, a novel BIPVT roofing panel (Fig. 1.12) was developed at Columbia University for solar energy harvesting. The purpose of the proposed FGM layer is to create a lightweight layer of solar roofing panel with a varying thermal conductivity in the thickness direction. The FGM layer gradually transits material phases from metal-dominated to polymer materials. Water tubes are embedded in the top part of the FGM layer, where the high aluminum concentration creates high thermal conductivity so that heat flow from the surface can be immediately transferred to water tubes in all directions, yet be insulated by the bottom part of the polymer layer. The BIPVT panel can be integrated into the building skin with water circulation, flow control, and heat storage and utilization systems. Solar energy is collected by BIPVT panels in the form of electricity and heat. Electricity can be used locally or transmitted to the

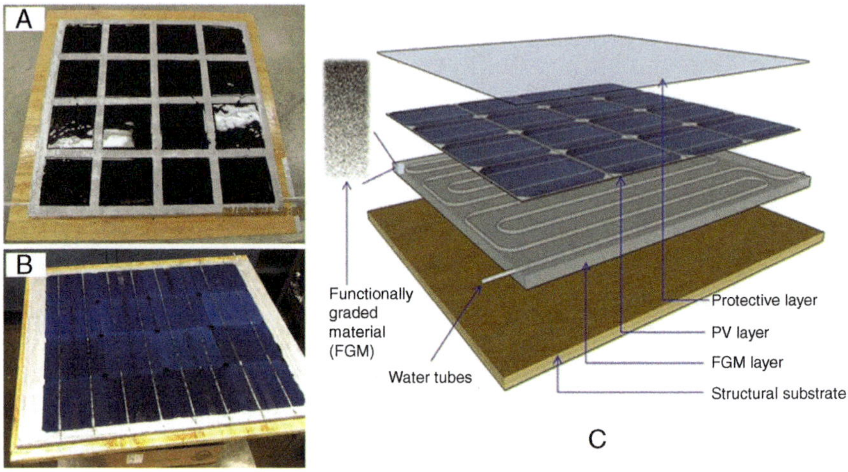

FIGURE 1.12 A novel building integrated photovoltaic-thermal (BIPVT) roofing panel: (A) functional graded material layer; (B) integrated solar panel, and; (C) the BIPVT layers for assembly. *BIPVT,* building integrated photovoltaic thermal.

grid, while the heat can be stored or directly used as an indoor or external heat supply such as floor heating, clean drying, and swimming pools.

The BIPVT panel is manufactured in the following procedure. When an FGM layer is fabricated with a thermal fluid circulation tubing system attached to the substrate in Fig. 1.12A, the aluminum-rich surface is polished and a stringed solar cell layer is attached to the surface for the encapsulation process in Fig. 1.12B. Infield application, a transparent protective waterproofing layer could be further mounted on the BIPVT panel to protect the power generating elements and underlying building materials from external environmental distress such as moisture migration, surface wear, and impact from dust, wind, storm, etc. A schematic illustration of the developed BIPVT panel is shown in Fig. 1.12C, where the FGM layer gradually transits material phases from a well-conductive side (aluminum dominated) attached to a PV solar cell to a highly insulated side (polymer materials) bonded to a structural substrate. The water tubes were embedded in the top part of the FGM layer, where the high aluminum concentration creates a high thermal conductivity so that heat can be immediately transferred to the water tubes from all directions, yet be insulated by the bottom part of the FGM layer and the thermal insulation plywood. The substrate provides support for mechanical loading and functions as thermal insulation for the building envelope. The multilayered solar panel was designed in such a way that layers with potentially shorter life expectancies can be easily replaced or removed from the design based on the sustainability measurement criteria, which can apply to other sustainable building materials and systems for building construction.

The FGM panel with embedded water tubes was fabricated by the combined vibration and sedimentation approach. The purpose of the designed double serpentine shape of the water tubes, as shown in Fig. 1.13A, is to remove the heat within the panel more uniformly, and thus it keeps the temperatures of the panel's surface much more uniform. When the FGM panel was cured and taken out from the vacuum oven, it was seamlessly glued to a structural substrate, such as fire-retardant plywood or lightweight concrete through a moisture-resistant epoxy adhesive, as shown in Fig. 1.13B. The integrated FGM panel and substrate were then sprayed by a thin layer of moisture-resistant resin, which reacts chemically with the natural moisture in the wood, creating a highly moisture-resistant substance similar to polyurethane. Due to the gradual change in the proportion of materials, only a small amount of aluminum powder is needed. This reduces the cost of the panel as the aluminum material is relatively expensive. The heat-absorbing layer and insulation layer in traditional PVTs can be replaced by one FGM layer that integrates the high thermal conductivity in the top part and low thermal conductivity in the bottom. Since the volume fraction of Al particle continuously varies in the thickness direction, the thermo-mechanical property distribution changes smoothly, and avoids the thermal stress concentration across layers, and increases the structural integrity and durability of the panels.

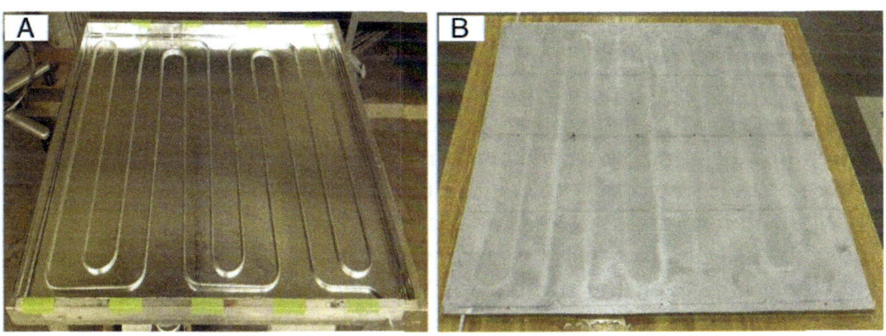

FIGURE 1.13 Embedment of water tube and integration of substrate: (A) double serpentine shape of water tubes in the casting mold; and (B) assembling of FGM panel with a structural substrate. FGM, functionally graded material.

A multifunctional roofing system can be easily assembled by using the proposed BIPVT panel, which is sized to be easily integrated with the industry-standard structural framing spacing of 304.8 mm on the center. Custom panels can also be easily developed for longer spans. The length of the panel can be either 1219 mm or 2438 mm, based on the weight and thickness requirements (1219 mm is illustrated for demonstration). The prototype design references the International Building Code (IBC 2015) for the roof assembly requirements for weather protection, such as flashings to prevent moisture penetrations at all roof interruptions and terminations. The proposed panelized system will employ a field-applied joint system while gasket design alternatives can also be developed. Both electrical and fluid connections will be made between the individual panels.

The manufacture of this panel will not be more complex than the existing combination of a traditional roof and PV panels. The proposed BIPVT panel can be integrated into a building's skin with relevant system components such as water circulation, flow control, and heat storage. The innovations of the proposed BIPVT technology are summarized as follows:

- Due to the ability to control temperatures through the water flow, the PV modules can work at lower temperatures in the summer and thus stably obtain a higher efficiency for PV utilization.
- The collected hot water can be directly utilized by water heating systems for indoor or external heat supply.
- Due to the temperature control on the roof, better thermal comfort in the building can be achieved in the summertime, cooler room temperature can be obtained, and the energy demand for cooling can be reduced.
- In winter, warm water flow can be used to remove frost or ice on the roof, clean solar panels, and thus restore and enhance solar energy utilization.

Note that traditional solar panels are rebuilt on the rooftop or attached to structural elements of the building skin, which is less than optimal. Firstly, the power generating elements of such panels are typically adhered to a structural substrate, supported by a structural framework that penetrates the building's waterproofing system. Secondly, traditional solar panels cannot shield the building skin from wind loads and conventional configurations necessitate redundancies as the panel substrate and frame must be designed to resist the same wind and snow loads as the building envelope. However, the present BIPVT is not a simple superposition of individual structural components and PV modules but provides a viable solution to significantly increase overall energy utilization efficiency while alleviating the disadvantages of a single approach.

This patent-pending holistic design meets multifunctional purposes, including insulation of thermal, water, and air flows and acoustic separation; structural integrity under the wind, rain, snow, and hail loading; mechanical strength for temperature and moisture-induced stresses; material integrity under aging, weathering, and deteriorating environments, and also its aesthetical appearance. Because the demand for heat and electricity in a building is supplementary, a thermal and electrical hybrid solar utilization system may reach a remarkable efficiency while achieving multifunctionality. The features, advantages, and benefits of this technology are summarized in Table 1.1.

TABLE 1.1 The features, advantages, and benefits of the building integrated photovoltaic thermal technology.

Feature	Advantages	Benefits
Photovoltaic cells	Energy harvesting for electricity supply	Reach acceptable electric efficiency 8%–15% for a temperature range of ˜ 85°C at a low cost
Water flow system	Solar heat harvesting and temperature control	Harvest 54% solar heat efficiency in the summertime and keep photovoltaic <45°C for better photovoltaic efficiency
Water flow control	Using water flow to control roof temperature	Defrost/deice roof surface for energy generation and reduce thermal strain for longevity
Functionally graded material layer	Continuous variation from metal-rich region to polymer materials for multifunctions	Transfer heat into the water flow through the metal-rich region but block the heat flux by the thermal insulating polymer material for thermal comfort
Fiber-reinforced lightweight concrete substrate	Using foamed postconsumer plastics as aggregate for lightweight concrete with sisal fiber as reinforcement	The lightweight (specific gravity ˜1) and good strength (compressive ˜1ksi) enable us to fabricate large solar panels (2′ by 8′) for excellent workability at a very low cost (˜0)
Building-integrated roof design	Integration of structural sheathing and solar energy collector into one element	Eliminate material redundancies for low cost, high performance, good workability, and easy maintenance
Modular design of the layered system	Thermal, acoustic, and water insulation combined with energy harvesting	Easy to recycle the useful components of the panel at the end of the lifetime or replace the damaged part during the service life
General design concept	Adaptive to future photovoltaic technology and materials	Different types of photovoltaic cells and materials can be integrated into the panel for life cycle benefits

1.8 Case study: Active Energy Building

The following project is a creative idea for smart and sustainable buildings by Anton Falkeis and Cornelia Falkeis-Senn architecture company (falkeis^{2}architects) based in Vienna, Austria. Considering that the cities are responsible for 75% of the world´s energy consumption, 80 of CO_2 emissions, and hosting more than 50% of the world´s population (Falkeis, 2015). Falkeis^{2}architects have developed and realized the first "active energy building."

The project successfully implements a building-integrated energy production system, including PVs tracking systems, and PCM climate wings (Figs. 1.14 and 1.15), which are part of a moving building envelope, producing solar power while also controlling the building climate. The building also demonstrates its high adaptability to changing spatial requirements, achieved by an extremely effective load-bearing structure with high-performance concrete (Fig. 1.16). Thus, this project works as a smart node in an interconnected network structure of existing buildings, sharing information as well as an energy surplus, delivering a contribution to the improvement of urban living conditions and energy demands. The goal is to make "active energy building" a prototype for a new urban and decentralized energy production system, providing a more flexible and resilient model of energy production as well as energy distribution.

FIGURE 1.14 PV tracking systems and PCM climate wings, part of a moving building envelope, harvest solar and interstellar radiation, producing solar power, and controlling the building climate (ACFNY, 2018). *PCM*, phase change materials; *PV*, photovoltaic.

FIGURE 1.15 The solar tracker is embedded in the Voronoi roof structure, lifting the PV solar wings out of the roof structure and positioning the panels toward the sun. *PV*, photovoltaic.

FIGURE 1.16 High-performance concrete.

1.8.1 Building innovation for architecture in motion

Some of the innovations in this "active energy building" project include acoustically active 3D light fields (Fig. 1.17), high-performance concrete with an outstanding reproduction quality, building-integrated energy production, and sustainable structures as well as "biological approaches" to building, structural design. Some of these technological inventions yielded a patent, like the movable building envelope, or were built as prototypes demonstrating new knowledge in the field of digital fabrication,

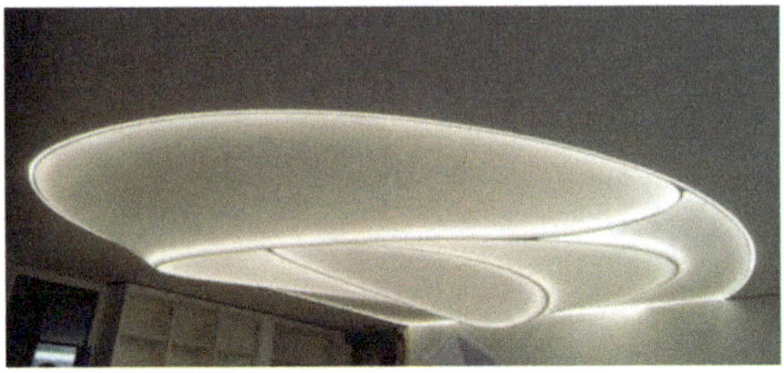

FIGURE 1.17 Acoustically active 3D light fields work as diffusors and absorbers.

such as the 3D freeform textile building envelope (Fig. 1.18)—working as a shading device—and the complex 3D Voronoi load-bearing structure (Fig. 1.19).

FIGURE 1.18 Working as a shading device, the 3D textile structure envelopes the main part of the west side of the building before penetrating the structure.

FIGURE 1.19 3D Voronoi load-bearing structure.

1.8.2 Structure design and optimization

This project is built according to the envisioned standards of climate change neutrality in 2050. Therefore, in its design, sustainability as a holistic concept has been applied in the massing of the building as well as to the construction of the load-bearing elements and the development of the energy technology. The design and distribution of the load-bearing elements allow for spatial reprogramming over the entire lifespan of the building without impacting load-bearing components. Fig. 1.20 shows

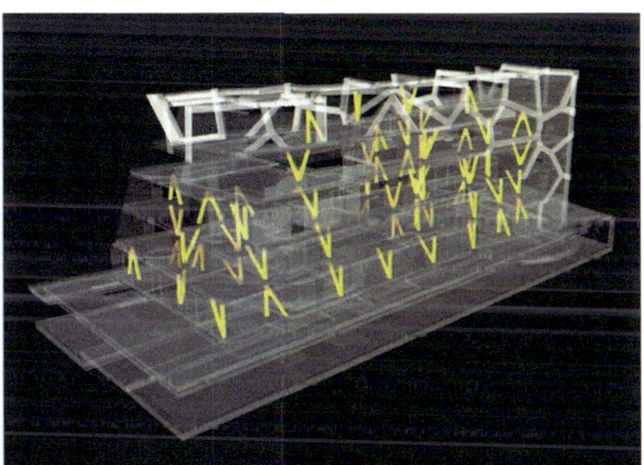

FIGURE 1.20 Structural development employing growth analogy, optimized by genetic algorithms, using V- and A-shaped double-column elements.

FIGURE 1.21 The building demonstrates its high adaptability to changing spatial requirements, achieved by an extremely effective load-bearing structure.

the structural development employing growth analogy, optimized by genetic algorithms including the V- and A-shaped double-column elements highlighted in Figs. 1.21 and 1.22 with extremely effective load-bearing performance.

FIGURE 1.22 The optimized distribution of columns influences the distribution of the program. By the development of characteristic rules, the dividing walls are affected by the placement of the columns, and the conventional plans are transformed into unexpected spatial sequences.

1.8.3 Voronoi load-bearing structure

For an active energy building, a high strength-weight ratio is critical for the material efficiency and building operation. A strong light-weight structure not only reduces the dead load of the building, but also provides easy control and high loading capacity of the movable parts. Inspired by the lightweight structure of a dragon-fly wing, 3D Voronoi load-bearing elements were developed to transfer load through in Fig. 1.23, which provides both optimal load transfer performance and aesthetic visual impact to the dynamic building design. The top story is completely wrapped by the energy skin supported by the 3D Voronoi load bearing

FIGURE 1.23 The development of the 3D Voronoi load-bearing elements is inspired by the lightweight structure of a dragon-fly wing (ACFNY, 2021).

FIGURE 1.24 The top story is completely wrapped by the energy skin, which features numerous cells housing PV- and PCM elements. The geometry hereby is derived from natural patterns that emerge when cells aggregate. *PCM*, phase change materials; *PV*, photovoltaic.

structure, which features numerous cells housing PV- and PCM elements (Fig. 1.24).

1.8.4 Energy design

The "active energy building" is an energy-independent structure that produces more energy than it consumes. Using renewable sources being, geothermal energy, and passive and active utilization of solar power. As for active solar systems, they use solar radiation to generate power and support climate regulation in the building, while the passive solar gains are part of the building's heating system. In this way, solar energy is implemented for heat, ventilation, and air conditioning of the building through the systems installed on the façade of the building as well as materials and systems that benefit from the latent heat storage properties (Falkeis et al., (2015)).

1.8.5 Improvement and optimization of passive solar gains

The architecture team has developed the design tool "solar erosion" to enhance passive solar gains. By inputting the maximum building volume according to the local building codes and transferring that number into a topographical structure with the criteria of maximizing sun exposure, the software does iterative processes to calculate the additional sun-exposed surfaces. Therefore, the sun elevation angles are treated as "eroding" surfaces and the final massing of the building can be manipulated to achieve the maximum available solar radiation.

This will result in a fanned-out topography on the east side and an inclined south side of the building, while "solar erosion" brings a deep canyon-like space on the west side of the building volume.

1.8.6 Active solar energy utilization

The "active energy building" is equipped with a movable envelope to harvest solar energy for power generation and heating, and cooling. In order to harvest the maximum solar irradiation, a building-integrated PV-tracking system is employed in addition to the roof and south side of the building that produce solar electricity. The supplementary systems are the use of e-mobility as temporary storage and the reservoirs of hydro power plants to store the surplus created at peak hours. Another innovative idea in this project is the use of "climate wings" in which, PCM is used on the western and eastern side of the building to store energy in the forms of latent heat storage units. Inside the wings, PCM capsules are mounted and when the wings are exposed to the sun, the heat is stored inside the wings, and when the wings are folded back and connected to the building's ventilation system, the stored energy is released.

Through studies done by Lucerne University of Applied Sciences and Arts, Switzerland, the described decentralized energy production systems have been evaluated by doing full-scale mock-up experiments and are now working as a movable building envelope producing green energy. This active energy building could work as the node for a network of connected structures, sharing information, energy surplus, and thus, managing a local cluster (Fig. 1.25).

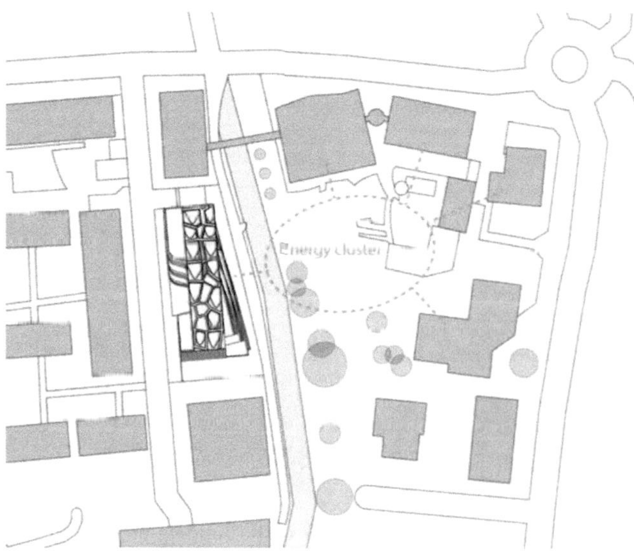

FIGURE 1.25 The building is a networked structure managing a local energy cluster.

1.8.7 Building integrated photovoltaic–tracker

The PV installation tracking system enhances solar power production by continuously following the sun (Fig. 1.26). All 13 PV solar wings have a double-axis tracking system that rotates the wings from east to west as well as moving up and down to match the zenith angle of the sun (Figs. 1.27 and 1.28). The solar tracker is embedded in the Voronoi roof structure and the

FIGURE 1.26 The new energy production systems have been scientifically evaluated by means of a full-scale mock-up (ACFNY, 2021).

FIGURE 1.27 The entire roof and south side of the building produce solar power.

FIGURE 1.28 Building integrated PV-tracking systems, developed for this project, increase the solar energy production by a factor of 2.9. *PV*, photovoltaic.

PV panels lie flat in the "Voronoi-pockets," completing the roof´s surface. When the sun is out in the morning, the tracking system starts moving by lifting the wings out of the roof structure and position the panels toward the sun. The angle and movement of the PV solar wings are controlled and programmed to achieve optimal actuating angles year-round. This active tracking system also monitors and collects the wind, radiation, and meteorological data by sensors and access to databases, and accordingly, responds to changes in the input. To avoid any damage to the tracking system in high wind speed occasions, the trackers move the PV solar panels back into the Voronoi roof structure to minimize the horizontal forces being absorbed by the structure

1.8.8 Phase change materials climate wings

As discussed earlier, the "active energy building" utilizes other innovative forms of energy management systems such as the use of PCMs. The PCMs as their name implies are materials that by going through a phase change, absorb or release energy. The latent heat in this process can be used to store excess energy and release it later when needed. More details on this topic can be found in Chapter 10.

Here in this building, PCMs are implemented in the wings that harvest solar and interstellar radiation for heating, cooling, and air conditioning (HVAC). The active energy building is equipped with seven wings in total, four of which dedicated to heating and three for cooling devices (Fig. 1.29). Twenty four square meters of heating-wing surface cover 10% of the heating requirements and 15 square meters of the cooling-wing surface cut out 15% of the required cooling energy.

FIGURE 1.29 Climate wing delivered on-site.

When the wings are exposed to the sunlight, they become hot and the excessive heat from the solar panels is transferred to the PCMs that are encapsulated inside the wings (Fig. 1.30). Thus, the PCMs start to melt due to the accumulated heat at 31°C, and as a result, the PCM inside the pouches store the excessive sun's heat energy by going through a phase-change process to a fluid state. The heating wings act as a recovering system and reduces the heating load of the building while lasting for days without recharging. While winter and also the transition times of spring and fall are the main usage periods of these wings, they are closed in summer and act mainly as thermal insulation to help the building from over-heating (Fig. 1.31). The arrangement of the profiles as a lamella-like structure ensures a high density of solar energy and temperature throughout the winter months. The heating wings are located on the western side of the building. At sunrise, the wings open up, exposing the PCM to the solar radiation, and at sunset, they close.

The three wings assigned for cooling purposes which are located on the eastern side of the building, work as an alternative to conventional

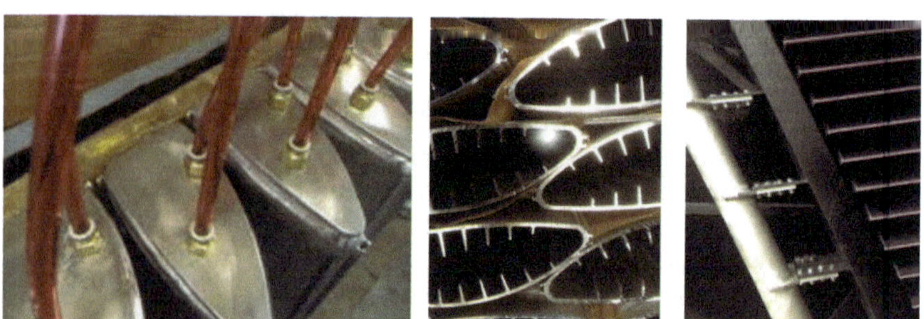

FIGURE 1.30 Testing of climate wing lamellas.

FIGURE 1.31 Full-scale mock-up evaluating the new energy technology.

FIGURE 1.32 The 3D freeform textile building envelope is built as a prototype, demonstrating new knowledge in the field of digital fabrication. The lamella-like structure is designed according to the solar radiation of the area and bridges 17 m without additional support.

cooling systems. They are open at night and use interstellar radiation to store the cold, therefore, freezing the PCM capsules at a temperature of 21°C changing the phase of the PCM from fluid back to solid. The air flows through these frozen wings and thus is cooled to be used for indoor cooling of rooms.

1.8.9 Textile building envelope

Some of the other innovation in this active energy building is used of textile patterns as part of the building envelope as well the interior lighting system. The 3D freeform textile structures which cover the main part of the west side of the building are used as a shading device.

Acoustically active 3D light fields work as diffusers and absorbers. The textile elements are stretched out of the building on the other side, continuing their flow through the canyon-like staircase (Fig. 1.32).

References

The exhibition catalogue from "falkeis²architects: active energy building" exhibition at Austrian Cultural Forum New York, 2018 (ACFNY, New York, NY).

Bejan, A., 2013. Convection Heat Transfer. John wiley & sons. DOI:10.1002/9781118671627.

Chen, F., Yin, H., 2016. Fabrication and laboratory-based performance testing of a building-integrated photovoltaic-thermal roofing panel. Appl. Energy 177, 271–284.

Chow, T.T., 2010. A review on photovoltaic/thermal hybrid solar technology. Appl. Energy 87 (2), 365–379.

Crabtree, G., Lewis, N., 2007. Solar energy conversion. Phys. Today 60, 37–42.

Daddazio, R.G., Kelly, H.M.Y., Garant, J., 2010. Final report of DOE STTR Phase I: hybrid building integrated solar energy system for photovoltaic, thermoelectric and heat utilization, H.M. Yin (PI). WAI, Department of Energy, DE-SC0003347.

Drury, E., Denholm, P., Margolis, R., 2010. Modeling the US rooftop photovoltaics market. National Renewable Energy Lab.(NREL), Golden, CO.

Dunnett, N., Kingsbury, N., 2008. Planting Green Roofs and Living Walls. Timber Press, Portland, OR.

EESI, 2006. https://www.eesi.org/papers/view/fact-sheet-solar-water-heating.

Falkeis, A., 2015. Thinking Out of the Urban Design Toolbox. In: Bast G., Carayannis E., Campbell D. (eds) Arts, Research, Innovation and Society. Arts, Research, Innovation and Society. Springer, Cham. https://doi.org/10.1007/978-3-319-09909-5_12.

Falkeis, A. et al., Façade element with latent heat storage device and method of air conditioning a building, 2015. EP 2896898. https://patents.google.com/patent/EP2896898A3/en.

Fthenakis, V., Lynn, P., 2018. Electricity from Sunlight. Wiley Online Library, John Wiley & Sons. DOI:10.1002/9781118963791.

Gregg, J.S.T.W., 2011. "Feed In Tariffs—The Devil Lies In The Details," https://www.shapingtomorrowsworld.org/2011/05/page/2.

Hasnain, S., 1998. Review on sustainable thermal energy storage technologies, part I: heat storage materials and techniques. Energy Convers. Manage. 39, 1127–1138.

Hegedus, S., 2006. Thin film solar modules: the low cost, high throughput and versatile alternative to Si wafers. Progress Photovolt. 14, 393–411.

Ibrahim, A., et al., 2011. Recent advances in flat plate photovoltaic/thermal (PV/T) solar collectors. Ren. Sustain. Energy Rev. 15 (1), 352–365.

IRENA, 2018. https://www.irena.org/Statistics/View-Data-by-Topic/Capacity-and-Generation/Country-Rankings.

Kant, K., et al., 2019. Analysis and design of air ventilated building integrated photovoltaic (BIPV) system incorporating phase change materials. Energy Convers. Manage. 196, 149–164.

Lewis, N.S., Crabtree, G., Nozik, A.J., Wasielewski, M.R., Alivisatos, P., Kung, H., ... Nault, R.M., 2005. Basic research needs for solar energy utilization. report of the basic energy sciences workshop on solar energy utilization, April 18–21, 2005. DOESC (USDOE Office of Science (SC)). https://www.osti.gov/biblio/899136.

Niachou, A., et al., 2001. Analysis of the green roof thermal properties and investigation of its energy performance. Energy Build. 33, 719–729.

NREL, 2020. Best Research-Cell Efficiency Chart. https://www.nrel.gov/pv/cell-efficiency.html.

Pao, F., US Patent No.: 11,114,976, Issue Date: 9/7/2021, USSN: 15/874,392, Modular Removable Building Integrated Thermal Electric Roofing System.

Preet, S., 2018. Water and phase change material based photovoltaic thermal management systems: a review. Ren. Sustain. Energy Rev. 82, 791–807.

SEIA, 2019. https://www.seia.org/solar-industry-research-data.

Timilsina, G.R., Kurdgelashvili, L., Narbel, P.A., Timilsina, G.R., 2011. A review of solar energy; markets, economics and policies (p. 5845). World Bank. https://openknowledge.worldbank.org/bitstream/handle/10986/3612/WPS5845.pdf.

U.S.P.I.R.S. Committee, Solar electric power: the US photovoltaic industry roadmap. 2001, Sandia National Laboratories.

Yang, D., et al., 2012. Simulation and experimental validation of heat transfer in a novel hybrid solar panel. Int. J. Heat Mass Transfer 55 (4), 1076–1082.

Yang, D., Yin, H., 2011. Energy conversion efficiency of a novel hybrid solar system for photovoltaic, thermoelectric, and heat utilization. IEEE Trans. Energy Convers. 26 (2), 662–670.

Yang, T., Athienitis, A.K., 2014. A study of design options for a building integrated photovoltaic/thermal (BIPV/T) system with glazed air collector and multiple inlets. Sol. Energy 104, 82–92.

Yin, H., et al., 2013. Design and performance of a novel building integrated PV/thermal system for energy efficiency of buildings. Sol. Energy 87, 184–195.

Yu, G., et al., 2020. A review of designs and performance of façade-based building integrated photovoltaic-thermal (BIPVT) systems. Appl. Therm. Eng. 182, 116081.

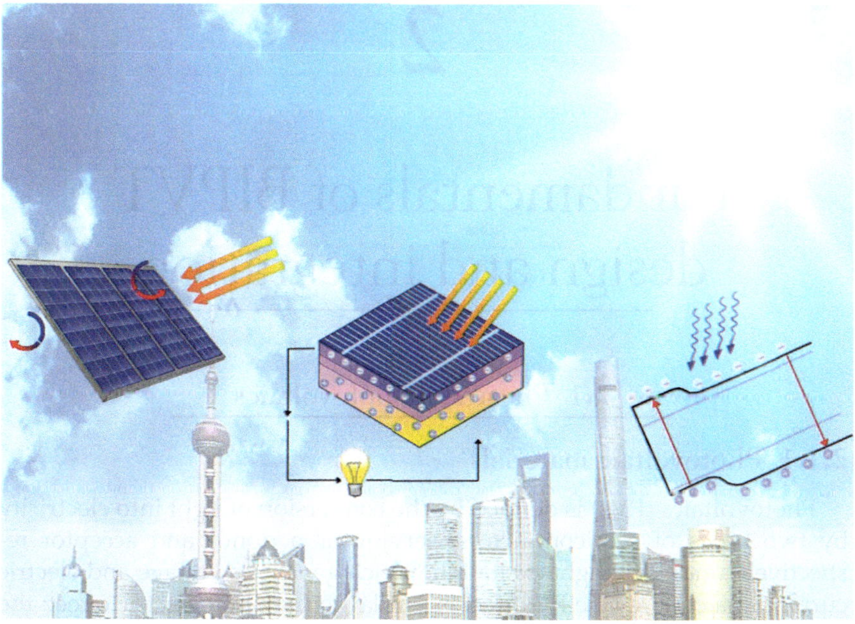

Skyscrapers are bathed under the strong sunshine with abundant energy, which often leads to overuse of electricity for the cooling, triggering blackout of the cities as well as the heat island effect. Photovoltaic systems on the building skin will cool the city by using solar energy. When sunlight strikes the surface of a photovoltaic (PV) module, most irradiation will be absorbed by the PV module with a minor portion reflected to the sky. Part of the absorbed energy is converted to electricity through the photovoltaic effect and the major part is wasted as heat. Depending on the PV efficiency, the unused solar energy will still heat the panel and be dissipated to the sky and indoor air by conduction, convection, and radiation. The PV modules are made of a group of solar cells that contain two semiconductors. Sunlight knocks electrons loose from atoms by the photovoltaic effect, which induces a voltage for electric current. At the atomic scale, photons of sunlight above a certain energy level (higher frequency) interact with the silicon atoms and form free electrons and holes drifting in a certain direction producing an electric current.

CHAPTER

2

Fundamentals of BIPVT design and integration

2.1 Physics of photovoltaics

2.1.1 Photovoltaic materials

Photovoltaics (PVs) is defined by the conversion of light into electricity by two types of semiconductors, serving as a donor and acceptor respectively under the light excitation, which generates voltage and electric current in a circuit. The PV effect is similar to the photoelectric effect: the absorbed light excites an electron or other charge carrier to a higher-energy state. However, the photoelectric effect usually refers to the electron ejected out of the material (usually into a vacuum); whereas, in the PV effect, the excited charge carrier remains within the material. In either case, an electric potential (or voltage) is produced by the separation of charges, and the light has to have sufficient energy to overcome the potential barrier for excitation.

A light beam can be considered as a collection of discrete wave packets, known as photons, which exhibit a certain level of energy proportional to the frequency of the light:

$$E = hf \tag{2.1}$$

where E is photon energy, $h = 6.62607015 \times 10^{-34}\ \mathrm{J \cdot s}$ the Planck constant, and f the frequency of the light. Here $f = c/\lambda$ in which $c = 299,792,458\ \mathrm{m/s}$ is the speed of light and λ is the wavelength.

To move an electron from its atomic binding, a certain energy is required to overcome the surface potential as:

$$W = hv_0 \tag{2.2}$$

where W is the threshold energy; v_0 the threshold frequency for given the band gap of a material. When $E > W$, electrons can be moved from the

Building Integrated Photovoltaic Thermal Systems: Fundamentals, Designs and Applications
DOI: https://doi.org/10.1016/B978-0-12-821064-2.00003-1

atom which produces a current. The maximum kinetic energy of the ejected electrons becomes

$$E - W = h(f - v_0) \tag{2.3}$$

which yields the stopping voltage as

$$V_0 = \frac{h(f - v_0)}{e} \tag{2.4}$$

where e is the elementary charge at $e = 1.602176634 \times 10^{-19}$ C. When the photon's energy level hf is less than the band gap hv_0, no interaction happens between the photon and the atom; when $hf > hv_0$, electrons (negatively charged) are excited and become moveable from the atoms. The extra energy will be eventually dissipated by heat. In the semiconductor of a solar cell, the electrons are only allowed to move in a single direction. The electronic structure of the PV materials is critical for energy generation. For example, silicon is the most popular solar cell semiconductor with a bandgap of $hv_0 = 1.1eV$. Commonly silicon incorporating small amounts of boron or phosphorus is used in different layers, which forms a p-n junction, in silicon PV cells.

The semiconductor chosen for a solar cell has to absorb as much of the solar spectrum as possible, therefore a low band gap is desirable. However, it is desirable to have as large a built-in voltage as possible which requires a larger band gap (discussed in the next Section). Therefore, there is an optimal band gap range between 1.0 and 1.7 eV, which makes an effective solar semiconductor. In this range, electrons can be freed without creating too much heat.

The photon energy of light varies according to the different wavelengths of light. The spectrum of sunlight, from infrared (IR) to ultraviolet, covers a range mainly of ~0.5–2.9 eV, which is corresponding to the wavelength range of 428–2,480 nm. For a single p-n junction, the efficiency is always much lower than 100%, as the energy content of photons above the bandgap will be wasted surplus re-emitted as heat or light and the energy content of photons below the bandgap will not be absorbed by the PV material at all. Different PV materials display different efficiencies and have different costs. Materials currently used for PV solar cells include monocrystalline silicon, polycrystalline silicon, amorphous silicon, cadmium telluride, copper indium selenide/sulfide, as well as organic materials and hybrid organic–inorganic materials.

2.1.2 Silicon solar cells

Solar cells are used to utilize solar energy and convert it to electricity. Using polycrystalline silicon (p-Si) solar cells as an example, highly pure p-Si ingots are afterward sliced into thin slices called wafers which form the base for the PVs cells. Silicon is a semiconductor and unlike conductors such as metals, it generally does not conduct electricity. However, under certain conditions, it can be made conductive. Silicon is an element located in group 14 of the period table, having 14 electrons in its atom. These electrons in the ground state are arranged in the configuration of [Ne] $3s^2 3p^2$, showing that there are four valence electrons in a silicon's atom structure. However, the silicon atoms are in a stable status and therefore, they are bonded together.

When an impurity such as phosphorus (P) with additional free electrons (five in this example) is added to the silicon layer, this process is called doping the silicon with a dopant (phosphorus). The result will be a layer with extra free electrons forming a negative charge, also known as the n-type layer. Although under light strike, the energy from the photons will cause some movement on the electron, the random movement will not result in any current. By doping another layer of silicon with a different dopant with less free electrons, for example, boron (B) with three valence electrons, we will have p-type doping. This will result in a layer of materials with a positive charge on this side, also known as "holes."

By attaching the two sides, some electrons from the n-type will start moving toward the p-type where there are holes and fill the holes, causing an area of "depletion zone" where there are no excessive electrons or holes. The electron migration in this process will result in a slightly negative charge on the p-type (bottom layer) and leave the n-type with a positive charge, thus, forming an electric field or built-in voltage. By having a negative layer on one side and a positive layer on the other side a p-n junction is formed, in which electrons and holes can only move in one direction, as shown in Fig. 2.1. Now, when the light strikes the thin highly doped top layer (n-type) of the PV cells, the high-energy photons of the incident light penetrate the depletion region and form electrons-holes again. The electric field that was present in the depletion zone attacks the electrons to the n-type layer and holes to the p-type layer, forming a high potential difference. By connecting these two regions through a load, we can see the flow of electrons and the recombination of electrons with the holes.

In an actual silicon cell, the electrons first are passed through the thin wires across the solar cell, known as the "fingers," and then are collected by the bus bars as illustrated in Fig. 2.2. From there, the top negative layer is connected to the bottom positive layer of the next cells by copper wires, forming a series connection. By putting these series of cells in series or

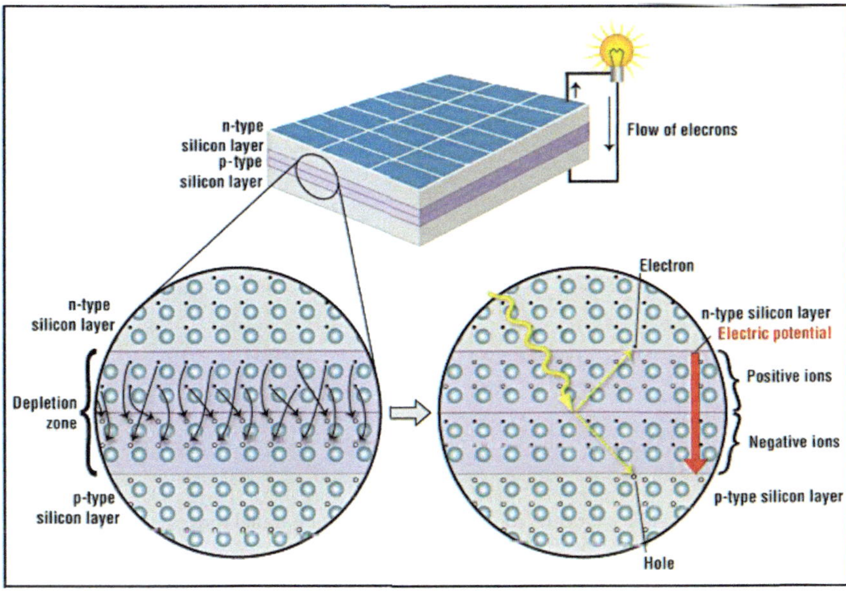

FIGURE 2.1 Schematic illustration of a silicon solar cell, showing the depletion zone in the p-n junction (courtesy of Anthony Fernandez) (Anthony Fernandez, 2021).

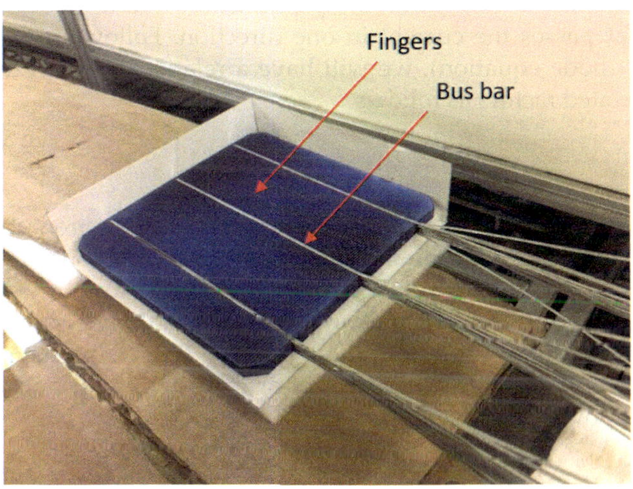

FIGURE 2.2 Fingers and Bus bars on the solar cell.

FIGURE 2.3 A set of solar cells connected in series and parallel forms a solar panel.

parallel to each other in Fig. 2.3, we will get a set of series and parallel cells, forming a solar panel that increases the voltage and the current values produced from each cell.

2.1.3 The efficiency of a photovoltaic cell

The depletion region turns the p-n junction into a diode, which is a device that passes the current in one direction. Following the Shockley equation (diode equation), we will have a relation between the current, and the related factors as follows

$$I = I_0\left[\exp\left(\frac{qV}{KT}\right) - 1\right] \tag{2.5}$$

where I_0 is the saturated reverse current, q is the electron charge, V is the across voltage of the diode, K is the Boltzmann's constant (1.3807×10^{-23} J/K), and T is the absolute temperature in Kelvin.

By drawing this equation, we can see the relation between the generated current and voltage, known as the I–V curve (Fig. 2.4). This graph is of significant importance as the product of these two values results in the generated power of the solar cell

$$P = VI \tag{2.6}$$

As one can see, for a single solar cell, the voltage is low, and typically might not exceed 0.6 volts; however, when the cells are grouped, this value increases to the operational level.

The efficiency of a PV cell can be evaluated from the ratio of the electricity generation to the solar energy flux density to the cell, which can be

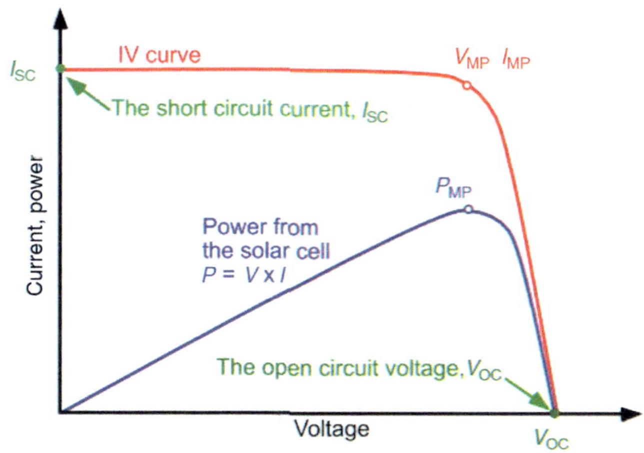

FIGURE 2.4 The voltage–current characteristic of a silicon cell (PVEducation 2021).

indicated by the energy of photons of sunlight. The energy content for each photon is related to the wavelength of the light as

$$E = \frac{hc}{\lambda} \tag{2.7}$$

where c is the light's velocity. Using the values of h and c, E can then be expressed in electron volts (eV) as

$$E\ (eV) = \frac{1.24}{\lambda(\mu m)} \tag{2.8}$$

where the wavelength λ is shown in micrometer (μm).

In reality, when an incident of light hits the surface of the solar panel, only photons with the same or greater energy of the bandgap of the semiconductor material can generate an electron-hole pair. The solar irradiation can be modeled as Planck's function as (Wieder, 1982)

$$B(\lambda) = \frac{2\pi hc^2}{\lambda^5\left(\exp\frac{hc}{\lambda k_B T} - 1\right)} \tag{2.9}$$

where the Boltzmann constant $k_B = 1.380649 \times 10^{-23}$ J/K, and the sun's temperature $T = \sim$6000K. Considering the relationship of $v = c/\lambda$, the spectrum of solar irradiation in wavelength can be written in frequency as

$$B(v) = \frac{c}{v^2}B\left(\frac{c}{v}\right) = \frac{2\pi hv^3}{c^2\left(\exp\frac{hc}{\lambda k_B T} - 1\right)} \tag{2.10}$$

where a multiplier of $|d\lambda/dv| = c/v^2$ is applied to the transformation between two spectra. By considering the photon energy $= hv$, the above equation can be rewritten as

$$B(E) = \frac{1}{h}B\left(\frac{E}{h}\right) = \frac{2\pi E^3}{h^3c^2\left(\exp\frac{E}{k_BT} - 1\right)} \tag{2.11}$$

The radiant flux on earth will be diluted by the sun–earth distance when it is emitted from the sun's surface as

$$B_E = f_w B \tag{2.12}$$

where $f_w = \left(\frac{R_{\text{sun}}}{D_{\text{earth}}}\right)^2$, in which the sun radius is $R_{\text{sun}} = 6.96 \times 10^8$m and $D_{\text{earth}} = 1.496 \times 10^{11}$m. Therefore, the incident spectral photon flux from the sun to the cell can be written as

$$\phi_0(E) = \frac{1}{E} f_w B(E) \tag{2.13}$$

Notice each photon at a different E level carries different energy. The total radiant flux density to the earth can also be expressed as (Wieder, 1982)

$$E_0 = \int_0^\infty E\phi_0(E)\,dE \tag{2.14}$$

which is the solar intensity external to the earth's atmosphere $E_0 = 1,353$ W/m^2. When the sunlight beam travels through the air mass (AM) of the atmosphere, the solar intensity will be attenuated by scattering and absorption, which is proportional to the relative path length. Because the atmosphere is a sphere around the earth, the sunlight beam's traveling path length will be equal to or higher than the atmosphere layer's thickness, AM 1.5G is commonly used for terrestrial power generation of solar panels. Therefore, the spectral photon flux onto the solar cell surface on earth will be further reduced, say $\phi(E)$. Fig. 2.5 shows the solar spectral radiation $f_w B(\lambda)$ of Planck function Eq. (2.9), extraterrestrial, and AM1.5G spectra. By replacing variables, $\phi(E)$ can be derived as well.

$$E_{\text{AM1.5}} = \int_0^\infty E\phi(E)dE \cong 1000\text{W/m}^2 \tag{2.15}$$

Although anti-reflective (AR) coatings are commonly used in solar cells, some parts of irradiation will be reflected to the sky at a reflective spectrum *R(λ)*. So the total reflection energy can be calculated by

$$E_{\text{ref}} = \int_0^\infty E\phi(E)\cdot R(E)dE \tag{2.16}$$

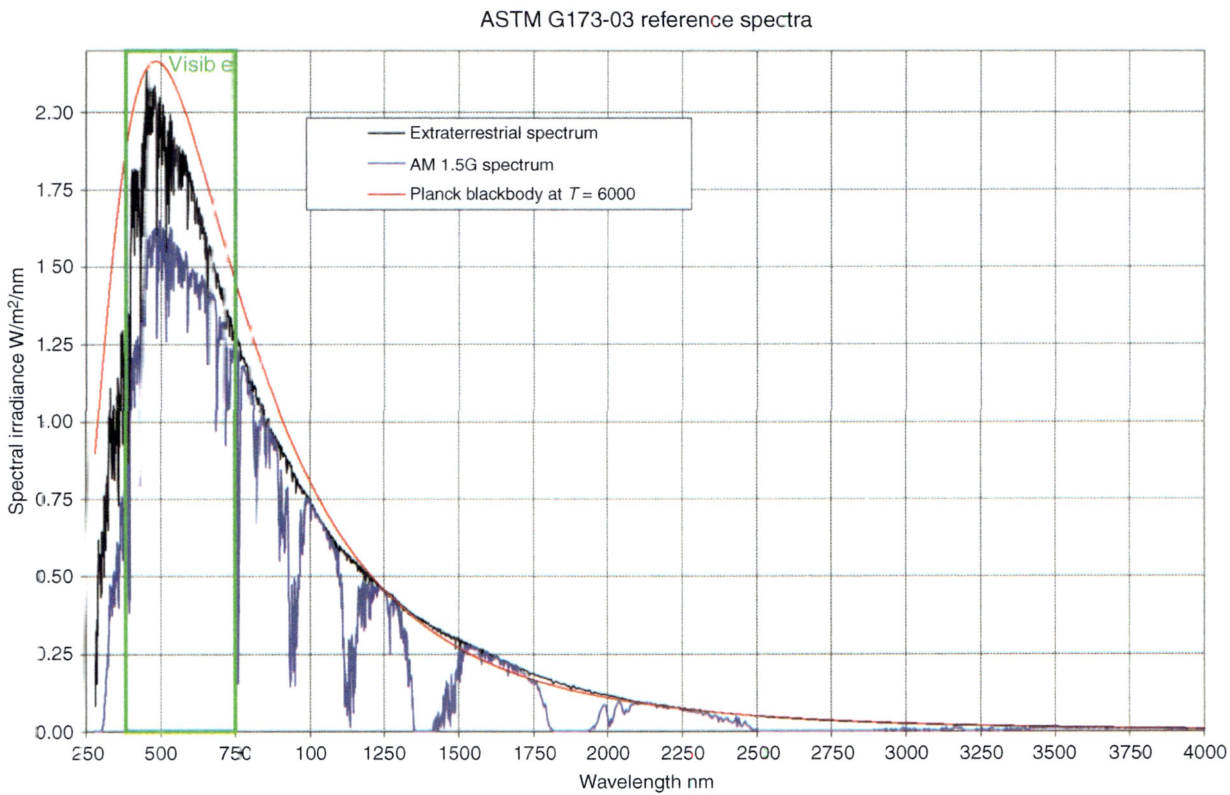

FIGURE 2.5 The solar irradiation distribution from the extraterrestrial spectrum, AM 1.5G spectrum, and the Planck Blackbody equation at T = ~6000 K.

The photons that arrive at the solar cell will not be absorbed completely. For example, the photons of the IR do not generally contribute to the formation of the electron-hole and generation of electricity in a silicon solar panel, and they will pass through the solar cell and be absorbed or reflected by the back sheet. Therefore, all photons solar cell captures for electricity generation are above the bandgap of silicon ($E_g = 1.1eV$) and one photon can only excite one pair of electron-hole. Therefore, the electric current generated can be written as

$$I_g = e\int_{E_g}^{\infty} \phi(E)\cdot[1 - R(E)]dE \tag{2.17}$$

Therefore, from the energy of $E_{AM1.5}$, the photons below the bandgap cannot be harvested, and the energy portion of photons above the bandgap is wasted as well.

Notice that some of the electron-hole pairs can meet and recombine, emitting a photon, following the second law of thermodynamics, which depends on the temperatures of the sun and the cell. Therefore, a fairly large portion of the electron-hole pairs in recombination will not contribute to electricity generation eventually. Thermal equilibrium can be reached when the cell is heated to a certain temperature. The recombination will lead to an emitted photon flux $\phi_e(E)$ changing with externally applied voltage V as

$$\phi_e(E, V) = \frac{2\pi E^2}{h^3c^2}\frac{1 - R(E)}{\exp\dfrac{E - eV}{k_BT_c} - 1} \tag{2.18}$$

where k_B is the Boltzmann constant, T_c the solar cell temperature, and the externally applied voltage V corresponding to the quasi-Fermi level splitting. The radiative recombination current density can be calculated as

$$I_r(V) = f_g e\int_{E_g}^{\infty}\frac{2\pi E^2}{h^3c^2}\frac{1 - R(E)}{\exp\dfrac{E - eV}{k_BT_c} - 1}dE \tag{2.19}$$

where $f_g = 2$ if the solar cell is free so that photons can be emitted through both the top and bottom and $f_g = 1$ if a perfect reflector is placed at the back so that photons can only be emitted through the top. Notice that a reflector under the solar cell can boost efficiency.

Then the short-circuit current density can be obtained

$$I_{SC} = I_g - I_r(0) \tag{2.20}$$

Assuming $R(E) = 0$, the I–V curve can be constructed by

$$I = I_{SC} + I_r(0) - I_r(V) \cong I_{SC} - I_r(0)\left(\exp\frac{V}{V_c} - 1\right) \tag{2.21}$$

where $V_c = KT_c/e$, and $\exp\frac{E-eV}{k_BT_c} - 1 \cong \exp\frac{E-eV}{k_BT_c}$ is used to simplify the equation.

From the above equation, one can draw an I–V curve and calculate $P = VI$, from which the maximum power output P_{max} can be calculated. Notice that some nonradiative recombination and generation processes also exist to slightly change the calculation (Rühle, 2016) but the error is very small. The electric power conversion efficiency of the solar cell η can be calculated by

$$\eta = P_{max}/E_{AM1.5} \tag{2.22}$$

The PV efficiency depends on the temperature of solar cells. At room temperature, the cap efficiency of a solar cell reaches a maximum value of 30%, which means that 70% of energy cannot be used for electricity generation but mostly turns to heat dissipation. This theoretical threshold is known as the Shockley–Queisser limit (Shockley and Queisser, 1961), which depends on the bandgap of the PV materials. The maximum is 33.16% for a PV material with a bandgap of 1.34 eV (Rühle, 2016).

There are multiple ways to exceed the Shockley–Queisser limit by breaking down the assumptions in the derivation. For example, tandem cells use multijunctions with different bandgaps to selectively harvest photons at different energy levels to reach a very high theoretical limit depending on the number of junctions—Bremner et al. predicted the maximum efficiency for a two junction tandem at 47% under the AM1.5G spectrum using the top cell with a bandgap of 1.63 eV and the bottom 0.96 eV (Bremner et al., 2008). Concentrated PV cells change the solar irradiation spectrum by light concentration, which leads to a different optimal bandgap and overall efficiency limit. Quantum dots also showed great potential to capture high-energy photons to generate more electron-hole pairs than the number of photons. Those technologies can be combined to get higher efficiency. As discussed in Chapter 1 (Fig. 1.3), the efficiency record reached 47.1% in the research phase. However, in commercial applications, efficiency should be balanced with the availability, simplicity, and durability of the PV cells for economic and technical considerations. Silicon cells will continue to dominate in this industry for many years.

When a PV cell heats up, electrons move faster and therefore the ability of the "built-in voltage" to separate the electrons and holes is reduced, and therefore the electricity generation efficiency decreases as well. Unfortunately, the solar cell temperature will commonly be higher than the ambient temperature because the dark panels are fully exposed to the sun with >70% of irradiation for heating the cell. Most solar panels have a temperature coefficient of efficiency reduction around −0.3% /°C to −0.5% /°C for every 1°C above 25°C. It is not always linear. When the temperature is higher than 45°C, this coefficient will be quickly increased or multiplied. optimized by the flow control. The sunlight strikes the solar panel surface and reflected a small portion. The absorbed energy will be harvested by PV cells, which exhibits an efficiency range of 15%–25% for silicon cells on the market today. The remaining energy will be transferred to heat, which is taken by the water flow and dissipated to the interior and exterior air as well as the structure body by convection, radiation, and conduction. Under the structural and temperature load, stress and deflection will be induced in the panel. In the following, the heat and mass transfer, the deflection and stress transfer, and energy dynamics within a BIPVT system under the service conditions will be introduced and the related analysis methods will be presented. The overall performance can be evaluated by a life cycle analysis.

2.2 Heat and mass transfer in BIPVT panels

A BIPVT panel typically contains multiple layers with coatings, glass, encapsulant, solar cell, structural substrate, pipes, and flow inside for thermal management as shown in Fig. 2.6. The solar energy is absorbed by PV cells with a minor part reflected to the sky. The heat and mass transfer will control the PV efficiency, building temperatures, and heat harvesting temperature, etc., and determine the performance of BIPVT systems. Because all these quantities are correlated with each other and

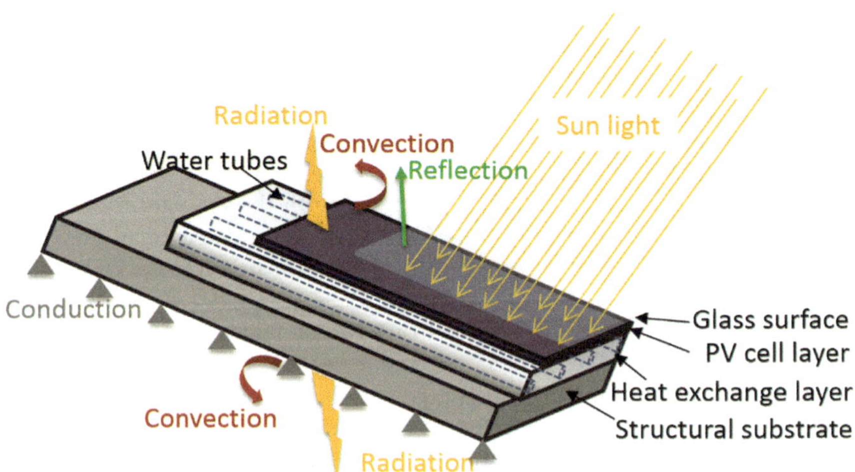

FIGURE 2.6 Schematic illustration of the structural configuration and energy flow of a BIPVT panel under solar irradiation (Yang and Yin, 2011).

depend on the sunlight and ambient environment, it is a challenging task to design, analyze, and optimize the BIPVT system for optimal performance.

2.2.1 The first law of thermodynamics and steady flow

According to the first law of thermodynamics, energy cannot be created or annihilated, but rather it can be converted from one form to another. The energy transfer can be passed through the state change with pressure, volume, and temperature as well as the mass and heat flow across the boundary. For a closed system in a domain Ω, the internal energy or energy content can be written as U_Ω, which is typically a function of pressure P and temperature T, as $U_\Omega(P, T)$. The internal energy can increase with the heat flow Q_Ω into the domain and decrease with the mechanical work W_Ω done by the system. Therefore, the changing of these three quantities shall follow the energy conservation as:

$$dU_\Omega = dQ_\Omega - dW_\Omega \tag{2.23}$$

which is the first law of thermodynamics for a closed system, where the sign of dW is negative because the energy is leaving the system. For example, an adiabatic process does not exhibit heat transfer, so $dU = -dW$; whereas an adynamic process does not have work exchange with the surroundings, so $dU = dQ$.

The concept can be extended to an open system by considering the domain Ω and its surrounding domain S. Because of the energy conservation, the heat and work exchange between two domains will be balanced with each other. Therefore one can write

$$dU_\Omega + dU_S = 0 \tag{2.24}$$

where subscripts show the domains of Ω and S. The combination of the above two equations also shows that the heat and work exchange causes the change of U_S. Notice that the absolute level of specific internal energy does not have to know in the above two equations and only the changes are taken into account for a thermodynamic process.

Considering a spatially inhomogeneous system with a steady flow in a water pipe, a small mass m of the flow, say $m = \rho \cdot A \cdot L$ with A—cross-sectional area of the pipe, L—the length of the cylinder, ρ—density, moves from entrance $\mathbf{x}^0$ to exit $\mathbf{x}^1$. Here the bold font means a vector or tensor variable. The total energy content at any location $\mathbf{x}$ can be written in terms of the internal energy, kinetic energy E_k, and potential energy E_p as

$$E = U + E_k + E_p \tag{2.25}$$

where $U = mU_m$, $E_k = \frac{1}{2}m\dot{\mathbf{x}}^2$ and $E_p = mgx_3$, in which $\dot{\mathbf{x}}$ is the velocity and x_3 the coordinate in the height direction. Eq. (2.23) can be rewritten as:

$$m \cdot d\left(U_m + \frac{1}{2}\dot{\mathbf{x}}^2 + gx_3\right) = dQ_\Omega - dW_\Omega \tag{2.26}$$

For this steady flow, integrating the change of the above quantities from $\mathbf{x}^0$ to $\mathbf{x}^1$ yields

$$\Delta U_m + \frac{1}{2}\Delta\dot{\mathbf{x}}^2 + g\Delta x_3 = \Delta Q_\Omega - \int_{x^0}^{x^1} dW_\Omega \tag{2.27}$$

where $\Delta(.)$ means the change of quantity from $\mathbf{x}^0$ to $\mathbf{x}^1$. The work exchange includes the work to be done by frictional force along the path as a negative quantity and by the pressure at the entrance $\mathbf{x}^0$ and the exit $\mathbf{x}^1$, which can be written as:

$$\int_{\mathbf{x}^0}^{\mathbf{x}^1} dW_\Omega = p^1\frac{m}{\rho^1} - p^0\frac{m}{\rho^0} - \int_{\mathbf{x}^0}^{\mathbf{x}^1} \mathrm{f}_x \cdot d\mathbf{x} \tag{2.28}$$

where p^1 and p^0 are the pressure at $\mathbf{x}^1$ and x^0, ρ^1, and ρ^0 the density at x^1 and x^0, respectively, f_x is the frictional force along the pipe cross-section while the mass travels along the pipe. Notice one can write $\frac{1}{\rho^1} = v_1$ and $\frac{1}{\rho^0} = v_0$ as the specific volume for a unit mass at x^0 and x^1, respectively. Therefore Eq. (2.27) can be rewritten as:

$$\Delta\left(U_m + pv + \frac{1}{2}\dot{\mathbf{x}}^2 + gx_3\right) = \frac{1}{m}\left(\Delta Q_\Omega + \int_{\mathbf{x}^0}^{\mathbf{x}^1} \mathbf{f}_x \cdot d\mathbf{x}\right) \tag{2.29}$$

where $h = U_m + pv$ is also the specific enthalpy. Therefore, the steady flow energy equation can be obtained as follows:

$$\Delta\left(h + \frac{1}{2}\dot{\mathbf{x}}^2 + gx_3\right) = \frac{1}{m}\left(\Delta Q_\Omega + \int_{\mathbf{x}^0}^{\mathbf{x}^1} \mathbf{f}_x \cdot d\mathbf{x}\right) \tag{2.30}$$

For liquid-based BIPVT, the thermal fluid can be considered incompressible. Due to the continuity and the closed-loop in the BIPVT, the effect of potential energy and kinetic energy will be relatively small, the power to produce pressure gradient and frictional force caused by the viscosity of the fluid will be the focus. Chapter 6 will conduct more analysis. When phase change occurs, the variation of enthalpy h and heat transfer will be the main consideration.

2.2.2 Heat transfer in a BIPVT panel

A BIPVT technology requires a thermal management system to monitor and control the heat flow and temperature distribution in BIPVT panels.

Heat or thermal energy is the kinetic energy of a distorted movement of the molecules of a body with a temperature above 0 K. The heat will flow from two bodies with a temperature difference from the hotter body to the cooler one. Usually, heat is denoted with the symbol Q (or q) as a form of energy and its unit is Joule (J). Heat transfer could be conducted in one or all of the three principal mechanisms of conduction, convection, and radiation (Jäger et al., 2016; Naraghi, 2020). Fig. 2.7 illustrates the three mechanisms. Radiation crosses space without a media; convection crosses the interface between a solid and fluid including liquid or gas, and conduction occurs through a volume of a matter.

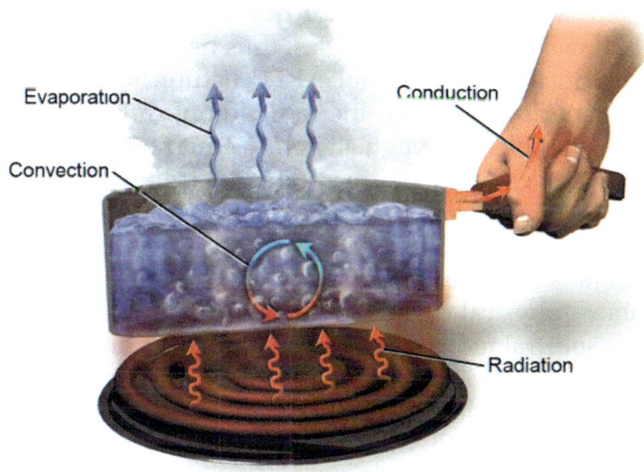

FIGURE 2.7 Three methods of heat energy transfer. Wikimedia, Mechanisms of Heat Transfer. https://commons.wikimedia.org/wiki/File:Heat_Transfer.png.

2.2.3 Thermal radiation

Every material with a temperature greater than the absolute zero (−273.15°C) emits this form of electromagnetic radiation. This thermal radiation is caused by the random motion of the particles in matter, which results in charge-acceleration or dipole oscillation with electromagnetic radiation. IR radiation from live animals and humans detected by an IR camera, cosmic microwave background radiation, and the glowing reddish or yellowish color from a hot metal are examples of thermal radiation.

In discussing the thermal radiation, some primary physical properties of a material are introduced:

- Emissivity ε: the effectiveness in emitting energy as radiation, or the fraction of the emitted radiation from the material.

- Absorptivity: the fraction of the absorbed incident radiation by the material.
- Reflectivity: the fraction of the incident radiation reflected by the material.
- Transmissivity: the fraction of the incident radiation passing through the material.

Based on the energy conservation, the sum of absorptivity, reflectivity, and transmissivity will be 100%. According to Kirchhoff's law of thermal radiation, in thermodynamic equilibrium, the emissivity is equal to the absorptivity. Therefore a material with good absorptivity is also a good emitter. A "black body" is a concept defined as a perfect emitter and absorber at all wavelengths and angles, thus having zero reflectivity and transmissivity. Such matter does not exist in nature, however, its definition makes the comparison between the radiation of different materials and formulating it possible. Although all bodies emit radiation, the maximum radiation at any given temperature is always attributed to the black body. Since a black body at its thermal equilibrium is not only emits the maximum radiation but also absorbs the maximum amount, the expectation is that it should look black since no light is reflected from its surface.

All the above physical constants are dependent on the wavelength of radiation. For spectral radiation, those physical constants can be provided in an average sense by the integral over the spectrum. For example, Eq. (2.9) provides the ideal blackbody radiation spectrum. The integral over the whole wavelength yields the Stefan-Boltzmann law, which describes the relationship between the power emitted from a black body and its temperature as

$$E_{\text{rad.}}^{\text{BB}}(T) = \sigma T^4 \tag{2.31}$$

where E is the total irradiant emittance (W/m^2), σ is the Stefan-Boltzmann constant $\sigma = \frac{2K_B^4\pi^5}{15h^3c^2} \approx 5.670 \times 10^{-8}$ W/m$^2K^4$ and T is the black body temperature.

For any object other than a black body that does not absorb all the incident radiation and accordingly, emits less energy than a black body ($\varepsilon < 1$), this equation becomes

$$E_e^{\text{BB}}(T) = \epsilon\sigma T^4 \tag{2.32}$$

When two surfaces exhibit different emissivity and temperature as ε_1 and ε_2, and T_1 and T_2 ($T_1 > T_2$) respectively, the heat will be transferred from T_1 to T_2 by the following equation

$$E_{\text{rad}}^{1-2} = \epsilon_{\text{eff}} \cdot \sigma \cdot \left(T_1^4 - T_2^4\right) \tag{2.33}$$

where the effective emissivity depends on ε_1 and ε_2 as $\varepsilon_{\text{eff}} = \left(\frac{1}{\varepsilon_1} + \frac{1}{\varepsilon_2} - 1\right)^{-1}$. For a BIPVT's working temperature commonly at 300 ± 50°K, the radiation wavelength is concentrated around 10 microns, which can pass the air very well. Therefore, a heated BIPVT panel will dissipate energy to the surrounding wall inside and to the sky through the above equation.

Notice that because the sky exhibits low temperature, if a coating exhibits a high reflectivity and a high emissivity, it cannot absorb solar irradiation but will emit heat to the sky, which leads to radiative cooling of the coating. This mechanism has been used in the smart window project which will be introduced in Chapter 10.

2.2.4 Heat convection

Heat convection is another means of energy transfer due to the mixing (diffusion) in fluids. On the microscopic scale, there is a relation between the kinetic energy of the molecules of fluid and their thermal energy. If there are heat flux and an increase in the temperature, the increased thermal energy will result in the movement and vibration of the molecules. In this way, the molecules of the fluid (gas or liquid) with higher kinetic energy move along from regions with higher kinetic energy to regions with lower energy. There are two types of convective heat transfer methods (Naraghi, 2020):

1) Free convection, in which heat transfer happens due to the temperature difference in the fluid. For example, air circulates because of differences in the buoyancy of a hot point with lower density to a cold point with higher density. This is why in winter, the hot air rises in the room and the cold air sinks near the floor.
2) Forced convection, where an external force is induced to enhance the convective heat exchange. This force can be exerted in either a laminar or turbulent flow. In the laminar flow regime, due to the lower energy exchange rate between the external particles and the fluid, the dominant factor responsible will be conduction in the cross direction, whereas in the turbulent flow, the increased heat transport is expected across the main direction of the flow.

Often, the overall convection is a product of the contribution of both free and forced components. The quantitative description of heat convection can be traced back to Newton's law of cooling: the rate of heat loss of a body is proportional to the difference in temperatures between the body and its surroundings while under the effects of a breeze. When the proportional coefficient is independent, or relatively independent, of the temperature difference between object and environment. For a solid with surface temperature T exposed to a fluid with temperature T_f, this linear relationship can be written as:

$$E_{\text{con.}} = h_c \cdot \left(T - T_f\right) \tag{2.34}$$

where h_c is the convective heat transfer coefficient, which depends on the physical configuration and properties of the fluid and the flow speed. The Nusselt number (Nu) is often used to characterize the relationship between heat transfer and fluid dynamics. It is a ratio of the convective to conductive heat transfer at the boundary in a fluid. Because the convection includes both advection caused by fluid motion and diffusion caused by heat conduction which will be elaborated in the next section, $Nu \geq 1$ where $Nu = 1$ means heat transfer by pure conduction. Because heating/cooling always causes density change of the fluid, which will lead to fluid motion, generally $Nu > 1$.

For a BIPVT panel exposed to indoor and outdoor air and containing a fluid pipe, h_c typically is small for indoor air because the air motion follows natural convection, but is larger for outdoor surface because of wind and breeze in an open space. For summer and winter, it exhibits different values due to the different flow nature in the cooling and heating season. For heat exchange with thermal fluid, h_c can be significantly higher due to the higher flow speed and specific heat of the fluid. Particularly when a phase change occurs in the fluid, h_c can increase at10^2 times, which lead to heat pipe technology in Chapter 6.

Notice that both radiation and convection occur on the surface or interface, when the BIPVT panel temperature distribution is studied, the radiation and convection will be considered as a boundary condition. It is often to combine convection and radiation by rewriting Eq. (2.33) as

$$E_{\text{rad.}}^{1-2} = h_{\text{rad.}} \cdot (T_1 - T_2) \tag{2.35}$$

where $h_{\text{rad.}} = \varepsilon_{\text{eff}}\sigma\left(T_1^1 + T_2^2\right)(T_1 + T_2)$.

For example, if the indoor air and wall are the same temperature of T_2, the radiative heat transfer can be superposed to the convective heat transfer by

$$h_{\text{eff}} = h_{\text{rad}} + h_{\text{con}} \tag{2.36}$$

2.2.5 Heat conduction

In heat conduction, kinetic energy is directly exchanged at the microscopic level between the particles of the materials at the boundary. Where there is a temperature difference (gradient) between two objects, heat flows from the object with a higher temperature to the other one with a lower temperature, until a thermal equilibrium is maintained (Figs 2.6 and 2.7).

In this form of heat transfer, energy is transported within a medium due to a temperature gradient by the transfer of kinetic energy between particles at the molecular level. The heat conduction relation was formulated by the French mathematician Joseph Fourier in 1822 and is known as Fourier's law of conduction

$$q = -k\nabla T \tag{2.37}$$

In which q is the local heat flux density (unit: W/m^2), k is the material's conductivity (W/m/K), and ∇T is the temperature gradient (K/m). If we integrate the above differential form over the material's oriented surface S, we will achieve the integral form of Fourier's law

$$\dot{Q} = -k \oiint_{S} \nabla T.dS \tag{2.38}$$

Here, $\dot{Q}$ is the heat flow rate from the volume V which causes the temperature change of the volume as $\frac{\partial Q}{\partial t} = -\rho c_p V \frac{\partial T}{\partial t}$, and dS is the surface area of the element (m^2) with an outward norm n of the surface plane. Using the Gauss' theorem, the surface integral can be written as a volume integral as follows:

$$\frac{\partial T}{\partial t} = \frac{k}{\sigma c_p} \frac{\iiint v^{\Delta^2 T dV}}{V} \tag{2.39}$$

When V approaches zero, the above equation can be reduced to

$$\frac{\partial T}{\partial t} = \alpha \nabla^2 T \tag{2.40}$$

Where $\alpha = \frac{k}{\rho c_p}$ is the diffusivity of the material. For a one-dimensional (1D) homogenous material between two points at constant temperature, the heat flow rate can be written as follows

$$\dot{Q} = -kA\frac{\Delta T}{\Delta x} \tag{2.41}$$

where Q is the flow of heat (J) with the cross-sectional surface area of A, temperature difference of ΔT and distance of Δx.

Therefore, for a uniform panel layer, with a thickness L and area A, subjected to a temperature difference ΔT, the heat flow rate can be written as

$$\dot{Q} = \frac{\Delta T}{R} \tag{2.42}$$

where $R = \frac{L}{kA}$ is the thermal conduction resistance with a unit of °C/W in the SI and (h-°F)/BTU in the English systems. The value of R for the unit area ($A = 1$, $R' = \frac{L}{k}$) is used to describe the resistance of insulation (known as R-value), which indicates the ability of a material in the building to resist heat flow through it. Therefore increasing the thickness can proportionally increase the R-value.

The thermal resistance addition concept is analogous to resistors in Ohm's law. Therefore, when several conducting layers lie between the cold and hot regions, assuming that all have the same cross-sectional area A, the total resistance for the series layers become

$$R_{\text{total}}^{\text{Series}} = R_1 + R_2 + R_3 + \cdots \tag{2.43}$$

However, the composite walls in most cases consist of multiple parallel layers and thus, the total resistance follows the parallel heat flow paths

$$R_{\text{total}}^{\text{Parallel}} = \left(\frac{1}{R_1} + \frac{1}{R_2} + \frac{1}{R_3} + \ldots\right)^{-1} \tag{2.44}$$

Similarly, U value, the reciprocal of R-value, is also widely used. For the series and parallel heat flow paths of layered materials, the above two equations can be rewritten as

$$U_{\text{total}}^{\text{Series}} = \left(\frac{1}{U_1} + \frac{1}{U_2} + \frac{1}{U_3} + \cdots\right)^{-1} \tag{2.45}$$

and

$$U_{\text{total}}^{\text{Parallel}} = U_1 + U_2 + U_3 + \cdots \tag{2.46}$$

In a BIPVT panel, the overall thermal resistance across the panel will follow the series case, whereas the heat transfer from one side to another will follow the parallel case. When conduction, radiation, and convection occurs at the same time from the hot roof surface at temperature T_h to the cold indoor air at temperature T_c, heat flux $\dot{E}$ can be written by the 1D heat transfer in Fig. 2.8

$$\Delta T = \frac{L}{k}\dot{E} + \frac{1}{h_{\text{rad}} + h_c}\dot{E} \tag{2.47}$$

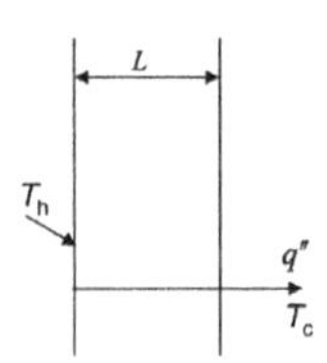

FIGURE 2.8 A one-dimensional heat transfer from a hot surface to a cold space.

where k is the thermal conductivity of the panel, h_{rad} and h_c are convective and radiative heat transfer coefficients. Therefore one can treat the configuration as a uniform material with an effective R value

$$R_{\mathrm{eff}} = \frac{L}{k} + \frac{1}{h_{\mathrm{rad}} + h_c} \tag{2.48}$$

Notice that h_{rad} is temperature-dependent. For BIPVT with a tube inside, the thermal fluid tubes will disturb the heat flow in the panel, which will make the 1D analysis invalid. More complex simulation and modeling work is required and can often be done with numerical methods. The effective U value or R-value can still be estimated in an average sense. Because the heat flow can be partially or fully intercepted by thermal fluid before it enters the indoor environment, the effective R-value will be changeable by the flow rate and temperature of the thermal fluid and can be extremely high. It is possible to have zero or negative heat flow into the indoor space. More analysis will be introduced in Sections 2.3 and 2.4.

2.2.6 The second law of thermodynamics and heat pump

In the operation of a BIPVT system, the heat is harvested from BIPVT at a certain temperature, say 35–70°C, and can be used for some purpose. However, this heat supply may not perfectly match the demands in all seasons. For example, on a hot summer day, the major need in a building is cooling. This medium-temperature heat supply may not be useful. However, cold thermal fluid or water is needed to cool down the PV cells. As an ideal case, the feeding temperature of the BIPVT keeps below 25°C, and leaving temperature keeps above 35°C. If the ambient temperature is close to or above 35°C, there is no natural way to dissipate the heat from the thermal fluid to the environment at ~35°C to cool down the liquid to <25°C. A heat pump can use external energy to take the heat away from the hot thermal fluid to a heat sink, i.e. the environment or heat storage, and cool it down to a temperature less than the heat sink temperature. Therefore, the thermal fluid circulation can be operated in a designed way.

Basically, a heat pump facilitates a way to dissipate heat from a heat source at a lower temperature to a heat sink at a higher temperature, or inversely to take heat from a heat source at a lower temperature to a heat sink at a higher temperature. This is not against the first law of thermodynamics but cannot be achieved by any of the three principal heat transfer mechanisms. External energy is required to make this unnatural process happen. The second law of thermodynamics can quantify the energy needs to achieve this goal.

Entropy, S, is a state variable in the Second Law of thermodynamics which is similar to enthalpy in the First Law of thermodynamics. The change in entropy, dS, is equal to the heat transfer dQ divided by the temperature T as:

$$dS = dQ/T \tag{2.49}$$

Given a thermodynamic process of a system interacting with the surrounding environment, the total entropy at the starting state S_0 and any following state S_1 should satisfy:

$$S_0 \leq S_1 \tag{2.50}$$

or

$$dS \geq 0 \tag{2.51}$$

which is the Second Law of Thermodynamics. When $S_0 = S_1$, the process is reversible and is called an isentropic process. For example, an adiabatic compression of an ideal gas that has no energy transfer by heat is an isentropic process although it may not exactly exist as motion always coupling with friction and viscous dissipation. In general, $S_0 < S_1$ and the process is irreversible. For example, two same masses with different temperatures T_1 and T_2 ($T_1 > T_2$) contact with each other, and reach an equilibrium temperature $\frac{T_1+T_2}{2}$ by a heat conduction Q from $1 \rightarrow 2$. The change of entropy is

$$\Delta S \cong -\frac{Q}{T_1} + \frac{Q}{T_2} = \frac{Q(T_1 - T_2)}{T_1 T_2} > 0 \tag{2.52}$$

where the thermodynamic process is approximated by one step. Strictly the temperature change can be divided into infinitesimal steps with heat flow dQ, which can be integrated along with the temperature change and will lead to the same conclusion.

When heat is used to generate mechanical work in a heat engine, or mechanical work is used to pump heat in heat pump, both using heat reservoirs at two temperatures, the Second Law of Thermodynamics states that a Carnot cycle reaches the greatest possible efficiency for any process between these two temperatures. Fig. 2.9 shows the Reversed Carnot Cycle for a heat pump, which follows the same principle of Carnot Cycle for heat engine but the opposite cycle direction.

In the Reversed Carnot Cycle shown in Fig. 2.9, four steps are described as follows:

1–2: an isothermal expansion, in which the refrigerant (two-phase mixture) is evaporated to state point 2. The latent enthalpy necessary for evaporation is supplied by the refrigerated space surrounding the evaporator.

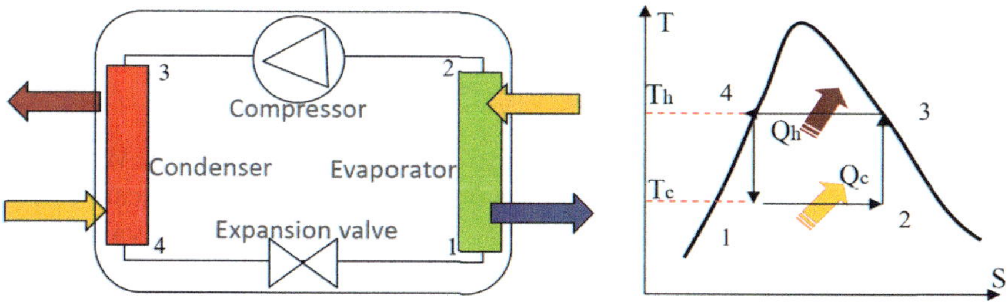

FIGURE 2.9 The Reversed Carnot Cycle for heat pump (left) and the temperature-entropy diagram (right).

The amount of heat transferred to the working fluid in the evaporator is called the refrigeration load. The heat absorbed is Q_c.

2–3: a reversible, adiabatic (isentropic) compression of the refrigerant. The two-phase mixture will heat up by the compression work from T_c to T_h, and reach the saturated vapor state 3.

3–4: an isothermal compression with heat rejection at T_h in which the working substance is de-superheated and then condensed to a saturated liquid state 4. During this process, the working substance rejects most of its energy to the condenser cooling water. The heat ejected is Q_h.

4–1: a reversible, adiabatic (isentropic) expansion in which the saturated liquid expand without heat input causes it to cool to T_c but the entropy remains unchanged.

The cycle includes two reversible isothermal steps 1–2 and 3–4 and two isentropic steps 2–3 and 4–1. Obvious the entropy changes in steps 1–2 and 3–4 should share the same value with the opposite sign, say $\Delta S = S_2 - S_1 = S_3 - S_4$. Then one can write

$$Q_c = \Delta S T_c \ \ and \ Q_h = \Delta S T_h \tag{2.53}$$

The overall work input to the system can be obtained by the conversation of energy in a cycle as

$$W = Q_h - Q_c \tag{2.54}$$

which can be confirmed by integral the P–V curve about the cycle. When the Reversed Carnot Cycle is used for heat pump, Q_h is the target and the coefficient of performance (COP) is defined by

$$COP_{\mathrm{HP}} = \frac{Q_h}{W} = \frac{T_h}{T_h - T_c} = 1 + \frac{T_c}{T_h - T_c} \tag{2.55}$$

Given T_c, the output temperature T_h is higher, the COP is lower. When the Reversed Carnot Cycle is used for refrigeration, Q_L is the target and the

COP is defined by

$$COP_R = \frac{Q_C}{W} = \frac{T_c}{T_h - T_c} \tag{2.56}$$

Given two heat reservoirs of T_h and T_c, $COP_{HP} = COP_R + 1$. When the temperature gap $T_h - T_c$ is small, say 30°C, for BIPVT application, the COP of Carnot cycle for either heating or cooling can reach a value ~10, which means 1 Joule of power can meet the heating or cooling need at 10 Joules. However, the Carnot Cycle is an ideal design for the upper limit of efficiency but is not practical. First, it is impossible to make every step reversible as they typically require a long time to process without internal friction or turbulence; secondly, it is hard to work on the mixture of liquid and vapor particularly in a compressor. Modern thermodynamic cycles typically modify the Carnot cycle for rapid and workable processes, while still reaching a high COP. It is practical for modern heat engines or pumps to reach 65%~75% of the Carnot efficiency.

A BIPVT system harvests both electricity and low-quality (temperature) heat. Using a portion of electricity, the temperature can be boosted to a higher one at high efficiency, so that higher-quality heat can be stored for energy generation and direct use.

2.3 Energy dynamics and modeling of BIPVT systems

A BIPVT system needs to address the energy harvesting efficiency, thermal management, thermal fluid circulation, and energy storage, and interfaces with the building's energy system in lighting, heating and cooling, water, and air circulation systems. Because BIPVT energy harvesting and building energy demands change with the weather, location, season, time, among other factors and can be affected by the thermal fluid flow, it is important to understand the energy dynamics in the BIPVT system for better integration with building energy system. The energy modeling of the BIPVT performance can be used for parameter optimization in design, building code compliance, and performance prediction and control of BIPVT in operation. It can be added to existing building energy modeling software (Beckman et al., 1994, DOE, 2020) for system performance optimization and control.

2.3.1 Electric modeling of BIPVT systems

For a single BIPVT panel, given solar irradiation, the power generation can be written as:

$$P = E.A_{\text{module}} \cdot \phi_{\text{cell}} \cdot \eta_{\text{cell}} \cdot \eta_{\text{inv}} \tag{2.57}$$

where:

P is the electrical power generated by the panel (W),
E is the solar radiation incident on PV module (W/m^2) which is affected by weather, shading, and reflection.
A_{module} is the area of the panel (m^2).
ϕ_{cell} is the fraction of surface area of PV cell in the panel.
η_{cell} is the PV cell conversion efficiency which depends on the module temperature.
η_{inv} is the conversion efficiency from direct current (DC) to alternative current (AC).

Here the power efficiency η_{cell} is calculated from the maximum value from the I–V curve, such as Eq. (2.21) in the physical analysis. Therefore, in actual application, the power output truly depends on how the load configuration matches with the panel's internal parameters.

However, in an actual PV module, although the I–V curve can be measured by the experiments, it is not easy to exactly predict it with a certain number of parameters. Among many models, the Four Parameter Model has been well accepted for its simplicity and accuracy. The four-parameter equivalent circuit model can be traced back to Townsend (1989) and Duffie and Beckman (2013), and was recently improved by Khezzar et al. (2014), which demonstrated high accuracy for mono-crystalline, multi-crystalline, and thin-film cells. The Four Parameter Model assumes that the slope of the IV curve is zero at the short-circuit condition with the equivalent circuit is shown in Fig. 2.10. The photocurrent I_L of solar module, the saturation current I_0, the series resistance R_s is given. In addition, the PV's I–V curve shape shall follow Eq. (2.21) with a scaling factor to be determined by curve fitting.

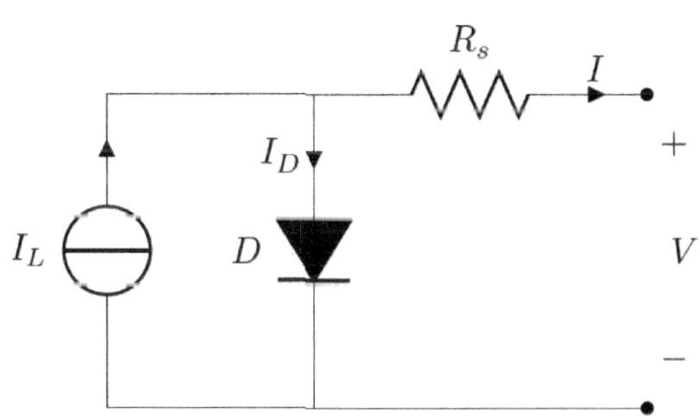

FIGURE 2.10 Equivalent circuit in the four-parameter model (Khezzar et al., 2014).

The current–voltage equation of the above circuit can be written as:

$$I = I_L - I_0\left(\exp\frac{V + IR_s}{a} - 1\right) \tag{2.58}$$

where a is the scaling factor to be determined by I–V curve shape, which can be corresponding to $\frac{N_cAKT_c}{q}$ in the reference (Khezzar et al., 2014).

Under operating conditions, when the V_{oc} open-circuit voltage, I_{sc} short-circuit current, I_m and V_m the voltage and current at the maximum power point (MPP) are measured, the four parameters can be determined by the following three equations for the three points on the I–V curve

$$I_{sc} = I_L - I_0\left(\exp\frac{0 + I_{sc}R_s}{a} - 1\right) \tag{2.59}$$

$$0 = I_L - I_0\left(\exp\frac{V_{oc} + 0}{a} - 1\right) \tag{2.60}$$

$$I_m = I_L - I_0\left(\exp\frac{V_m + I_m R_s}{a} - 1\right) \tag{2.61}$$

and the additional condition at the MPP (V_m, I_m),

$$\frac{dP}{dV}|_{V=V_m} = V_m\frac{dI}{dV}|_{V=V_m} + I_m = 0 \tag{2.62}$$

where $\frac{dI}{dV}$ can be derived from $\frac{dI}{dV} = -\left(\frac{I_0}{a}\exp\frac{V+I\,R_s}{a} + \frac{R_sI_0}{a}\exp\frac{V+I\,R_s}{a}\frac{dI}{dV}\right)$, i.e.

$$\frac{dI}{dV} = -\frac{\frac{I_0}{a}\exp\frac{V + I\,R_s}{a}}{1 + \frac{R_sI_0}{a}\exp\frac{V + I\,R_s}{a}} \tag{2.63}$$

Consider the difference between the photo-generated current I_L and the short-circuit current I_{sc} is generally very small and $exp\ \frac{V+IR_s}{a} \gg 1$ when $V > 0$, the above four equations can be simplified as

$$I_{sc} = I_L \tag{2.64}$$

$$0 = I_L - I_0\exp\frac{V_{oc}}{a} \tag{2.65}$$

$$I_m = I_L - I_0\exp\frac{V_m + I_m R_s}{a} \tag{2.66}$$

and

$$I_m - \frac{\frac{I_0}{a}\exp\frac{V_m + I_m R_s}{a}}{1 + \frac{R_sI_0}{a}\exp\frac{V_m + I_m R_s}{a}}V_m = 0 \tag{2.67}$$

where $exp\frac{V_m+I_m R_s}{a} = \frac{I_{sc}-I_m}{I_0}$. Therefore, the four parameters can be explicitly written by I_{sc}, V_{oc}, I_m, and V_m as follows

$$I_L = I_{sc};\ I_0 = I_{sc}exp\left(-\frac{V_{oc}}{a}\right) \tag{2.68}$$

$$R_s = \frac{a\ln\left(1-\frac{I_m}{I_{sc}}\right) + V_{oc} - V_m}{I_m} \tag{2.69}$$

$$a = \frac{2V_m - V_{oc}}{\frac{I_m}{I_{sc}-I_m} + \ln\frac{I_{sc}-I_m}{I_{sc}}} \tag{2.70}$$

Therefore, once the four parameters are obtained from the three measurement conditions of open circuit, short-circuit, and maximum power conditions, the I–V function or curve can be defined as.

$$I = I_{sc}\left(1 - \exp\frac{V - V_{oc} + IR_s}{a}\right) \tag{2.71}$$

Because in the operational situation, the BIPVT module performance depends on temperature and irradiation level, the I–V curve will be different. For example, researchers from Columbia University have measured the I–V and P–V curves for an amorphous silicon cell at different temperatures and irradiations, respectively (Lin et al., 2020).

In Fig. 2.11, the I–V curves under 1263 W/m^2 solar irradiation and different temperature condition are plotted, which shows that the temperature influences more on the open-circuit voltage output but has a negligible impact on the short-circuit current output. The corresponding

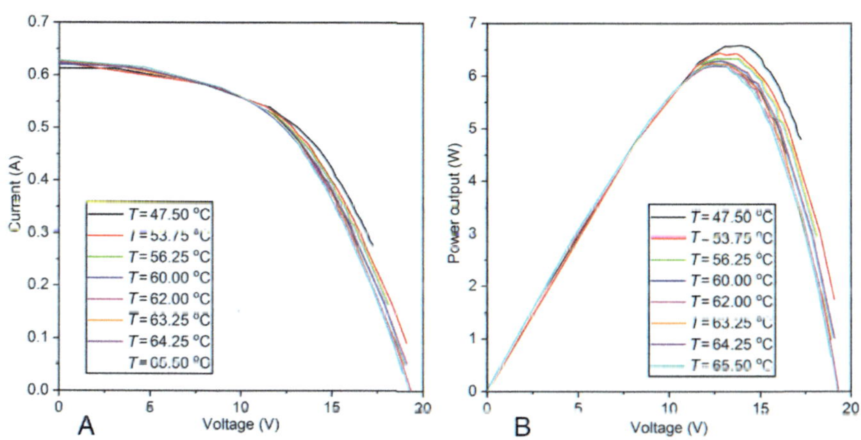

FIGURE 2.11 (A) Current and voltage relationships and (B) power output and external resistance relationships, under 1263 W/m^2 solar irradiation and different temperature conditions (Lin et al., 2020).

relationships between the power output to the external resistance are shown in Fig. 2.11A from which it is observed that the power outputs lie between 6 and 6.5 W and that the ~20°C temperature difference will generate approximately 6% performance compromise. The optimal power output points and the corresponding external resistances are captured from Fig. 2.11B and plotted versus the different temperature conditions in Fig. 2.12. The linear relationships to the temperature are observed for both the optimal power output and the external resistance.

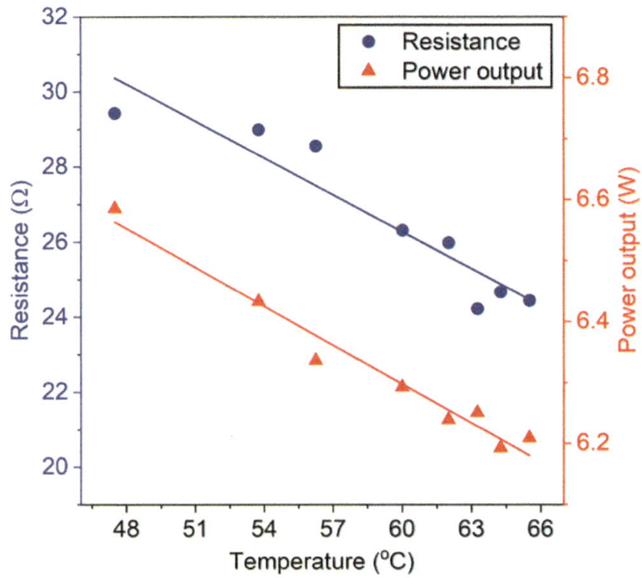

FIGURE 2.12 Temperature influence on the optimal power output and external resistance under 1263 W/m^2 solar irradiation (Lin et al., 2020).

Similar tests are carried out under different solar irradiation levels from 859 to 1709 W/m^2 after the temperature stabilized. The I–V curves are plotted in Fig. 2.13A, and it is observed that the solar irradiation has a significant influence on the short-circuit current, while almost no influence on the open-circuit voltage. The corresponding power output to the external resistance is given in Fig. 2.13B, from which a clear linear trend is observed between the optimal power output and the solar irradiation. The relationships between power output and external resistance under different solar irradiations are plotted in Fig. 2.14. Therefore, the best power output and the corresponding resistance can be calculated for all cases, given the solar irradiation and the temperature condition, with the help of bilinear interpolation and extrapolation of the current test results.

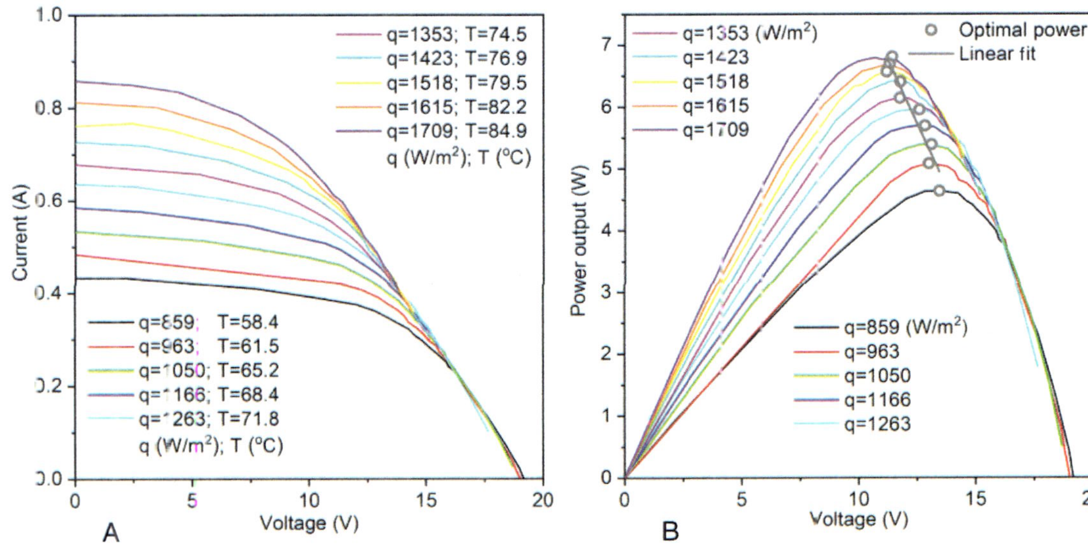

FIGURE 2.13 Current and voltage relationships (A) and power output and external resistance (B) under different solar irradiation levels (Lin et al., 2020).

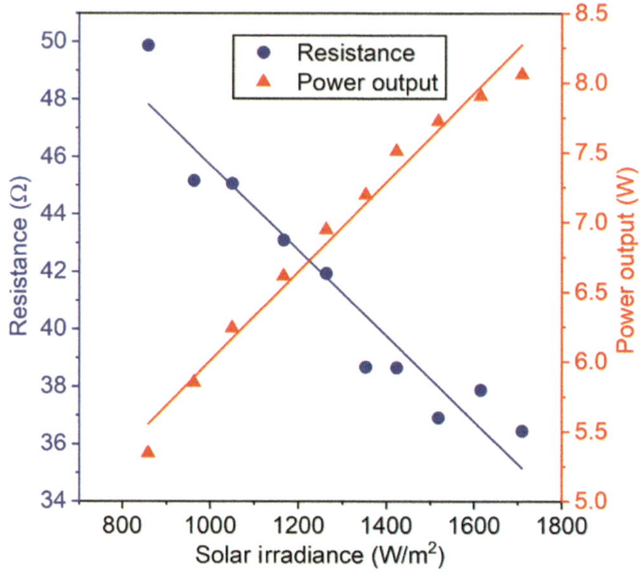

FIGURE 2.14 Solar irradiation influence on the optimal power output and external resistance (Lin et al., 2020).

Therefore, the actual electric power output will change with the load or external resistance. To yield the maximum power, the external resistance shall match with

$$R = \frac{V_m}{I_m} \tag{2.72}$$

Otherwise, the power output will be lower. Once the reference test is performed at a standard temperature T_{ref} and irradiation E_{ref} to obtain I_{sc}^{ref}, V_{oc}^{ref}, I_m^{ref}, and V_m^{ref}, the I–V curve for different operational conditions can be obtained by modeling the change of those parameters. Given a different temperature T and irradiation E, one popular way to generate the parameters is given as follows (Khezzar et al., 2014):

$$I_{sc} = I_{sc}^{ref}\frac{E}{E_{ref}} + \mu_{I_{sc}}(T - T_{ref}) \tag{2.73}$$

$$V_{oc} = V_{oc}^{ref} + a\frac{T}{T_{ref}}\ln\frac{E}{E_{ref}} + \mu_{V_{oc}}(T - T_{ref}) \tag{2.74}$$

$$I_m = I_m^{ref}\frac{E}{E_{ref}} + \mu_{I_{sc}}(T - T_{ref}) \tag{2.75}$$

$$V_m = V_m^{ref} + a\frac{T}{T_{ref}}\ln\frac{E}{E_{ref}} + \mu_{V_{oc}}(T - T_{ref}) \tag{2.76}$$

Notice that there are other models to generate new I–V curves. The applications of those models shall be cautious for different cell types and energy levels. Validation and calibration of the I–V models are necessary given a new material or application. When multiple BIPVT modules are connected in series or parallel, the I–V curves for each module should be matched to avoid mismatch losses. The performance of an array of identical modules is assumed to be linear with the number of modules in series and parallel. The inverter efficiency is applied linearly to take into account the efficiency loss in the DC–AC conversion. The inverter capacity forms a limit for power production from a PV generator.

2.3.2 Thermal modeling of BIPVT systems

A BIPVT system combines solar electric cells and thermal working fluid to collect both electricity and heat. It is not a simple superposition of the two components for energy harvesting, their efficiencies and capacity depend on each other in a complex way. First, based on energy conservation, if one way uses more, the energy for another one will be less. Secondly, if more heat is taken from the BIPVT panel, the lower temperature will lead to an increase in PV efficiency. Thirdly, to reach a lower temperature, the thermal fluid flow rate should be high, which lead to an increase pumping power and yield lower quality heat. Therefore, the temperature control and heat flow from the BIPVT system are critical to maintaining a safe operational mode with high efficiency.

Thermal management can be conducted through a thermal fluid flow with either air or liquid or phase change materials (PCMs). The advantage of using thermal fluid is the flexibility to control the fluid circulation rate adaptive to the environment, but more components are required including a fluid pump, pipe/duct system. Using air, it can be integrated into the heating, ventilation, and air condition air system loop; whereas using water, it can be connected with a water heating system for preheating water. However, a similar analysis procedure can be used for both systems. In the following, a water-based BIPVT system is used (Yang et al., 2012) to demonstrate the heat harvesting and panel cooling, as shown in Fig. 2.15. To monitoring the temperature change, eight observation points are provided on the solar panel to evaluate the temperature distribution, which also indicates the direction of water flow. The point number sequence follows the water flow directions, and point 8 stands for the outlet water temperature.

The energy flow rate into the panel upper surface as heat can be written as

$$\dot{q}_{in} = E(1 - R_{cell})(1 - \phi_{cell} \cdot \eta_{cell}) - \frac{T_{surf\,1}^4 - T_{sky}^4}{R_{rad\,1}} - \frac{T_{surf\,1} - T_{air}}{R_{conv\,1}} \tag{2.77}$$

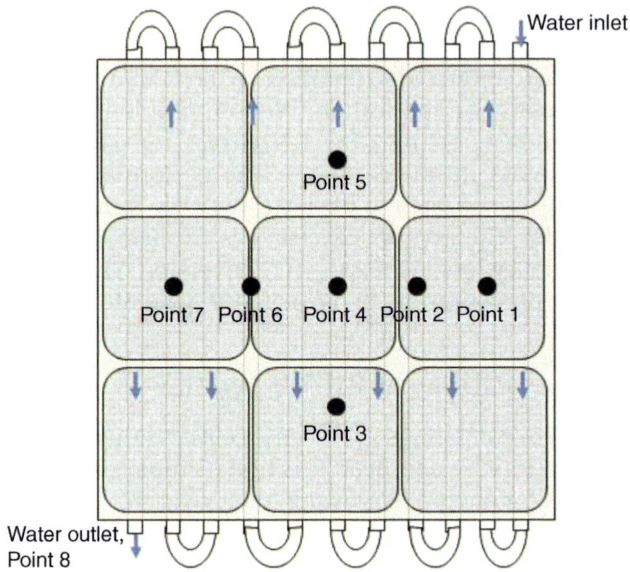

FIGURE 2.15 Thermal modeling of a photovoltaic thermal panel to harvest heat by water flow (Yang et al., 2012).

where:

$\dot{q}_{in}$ is the heat flow intensity ($\frac{J}{s.m^2}$) into the panel,
E is the solar radiation incident on PV module (W/m^2) which is affected by weather, shading, and reflection,
R_{cell} is the reflectivity of the surface area of PV cell in the area,
ϕ_{cell} is the fraction of surface area of PV cell in the area,
η_{cell} is the PV cell conversion efficiency which depends on the module temperature,
$T_{surf\,1}$ is the upper surface temperature facing to the sky,
T_{sky} is the temperature of the sky,
$R_{rad\,1}$ is the radiative resistance of the upper surface material of the BIPVT panel,
T_{air} is the ambient temperature of the air,
$R_{conv\,1}$ is the convective resistance of the upper surface material of the BIPVT panel.

The energy flow rate from the panel lower surface can be written as:

$$\dot{q}_{out} = \frac{T_{surf2}^4 - T_{indoor}^4}{R_{rad\,2}} + \frac{T_{surf2} - T_{indoor}}{R_{conv2}} \tag{2.78}$$

where:

$\dot{q}_{\text{out}}$ is the heat flow intensity ($\frac{J}{s.m^2}$) out from the panel to indoor air,
T_{surf2} is the lower surface temperature facing to the sky,
T_{indoor} is the indoor temperature,
$R_{\text{rad}}2$ is the radiative resistance of the lower surface material of the BIPVT panel,
R_{conv} is the convective resistance of the lower surface material of the BIPVT panel.

For a steady heat conduction problem with a steady flow rate, the temperature field becomes stable, the heat collected by the thermal fluid $\dot{q}_{\text{harvest}}$ will be constant as

$$\dot{q}_{\text{harvest}} = \dot{q}_{\text{in}} - \dot{q}_{\text{out}} \tag{2.79}$$

Scaling up the heat flow rate over the whole roof, the total heat flow rate can be written as $\dot{Q} = \dot{q}A$ with A being roof area. One can write the work fluid temperature change related to the water flow rate as

$$T_{\text{out}} = T_{\text{in}} + \frac{\dot{Q}_{\text{harvest}}}{\dot{m}c_p} \tag{2.80}$$

where: T_{out} is the temperature of the working fluid leaving the BIPVT system,

T_{in} is the temperature of the working fluid entering the BIPVT system,
$\dot{m}$ is the entire mass flow of the working fluid through the PV/T,
c_p is the specific heat of the working fluid.

On average, one can assume the effective collector temperature T_{BIPVT} is controlled by the T_{in} and T_{out}, for example, taking $T_{\text{BIPVT}} = \lambda T_{\text{out}} + (1 + \lambda)\, T_{in}$. By assuming the upper and lower surface temperatures at the average temperature, say $\lambda = 0.5$, one can calculate T_{out} by given $\dot{m}$, E, and T_{in} as well as panel configuration and other physical parameters.

This is a very rough estimation because a steady heat conduction state never happens on the roof as solar irradiation E changes with time and the water flow rate needs to change accordingly. To make an automatic control system, a transient heat conduction problem should be considered to predict the temperature change with the flow rate. In addition, the water temperature from the inlet to outlet gradually increases with the heat transfer along the pipeline, the temperature field will significantly change with spatial location as well.

To accurately predict the BIPVT performance, a transient boundary value problem should be set up. Numerical methods can be used to predict the performance with any given initial condition. The heat flow through the upper and lower surface will be used as boundary conditions. By discretization of the BIPVT into small elements, heat conduction in the solid can be modeled. The heat transfer in the fluid will be very complex at microscale due to the possible turbulence in the serpentine configuration, it can be simplified by a 1D element with temperature T_f changing with time and location. The pipe inner surface temperature can be averaged with the nodes around the cross-section so that the convective heat transfer equilibrium equation can be written as (Hanby et al., 2002):

$$Mc_p\frac{dT_i}{dt} = \dot{m}c_p(T_{f,i-1} - T_{f,i}) - h_f A_i(T_{f,i} - T_{w,i}) \tag{2.81}$$

where:

M is the mass in the element,
T_i is the temperature on the element,
h_f is the convective heat transfer coefficient,
A_i is the convective heat transfer area along the wall,
T_f and T_w are the average temperature of the fluid and pipe wall, respectively.

Here h_f can be used to take into account the fluid rate, fluid physical properties, and surface materials. Then the heat transfer and temperature field in the panel can be simulated and the energy efficiency can be optimized by control the water flow rate.

We have conducted a finite element simulation of the BIPVT panel (Yang et al., 2012) in Fig. 2.15 and validated it by physical experiments. Fig. 2.16 shows the equilibrium temperature space distribution in the panel without and with water flow. The red color and blue color stand for high and low temperature, respectively. Before the water flow starts, the whole panel surface is uniformly heated up, as depicted in Fig. 2.16A. When the water flow was introduced, the overall panel surface temperature decreased and exhibited a rising trend from the inlet to the outlet position.

The temperature variation with the temperature of the seven points on the panel is also simulated and compared to the experimental results, as shown in Fig. 2.17. Basically, the FEM simulation results well fit the experimental results. The fluctuation during the temperature falling region after the water flow is introduced due to the variations of the water flow rate.

2.3.3 Water pump design of BIPVT system

The thermal fluid circulation in a BIPVT system commonly requires additional power for flow rate control. A water pump or air pump is required. This section will use a water pump as an example. In the pump design, pressure, friction, and flow are three important characteristics. Pressure is the driving force to move the fluid. Friction slows down fluid

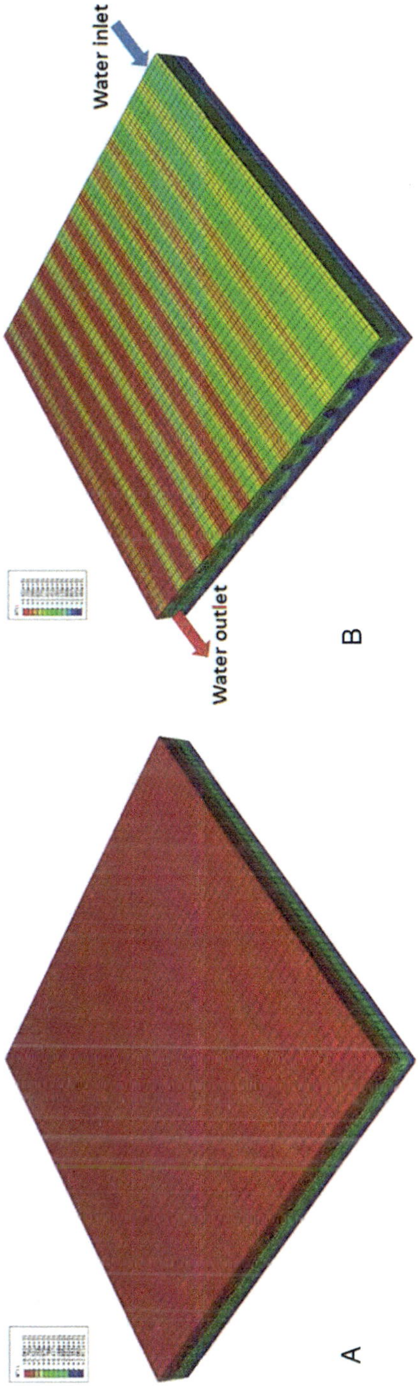

FIGURE 2.16 Temperature space distribution of the panel at (A) without water flow, (B) with water flow (Yang et al., 2012).

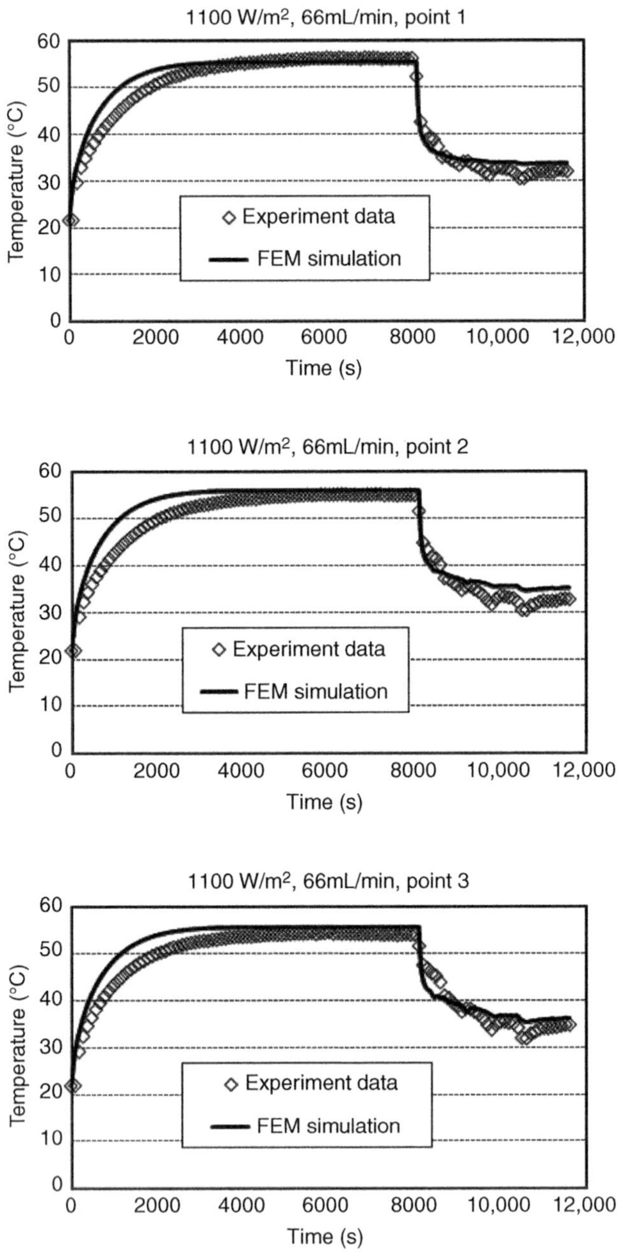

FIGURE 2.17 Finite element method (FEM) simulation and experimental data of the temperature distribution under 1100 W/m^2 irradiation and 66 mL/min water flow rate (Yang et al., 2012).

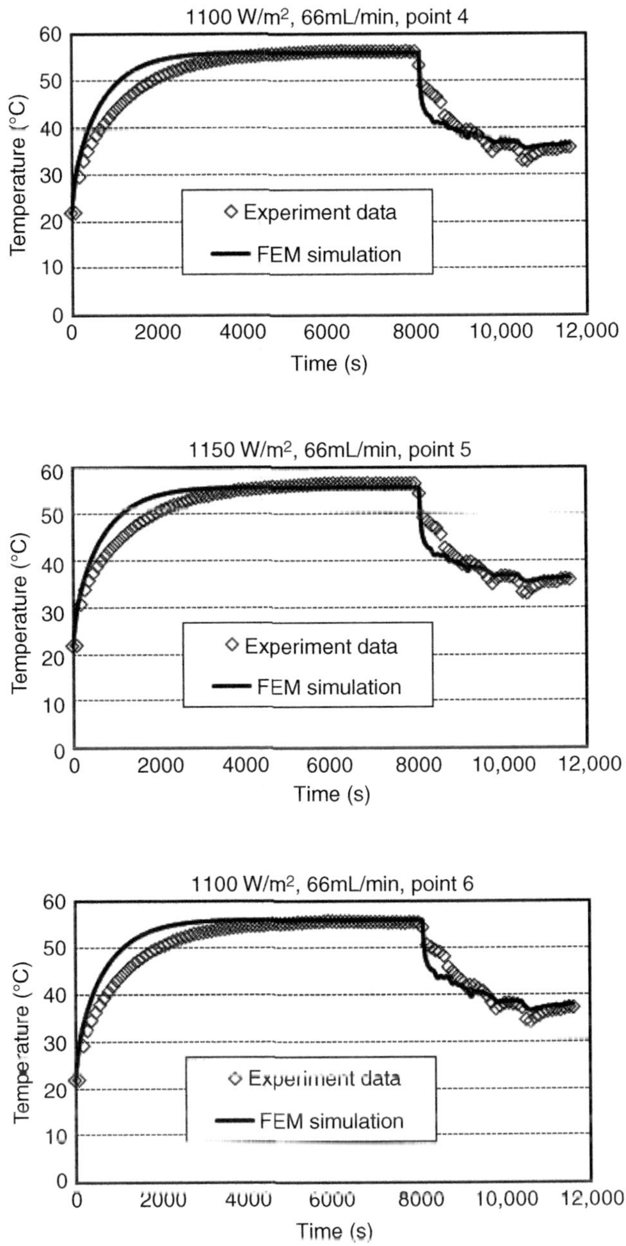

FIGURE 2.17, cont'd.

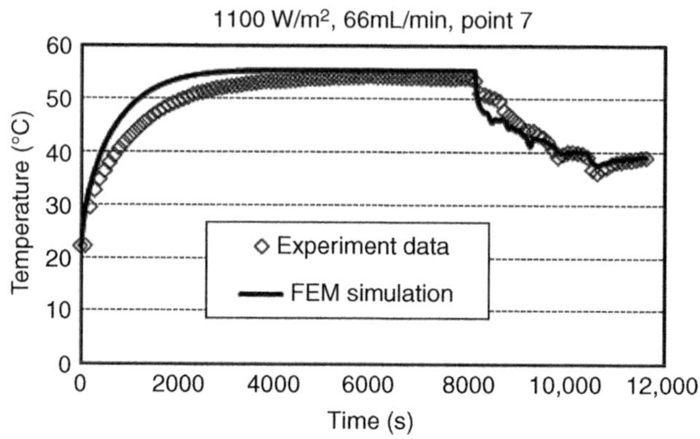

FIGURE 2.17, cont'd.

particles and dissipates energy. The flow rate will control the heat transfer rate as well as output temperature.

Because most BIPVT systems is a closed-loop system, in which the thermal fluid will be repetitively used for heat exchange. Therefore, the water head by elevation different will be balanced by itself. A water pump is used to produce enough pressure to overcome the pressure drop caused by friction.

Friction is the force that resists the movement of objects. In fluids, friction occurs between fluid layers that are traveling at different velocities within the pipe. In Chapter 6, friction in a Newtonian fluid is formulated. It depends on the viscosity and flow speed for laminar flows. The head losses are proportional to fluid velocity. When the fluid velocity is higher, such as $Re > 2500$, the flow becomes unstable, varying with time as vortices within the flow form and vanish randomly. This domain of flow is not well modeled, nor are the details well understood. When $Re > 4000$, it gets in turbulent flow, and losses are proportional to the square of the fluid velocity, V^2, which is very sensitive to the roughness of the pipe surface. The turbulent flow consumes more energy and promotes the heat transfer rate but also generates heat. For heating application, it may produce a positive effect. However, BIPVT systems mostly use cooling for temperature control of roof panels. Therefore, laminar flow is favorable with a $Re < 2000$.

In BIPVT systems, the pipelines are in serpentine configuration with manifolds for water distribution. More friction losses, known as "minor loss," can be caused by bends, couplings, valves, or transitions in hose or pipe diameter, or objects intruded into the fluid flow, which can be found in engineering manuals and building codes.

In general, the water pump power for a BIPVT system is fairly low, so its contribution to water temperature change is ignored. However, for applications with high heat transfer rates, it is necessary to consider this effect.

2.3.4 Stress and deflection analysis of BIPVT panels

BIPVT panels are supported by building frame and subjected to various environmental, dead, and live loads. Stress and deformation will be induced by those loads, which may lead to cell cracking, panel delamination, permanent deformation, or panel failure. Therefore, it is critical to predict the stress and deflection of BIPVT panels in the structural design phase, particularly when very large modules are often used in modern structures. Compared with the conventional building applied PV system, which is attached to and supported by the existing roof, the BIPV system is a part of the building structure and will transfer the load to the building framework and the foundation. The deflection and load capacity of BIPVT panels will directly impact the functionality, appearance, and safety of the building. However, most research on the BIPVT system has been focusing on the energy-related aspects, such as electrical and thermal behavior, but the structural behavior has not been extensively studied, and by today there is no specific building code about the PV modules applied to the BIPV or BIPVT system yet (Li et al., 2020). Although IEC 61215 (Commission, I.E.t., IEC 61215 2005) has addressed the mechanical behavior of a single module for the strength and toughness under an accelerated testing environment, it does not address the structural loads yet. Engineers typically use existing building codes for general building materials on BIPVT design (Lintag, 2017). Because BIPVT panels exhibit unique material and structural properties including stiff and brittle glass and PV cell, compliant but ductile encapsulant, strong and tough substrate, with different thicknesses, it is very challenging to model the stress transfer through the panel, particularly considering the extremely high stiffness ratio of glass/PV cell to encapsulant (10^3) and thickness ratio of glass to PV cell ($10^?$). Layer-wise composite models (Naumenko and Eremeyev, 2014; Eisenträger et al., 2015; Schulze et al., 2012) were often used to study the stress and deflection by treating a panel as two stiff skin layers and a soft core, which fits the double-glass modules very well but does not address the stress on solar cells. In actual BIPVT in roof applications, solar panels are mounted onto roof rafters or purlins with thermal fluid pipe integrated or attached (Fig. 2.18). In general, because of the tilted angle θ of the roof pitch, the surface load on the roof needs to be decomposed for structural analysis. Three types of loads are considered: vertical load is perpendicular to the

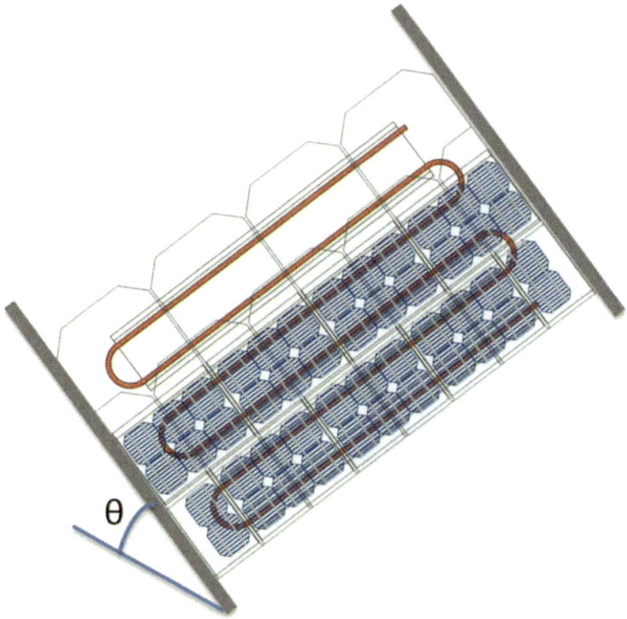

FIGURE 2.18 A BIPVT panel mounted on rafters with multiple layers of glass, EVA, back sheet, and substrate. *EVA*, ethylene-vinyl acetate.

panel surface, the in-plane shear load is parallel to the roof, and temperature load in the panel. The combination of them can be used to solve most practical loading cases in the design, such as the dead load of the self-weight, the live load of installer, wind, and snow, as well as temperature gradient or simple temperature change of the panel. Because there are different types of BIPVT configurations, such as single glass, double glass, and various supporting methods, such as simply support by four edges, two opposite edges, multiple points, this section is not targeting every problem but provides an introduction to the general analysis procedure for three basic load conditions. A simple support by two opposite edges is considered for simplicity.

2.3.4.1 Vertical load

For a simply support BIPVT panel under a uniform distributed load in Fig. 2.19A, it can be simplified to a beam with multiple layers shown in Fig. 2.19B, including the glass, ethylene-vinyl acetate (EVA), and the substrate with a pipe inside. If classic beam theory is used, it could be considered as a composite beam under the plane assumption that every point on the same cross-section will remain on the same plane. Therefore, the deformed shape can be shown in Fig. 2.19C.

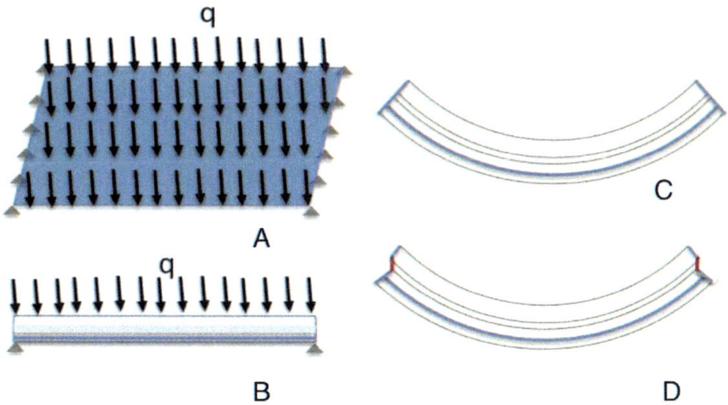

FIGURE 2.19 A simply supported BIPVT panel under a uniform load in (A) is simplified to a beam in (B). Classic composite beam theory predicts the deformation in (C) but the actual deformation is in (D) with a compliant EVA layer releasing the stress in the glass and substrate layers. *EVA*, ethylene-vinyl acetate.

If the deformation or deflection is too big, it may cause leakage problems or aesthetic issues. When the stress is too big, the fragile PV cell may exhibit micro-cracking first reducing the energy harvesting efficiency. Moreover it may cause the weak part to fracture and even the total module to break down. To formulate the problem, let the span between the two supports be L. The cross-section of the beam is shown in Fig. 2.20. The width is b, and the thicknesses of glass, EVA, and substrate layers are t_g, t_e, and t_s, respectively.

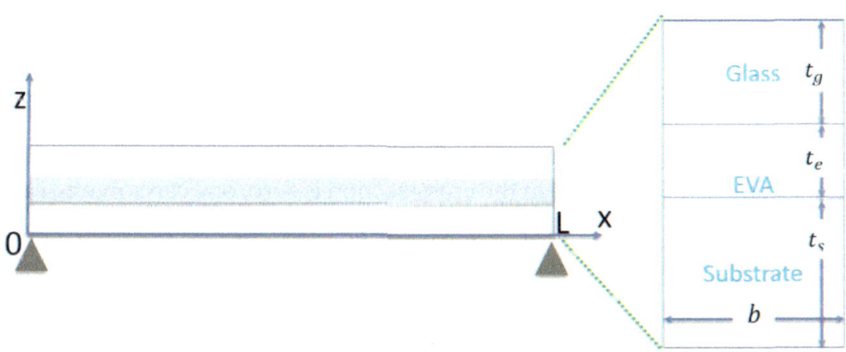

FIGURE 2.20 The coordinate of the beam bending problem with the cross-section at the right.

For any point in the beam, the equilibrium equation in absence of body force for the plane stress problem can be written

$$\sigma_{xx,x} + \sigma_{zx,z} = 0 \tag{2.82}$$

and

$$\sigma_{xz,x} + \sigma_{zz,z} = 0 \tag{2.83}$$

where σ is the stress tensor. By the integral of $\int_0^{t_s+t_e+t_g} bz(\sigma_{xx,x} + \sigma_{zx,z})dz = 0$, one can obtain

$$-\frac{dM}{dx} = V \tag{2.84}$$

where the moment $M = \int_0^{t_s+t_e+t_g} bz\sigma_{xx}dz$, and the shear force $V = -\int_0^{t_s+t_e+t_g} b\sigma_{xz}dz$. By the integral of $\int_0^{t_s+t_e+t_g} b(\sigma_{xz,x} + \sigma_{zz,z})dz = 0$, one can obtain another equilibrium equation:

$$\frac{dV}{dx} = -q \text{ or } \frac{d^2M}{dx^2} = q \tag{2.85}$$

The deflection of the beam is written as $w(x)$. Using the classic beam's assumption, one can imagine there is a neutral axis (NA) with $u_x(x) = 0$. Then the points in a cross-section vertical to the NA will remain vertical with no shear strain, so one can write

$$u_z = -w(x); \; u_x = w'(z - z_{NA}) \tag{2.86}$$

Then strain ε_{xx} is written

$$\varepsilon_{xx} = w''(z - z_{NA}) \tag{2.87}$$

The stress σ_{xx} is written

$$\sigma_{xx} = E(z)w''(z - z_{NA}) \tag{2.88}$$

where

$$E(z) = \begin{cases} E_s & x \le t_s \\ E_e & t_s < x \le t_s + t_e \\ E_g & t_s + t_e < x \le t_s + t_e + t_g \end{cases} \tag{2.89}$$

Assume the total stress in the cross-section is zero, one can solve z_{NA} as

$$z_{NA} = \frac{1}{2}\frac{E_s t_s^2 + E_e\left(t_e^2 + 2t_s t_e\right) + E_g\left(t_g^2 + 2t_g t_e + 2t_g t_s\right)}{E_s t_s + E_e t_e + E_g t_g} \tag{2.90}$$

The moment can be written in term of stress as

$$M = \int_0^{t_s+t_e+t_g} bz\sigma_{xx}dz = \int_0^{t_s+t_e+t_g} b(z - z_{NA})\sigma_{xx}dz = \overline{EI}w'' \tag{2.91}$$

where

$$\overline{EI} = \int_0^{t_s+t_e+t_g} E(z)b(z - z_{NA})^2 dz \tag{2.92}$$

Therefore, one can obtain the governing equation of the deflection as

$$\overline{EI}\frac{d^4w}{dx^4} = q \tag{2.93}$$

Using the boundary conditions

$$w(0) = w(L) = 0; M(0) = M(L) = 0 \tag{2.94}$$

one can solve the deflection as

$$w = \frac{qx}{24\overline{EI}}(L^3 - 2Lx^2 + x^3) \tag{2.95}$$

However, in the classic beam analysis, the three assumptions may produce inaccuracy to this unique problem for such a thin layered material with a soft core:

1. There exists a cross-sectional plane over the whole beam during the deformation.
2. The cross-sectional plane is always perpendicular to the NA.
3. The resultant force along the x-axis on the cross-sectional plane is zero.

Actually, to maintain a cross-sectional plane, it requires the soft core zone of the EVA to carry a large shear load which EVA will not provide, so the glass and substrate surfaces will exhibit a mismatch, as shown in Fig. 2.19D, which depends on the stiffness of the EVA. For an extreme case, if EVA produces zero shear resistance, the glass and substrate will perform as individual beams to carry the load, which can be much more flexible than the composite beam. This issue will be revisited in the next Section.

The second assumption has been well studied in the literature because it produces a zero shear strain, which is contradictory to the equilibrium of shearing force in the z-direction and can be addressed by Timoshenko's beam theory or other high-order beam theory. Notice that although the first two assumptions will exhibit some disparities from actual deformation in beam bending, the issue is mainly concentrated in the neighborhood of the support ends. In the end zones, the shear force transfer may distort the beam in 3D and the plane assumption becomes inaccurate. Therefore, for short or thick beams, the accuracy can be considerably sacrificed. However, for BIPVT panels with the ratio of length-to-thickness higher than 100, because the package of EVA can keep fairly good integrity of the composite structure, the effects of the end zone will not be severe, so that the classic beam theory can still provide acceptable accuracy with the two above assumptions.

The third assumption has often been ignored but may play a significant role in large BIPVT installation. The supports at the end provide not only vertical force but also an in-plane tensional force to keep the panel or beam in place without a gap induced between panels. In Fig. 2.21, the problem can be simplified as membrane forces in the plane to balance the vertical load.

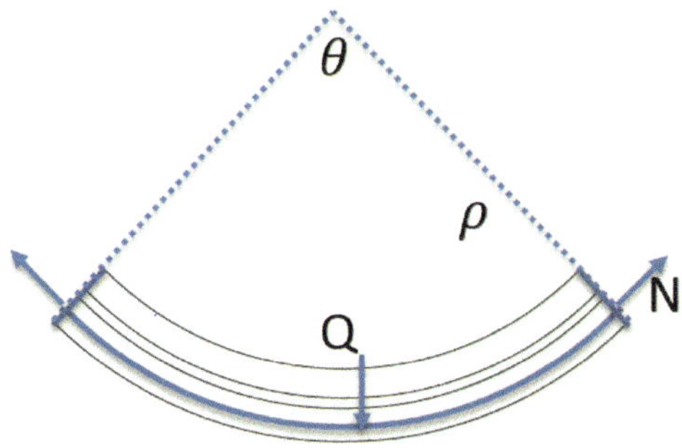

FIGURE 2.21 The curved beam caused by bending is balanced by the in-plane force.

Consider the length of the beam section is $\rho\theta$, so $Q = \rho\theta q$. If the force is balanced by membrane force N only, one can write the equilibrium as

$$2N.\sin\frac{\theta}{2} = Q \tag{2.96}$$

or

$$N = \rho q \tag{2.97}$$

Considering $\rho = \frac{1}{w''}$, if the in-plane support force N is known, the effect of the in-plane force on the equilibrium of beam bending can be equivalently considered as a distributed force of $-Nw''$. Here N can be solved by the membrane strain caused by the length change of the beam during bending, such as

$$N = \int_0^{t_s+t_e+t_g} E(z) b \varepsilon_{xx}^m dz \tag{2.98}$$

where the membrane strain caused by the deformed length of the neutral axis can be written as

$$\varepsilon_{xx}^m = \frac{L_{NA} - L}{L} = \frac{\int_0^L \sqrt{1 + w'^2} dx - L}{L} \tag{2.99}$$

By superposition of existing load and the equivalent load, Eq. (2.93) can be rewritten with replacing q with $q - Nw''$ as:

$$\overline{EI}\frac{d^4w}{dx^4} + N\frac{d^2w}{dx^2} = q$$

Using this new governing equation, we can justify the solution from the classic beam theory for a more accurate solution. Notice that because the membrane force N depends on w, this is a nonlinear differential equation, which can be solved by the numerical method. One numerical procedure can be as follows:

1. Let $w_0'' = 0$ and $N_0 = 0$, solve for the boundary value problem for $w_1(x)$.
2. Determine the in-plane force N_1 and $w_1''(x)$ by the deformed configuration.
3. Solve $q - N_1 w_1''$ the governing equation for $w_2(x)$.
4. Repeat steps 1–3 until $N_i - N_{i-1} \rightarrow 0$.

In this way, the effect of the in-plane support force can be predicted. This procedure can be extended to point load in the beam with other boundary conditions. For a BIPVT panel with arbitrary load, such as a human walk on the roof, or for other boundary conditions, such as simple supports for all edges, the plate theory is necessary. The beam analysis procedure can be extended to the plate theory as well. Some detailed analyses will be published in near future. In addition, although the simple supports assume that the vertical and horizontal displacements are zero, in reality, the support may not completely constrain the horizontal displacement, which cause the Eq. (2.99) to be inaccurate. The BIPVT panel is usually attached to the framework by polymetric adhesive which exhibits a certain rigidity to constrain the horizontal displacement by the shearing force. A linear-spring support can be used to simulate the horizontal motion of the ends.

2.3.4.2 *In-plane shear and temperature load*

As the BIPVT panel is fixed at the bottom edge, when the in-plane load is applied to the upper surface, such as by self-weight, snow, or wind, it will transfer from glass, EVA, and substrate to the rafter among other support structures through shear stress between layers. If the shear resistance is less than the shear load, delamination may happen. For uniformly distributed shear load on the glass surface, because the stiff glass and substrate do not deform as EVA, it can distribute the shear force uniformly to EVA and solar cell. It is relatively simple to deal with the uniform shear force transfer.

However, when there is a mismatch in-plane force or deformation, nonuniform shear stress will be induced to maintain the deformation compatibility and structural integrity of the layer-wise system. For example, BIPVT panels are constantly exposed to temperature change. For a free stand uniform plate, temperature change with ambient temperature can induce uniform thermal expansion or contraction. When a heat flow goes through it, a temperature gradient across the thickness causes a pure bending deformation due to the linear thermal strain across the thickness. Both cases can be free of stress without any constraint from the boundary.

However, for a multilayered BIPVT panel subjected to any temperature change, because of the mismatch of thermal expansion coefficients between material layers, mismatch thermal strain may be caused. To maintain the structural integrity, the deformation has to be continuous and compatible between layers, and thermal stress will be induced.

2.3.4.3 *Shear lag model*

The shear lag model has been developed to solve this type of problem (DILLARD, 2002).

Fig. 2.22 shows the configuration of the stress transfer in a multilayered solar panel subjected to a uniform temperature change. Here the material of each layer of the BIPV module is elastic isotropic. Because EVA is much more compliant than other layers, it is assumed that shear deformation is only in the EVA layer and no bending is considered in the model. Therefore, the normal stress in the glass and back layer or substrate will change along x through shear stress transfer in EVA.

By taking an infinitesimal section of the model, the stresses applied to the section are shown in Fig. 2.22B. The normal stresses along *x*-axis direction in back layer and front glass are written as σ_1 and σ_2, respectively. The shear stresses in the EVA layer EVA are written as τ. *The* young's modulus E, Poisson's ratio v, thermal expansion coefficient α, and thickness t for back layer cell and glass are indicated with subscripts 1 and 2 respectively. The EVA layer exhibits shear modulus G and thickness h. Other parameters are shown in Fig. 2.22B.: γ is shear strain, δ is displacement, and L is length of the module along the shear force direction.

Applying equilibrium equation on the back layer along x-direction:

$$\frac{d\sigma_1}{dx}dx.t_1 + \tau dx = 0 \tag{2.100}$$

or

$$\frac{d\sigma_1}{dx}t_1 + \tau = 0 \tag{2.101}$$

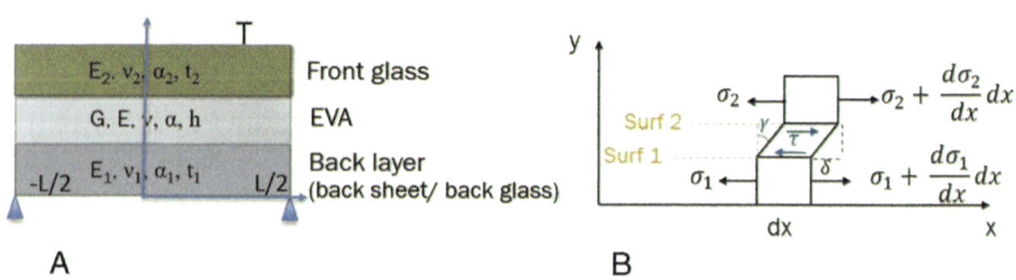

FIGURE 2.22 Schematic illustration of the shear lag model: (A) a multilayer solar panel with the back layer fixed and subjected to a shear load on the front glass; (B) free body diagram of an infinitesimal section with the shear force in the x-direction.

Similarly, for the glass layers, one can write

$$\frac{d\sigma_2}{dx}t_2 - \tau = 0 \tag{2.102}$$

where τ_0 is the uniform shear stress with $\tau_0 = \frac{T}{L}$. Notice that the shear stress in the back layer and glass layer is in the opposite direction from the one in the EVA layer.

The stress and strain relation for each above layer is written:

$$\varepsilon_i = \frac{\sigma_i}{E_i} + \alpha_i T, \quad i = 1, 2 \tag{2.103}$$

and

$$\gamma = \frac{\tau}{G} = \frac{\delta}{h} \tag{2.104}$$

The compatibility equations are:

$$d\delta = \varepsilon_2 dx - \varepsilon_1 dx \tag{2.105}$$

Using $\frac{d\tau}{G} = \frac{d\delta}{h}$, one can write

$$\frac{h}{G}\frac{d\tau}{dx} = \varepsilon_2 - \varepsilon_1 \tag{2.106}$$

or

$$\frac{h}{G}\frac{d^2\tau}{dx^2} = \frac{d\varepsilon_2}{dx} - \frac{d\varepsilon_1}{dx} = \frac{1}{E_2}\frac{d\sigma_2}{dx} - \frac{1}{E_1}\frac{d\sigma_1}{dx} \tag{2.107}$$

Using the equilibrium equation, one can write the governing equations as

$$\frac{h}{G}\frac{d^2\tau}{dx^2} = \frac{1}{E_2}\frac{\tau}{t_2} + \frac{1}{E_1}\frac{\tau}{t_1} \tag{2.108}$$

or

$$\frac{d^2\tau}{dx^2} = \left(\frac{G}{E_1 t_1 h} + \frac{G}{E_2 t_2 h}\right)\tau \tag{2.109}$$

The solution of the shear stress distribution is

$$\tau(x) = A\cosh(wx) + B\sinh(wx) \tag{2.110}$$

where $w = \sqrt{\frac{G}{E_1 t_1 h} + \frac{G}{E_2 t_2 h}}$. Integrating the equilibrium equations yields

$$\sigma_1 = -\frac{1}{wt_1}(A\sinh(wx) + B\cosh(wx)) + C \tag{2.111}$$

$$\sigma_2 = \frac{1}{wt_2}(A\sinh(wx) + B\cosh(wx)) + D \tag{2.112}$$

Using the boundary conditions $\sigma_i\left(x = \pm\frac{L}{2}\right) = 0, \quad i = 1$ and 2, one can determine the constants A and B as well as the normal stress distributions as follows.

$$A = 0 \tag{2.113}$$

$$-\frac{1}{wt_1}B\cosh\left(\frac{wL}{2}\right) + C = 0 \tag{2.114}$$

$$\frac{1}{wt_2}B\cosh\left(\frac{wL}{2}\right) + D = 0 \tag{2.115}$$

Using $\frac{h}{G}\frac{d\tau}{dx} = \varepsilon_2 - \varepsilon_1$, one can obtain

$$D + C + (\alpha_2 - \alpha_1)T = 0 \tag{2.116}$$

So one can obtain

$$B = \frac{wt_2t_1(\alpha_1 - \alpha_2)T}{(t_1 - t_2)\cosh\left(\frac{wL}{2}\right)}; C = \frac{t_2(\alpha_1 - \alpha_2)T}{(t_1 - t_2)}; D = \frac{t_1(\alpha_2 - \alpha_1)T}{(t_1 - t_2)} \tag{2.117}$$

2.3.4.4 *Two-dimensional stress transfer model*

The shear lag model can predict the stress distribution in the plane direction, but cannot produce the stress or strain distribution along with the thickness or cross-plane as a single layer is assumed for each material layer. To study the stress transfer in a general layered material, the Columbia team has developed a two-dimensional stress transfer model (Yin and Prieto-Muñoz, 2013). When a layered material has mismatched strain in one layer from others, it can be treated by superposition of a uniformed strain field and a single layer subject to a load.

Without loss of generality, consider a multilayered material system containing n layers of length 2λ. Each layer has a thickness h_i, Young's modulus E_i, and Poisson's ratio ν_i with $i = 1, 2, \ldots, n$. They are fully bonded to their neighboring layers, as illustrated in Fig. 2.22. When one layer, say layer 1, is subjected to a force P, the force will be transferred to the other layers through the fully bonded interfaces. Although the ends of the other layers are free, the normal stress will increase from the ends and reach a maximum at the center, which may produce an open-mode fracture in the middle. Because the width of the system is much larger than the thickness, our formulation will be based on the plane strain problem. The coordinate system is set up with the x-axis along with the interface, and the y-axis vertical.

Because the load P is along the x-direction, the displacement in the y-direction should be much smaller than that in the x-direction. Following the assumption made in the shear-lag theory, we assumed that all points in one layer normal to the y-direction remain in the same plane after deformation or at least that $|u_{y,x}| \ll |u_{x,y}|$ (Yin et al., 2007, Nairn and Mendels, 2001). Therefore, we can approximately write:

$$u_{y,x} \cong 0. \tag{2.118}$$

This is the so-called plane assumption, which will be used to simplify the shearing strain and stress. The constitutive law is written as

$$\begin{cases} \sigma_{xx} = \dfrac{E_i}{(1+\nu_i)(1-2\nu_i)}\left[(1-\nu_i)u_{x,x} + \nu_i u_{y,y}\right], \\ \sigma_{yy} = \dfrac{E_i}{(1+\nu_i)(1-2\nu_i)}\left[\nu_i u_{x,x} + (1-\nu_i)u_{y,y}\right] \\ \tau_{xy} = \dfrac{E_i}{2(1+\nu_i)}u_{x,y}, \end{cases} \tag{2.119}$$

where the subscript $i = 0, 1, \ldots, n$ denotes the material layers. Substituting the above equations into the equilibrium equations in the absence of body force, i.e.,

$$\begin{cases} \sigma_{xx,x} + \tau_{xy,y} = 0, \\ \tau_{xy,x} + \sigma_{yy,y} = 0, \end{cases} \tag{2.120}$$

we obtain the governing equations as follows:

$$\frac{1-v_i}{1-2v_i}u_{x,xx} + \frac{1}{2}u_{x,yy} = 0, \tag{2.121}$$

$$(1-v_i)u_{y,yy} + \frac{1}{2}u_{x,xy} = 0. \tag{2.122}$$

Notice that Eq. (2.121) is a decoupled partial differential equation of u_x. Once u_x is obtained, we can solve for u_y through Eq. (2.122). The solution of u_y can be used to evaluate the accuracy and applicability of the assumption in Eq. (2.118). However, because the upper and lower surfaces are free, one can assume $\sigma_{yy} = 0$, or $u_{y,y} = \frac{\nu_i}{1-\nu_i}u_{x,x}$, so that u_y becomes an insignificant value in the analysis. Only u_x is derived and used in the following analysis.

In the x-direction, the general solution for Eq. (2.121) can be obtained by the method of separation of variables by using the following simplification

$$u_x(x,y) = X(x)Y(y) + \varepsilon_x^u x, \tag{2.123}$$

where the unknown constant ε_x^u is introduced as a uniform strain caused by the force P. Plugging this assumed solution into the governing equation (i.e., Eq. (2.121)) produces

$$\frac{X''Y(1-\nu_i)}{(1-2\nu_i)} + \frac{XY''}{2} = 0. \tag{2.124}$$

Two independent equations can be set equal to an independent constant c_i

$$\frac{X''}{X} = -\frac{Y''}{Y}\frac{(1-2\nu_i)}{2(1-\nu_i)} = c^2. \qquad (2.125)$$

Two ordinary differential equations can be extracted from Eq. (2.125) such that

$$X'' - c^2X = 0;\ Y'' + 2c^2\frac{1-\nu_i}{1-2\nu_i}Y = 0. \qquad (2.126)$$

The general solutions for the ordinary differential equations (2.126) are written as

$$\begin{cases} Y(y) = A_i \sin d_i y + B_i \cos d_i y, \\ X(x) = C_i \sinh cx + D_i \cosh cx, \end{cases} \qquad (2.127)$$

where the constant d_i is given as

$$d_i = c\sqrt{\frac{2(1-\nu_i)}{1-2\nu_i}}. \qquad (2.128)$$

Therefore, the general solution for u_x in Eq. (2.123) can be rewritten as

$$u_x(x,y) = (A_i \sin d_i y + B_i \cos d_i y)(C_i \sinh cx + D_i \cosh cx) + \varepsilon_x^u x. \qquad (2.129)$$

The parameters A_i, B_i, C_i, and D_i and ε_x^u in the general solutions of Eq. (2.129) are different for each layer. For each layer, a series form solution may be needed to satisfy the boundary conditions exactly. In that case, a set of parameters for the asymptotic solution may be used for each layer. However, in this paper, we will demonstrate that a closed-form solution with one function has already reached an acceptable accuracy for a symmetric three-layer system (Fig. 2.23).

For boundary conditions in terms of stresses, u_x and $u_{y,y} = \frac{\nu_i}{1-\nu_i}u_{x,x}$ are used to express the stresses through Eq. (2.119), and thus the constants in

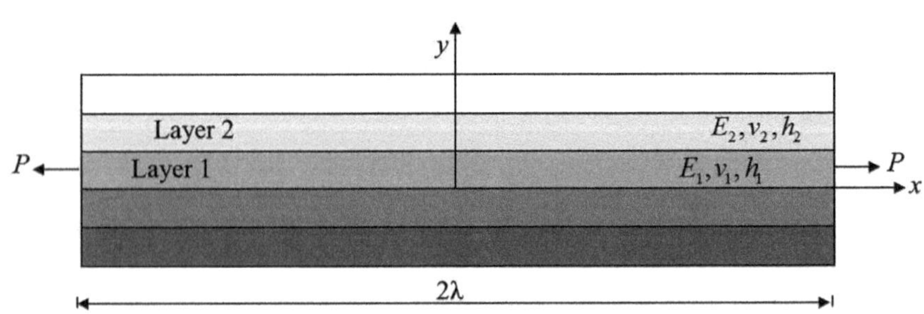

FIGURE 2.23 Schematic illustration of stress transfer in a multilayered material system (Yin and Prieto-Muñoz, 2013).

Eq. (2.129) can be solved by the boundary conditions and the interfacial continuities with combined terms of u_x.

Applying this theory to the BIPVT with dual glass layers and a soft core of EVA, shown in Fig. 2.24, two glass layers (thickness h_2, length 2λ, Young's modulus E_2, Poisson's ratio ν_2) fully bonded symmetrically to the EVA layer (thickness $2h_1$, Young's modulus E_1, Poisson's ratio ν_1). Typically, EVA has a higher thermal expansion coefficient than glass. In winter, the mismatch of the thermal strain will produce tensile stress in the EVA layer. The overall deformation can be the superposition of a uniform contraction and a tensile load on the EVA as shown in Fig. 2.24. The coordinate system is set up with the origin 0 at the center of layer 1. The x-axis is along the loading direction, and the y-axis is in the upward direction. Based on the symmetry, a quarter of the system can be used to represent the entire coating system.

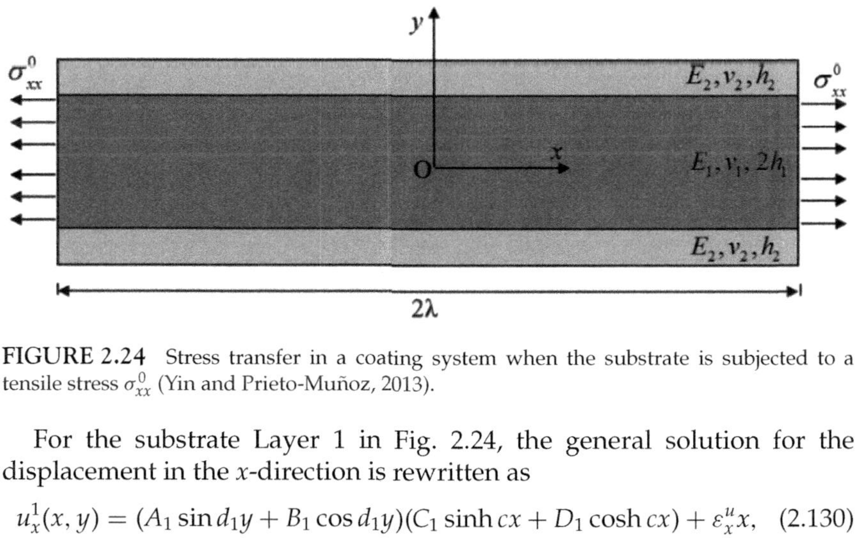

FIGURE 2.24 Stress transfer in a coating system when the substrate is subjected to a tensile stress σ_{xx}^0 (Yin and Prieto-Muñoz, 2013).

For the substrate Layer 1 in Fig. 2.24, the general solution for the displacement in the x-direction is rewritten as

$$u_x^1(x, y) = (A_1 \sin d_1 y + B_1 \cos d_1 y)(C_1 \sinh cx + D_1 \cosh cx) + \varepsilon_x^u x, \quad (2.130)$$

where, as shown in Eq. (2.128),

$$d_1 = c\sqrt{\frac{2(1-\nu_1)}{1-2\nu_1}}. \quad (2.131)$$

The symmetric conditions along the x and y-axes produce the following boundary conditions:

$$u_x|_{x=0} = 0, \quad \tau_{xy}|_{y=0} = 0 \quad (2.132)$$

Applying the above boundary conditions to Eq. (2.130) reduces the displacement to

$$u_x^1(x, y) = F_1 \cos d_1 y \sinh cx + \varepsilon_x^u x, \quad (2.133)$$

where the leading coefficient F_1, is a constant to be determined from the remaining boundary conditions, and the superscript 1 represents the displacement for Layer 1.

Similarly to the derivation in Eqs. (2.130)–(2.133), we can simplify the x-directional component of the displacement in the coating as

$$u_x^2(x, y) = F_2 \cos d_2(h_1 + h_2 - y) \sinh cx + \varepsilon_x^u x, \tag{2.134}$$

where the constant d_2 satisfies

$$d_2 = c\sqrt{\frac{2(1 - \nu_2)}{1 - 2\nu_2}}. \tag{2.135}$$

Notice that the displacements of u_y^1 and u_y^2 can be derived as well, however, because their values are much smaller and may fall into the range that this model may not accurately predict. Since a fully bonded interface is considered, interfacial continuity conditions are required at $y = h_1$:

$$u_x^1|_{y=h_1} = u_x^2|_{y=h_1}; \; \tau_{xy}^1|_{y=h_1} = \tau_{xy}^2|_{y=h_1} \tag{2.136}$$

Then, it is obtained

$$F_1 \cos d_1 h_1 = F_2 \cos d_2 h_2 \tag{2.137}$$

$$-\frac{E_1}{2(1 + v_1)} F_1 d_1 \sin d_1 h_1 = \frac{E_2}{2(1 + v_2)} F_2 d_2 \sin d_2 h_2 \tag{2.138}$$

Notice that the approximation of the shearing stress in Eq. (2.119) is used. Division of both sides of Eq. (2.138) by Eq. (2.137), respectively, yields

$$\frac{E_1}{1 + v_1} d_1 \tan d_1 h_1 = -\frac{E_2}{1 + v_2} d_2 \tan d_2 h_2. \tag{2.139}$$

Using the definitions of Eq.(2.131) and Eq.(2.135), we can calculate the constant c, and then d_1 and d_2 from Eq. (2.139). The constant c changes with the elastic constants and thickness, while the fracture spacing has no effects on c. There are many solutions for constant c because of the periodicity of the tan ( ) function, which will make a series form solutions for this problem. For simplicity, we choose the first root to demonstrate this theory. To solve for c, we define a parameter

$$\xi = \frac{d_2 h_2}{d_1 h_1} = \sqrt{\frac{(1 - 2\nu_1)(1 - \nu_2)}{(1 - \nu_1)(1 - 2\nu_2)}} \frac{h_2}{h_1}. \tag{2.140}$$

Notice that Poisson's ratios of the coating and the substrate are assumed to be less than 0.5. Otherwise, the effective stiffness is Eq. (2.119) will be infinite. If incompressible materials are used in the coating, the plane strain assumption in this theory may be invalid and thus the formulation would need to be reconstructed. Based on the mechanics of the coating system, the first root of c can be obtained within a range of $0 < d_1h_1 < \pi$, which makes the shearing stress along with the interface negative. More specifically, c can be obtained through the following three conditions.

When $\xi > 1$, c is the smallest value that satisfies $0 < d_1h_1 < \pi/2$. Typically, a thick coating on a thin substrate will fall within this category. When $\xi \to \infty$, we can use $d_2h_2 \cong \pi$ to approximately solve for $c = \frac{\pi}{h_2}\sqrt{\frac{1-2\nu_2}{2(1-\nu_2)}}$.

When $\xi = 1$, $d_1h_1 = d_2h_2 = \pi/2$ is used to solve for $c = \frac{\pi}{2h_1}\sqrt{\frac{1-2\nu_1}{2(1-\nu_1)}} = \frac{\pi}{2h_2}\sqrt{\frac{1-2\nu_2}{2(1-\nu_2)}}$. A representative case is when the coating and the substrate have the same elastic constants and the coating is half the thickness of the substrate.

When $\xi < 1$, we impose that c is the smallest solution meeting the constraint where $\pi/2 < d_1h_{1\&}$#x2005;$< \pi$. Typically, a thin coating deposited on a thick substrate will fall in this category. When $\xi \ll 1$, we can assume $d_1h_1 \cong \pi$, and c can be obtained as $c = \frac{\pi}{h_1}\sqrt{\frac{1-2\nu_1}{2(1-\nu_1)}}$.

Fig. 2.25 illustrates the relationship between the first root of c and ξ. With the increase of h_2 from $0.01h_1$ to 100 h_1, the value of c is illustrated in the log coordinate of ξ for the three cases of $E_1/E_2 = 100,\ 1,\ 1/100$. With the increase of ξ, the value of c decreases from a certain value to zero. Notice that c has a dimension of "length^{-1}," so that its value depends on the length scale of h_1 and h_2.

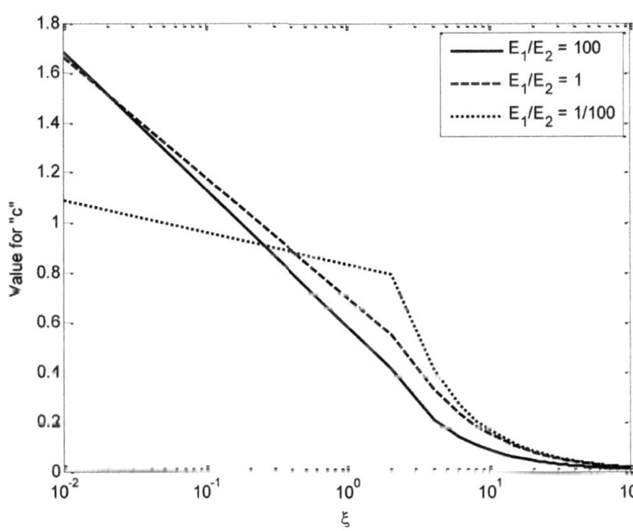

FIGURE 2.25 The calculation result of c based on varying values of ξ with $h_1 = 1$, $v_1 = v_2 = 0.3$ (Yin and Prieto-Muñoz, 2013).

In addition, Eq. (2.137) provides the relation between F_1 and F_2 as

$$F_2 = F_1 \frac{\cos d_1 h_1}{\cos d_2 h_2} \tag{2.141}$$

Therefore, if the constants F_1 and ε_x^u can be determined, the closed-form solution of the displacement field in the coating system will be obtained, and thus the strain and stress fields can be derived accordingly. We will use the principle of stationary potential energy to determine these two constants.

Because the stresses within the coating system are caused by the load σ_{xx}^0 applied in the substrate and no external load is applied in the y-direction, we assume that the stress $\sigma_{yy} \ll \sigma_{xx}$. Again, this assumption may not be accurate near the edge but is reasonable in the zone away from the edge. Using this assumption, the stress σ_{xx} can be approximated from Eq. (2.119) as

$$\sigma_{xx} \cong \frac{E_i u_{x,x}}{1 - \nu_i^2}, \quad (i = 1, 2). \tag{2.142}$$

Notice that σ_{xx}^0 can be taken as the average stress of the resultant force distributed in the cross-section of the substrate. The Saint–Venant principle will guarantee the accuracy for the zone away from the edge.

The strain energy density of a material point in the coating system is from two stress components σ_{xx} and τ_{xy}. Using the third equation in Eq. (2.119) and Eq. (2.142), we can obtain the strain energy in the coating system written as

$$U = 2\int_0^{h_1}\int_{-\lambda}^{\lambda} \frac{E_1\left(u_{x,x}^I\right)^2}{2(1-\nu_1^2)} + \frac{E_1\left(u_{x,y}^I\right)^2}{4(1+\nu_1)} dxdy + 2\int_{h_1}^{h_1+h_2}\int_{-\lambda}^{\lambda} \frac{E_2\left(u_{x,x}^{II}\right)^2}{2(1-\nu_2^2)} + \frac{E_2\left(u_{x,y}^{II}\right)^2}{4(1+\nu_2)} dxdy. \tag{2.143}$$

Substituting Eqs. (2.133) and (2.134) into the above equation, we can obtain

$$U = MF_1^2 + 2N\varepsilon_x^u F_1 + L(\varepsilon_x^u)^2. \tag{2.144}$$

where

$$M=\frac{c^2}{2}\left(\frac{\sinh(2c\lambda)}{2c}+\lambda\right)\left(\frac{E_2h_2\cos^2(d_1h_1)}{(1-\nu_2^2)\cos^2(d_2h_2)}+\frac{E_1h_1}{1-\nu_1^2}\right)$$
$$+\frac{c^2}{2}\left(\frac{\sinh(2c\lambda)}{2c}-\lambda\right)\left[\frac{E_2h_2(1-\nu_2)\cos^2(d_1h_1)}{(1+\nu_2)(1-2\nu_2)\cos^2(d_2h_2)}+\frac{E_1h_1(1-\nu_1)}{(1+\nu_1)(1-2\nu_1)}\right]$$
$$+\frac{c^2}{4}\left(\frac{\sinh(2c\lambda)}{2c}+\lambda\right)\left[\frac{E_1\sin(2d_1h_1)}{d_1(1-\nu_1^2)}-\frac{E_1(1-\nu_1)(1-2\nu_2)\sin(2d_1h_1)}{d_1(1+\nu_1)(1-2\nu_1)(1-\nu_2)^2}\right], \tag{2.145}$$

and

$$N=2E_1\left(\frac{1}{d_1(1-\nu_1^2)}-\frac{d_1}{d_2^2(1+\nu_1)(1-\nu_2)}\right)\sinh(c\lambda)\sin(d_1h_1), \tag{2.146}$$

and

$$L=2\left(\frac{E_1h_1}{1-\nu_1^2}+\frac{E_2h_2}{1-\nu_2^2}\right)\lambda. \tag{2.147}$$

The work done by the external tensile load is written as

$$W=2\int_0^{h_1}\sigma_{xx}^0u_x^1(\lambda,y)dy=2\sigma_{xx}^0h_1\lambda\varepsilon_x^u+\frac{2\sigma_{xx}^0\sin(d_1h_1)\sinh(c\lambda)}{d_1}F_1. \tag{2.148}$$

Using the principle of the stationary potential energy $\Pi=U-W$, we can write

$$\frac{\partial\Pi}{\partial F_1}=0 \text{ and } \frac{\partial\Pi}{\partial\varepsilon_x^u}=0 \tag{2.149}$$

or

$$2MF_1+2N\varepsilon_x^u=\frac{2\sigma_{xx}^0\sin(d_1h_1)\sinh(c\lambda)}{d_1}, \tag{2.150}$$

and

$$2NF_1+2L\varepsilon_x^u=2\sigma_{xx}^0h_1\lambda. \tag{2.151}$$

Solving the above two equations, we can obtain

$$\begin{pmatrix}F_1\\ \varepsilon_x^u\end{pmatrix}=\begin{pmatrix}M & N\\ N & L\end{pmatrix}^{-1}\begin{pmatrix}\sigma_{xx}^0/d_1\sin(d_1h_1)\sinh(c\lambda)\\ \sigma_{xx}^0h_1\lambda\end{pmatrix}. \tag{2.152}$$

Therefore, a closed-form solution for a coating system is obtained with the combination of displacements in Eqs. (2.133), (2.134, 2.142), and (2.151).

2.3.4.5 A hybrid stress transfer model

The shear lag model provides the normal stress in the stiff layer and shear stress in softcore as a 1D solution on x. The 2D stress transfer model can provide the elastic fields with higher accuracy. However, both models ignore an interesting physical behavior for the curling of layered material induced by the mismatch normal strain, which has been firstly studied by Stoney. For example, Fig. 2.26 shows the layout and deformation pattern of two types of solar panels under a temperature change. In both design schemes, the EVA is selected as an adhesive to bond the layers. Since the Tedlar and EVA are considered as nonmechanical layers due to their very small stiffness compared to glass, the models can be simplified into a glass substrate and PV cell overlay with a fully bonded interface. Therefore, the following discussion will focus on the glass-cell-glass (GCG) and glass-cell-Tedlar (GCT) models, in which only two material phases are involved. Considering the geometric and material properties, the GCT system shows curvature under the thermal load, while the GCG system remains flat due to the symmetry.

The 2D model in the last section has exhibited very good modeling results for GCG. However, it is not accurate for GCT. To interpret the residual stress transfer and distribution in an overlay-substrate structure, Stoney proposed a theory that assumed a uniform uniaxial residual stress in the overlay, which will produce a moment and in-plane stress. This theory could be applied to situations where a thin film layer is deposited on a thick substrate so that the bending stiffness of the thin film is ignored, and thus the residual stress is distributed uniformly. However, in Stoney's

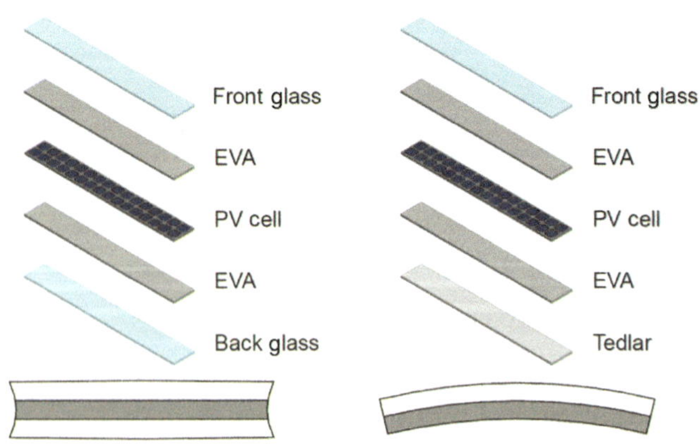

FIGURE 2.26 The layout and deformation pattern of two types of solar panels under a temperature change: (left) a symmetrical configuration of dual-glass module and (right) an unsymmetrical configuration.

theory, shear stress is not considered. It basically extended Bernoulli-Euler's classic beam theory to this problem. Therefore, it is not sufficient to provide an accurate result.

The Columbia team recently integrated Stoney's theory into the 2D stress transfer model and thus predicted the deformation caused by both shearing and bending mechanisms. The GCT can be simplified in Fig. 2.27. A half is considered for symmetry. A stress $\bar{\sigma}^0$ is applied to simulate the mismatch strain.

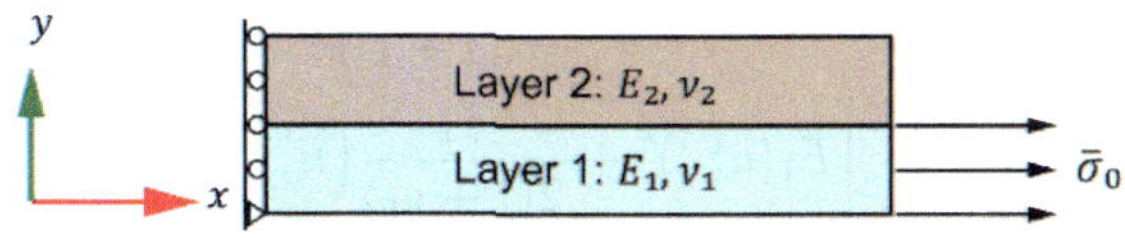

FIGURE 2.27 Schematic illustration of the two-dimensional bilayered model.

The displacement is written in terms of two parts by the pure bending and shear transfer:

$$\begin{cases} u_x^i = u_x^{si} + u_x^{bi} \\ u_y^i = u_y^{si} + u_y^{bi} \end{cases} \tag{2.153}$$

where the superscripts b and s denote the bending and shear transfer, respectively, and $i = 1, 2$. The trial function in pure bending is determined from the conventional beam theory as:

$$\begin{cases} u_x^{bi} = \bar{\kappa} x (y - y_{NA}) \\ u_y^{bi} = -\frac{1}{2} \kappa_i x^2 \end{cases} \tag{2.154}$$

$$\begin{cases} \sigma_{xx}^{bi} = E_i \bar{\kappa} (y - y_{NA}) \\ \sigma_{yy}^{bi} = 0 \\ \sigma_{xy}^{bi} = 0 \end{cases} \tag{2.155}$$

where the parameter $\bar{\kappa}$ indicates the curvature and y_{NA} indicates the location of the neutral axis. The location of the neutral axis is determined by its definition that the axial force is zero:

$$N = \int_0^{h_1} E_1 \bar{\kappa} (y - y_{NA}) dy + \int_{h_1}^{h_1+h_2} E_2 \bar{\kappa} (y - y_{NA}) dy = 0 \tag{2.156}$$

$$y_{NA} = \frac{E_1 h_1^2 + E_2 h_2^2 + 2E_2 h_1 h_2}{2(E_1 h_1 + E_2 h_2)} \tag{2.157}$$

As for the trial function featuring shear transfer, u_y^{si} is mainly caused by the deformation mismatch between layers. It is hypothesized that the vertical displacement in the shear transfer is only a function of y coordinate

(Nairn and Mendels, 2001) because the horizontal layer keeps the plane approximately during the deformation. Therefore, u_x^{si} will be from the last section (Zhang et al., 2021).

Using the displacement field, the corresponding stress components σ_{ij}, the strain energy U, and the work done by the external load W are determined as:

$$\sigma_{xx}^{i} \cong \sigma_{xx}^{bi} = E_i u_{x,x}^{bi} \tag{2.158}$$

$$\sigma_{xy}^{i} = \frac{E_i}{2(1+\nu_i)}\left(u_{x,y}^{i} + u_{y,x}^{i}\right) \cong \frac{E_i}{2(1+\nu_i)}\left(u_{x,y}^{i} + u_{y,x}^{bi}\right) \tag{2.159}$$

$$U = \frac{1}{2}\int_0^{h_1} \mathrm{d}y \int_{-\lambda}^{\lambda}\left[E_1\left(u_{x,x}^{1}\right)^2 + \frac{E_1}{2(1+\nu_1)}\left(u_{x,y}^{1} + u_{y,x}^{b1}\right)^2\right]$$

$$\mathrm{d}x + \frac{1}{2}\int_{h_1}^{h_1+h_2} \mathrm{d}y \int_{-\lambda}^{\lambda}\left[E_2\left(u_{x,x}^{2}\right)^2 + \frac{E_2}{2(1+\nu_2)}\left(u_{x,y}^{2} + u_{y,x}^{b2}\right)^2\right]\mathrm{d}x \tag{2.160}$$

$$W = 2\sigma_0 \int_0^{h_1} u_x^1(\lambda, y)\mathrm{d}y \tag{2.161}$$

Rearranging Eq. (2.159), it produces:

$$U = T_{11}\bar{\kappa}^2 + 2T_{12}\bar{\kappa}\bar{\varepsilon} + 2T_{13}\bar{\kappa}F_1 + T_{22}\bar{\varepsilon}^2 + 2T_{23}\bar{\varepsilon}F_1 + T_{33}F_1^2 \tag{2.162}$$

where

$$T_{11} = \frac{E_1 h_1 \lambda}{3}\left(h_1^2 - 3h_1 y_{\mathrm{NA}} + 3y_{\mathrm{NA}}^2\right) + \frac{E_2 h_2 \lambda}{3} \times \left(3h_1^2 + 3h_1 h_2 - 6h_1 y_{\mathrm{NA}} + h_2^2 - 3h_2 y_{\mathrm{NA}} + 3y_{\mathrm{NA}}^2\right) \tag{2.163}$$

$$T_{12} = \frac{E_1 h_1 \lambda}{2}(h_1 - 2y_{\mathrm{NA}}) + \frac{E_2 h_2 \lambda}{2}(2h_1 + h_2 - 2y_{\mathrm{NA}}) \tag{2.164}$$

$$T_{13} = 0 \tag{2.165}$$

$$T_{22} = E_1 h_1 \lambda + E_2 h_2 \lambda \tag{2.166}$$

$$T_{23} = 0 \tag{2.167}$$

$$T_{33} = \frac{E_1 d_1}{32(1+\nu_1)c}[\sinh(2c\lambda) - 2c\lambda][2d_1 h_1 - \sin(2d_1 h_1)] + \frac{E_2 d_2[\cos(d_1 h_1)]^2}{32(1+\nu_2)c[\cos(d_2 h_2)]^2}[\sinh(2c\lambda) - 2c\lambda][2d_2 h_2 - \sin(2d_2 h_2)] \tag{2.168}$$

The work done by the external load is determined with the same approach:

$$W = 2u_1\bar{\kappa} + 2u_2\bar{\varepsilon} + 2u_3F_1 \tag{2.169}$$

where

$$u_1 = \frac{1}{2}\sigma_0\lambda\left(h_1^2 - 2h_1 y_{\text{NA}}\right) \tag{2.170}$$

$$u_2 = \sigma_0\lambda h_1 \tag{2.171}$$

$$u_3 = \frac{\sigma_0 \sinh(c\lambda)\sin(d_1 h_1)}{d_1} \tag{2.172}$$

The potential energy can be calculated as:

$$\Pi = U - W \tag{2.173}$$

The principle of stationary potential energy requires $\partial\Pi/\partial q_i = 0$, where q_i represents the unknown parameters. Therefore, a set of linear equations are produced from the requirements, and the corresponding parameters are determined from:

$$\begin{Bmatrix} \bar{\kappa} \\ \bar{\varepsilon} \\ F_1 \end{Bmatrix} = \begin{bmatrix} T_{11} & T_{12} & T_{13} \\ T_{12} & T_{22} & T_{23} \\ T_{13} & T_{23} & T_{33} \end{bmatrix}^{-1} \begin{Bmatrix} u_1 \\ u_2 \\ u_3 \end{Bmatrix} \tag{2.174}$$

After the calculation of $\bar{\kappa}$, $\bar{\varepsilon}$, and F_1, the analytical solution is obtained as follows:

$$\begin{cases} u_x^i = \bar{\kappa}x(y - y_{\text{NA}}) + F_i cos\left[d_i(y - y_i)\right]sinh(cx) + \bar{\varepsilon}x \\ u_y^i = -\dfrac{1}{2}\bar{\kappa}x^2 \end{cases} \tag{2.175}$$

and

$$\begin{cases} \sigma_{xx}^i = E_i\bar{\kappa}(y - y_{\text{NA}}) + E_i F_i c cos\left[d_i(y - y_i)\right]cosh(cx) + E_i\bar{\varepsilon} \\ \sigma_{xy}^i = -\dfrac{E_i}{2(1+\nu_i)}F_i d_i sin\left[d_i(y - y_i)\right]sinh(cx) \end{cases} \tag{2.176}$$

2.4 Life cycle analysis of BIPVT systems

2.4.1 Economic metrics

The United Nations definition of sustainable development is "how to meet the needs of the present generation without compromising the ability of future generations to meet their needs." In order to achieve this goal, the engineers need to design and redesign processes and products with full knowledge of their environmental impacts on Earth's resources such as minerals, fuels, and water. One way to do so is to consider the life cycle assessment (LCA) of the products. Life cycle assessment, also called life cycle analysis, is the methodology to assess the environmental impacts of a product, process, or activity from the very first stem of extracting raw materials to the last step which is the end of life of that product (from cradle to grave). A schematic overview of the LCA method is shown in Fig. 2.28. These assessments may involve:

1. Life cycle inventory (LCI): creating an inventory of material and energy use and environmental emissions.
2. Life cycle impact assessment: evaluating and quantifying the potential environmental and human health impacts of uses and emissions.
3. Interpretation & Reporting: identify and conclude the results of LCI and LCA and offer recommendations

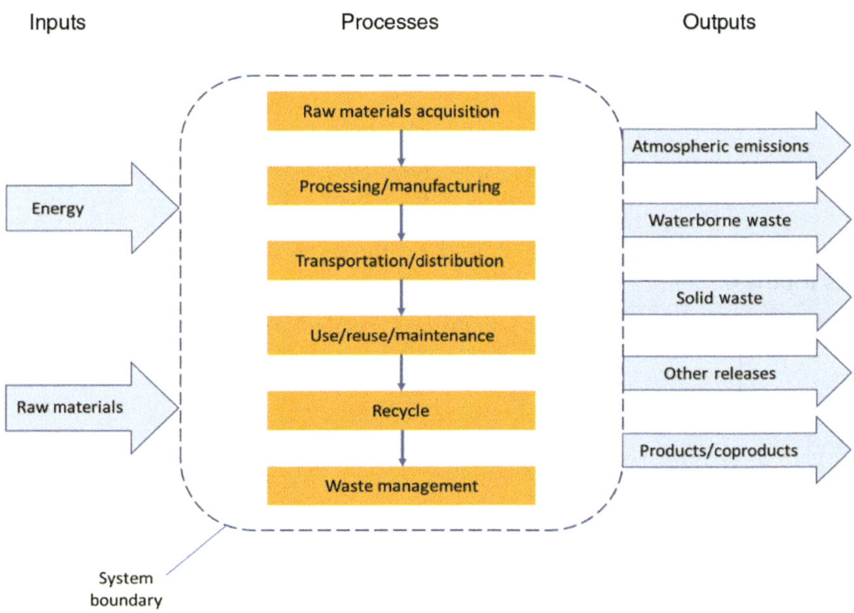

FIGURE 2.28 Overview of a life cycle assessment of a product. Source: redrawn from Fava (1991).

An important concept in the life cycle assessment is the calculation of the energy payback time (EPBT), which is defined as the period required for a renewable energy system to generate the same amount of energy that was used to produce the system itself. The EPBT is calculated as (Fthenakis et al., 2011):

$$EPBT = \frac{(E_{\text{Mat}} + E_{\text{Manuf}} + E_{\text{Tran}} + E_{\text{Inst}} + E_{\text{EOL}})}{\left(\left(\frac{E_{\text{Agen}}}{\eta_G}\right) - E_{Aoper}\right)} \tag{2.177}$$

where

E_{Mat}: primary energy demand to produce materials comprising PV system.
E_{Manuf}: primary energy demand to manufacture PV system.
E_{Tran}: primary energy demand to transport materials used during the life cycle.
E_{Inst}: primary energy demand to install the system.
E_{EOL}: primary energy demand for end-of-life management.
E_{Agen}: annual electricity generation.
E_{Aoper}: annual energy demand for operation and maintenance in primary energy terms.
η_G: grid efficiency, the average primary energy to electricity conversion efficiency at the demand side.

Energy return on investment (EROI) or also sometimes called energy returned on energy invested (EROEI), is the other important factor in characterizing the economical efficiency of an energy source, which is defined as the ratio of the total electricity produced during the lifetime of the PV system to the expended exergy used to obtain that system (Fthenakis et al., 2011)

$$EROI = \frac{Lifetime}{EPBT} = \frac{T.\left(\left(\frac{E_{Agen}}{\eta_G}\right) - E_{Aoper}\right)}{\left(E_{Mat} + E_{Manuf} + E_{Tran} + E_{Inst} + E_{EOL}\right)} \tag{2.178}$$

where T is the expected lifetime of the PV panels.

Although the oil has had a very high EROI almost all over the world, there has been a declining trend in this value. The EROI of US oil has declined from 100:1 in 1930 to 40:1 in 1970 and about 14:1 in 2000. According to a 2014 report by the Royal Society of London (Murphy, 2014), the EROI for oil has fallen below a ratio of 15 to 1 globally, and in the United States where the use of unconventional extraction techniques such as fracking is used, this ratio is even lower to about 10 to 1 (Fig. 2.29).

However, in recent years, newer technologies and improved solar cell conversion efficiency, expansion of manufacturing facilities in the world, and a better supply chain have resulted in increasing the solar energy EROI. A study done by the Imperial College of London researcher (Koppelaar, 2017) showed that as of 2015, the solar energy's return on investment ratio was 14 to 1 and rising. Therefore, by investing in solar electricity production, we will get back more energy for each unit invested in it. Based on the growing electrification trend and the efforts of the governments to invest in renewable energy resources in addition to the technological advancements in production, module efficiency, and installation, we should expect even higher EROI in the coming years.

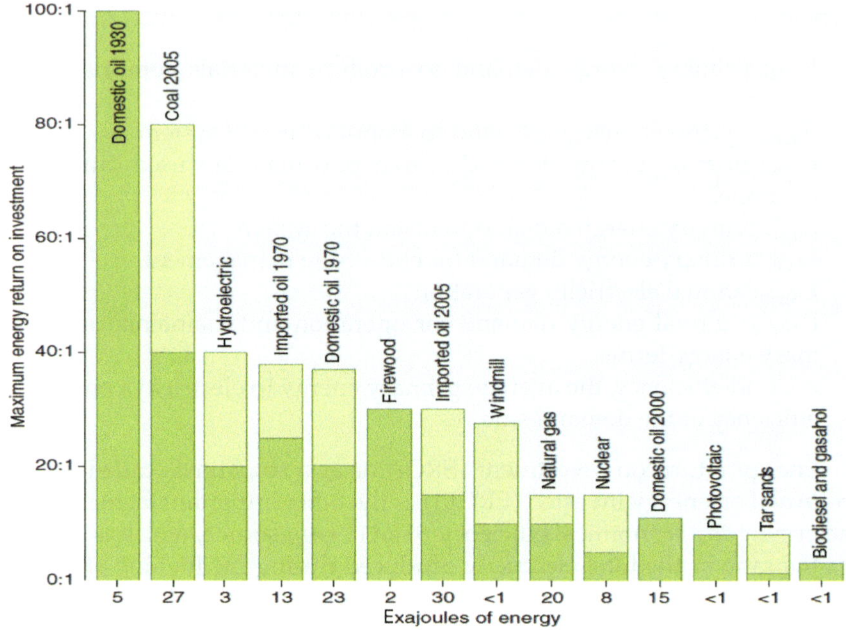

FIGURE 2.29 EROI for some common resources of energy, as of 2009 (Hall and Day, 2009). *EROI*, Energy return on investment.

2.4.2 LCA functional unit

A key element for doing LCA of a system or material is the correct choice of the functional unit. The functional unit is a reference to relate inputs and outputs and is used as a measure of the function of the system studied (ec.europa.eu, 2005). A precise and comparable definition of the unit is essential for the study as a reference. For example, the functional unit for a carton of milk could be a 0.5-gallon paper carton of pasteurized milk for sale and use in the continental United States. In a study, to perform a correct comparison, the functional unit has to be the same for all the compared product systems. The unit is identified based on the project, collected data, market segment, product properties, etc. (ec.europa.eu, 2005; Consequential LCA).

2.4.3 Boundary definition

The system boundaries are for determining the unit processes that will be included in the study. The system boundary is subjective and should be defined in accordance with the stated goal of the study (Wikipedia, 2021). The system boundaries can be based on geographical area, time horizon, boundaries between technological systems and nature, and the boundaries between the current life cycle and the related life cycles of other technical systems (ec.europa.eu, 2005).

2.4.4 LCA performance of BIVPT systems

After defining the functional unit and the system boundary, the first step to identify all the raw materials and processes involved in the production of a BIPVT system until the final day of the lifetime of the panel. In other words, we recognize and calculate all the energy, exergy, and materials used from cradle to grave in the production of an individual BIPVT system. There are many databases that could be used to search for these detailed elements including EcoInvent, Franklin, NREL LCI. For PV life-cycles, the IEA PVPS Task12 LCI database is recommended (Fthenakis et al., 2011). This task can be done using commercial software such as SimaPro. Next, from the Eqs. (2.176) and (2.177), the EPBT and EROI can be calculated.

In 2010, Agrawal and Tiwari performed an LCA study of a rooftop BIPV and BIPVT system and compared different PV technologies in terms of their annual output versus cost, overall efficiency, and exergy (Agrawal and Tiwari, 2010). They found that a BIPVT with mono-crystalline silicon (a-SI) produces the maximum annual electrical output (15,131 kWh), whereas the amorphous silicon (a-Si) BIPVT produces the minimum output (6066 kWh), due to the higher conversion efficiency of the c-Si BIPVT. Another aspect of doing an LCA study could be calculating the ecotoxicity and waste produced from the production of a system or goods. For example, in a research work done by Lamnatou et al. on an environmental LCA study of a BIPVT (Lamnatou et al., 2019), the authors concluded that recycling steels in the BIPVT can reduce the environmental impact from 47% to 85%. They also showed that for every m^2 area of a thermal absorber in a BIPVT, 0.34 tons of CO_2 is released in the production and transport processes.

2.5 Case study: Development of a BIPVT panel with a foamed aluminum substrate

2.5.1 Concept and design outline

The current increasing energy demand and the goal to meet the Paris Agreement to keep the global average temperature increase below 2°C require new and innovative methods in energy management to be implemented. For example, in the building section, heating, ventilation, and air condition account for almost 35% of the total energy consumption. Any savings in this area could be significant on the large scale. In addition, the depleting fossil fuel reserves have urged the governments to look for more renewable and eco-friendly energy resource alternatives. The growing technologies and governmental subsidies have facilitated the transition from fossil fuels to renewable energies such as solar energy, which has taken a sharp peak in recent years. Solar energy is an abundant source of energy, and with newer technologies in the semiconductor industry, applications of silicon solar cells have grown substantially since their invent in 1954 at the Bell Labs (Figs 2.28 and 2.29).

Among different solar cell technologies, polycrystalline silicon accounts for more than 60% of the market share. In silicon solar cells, when an incident light hits the surface of the semiconductor, the photons that have high energy equal to or larger than that of the energy difference between the valence and the conductance band, excite the electrons, and form a current inside the cell. However, the weak photons and also the excessive energy of the strong photons turn into heat and increase the surface temperature of the solar module. Thus, the solar energy absorbed by the PV panel is partially converted to electricity and the rest is dissipated in the panel in the form of heat. Consequently, one issue with the conventional silicon solar panels has always been the reduction of conversion efficiency with the increase in the surface temperature of the panel. Mainly, the open-circuit voltage of the cell decreases with temperature increase (Fig. 2.30) (Fthenakis and Lynn, 2018). Thus, as the PV panels are exposed to the

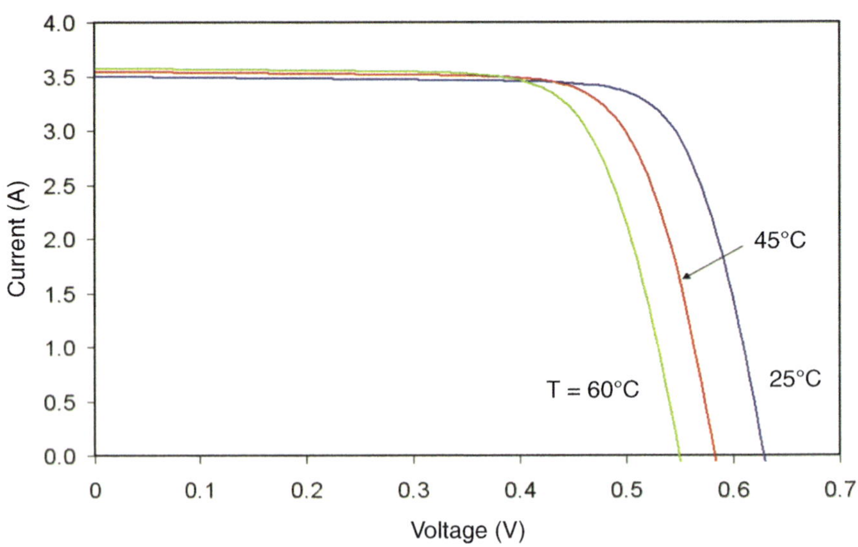

FIGURE 2.30 Effect of temperature on the I–V curve (Haber and Farkas, 2011).

sunlight and become hotter, their output power which is the product of the current and voltage decreases with time. Laboratory experiments show that just 30 minutes of sunlight with an intensity of 1000 W/m^2 increases the surface temperature of the panel from room temperature (25°C) to almost 85°C. Under standard testing conditions, for every degree rise in temperature, the efficiency decreases by about 0.40%–0.50 % (Natarajan et al., 2011). Consequently, not only the panel will be less efficient and generate less power (by as much as 0.5% Watt per degree Celsius), but also the consistent heat exposure accelerates the aging of the cells and shortens their lifespan. This brings the concept which is basically collecting heat from the solar panels and utilizing it for space heating or domestic hot water use. These hybrid systems are called photovoltaic thermal (PVT) collectors, and not only they produce solar electricity, but also heat with a higher total energy efficiency.

This project is an improvement of the design concept and method of a recent work published by Chen et al. (2018). Based on research work from Chen & Yin (2016), it is possible to reduce the surface temperature of a solar panel and thus, maintain its efficiency at the optimal level. Herein, a new design is presented for the solar panels using a hydronic loop system on the backside of the panel. Using this integrated design, the temperature of the panel can be regulated during usage and the harvested heat could be used for warm water for application inside the building. In this design, aluminum tubes that are cast inside a foamed aluminum layer are integrated on the backside of the solar panel and fluid with high heat capacity (i.e., water) circulates inside the tubes in a closed loop. The fluid absorbs the heat when it comes in contact with the panels and thus, reduces the surface temperature of the cells and makes them cooler. Therefore, the fluid is acted as both a heat sink and a heat collector. The heated fluid is then transferred to the facility room of the building via a pump to heat a water tank indirectly. As a result, the temperature of the panel can be regulated during the operation time, and the harvested heat could be used for partially warmed water to be used for applications in the building. This system not only maintains the efficiency of the solar panels at their feasible peak but also harvests heat and reduces the total energy consumption of the building. In this work, by using a combination of a simulations approach and experimental study, the proposed hypothesis is examined. The results show that the designed hydronic system can successfully cool down the surface of the solar panel and improve its efficiency approximately to the value at the initial temperature state.

The components of this design are shown in Fig. 2.31. A protective layer, which is glass, is placed on top, and an EVA film is used to sandwich the PV layer. A foamed aluminum layer with tubes cast inside the layer is integrated on the backside of the solar panel. The tubes are made of aluminum and thus, are highly conductive.

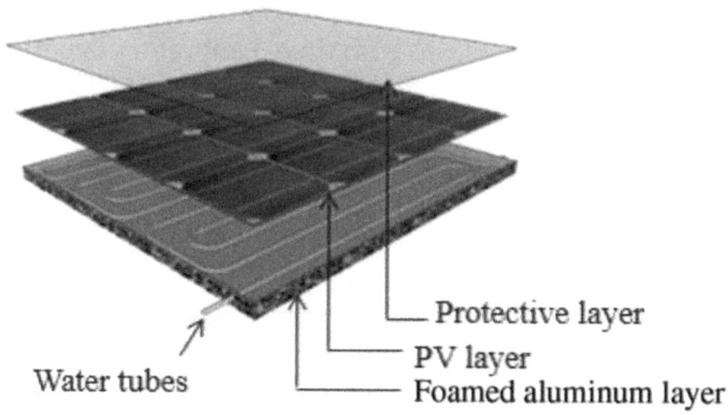

FIGURE 2.31 Schematic illustration of the hydronic system.

An absorbing fluid with high heat capacity which could a mixture of water and glycol is circulated inside the tubes in a closed loop. When water is exposed to heat, it expands, and its temperature increases, thus, reducing the surface temperature of the cells and making them cooler. The heated fluid could then be transferred to the facility room of the building via a pump to indirectly heat a water tank, which could be used for indoor heated water usage. Therefore, the water is acted as both a heat sink and a heat collector. Providing a large surface area for heat exchange while being lightweight, relatively high thermal conductivity, resistance to thermal shocks, high pressures, high temperatures, and the recyclability of the aluminum make it the right choice as the back support to embed this system. The foamed aluminum layer which encapsulates the tubes has open pores that can be filled with PCMs that absorb and release heat with a time latency. Therefore, the daily temperature fluctuation of the building façade will significantly be lowered, which per se has a role in the total energy consumption of the building. The foamed aluminum layer is assumed to be homogenous, with a consistent thermal conductivity. The applications of the foamed layer could be the implementation of PCM in the pores to control the temperature shifts due to the disputed heat.

2.5.2 Modeling and simulation

The simulation work has been done in two different approaches: 1) coupling finite element method (FEM) with solid-fluid interaction (ABAQUS) and coupling finite volume method (FVM) and FEM (Fluent and ANSYS). The first approach involves FEM to solve both solid and liquid parts in ABAQUS, the transient simulation results have been shown in the poster competition in Energy Conference International – Energy Harvesting from Infrastructure and Ocean Systems (EHIOS) in 2019. The second approach couples the FVM with FEM, which provides the steady-state thermal solution. As a well-known method, FEM will not be introduced on the following page. In this section, updating algorithms and convergence will be illustrated briefly.

2.5.3 Governing equations

For incompressible flow, the basic equations are continuity and momentum conservation equations.

$$\frac{\partial \rho}{\partial t} + \nabla \cdot (\rho\, \vec{v}) = S_m \tag{2.179}$$

The above equation is the general form of the mass conservation equation, and the source S_m is the mass added to the continuous phase from the dispersed second phase (i.e., due to the vaporization of liquid droplets). For the momentum conservation equation:

$$\frac{\partial}{\partial t}(\sigma \vec{v} + \Delta \cdot (\sigma \vec{v}\vec{v}) = -\Delta_p + \delta \cdot (\underline{\tau}) + \sigma \vec{g} + \vec{F} \tag{2.180}$$

In which

$$\underline{\tau}\mu\left[(\Delta\vec{v} + \Delta\vec{v}^{T} - \frac{2}{3}\Delta \cdot vecvI\right] \tag{2.181}$$

and,

$$\nabla \cdot \vec{v} = \frac{\partial v_i}{\partial x_i} \tag{2.182}$$

where p is the static pressure, $\underline{\tau}$ is the stress tensor, and $\rho\ \vec{g},\ \vec{F}$ are gravitational body force and external body forces respectively. In the stress equation, μ is the molecular viscosity, I is the unit tensor, and the second term on the right-hand side (RHS) is the effect of the volume dilation. The application of the working fluid is to transfer heat from the system; therefore, the energy equation is included to measure the thermal quantities and verify the energy balance (conservation). The flow of thermal energy from matter occupying one region in space to matter occupying a different region is known as heat transfer (Moukalled et al., 2016). Heat transfer could happen in three main ways: (1) conduction, (2) convection, (3) radiation. As for simulation of heat transfer by fluid, the physical models usually contain thermal conduction and thermal convection only. The energy equation is shown as:

$$\frac{\partial}{\partial t}(\rho E) + \nabla \cdot (\vec{v}(\rho E + p)) = \nabla \cdot \left\{k_{\mathrm{eff}} \nabla T - \sum\nolimits_j h_j \vec{J_j} + (\tau_{\mathrm{eff}} \cdot \vec{v})\right\} + S_h \tag{2.183}$$

where k_{eff} is the effective conductivity, and $\vec{J_j}$ is the diffusion flux of species j. The first three terms on the RHS represent energy transfer due to the conduction, species diffusion, and viscous dissipation respectively. S_h includes the volumetric heat sources in the system. And internal energy, $E = h - \frac{p}{\rho} + \frac{v^2}{2}$, and h is the sensible enthalpy.

2.5.4 Fluid–solid interface cosimulation

As the software solves the fluid domain by FVM, it also solves the solid domain by FEM. The major reason for the two different methods is that FVM generally takes more time and computing resources to solve a nonflow problem. As for the system coupling, it is necessary to understand how the coupling process works so that one can avoid making conceptual mistakes.

The external surfaces of the working fluid and the inner surfaces of the water pipes are the fluid-solid interface (FSI) and it requires thermal coupling (static load is not important this time). In general, for solid to solid interactions, the thermal conductance is highly important, which ensures that the heat flux through two parts is exactly the same, which guarantees consistency. However, to solve the thermal interaction between the fluid and solid, there are three necessary quantities, which are needed to be updated in each time step (for steady-state, it means data transfer difference iterates to zero).

2.5.5 Simulation geometry and material properties

The simulation geometry includes working fluid, water pipes, foamed aluminum substance, and thin-film PV module, and the EVA (Fig. 2.32).

The dimensions of the foamed aluminum box, water pipes, and the layers are shown in Table 2.1.

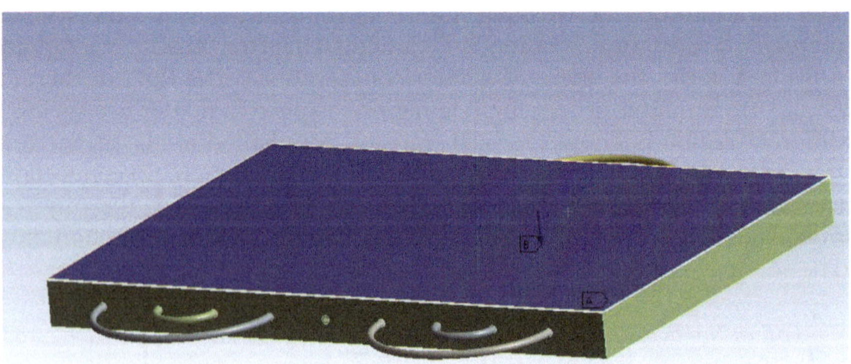

FIGURE 2.32 Dimensions of simulation.

TABLE 2.1 Dimensions and thermal conductivity of the different layers.

Name of the layer	Length (mm)	Width (mm)	Thickness (mm)	Thermal conductivity (W/m K)
Photovoltaic module	400	300	0.4	130
Ethylene-vinyl acetate layer	400	300	0.4	1.3
Box	400	300	30	18

TABLE 2.2 Material properties of the water.

Material	Density (kg/m^3)	Dynamic viscosity (*Pa.s*)	Thermal conductivity (W/m K)
Water	1000	8.9×10^{-4}	0.59

The inner radius of the water pipe is 1.6 mm, and the outer radius is 2.5 mm. The material of the water pipes is aluminum with thermal conductivity of 237 W/m K. The cross-section of the working fluid is the same as the inner section of the water pipe. Due to the convenience in the experiment, the working fluid is water, and the properties of it are listed in Table 2.2.

2.5.6 Boundary conditions

Due to the complexity of the boundary conditions, an insulated boundary condition has been applied to the bottom, the external surface of water pipes, and all sides of the solid part including the two layers. The top surface of the solid part is subjected to light at the intensity of 800, 1000 W/m^2. If converted to heat flow rate:

$$q = 800 \text{ or } 1000 \cdot 0.3 \cdot 0.4 = 96 \text{ } or \text{ } 120 \text{ W}$$

The air temperature near the top surface may create heat convection by Newton's cooling law, however, in the simulation; only one boundary condition can be prescribed. Therefore, the heat convection and thermal radiation at the top surface are neglected.

The interaction between solid and fluid surfaces has been illustrated in the previous section. The data transfer includes heat transfer coefficient, convection reference temperature, and temperature. Besides that, the inner surface of the water pipe serves as nonslip boundary conditions for the fluid, where the velocity of the boundary layer is zero due to Newton's viscous assumptions. For ideal boundary conditions, the incoming heat flow rate should be between the outcoming heat transfer rate, which means the temperature difference between the inlet and outlet fluid is a key factor. To match the experiment data, 293 K is selected for the inlet and

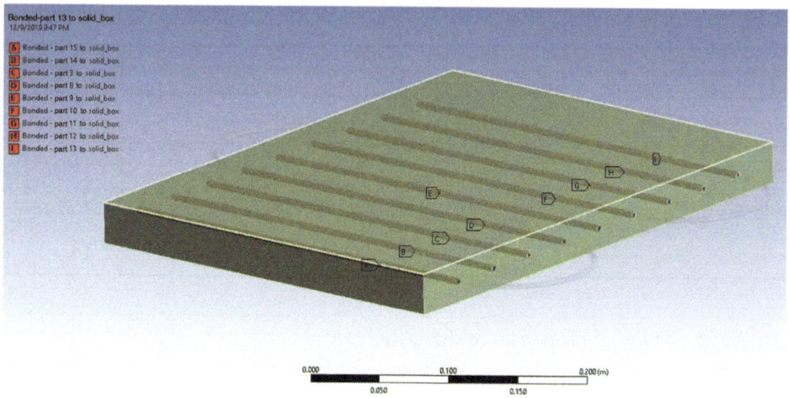

FIGURE 2.33 Interaction definition area of the foamed aluminum box and water pipes.

the quantum of the fluid is controlled based on the average velocity of the inlet due to a fixed pipe cross-section. The outlet boundary condition is set to be pressure gradient-free as a natural outflow condition.

Finally, the interaction between the foamed aluminum box and the water pipes should be studied as the relationship is not known. Therefore, a parametric study will be performed for the thermal conductance between the two surfaces (Fig. 2.33).

2.5.7 Steady–state thermal solution results

For the thermal conductance between the water pipes and foamed aluminum box, we select the value of 120 W/m^2. The temperature distribution of the aluminum box at 7 selected nodes are shown in Fig. 2.34.

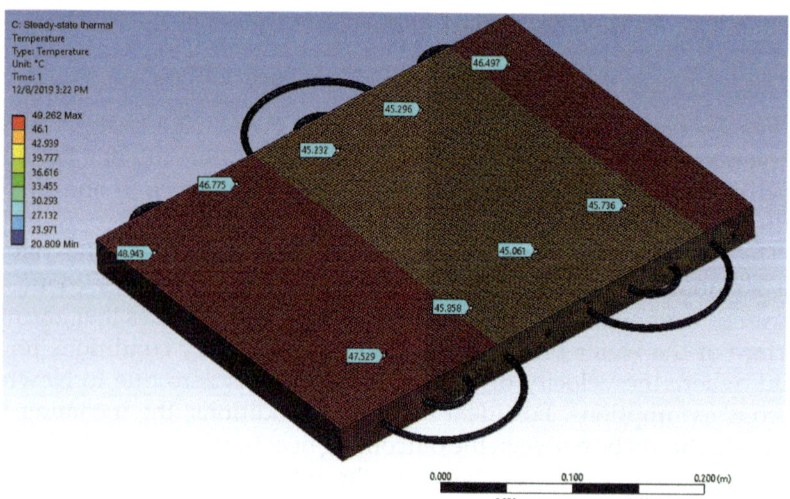

FIGURE 2.34 Temperature of the foamed aluminum box with seven picked nodes.

The above results are close to the experiment explained later. The heat flux at the top surface is 800 W/m^2. The temperature difference of inlet and outlet is 10.3 K. Calculating with the maximum heat transfer rate equation,

$$Q\rho C_p \Delta u = 1.6^2 \cdot \pi \cdot 342 \cdot 1000 \cdot 1000^{-3} \cdot 4182 \cdot 10.2 = 117.333\ W > 800 \cdot 0.12 = 96$$

After performing the surface integral along the interaction surface of the fluid, the entire heat flow rate inside the water fluid is 95.2 W, which is reasonable as some convergence tolerance exists. As the flow is incompressible, the outlet mass flow must equal the inlet mass flow, and they are equal after verification in the software.

In Fig. 2.35, the temperature of the working fluid gradually increases along the path, while the highest temperature on the top surface of the solid box is at the edge, not the outlet area. This could be explained as heat conduction inside the solid panel, which causes the lowest temperature to exist at the center. Based on the definition of thermal conductance:

$$C = \frac{k}{l} \tag{2.184}$$

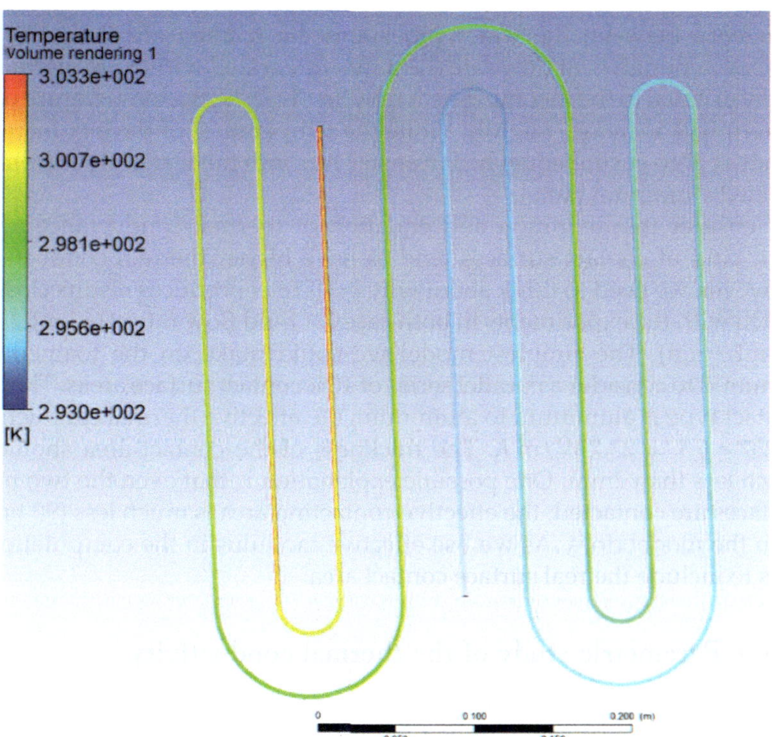

FIGURE 2.35 Temperature of the water fluid.

where k is thermal conductivity and l is the thickness of the heat path. The thermal conductivity for the panel is 18 W/m K, thus based on the parallel series, $l = \frac{18}{100} = 0.18$ m. This value is not reasonable as the entire thickness of the system is 30 m. A parametric study about thermal conductance is shown in the next subsection.

2.5.8 Parametric study of the thermal conductance of interfaces

Here, three case studies are made for thermal conductance, and we change the parameter as 80, 120, 150W/m^{2}K.

The change in thermal conductance produces significant temperature variations. The highest temperature of the two figures is 58.7 °C and 45.3 °C, respectively. And the temperature distribution seems to be more uniform than the right one with higher thermal conductance (Fig. 2.36). However, the temperature difference between the inlet and outlet of the fluid keeps the same. As derived from the energy balance equation in the above sections, the entire income heat flow rate should be the same as the heat transfer rate of the fluid, which means the total energy transferred in the fluid part is always the same. The only change is the temperature difference between the water pipes and the foamed aluminum box to generate enough heat flow rate (heat flux integrated within the interaction surface) equal to the incoming heat flow. In Fig. 2.37, the temperature of the water pipes increases slightly, while the temperature of the box increases about 11 K to ensure equal heat transfer rate into the water pipes from the foamed aluminum panel.

Consider the definition and equation of thermal conductance again, such kind of contact surfaces tend to have higher thermal conductance. However, we need to think about why $l = 0.18$ m produces results closer to match with the experiments in both cases of fluid flow rates (165 mL/min, 42 mL/min). The simplest model we could make on the foamed aluminum is to consider a parallel series of 10% contact surface areas. Then the contact type is aluminum to aluminum, the effective thermal conductivity is $237 \cdot 0.1 = 23.7$ W/m K. The thickness of the contact area should be much less than 2mm. One possible explanation is that even the two metal surfaces are contacted, the effective contacting area is much less (90 times) than the model does. As we use effective modulus in the computation, it fails to include the real surface contact area.

2.5.9 Parametric study of the thermal conductivity

Due to the uncertainty of the thermal conductivity of the foamed aluminum. Calmidi considered the thermal conductivity as a function of the volume fraction (Yang and Yin, 2011), and with 90% of air fraction, it yields

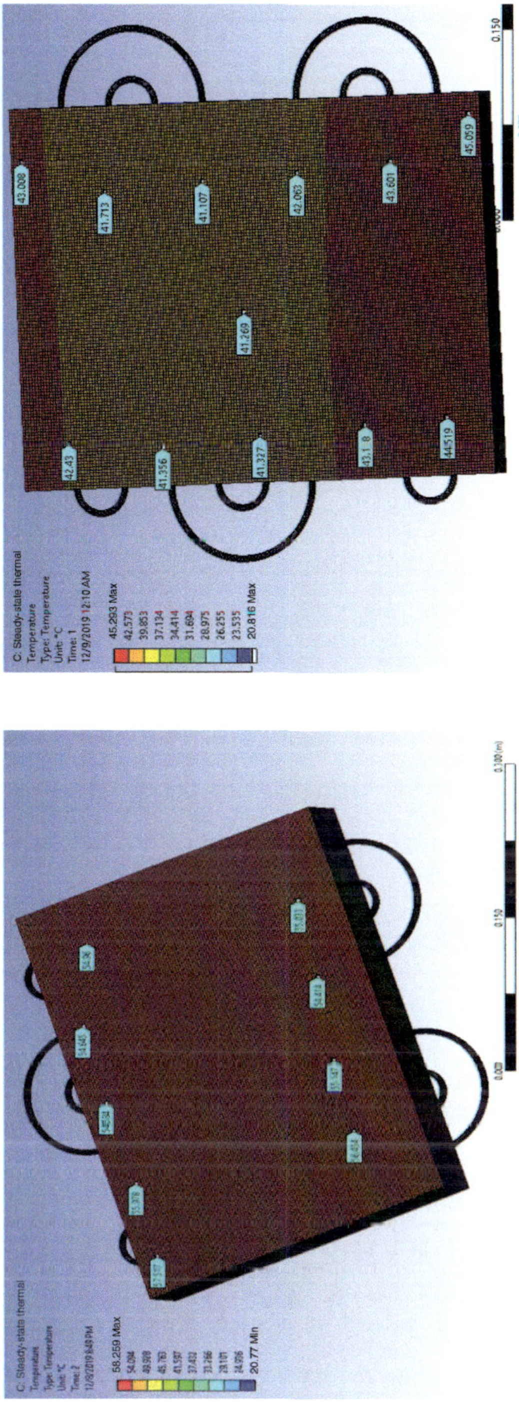

FIGURE 2.36 Temperature distribution of the solid panel (A), $C = 80\ W/m^2\ K$, (B) $C = 150\ W/m^2\ K$.

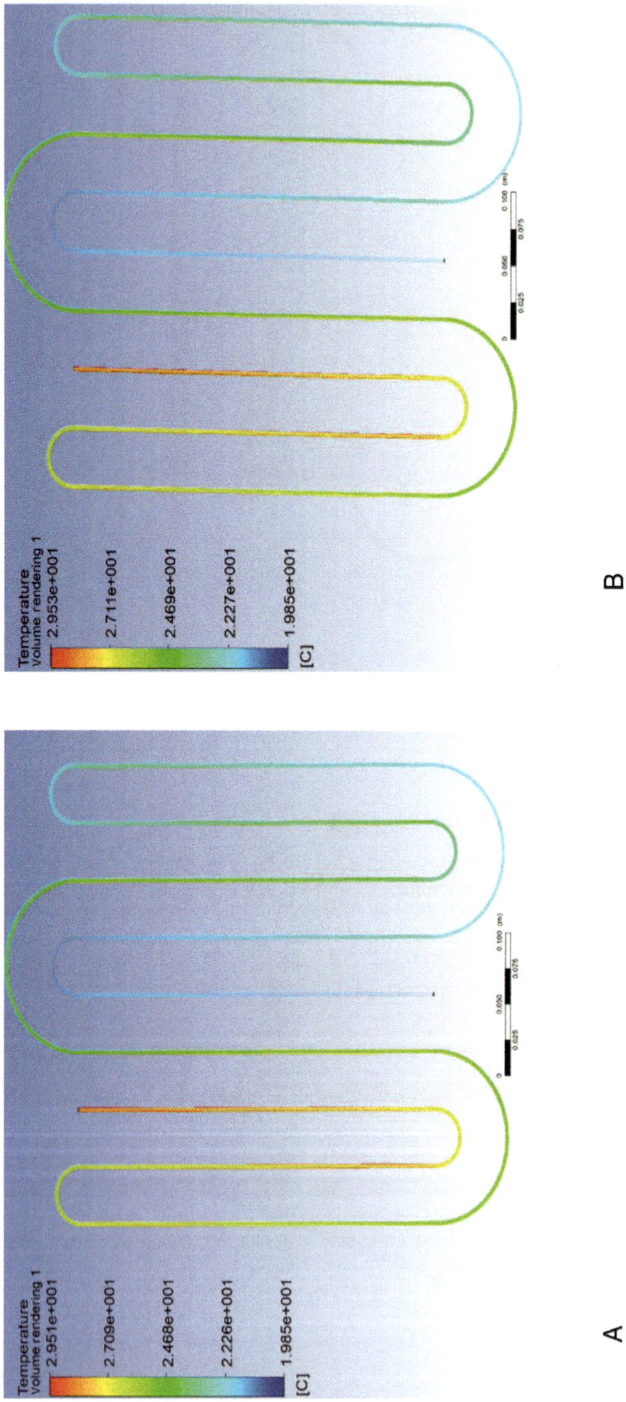

FIGURE 2.37 Temperature distribution of the water. (1) left one, $C = 80\ \mathrm{W/m^2\,K}$,(2) right one, $C = 150\ \mathrm{W/m^2}$.

18 W/m K. However, we still want to launch a parametric study on it to observe whether it is a sensitive factor in the model. Two case studies were done as $k = 180,\ 18$ W/m K (Fig. 2.38).

In Fig. 2.39, the temperature change in the foamed aluminum panel is 4 K. Compared to the 10 times difference of the thermal conductivity, the temperature change is not sensitive to the thermal conductivity change. Therefore, even the thermal conductivity of foamed aluminum is important, it is of less significance compared with the thermal conductance of the interaction surfaces.

2.5.10 Experiments

As mentioned before, there is a reverse correlation between the surface temperature of the panel and its output power (efficiency). Therefore, any tests that are done on the improvement of the efficiency of the solar panel have to include both the temperature change in the course of time and the corresponding effect on the output power of the panel. In order to simulate the sunlight on the panel, a solar lamp tester was used, in which the irradiation of the light can be adjusted to mimic the sunlight intensity at different weather conditions and locations. Under the standard test conditions with cell temperature of 25°C, an irradiance of 1000 W/m^2 with an AM of 1.5 (AM 1.5), water flow rate of 165 mL/min was examined and the changes in the I–V curve of the solar panel is measured. Also, the effect of the hydronic system will be compared with a regular panel without this cooling technology. Thus, another panel with the same cell specifications and geometry of 40 cm (15.7″) by 30 cm (11.8″) will be chosen and the short-circuit current, open-circuit voltage, and the power at different temperatures will be measured. The results will be compared with the results of the regular panel (I_{sc} 5.2 A, V_{oc}= 3.7 V, and P_{max} is 14W) to see how much improvement in the power is yielded. In Fig. 2.31, the designed solar panel with the temperature regulation system is shown (left). Nine thermocouples were used to collect the data during different times with and without the introduction of the water. Then the results are mapped for almost 83 minutes showing how the panel's temperature changes over time and with the cooling property of the hydronic system. There is another thermocouple that measures the ambient temperature as a reference point. The position of the thermocouples was chosen in a way to cover the whole area of the panel, but to make sure the distribution of the double serpentine tubes is covered. In this way, the temperature profile of the test can show how evenly the temperature is distributed after the flow of the water in the tubes begin. This is of significant importance since if there is a high-temperature difference between various points in the panel, not only the panel will not efficiently cool down all over, but also the resulting thermal stress mismatch at different points will accumulate and cause microcracks in the future (Figs 2.40 and 2.41).

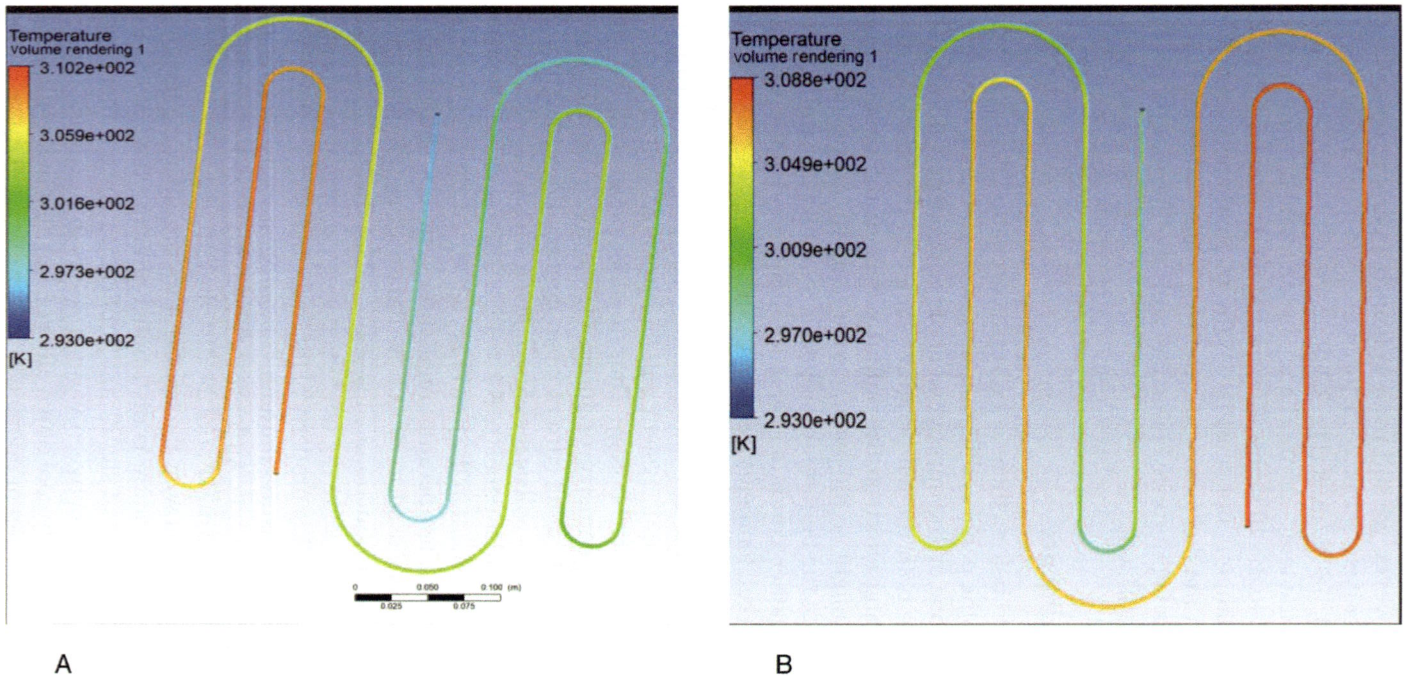

FIGURE 2.38 Temperature distribution of the water (1) left one, k = 180W/m K,(2) right one, k = 18W/m K.

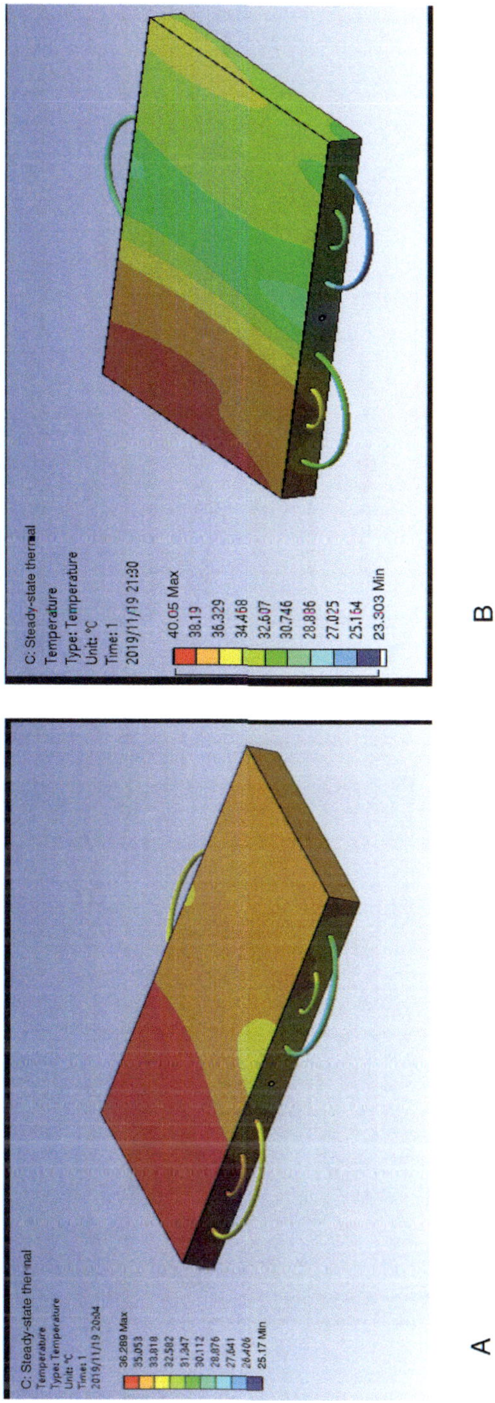

FIGURE 2.39 Temperature distribution of the foamed aluminum panel (A) k = 180W/m K, (B) k = 18W/m K.

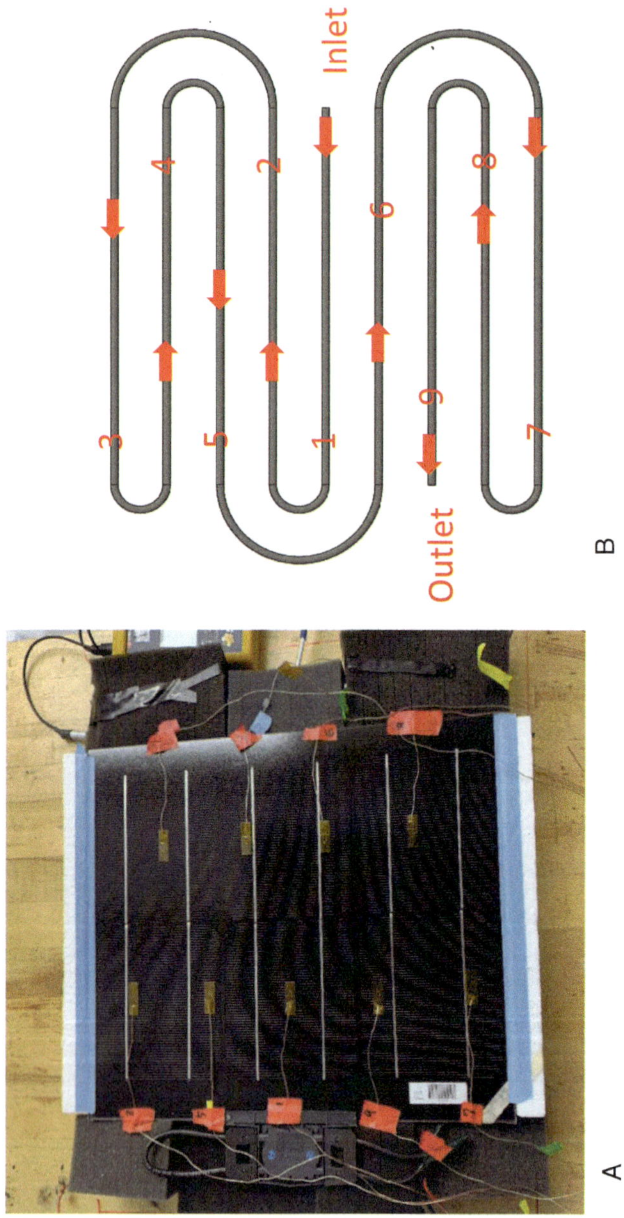

FIGURE 2.40 (A) The prototype of the BIPVT design, (B) schematic map of the thermocouples.

FIGURE 2.41 Testing the solar panel under the simulated light with an intensity of 1000 W/m^2.

2.5.11 Results and discussion

The tested results from the thermocouples and the I–V curve tracer are shown in Figs 2.42 and 2.43. In the beginning, before starting the solar lamp, the panel is at room temperature. As time passes, the surface temperature of the panel increases reaching the highest point of almost 84°C. At this point, the open-circuit voltage (V_{oc}) is 3.46 V and the short-circuit current (I_{sc}) is 5.18 A, and the MPP is read at 12.1 W. However, soon after the solar simulator starts, the surface temperature of the panel begins to rise, reaching about 84°C after about 50 minutes. At this point, the open-circuit voltage has decreased value to 2.88 V while the short-circuit current

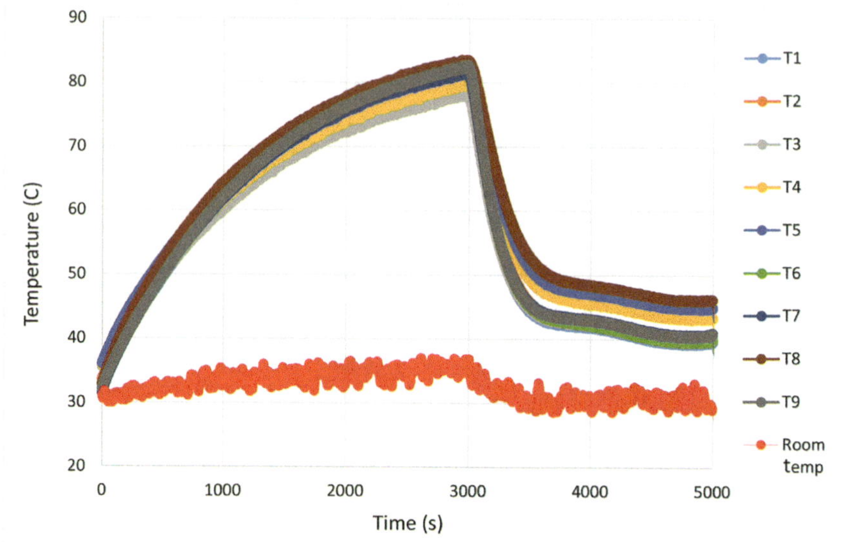

FIGURE 2.42 Temperature distribution of panel over time before and after water introduction.

increased to 5.27A. We can see that the MPP has significantly decreased to 8.89 W due to the temperature increase. When the water is introduced to the system, a sudden decay is observed in the temperature of the panel down to around 37°C. The power output gradually starts to increase and at the highest point, it reaches 11.7 W. It is observed that the V_{oc} improved to 3.39 V while the I_{sc} value is 5.21 A. It is observed that while the temperature increase had a noticeable effect on the open-circuit voltage, the short-circuit current slightly changed, confirming that the temperature results in less voltage of the panel and according to power.

As for the temperature distribution of the double serpentine hydronic system, it can be seen that the readings from the thermocouples are all within an accepted deviation 8°C difference after the water is introduced to the system. The highest temperature can be seen at T8 (46.1°C) which is near the outlet of the pipe, and the lowest reading is at T6 (38.3°C) near the center point of the panel. The values are in accordance with expectation, especially for T8 since this where the heat is carried away to the outlet and the water should be at its highest point after circulating and collecting heat from the surface of the panel. It is speculated that with a different design, i.e., single serpentine or zig-zag, different values of temperature difference will be observed, which requires more design parametric analysis.

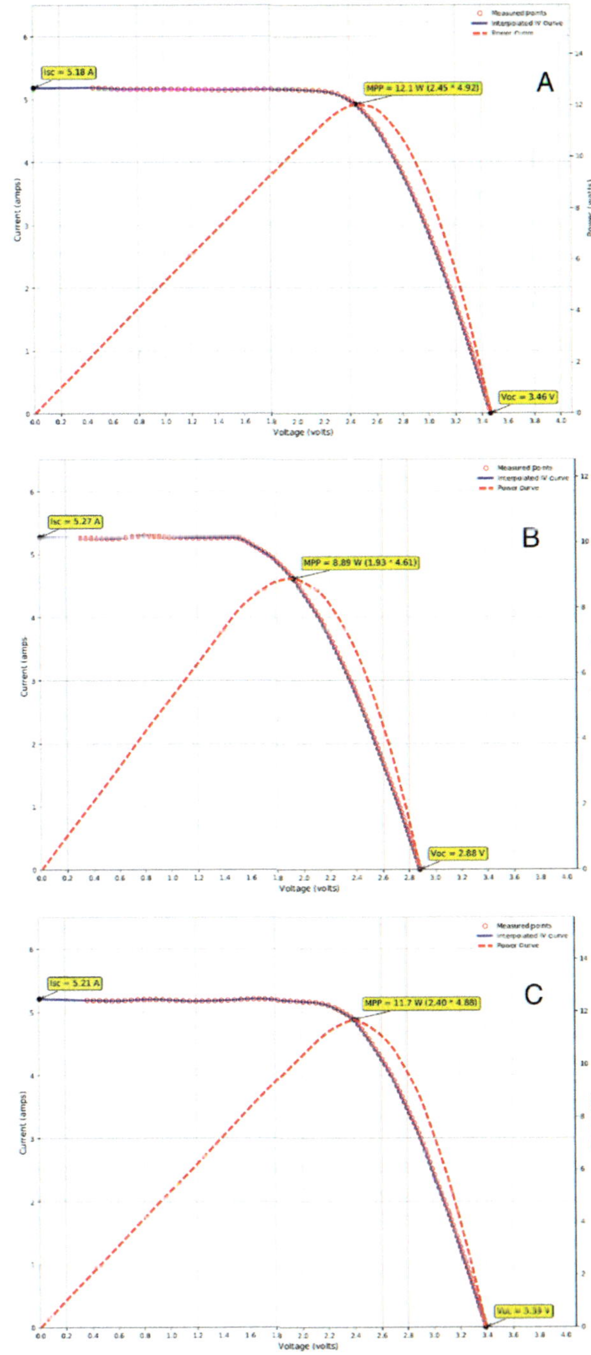

FIGURE 2.43 Current versus voltage and the output power in the BIPVT at: (A) initial state (32°C) without the water system; (B) final state (81°C) without the water system; (C) cooled state (36°C) with the water system.

2.5.12 Conclusions

In this work, the feasibility of using a temperature-regulated system on solar panels was examined. A hydronic system consisting of aluminum water tubes inside a foamed aluminum layer was designed and adhered to the backside of the solar panel. The water flow in the tubes contributed to cooling the surface temperature of the panel and thus, increasing the output power and efficiency. Results from the thermocouple reading and the measured I–V curves proved this hypothesis. An almost 27% increase in the efficiency of the panel was observed after the water was introduced into the hydronic system, increasing the V_{oc} from 2.88 to 3.46 V. Future works can include examining different water flows and water tube size to find the optimum design for the panel.

References

Agrawal, B., Tiwari, G., 2010. Life cycle cost assessment of building integrated photovoltaic thermal (BIPVT) systems. Energy Build. 42 (9), 1472–1481.

Anthony Fernandez, A.C.S., 2021. P-N junction, ChemMatters, Issues, Archive, 2013–2014, How a Solar Cell Works. https://www.acs.org/content/acs/en/education/resources/highschool/chemmatters/past-issues/archive-2013-2014/how-a-solar-cell-works.html.

Beckman, W.A., et al., 1994. TRNSYS the most complete solar energy system modeling and simulation software. Ren. Energy 5 (1-4), 486–488.

Bremner, S., Levy, M., Honsberg, C.B., 2008. Analysis of tandem solar cell efficiencies under AM1. 5G spectrum using a rapid flux calculation method. Prog. Photovolt. Res. Appl. 16 (3), 225–233.

Commission, I.E.t., IEC 61215:2005, Crystalline silicon terrestrial photovoltaic (PV) modules-design qualification and type approval. IEC 61215, 2nd Edition, April 2005, Geneva.

Consequential LCA, https://consequential-lca.org/clca/the-functional-unit/define-the-functional-unit/.

DOE, 2020. Departement of Energy's, Building Technology Office. "Energy Plus Homepage". This is a software by DOE, so no more details can be provided.

Duffie, J.A., Beckman, W.A. (Eds.) Solar Engineering of Thermal Processes, 2nd ed., John Wiley and Sons, New York, 1991. 919 pages.

ec.europa.eu, 2005. Life Cycle Assessment (LCA) https://ec.europa.eu/environment/life/project/Projects/index.cfm?fuseaction=home.showFile&rep=file&fil=ECOIL_Life_Cycle.pdf.

Eisenträger, J., et al., 2015. Application of the first-order shear deformation theory to the analysis of laminated glasses and photovoltaic panels. Int. J. Mech. Sci. 96, 163–171.

Fava, J.A., Society of Environmental Toxicology and Chemistry. A technical framework for life cycle assessment: workshop report, August 18–23, 1990. SETAC.

Fthenakis, V., et al., Methodology guidelines on life cycle assessment of photovoltaic electricity. IEA PVPS Task, 2011, 12.

Fthenakis, V.M., Lynn, P.A., 2018. Electricity from sunlight: Photovoltaic-systems integration and sustainability. John Wiley & Sons.

Haber, I.E., Farkas, I., 2011. Combining CFD simulations with blockoriented heatflow-network model for prediction of photovoltaic energy-production. J. Physics: Conf. Ser. 268 (1) 012008.

Hall, C.A., Day, J.W., 2009. Revisiting the limits to growth after peak oil: in the 1970s a rising world population and the finite resources available to support it were hot topics. Interest faded—but it's time to take another look. Am. Sci. 97 (3), 230–237.

Hanby, V.I., et al., 2002. Modeling the dynamic response of conduits. HVAC&R Res. 8 (1), 1–12.

Jäger, K.D., Isabella, O., Smets, A.H., van Swaaij, R.A., Zeman, M. Solar energy: fundamentals, technology and systems. UIT Cambridge, 2016.

Khezzar, R., Zereg, M., Khezzar, A., 2014. Modeling improvement of the four parameter model for photovoltaic modules. Sol. Energy 110, 452–462.

Koppelaar, R.H., 2017. Solar-PV energy payback and net energy: meta-assessment of study quality, reproducibility, and results harmonization. Ren. Sustain. Energy Rev. 72, 1241–1255.

Lamnatou, C., Smyth, M., Chemisana, D., 2019. Building-integrated photovoltaic/thermal (BIPVT): LCA of a façade-integrated prototype and issues about human health, ecosystems, resources. Sci. Total Environ. 660, 1576–1592.

Li, Y., et al., 2020. Mechanical analysis of photovoltaic panels with various boundary condition. Renewable Energy 145, 242–260.

Lin, Q., et al., 2020. Design and experiment of a sun-powered smart building envelope with automatic control. Energy Build. 223, 110173.

Lintag, J.E.a.L., M., Worcester Blackstone visitor Center Phase 2 – Glass Canopy Structural Calculations. Aesthetic Green Power Corp., 2017.

Murphy, D.J., 2014. The implications of the declining energy return on investment of oil production. Philos. Trans. R. Soc., A 372 (2006), 20130126.

Nairn, J.A., Mendels, D.A., 2001. On the use of planar shear-lag methods for stress-transfer analysis of multilayered composites. Mech. Mater. 33 (6), 335–362.

Naraghi, M.H., 2020. Energy Dynamics of Green Buildings. Linus Learning. ISBN 13: 978-1-60797-780-3.

Natarajan, S.K., et al., 2011. Numerical investigations of solar cell temperature for photovoltaic concentrator system with and without passive cooling arrangements. Int. J. Therm. Sci. 50 (12), 2514–2521.

Naumenko, K., Eremeyev, V.A., 2014. A layer-wise theory for laminated glass and photovoltaic panels. Compos. Struct. 112, 283–291.

Pocius, A.V., Dillard, D.A., 2002. Adhesion science and engineering: surfaces, chemistry and applications. Elsevier.

PV Education, 2021 I-V curve. https://www.pveducation.org/pvcdrom/solar-cell-operation/iv-curve.

Rühle, S., 2016. Tabulated values of the Shockley–Queisser limit for single junction solar cells. Sol. Energy 130, 139–147.

Schulze, S.-H., et al., 2012. Analysis of laminated glass beams for photovoltaic applications. Int. J. Solids Struct. 49 (15-16), 2027–2036.

Shockley, W., Queisser, H.J., 1961. Detailed balance limit of efficiency of p-n junction solar cells. J. Appl. Phys. 32 (3), 510–519.

Townsend, T.U. A method for estimating the long-term performance of direct-coupled photovoltaic systems (Doctoral dissertation), University of Wisconsin. 1989.

Wieder, S., 1982. Introduction to Solar Energy for Scientists and Engineers. In: 1982-01-01OSTI, United States.

Wikipedia, 2021. Life Cycle Assessment. https://en.wikipedia.org/wiki/Life-cycle_assessment.

Yang, D., et al., 2012. Simulation and experimental validation of heat transfer in a novel hybrid solar panel. Int. J. Heat Mass Transfer 55 (4), 1076–1082.

Yang, D., Yin, H., 2011. Energy conversion efficiency of a novel hybrid solar system for photovoltaic, thermoelectric, and heat utilization. IEEE Trans. Energy Convers. 26 (2), 662–670.

Yin, H.M., Buttlar, W.G., Paulino, G.H., 2007. Simplified solution for periodic thermal discontinuities in asphalt overlays bonded to rigid pavements. J. Trans. Eng., ASCE 133 (1), 39–46.

Yin, H.M., Prieto-Muñoz, P.A., 2013. Stress transfer through fully bonded interface of layered materials. Mech. Mater. 62, 69–79.

Zhang, Y., Lin, Q., Yin, H., 2021. Thermoelastic modeling of layered composites considering bending and shearing effects. J. Eng. Mech. 147 (7), 04021034.

Fossil fuel has powered the earth as the primary energy in the past centuries but created challenging problems for the sustainable development of human civilization on earth: the land devastation by mining, greenhouse emission, and air pollution during burning, climate change, and ecosystem disorder by global warming, which directly causes the change of the earth's surface by desertification. Now the desert and grass are competing for the land. This competition cannot last long as fossil fuels are limited, and it is hard for humans to survive in the desert. Nature creates a solution already: turn sand into silicon cells, power the earth with solar energy, reduce greenhouse gas emissions, rein the global warming, and heal the earth again for sustainable development. The "dove" solar farm will replace fossil-based electric plants and power the onsite silicon manufacturing (under the solar roof), which forms a self-sustained circle: harvest electricity by silicon solar cells—power silicon production from sand—use silicon for solar cell manufacturing. This circle does not require extra resources but generates electricity and solar cells toward the win of the green land over the desertification.

CHAPTER

3

Solar cell manufacture and module packaging

3.1 Introduction

In order to maximize the efficiency of a single junction solar cell, the following aspects shall be considered: (1) increase the light absorption balanced by the built-in voltage, which is characterized by the Shockley–Queisser limit; (2) minimize the forward bias dark current through the diode of the p-n junction; (3) maximize the electricity utilization by matching the internal resistance for minimized resistive loss. A semiconductor at a bandgap at 1.34 eV may exhibit the highest light absorption efficiency. Heterojunction or multijunction cells may overcome the Shockley–Queisser limit. However, the actual solar cell design is subjected to some constraints, such as material availability and price, service workability and durability, environmental toxicity, and safety. As solar energy generation requires a large area to make a sizeable application, the actual adoption of solar cells will not only rely on efficiency, the life-cycle cost and availability become key factors.

Although silicon (Si) is not ideal to reach the maximum of the Shockley–Queisser limit, it has been the dominant solar cell material, and currently occupies more than 90% of the total PV market revenue. As silicon is in such an abundance everywhere in the world, it is wise to be used as an energy generation resource. Notice that silicon is in general dominant in the semiconductor industry such as integrated circuits, transistors, and printed circuit boards. Crystalline silicon (c-Si) is one of the most widely used semiconductor materials in photovoltaic (PV) technology to manufacture solar cells.

The first silicon solar cell can be traced back to Ohl's patent in 1941 (Ohl, 1946) when the understanding of a dopant was very limited, and thus very low efficiency was obtained. The first practical photovoltaic cell was demonstrated with an efficiency of 6% in 1954 by Chapin et al. at Bell Laboratories (Chapin et al., 1954). Four years later, silicon solar cells were

Building Integrated Photovoltaic Thermal Systems: Fundamentals, Designs and Applications
DOI: https://doi.org/10.1016/B978-0-12-821064-2.00012-2

used in the early satellite - Vanguard I - that achieved the highest altitude of any man-made vehicle to that time (Vanguard I - the World's Oldest Satellite Still in Orbit, 2008). Considering conventional batteries used to power another transmitter on board Vanguard lasted only 20 days, the silicon solar cells used renewable energy and helped Vanguard I achieve all research goals. Since then, silicon solar cells have been commonly used in space applications. However, the high price prevents its applications to the building until the 1973 oil crisis. Adjusting for the inflation, the solar electricity price reduces from $96 per watt in the 1970s to $0.68 per watt in 2016 (Wikipedia, 2021). Silicon-based solar technology has become mature and driven a rapidly growing industry.

Let us start by looking at how silicon is produced. Silicon is extracted from silica or silicon dioxide (SiO2) which is the major component of sand. It can be extracted at high temperatures of ~2000°C by a reaction with carbon in an electrode arc furnace, which results in metallurgical grade silicon (MGS) containing some metal impurities such as iron, aluminum, etc., and cannot be directly used for solar cell yet before a further refining process. The most common refining process today in the industry is the Siemens process. Recently, Schmid has come up with a new proprietary process called Schmid Silicon Technology (SST) (Schmid Silicon Technology (SST), 2020) that claims to be less energy-intensive and costly compared with the Siemens process.

3.2 Silicon refining process

3.2.1 The Siemens process vs Schmid process

Silicon purification and production is the costliest part of the whole operations as compared to the wafer, solar cells, and module manufacturing, and the capital outlay would be in the order of billions. Initially, there were only four countries (Germany, the USA, Italy, and Japan) that had the manufacturing technology of solar cell silicon. In recent years, numerous other countries have become producers of solar cells including China, Taiwan, South Korea, Saudi Arabia, United Arab Emirates, Canada, Norway, Spain, Iceland, South Africa, India, and Brazil. Even though countries such as Norway, Iceland, and Taiwan are not vast countries and have land limitations, they have taken great actions on silicon production and renewable energy which could be efficacious for the future of the world to be less dependent on fossil fuel.

On silicon processing, there are three different procedures:

- **i.** Siemens process, which has been the dominion for years and still is,
- **ii.** silane/fluidized bed reactor (FBR) process; and most recently, and
- **iii.** SST process.

The Siemens process involves three mains steps:

1. Silicon is converted to trichlorosilane (TCS) in an FBR through the reaction (Si + 3HCl → SiHCl3 + H2); this removes the impurities but additional purification is required (Fig. 3.1A). There are two processes to generate the TCS through metallurgical silicon (SiMGS):
   (a) Low-pressure process: a pressure of 1–5 bars is applied at a temperature range of 300–350°C for hydrochlorination on impure metallurgical silicon (mgSi) by hydrogen chloride

$$Si(MGS) + 3HCl \rightarrow SiHCl_3 + H_2$$

   and a thermal converter is used to convert silicon tetra chloride (STC) to TCS at 1300°C as $SiCl_4 + H_2 \rightarrow SiHCl_3 + HCl$
   (b) High-pressure process: a pressure of 20–35 bars is applied at a high temperature (550°C) for hydrochlorination of MGS and STC at one one-step operation as

$$Si(MGS) + 3SiCl_4 + 2H_2 \rightarrow 4SiHCl_3$$

2. The TCS, which boils up to 32°C is then distilled in order to reach the required purity.

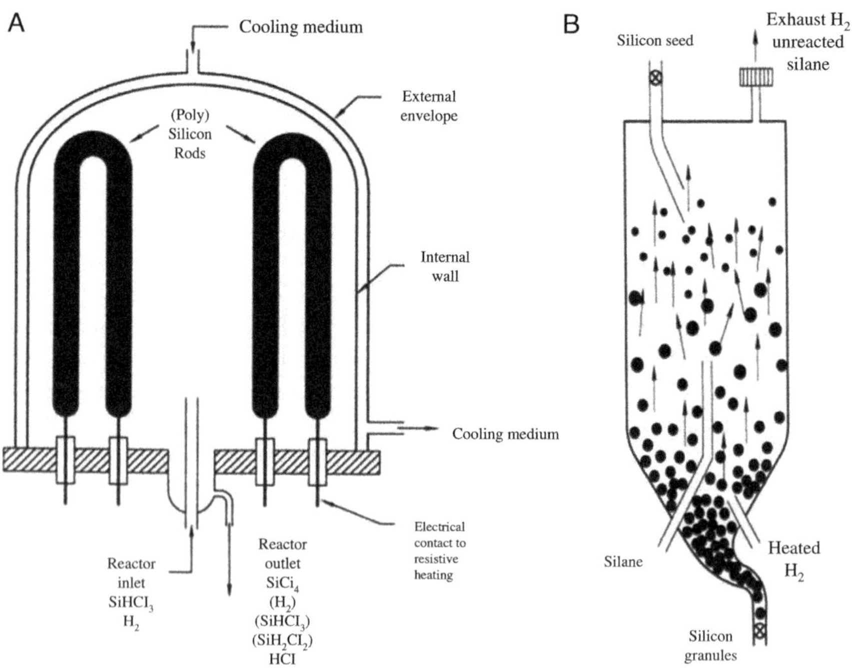

FIGURE 3.1 TCS/Siemens and Silane/FBR process (Bye and Ceccaroli, 2014). *FBR*, fluidized bed reactor; *TCS*, trichlorosilane.

3. The TCS is broken down in reducing atmosphere at around 900–1000°C via the reaction $SiHCl_3 + H_2 \rightarrow Si + 3HCl$, so that Si atoms from the vapor are deposited on silicon starting seed, usually a cylindrically arranged array of thin Si rods.

The high purity TCS will then vaporized and put into the deposition reactor. The gas will decompose onto the surface of the heated silicon seed rods, heated around 1100°C building rods of high purity silicon (Safarian et al., 2012).

$$2SiHCl_3 \rightarrow SiH_2Cl_2 + SiCl_4$$

$$SiH_2Cl_2 \rightarrow Si + 2HCl$$

$$SiHCl_3 + H_2 \rightarrow Si + 3HCl$$

$$SiHCl_3 + HCl \rightarrow SiCl_4 + H_2$$

The Siemens process consumes high energy which is a concern. However, in recent years the process has made advancements reducing the energy consumption and productivity increase especially considering the growth in demand.

This chemical vapor deposition (CVD) process produces a thick rod of highly pure silicon. Dopants such as Boron (B), Phosphorus (P), and Arsenic (As) at the required concentrations to produce p-type (B) or n-type (As, P) solar cells are introduced by parallel reactions involving BH, AsH, or PH. After deposition, the rods are removed from the reactor and broken into small polysilicon pieces. Afterward, the silicon is melted as starting material for the growth of large single-crystal or multicrystalline blocks.

As for the second procedure, the silane/fluid bed reactor process is a challenger to the Siemens process because it has a big reduction of energy consumption (Fig. 3.1B). The process initially uses silane as the feed gas, then it can be converted to silicon and with hydrogen gas as the byproduct according to the equation shown below.

$$SiH_4 \rightarrow Si + 2H_2$$

The main benefit of the silane/FBR process is its simplicity and energy efficiency. However, it is a slower process with less productivity. In a batch cycle, the Siemens process would take 60–150 hours and the silane/FBR process would take 60–120 days.

The third method, SST Process includes:

1. Hydrochlorination: (a) conversion of STC to TCS
(b) The synthesis of mgSi with HCl. The process could save around 70% of energy.

2. Disproportionation: the process is based on catalytical disproportionation of TCS to monosilane (25%) and STC (75%) in one process step and afterward, the monosilane is being purified where nitrogen, hydrogen, and oxygen residuals will be removed.
3. MS-CVD (monosilane-chemical vapor deposition): The Schmidt Silicon Technology, CVD has a great number of advancements compared to other similar systems in the market today.

According to the Schmid process of the high-end polysilicon and monosilane, recycling of the immediate product such as TCS and hydrogen is achieved by closed-loop. In the first step, the hydrochlorination, TCS is produced from feed materials of mgSi, STC, and Hydrogen (H_2) in a FBR. In the second step, the catalytic disproportionation, TCS is then converted to monosilane and STC. In the last process step, the MS-CVD deposition (Monosilane CVD), the monosilane gas is heated, and high-purity polysilicon deposits on heated rods, after which high-end polysilicon can be obtained. The following advantages have been listed for this process:

1. >25% STC conversion rate.
2. Automated solids treatment and removal with 95% solid content.

One step disproportionation:

1. Reactive distillation saves one reactor and one column.
2. Significant savings in energy consumption.
3. Operation with dichlorosilane (DCS) or TCS or any combination of the two.

Monosilane CVD:

1) >300 metric ton (MT)/a/reactor.
2) Power consumption less than 30 kWh/kg.

The entire process: superior product quality range 9–11N+ at lowest energy consumption

To learn more about the Schmid process, you can refer to the provided references (Ohl, 1946; Chapin et al., 1954; Vanguard I - the World's Oldest Satellite Still in Orbit, 2008).

As more advancements are made on these processes, it will lower the cost of solar technology and improve performance. Currently, the most common technology for solar cell manufacturing is (1) the monocrystalline or Czochralski (CZ) single crystalline, and (2) the directional solidification or cast, multicrystalline cells. In the past, the CZ dominated the market but in recent years, the multicrystalline cells are made more commercially available as the cost has outperformed the monocrystalline cells. Most of the wafer substrates used in production facilities have dimensions relating to monocrystalline silicon standard at 5 inches or 6 inches (125 mm or 156 mm) which in turn influenced standards of wafer carriers in automating packaging and module manufacturing. However, in order to increase the power density of the modules, wafers are shaped in squares or pseudo-square in the case of monocrystalline silicon. The round corners of the cylinders are cut off which reduces the surface area by 2%–5% as compared with a full square. A schematic of the manufacturing processes of silicon from sand is shown in Fig. 3.2.

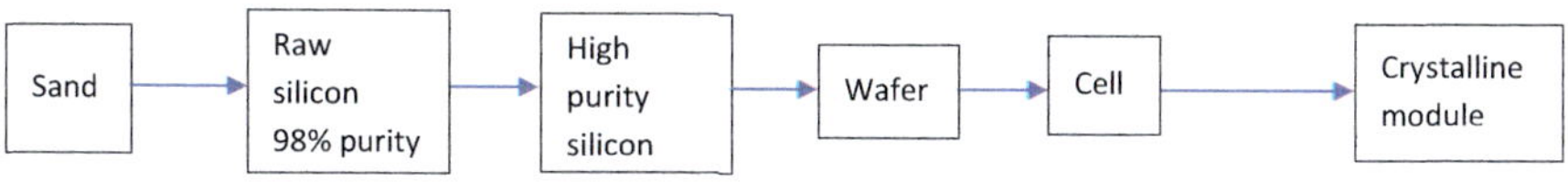

FIGURE 3.2 Sand to single-crystalline module.

The latest projection is that the price of silicon and polysilicon ingots will slowly decrease. However, the production cost of mono silicon is still relatively high. The Siemens and the FBR processes remain the main technologies for the production of polysilicon. Though the FBR consumes less electricity, the Siemens process will continue to the mainstream. Other technologies such as the upgraded metallurgical grade silicon (UMG-Si) or direct wafer technologies are going to be available but are not anticipated to yield a significant cost advantage. On the other hand, with the introduction of the diamond saw from the slurry-based wafer, there has been a significant improvement in terms of wafer process stability and cost reduction. According to the international roadmap technology of photovoltaic, the wafer accounts for 52% of the cell cost which is quite substantial. Looking ahead, the PV industry will continue to consider reducing the thickness of the cell in order to optimize the use of silicon.

3.2.2 Sawing ingots to wafers

Once high purity silicon is obtained by the refining process, often at 99.9999999% (9N) for advanced silicon PV cells, it can be melt at 1415°C in a large crucible. Let it cool slowly to form a large crystal. Very large grains (>1 cm) can be obtained by controlling the cooling rate. A large ingot, containing a large number of grains, needs to be cut to smaller bricks first and then be further cut to wafers for solar cell processing. Depending on the grain size, different types of wafers can be obtained for different types of silicon cells, such as single-crystalline, multicrystalline, polycrystalline, or amorphous (Fig. 3.3). For example, single crystalline wafers made by the CZ process exhibit a crystal size of ~10 cm.

In the past, the standard for silicon wafers thickness was about 200–250 μm. However, in recent years, this number has been reduced to the range of 120–160 μm, resulting in a much thinner wafer. This push from the silicon manufacturers and the solar industry is mainly done to optimize

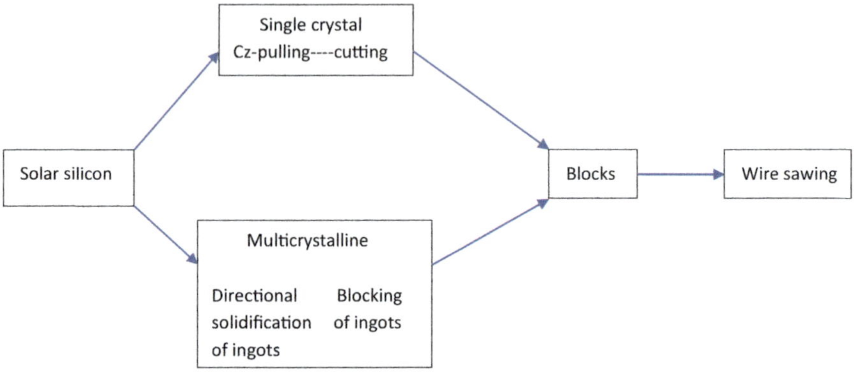

FIGURE 3.3 Single and multicrystalline solar cell processes.

the use of silicon, saving the cost of the materials, and make the final product more price-competitive. As for the sawing systems, the industry mainly has switched from loose abrasive slurry in Fig. 3.4A which is costly and has a high negative environmental impact, to fixed abrasive diamond wire sawing in Fig. 3.4B, which is more economical and environmentally friendly. Major machinery manufacturers are for this saw are Applied Materials Inc in Santa Clara, California, and Komatsu Ltd, Japan.

After the wafers come out of the sawing machines, they need to be ground and chamfered in order for the edges to become smooth which eventually results in improving the quality of the end product and increase the manufacturing capacity as a means of reducing costs.

3.3 Silicon solar cell production

3.3.1 Silicon cell production procedure

After the cells are ground and polished, the following steps are applied to the silicon cells:

1. Front surface texturing.
2. Optimized emitter surface concentration and doping profile.
3. Effective front surface passivation.
4. Fine line front electrode.
5. Front electrode passivation:
    - Point contact.
    - Deep and highly doped emitter under contact.
    - Metal insulated semiconductor contact.
6. Thin base.
7. Back surface passivation.

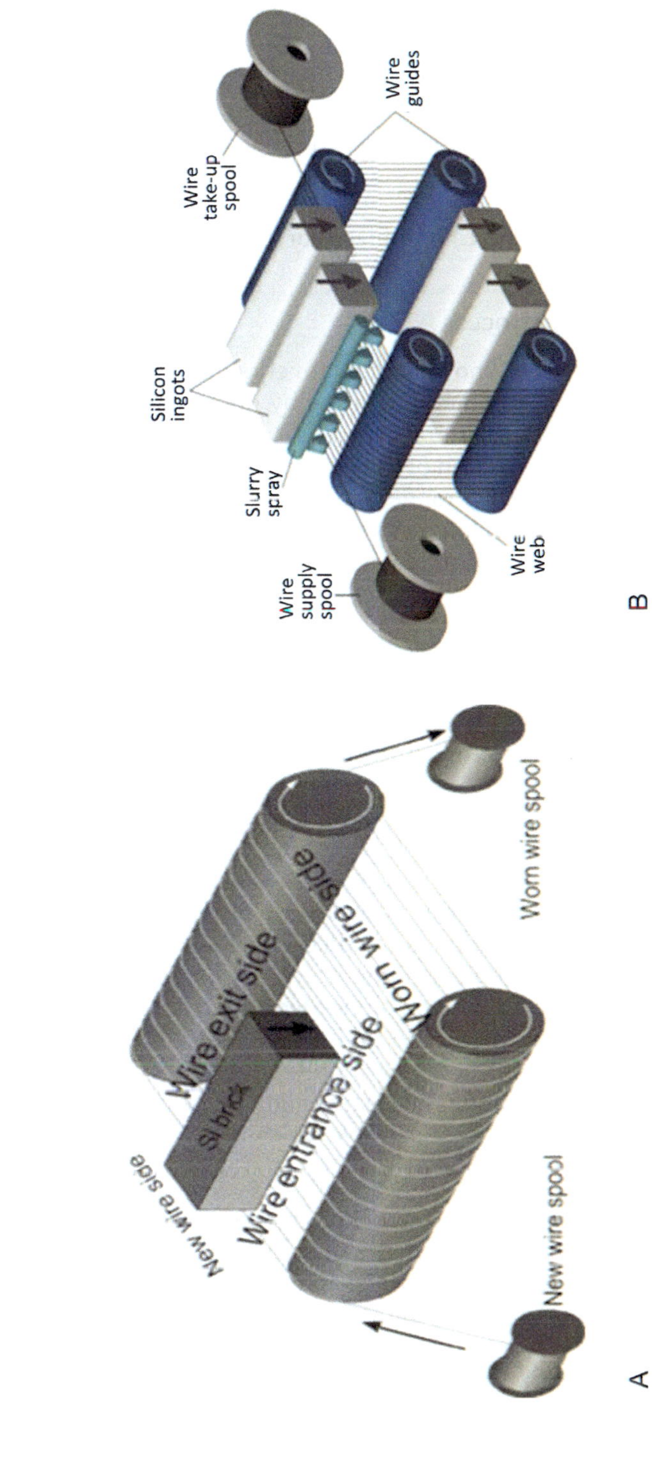

FIGURE 3.4 (A) Loose abrasive saw (*LAS*); and (B) fixed diamond wire saw (*DWS*), for cutting the ingots to wafers (Kumar and Melkote, 2018).

- Oxide and/or nitride passivation + local back surface field (BSF).
- Floating junction structure.

8. Back electrode passivation:
   - Point contact.
   - Deep back surface diffusion under the contact.
9. Back reflector.
10. Back surface texture.
11. Anti-reflective coating (ARC) optimized for encapsulation.

3.3.2 Screen printing process

This process has been well known and relatively simple. Nowadays, it is highly automated in solar cell manufacturing by numerous corporations. These systems could handle the front surface texturing and Al-alloyed BSF in the initial automated generation of the screen printing of cell processing where the conversion efficiency range has been in the order of 13%-15% throughout the earlier years. However, in recent years especially companies in Japan and Europe have made advancements in developing new types of paste formulation fabrication on finer line screens to modernize the screen-printing technology process. These types of silver pastes have additives that would allow at the firing process, selectively dissolve of silicon dioxide, and the ARC (ARC layers of TiOx (Tungsten Oxide) or SiNx (Silicon Nitride) while avoiding deep penetration into a sizable portion of the silicon. This will permit a simpler process in firing through passivating ARCs.

The advanced paste and the metal screen enable the screen to print the lines narrower in the range of 50–60 μm with a better aspect ratio. Therefore, the lines can be reduced which resulted in a more lightly doped emitter that would have a shorter wavelength response. The concern in nonoptimal surface passivation is handled using a selective structure emitter or a shallow homogenous emitter.

The phosphorous paste applied on a one-step selective emitter process in selective screen printing has made the pilot line efficiencies to 17.9% on 10 Ohm cm CZ (100 cm^2) and 16.8% poly-Si (100 cm^2). In a double diffusion and etch-back situation, and conversion efficiency of 17.0% on a polysilicon 156 cm^2 solar cell has been able to achieve. The features of the screen-printed solar cells are listed below:

- Isotopic front surface texturing independent of grain orientation.
- Reduced total front shadowing losses down to 6%.
- Shallow emitter with an optimal profile and front surface passivation.
- Efficient bulk passivation based on thermal treatment of hydrogen-rich silicon nitride.
- Back Surface Field (BSF).

The alignment of the front contacts on top of the highly doped n++ region is done automatically by the last generation of screen printers equipped with digital cameras. The movement of the table and squeeze is driven by linear motors control by the central computer.

The simplified shallow homogeneous process retains many advanced features described above. An independently confirmed efficiency of 16.8% has been obtained on 200 μm thick multi-Si substrates. The applied process sequence on the cross-section of such a cell is shown in Fig. 3.5. All the processing steps can be easily transferred to production lines.

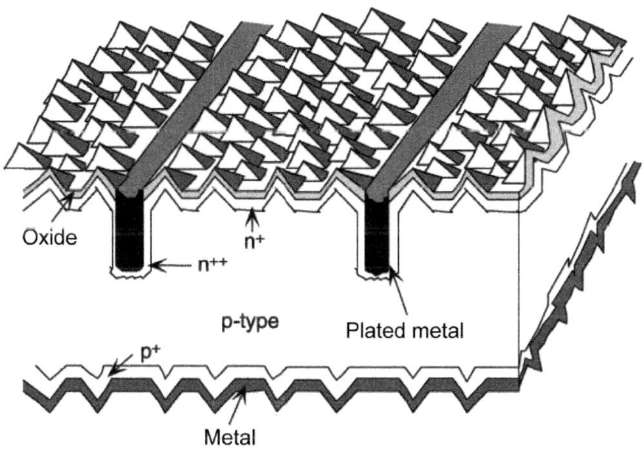

FIGURE 3.5 Cross-section of an advanced thin screen-printed cell (with permission from Elsevier) (McEvoy et al., 2012).

A process that combines screen printing and grooving has demonstrated efficiencies above 17% on 100 cm^2 multicrystalline solar cells. This screen-printing process is also well suited for larger cell sizes with efficiencies above 16%.

The processing sequence for a thin screen-printed multicrystalline silicon solar cell based on a shallow emitter is shown below.

Steps Process description:

1. Saw damage etching + isotopic texturization in one single step.
2. Shallow n+ diffusion covers the whole front surface (70 Ohm/sq.
3. Glass removal.
4. Edge junction isolation.
5. PECVD SiNx as front ARC and hydrogen source.
6. Printing + drying of adapted front and rear contacts.
7. Cofiring of both contacts:
   (i) Creation of contact to highly resistive emitter with low contact resistance by firing through SiNx.

(ii) Surface and bulk passivation by hydrogen release from SiNx.
(iii) Creation of Al-BSF for rear surface passivation without wafer bending.

3.3.3 Back contact cells

Another configuration of the rear contact cells is interdigitated back contact (IBC) cells. The main reason for looking at the back contact cells is that they are aesthetically pleasing and have a higher efficiency, which makes them very suitable for the building integrated photovoltaic (BIPV) systems. In addition, the IBC solar cell is particularly suitable for high concentration since both metal contacts are made on the backside of the cell as shown in Figs 3.6 and 3.7. Consequently, there is no shadowing effect on the front side. The trade-off between the shadowing effect and the series resistance is eliminated, which is particularly interesting for concentration applications.

In addition, the design of the front side (optical side) and the backside (electrical side) can easily be optimized separately. In a conventional solar cell, the doping of the front side emitter is a trade-off between series resistance and efficiency; a lighter doped front side emitter would improve the quantum efficiency and reduced the emitter saturation current density, but it would increase the series resistance of the cell. On the other hand, in IBC solar cells, the front side could be optimized for maximum quantum efficiency, reduced recombination, and reduced sublinearity, without compromising the series resistance of the cell.

1. Long recombination lifetime in the bulk after the solar cell fabrication. The diffusion length must be at least 5 times longer than the solar cell thickness. Thus, float-zone, high-resistivity, n-type substrates with carrier lifetime greater than 1 ms are generally preferred. The typical solar thickness is around 120–150 μm.
2. Low front surface recombination velocity. The front surface must be passivated with a thin thermally grown, silicon dioxide layer, grown in a very clean environment.

The typical process flow to manufacture IBC solar cells is summarized in Fig. 3.8. Though there may be some variations that would be subject to the manufacturer. Generally, it requires six photolithography steps, while additional photolithography steps may also be required.

1. Inverted pyramid texturization instead of random texture.
2. Local thinning instead of uniform wafer thinning.

As to the stability of the cells, like any high-efficiency silicon solar cell, the IBC is subject to efficiency degradation. If the float-zone wafer is used, the degradation is limited to the silicon-silicon dioxide interface.

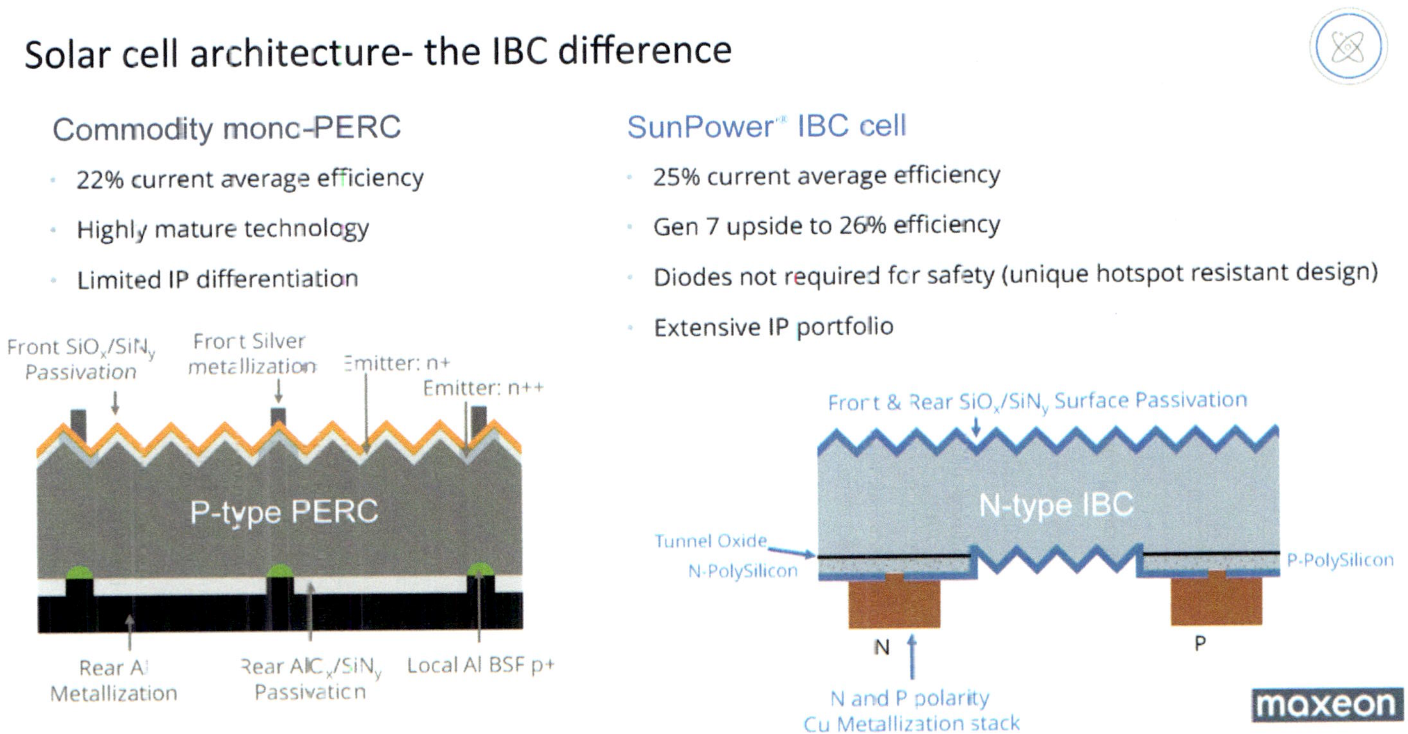

FIGURE 3.6 Sunpower and conventional solar cell cross-section (with permission from Maxeon) (MaxeonCells, 2020).

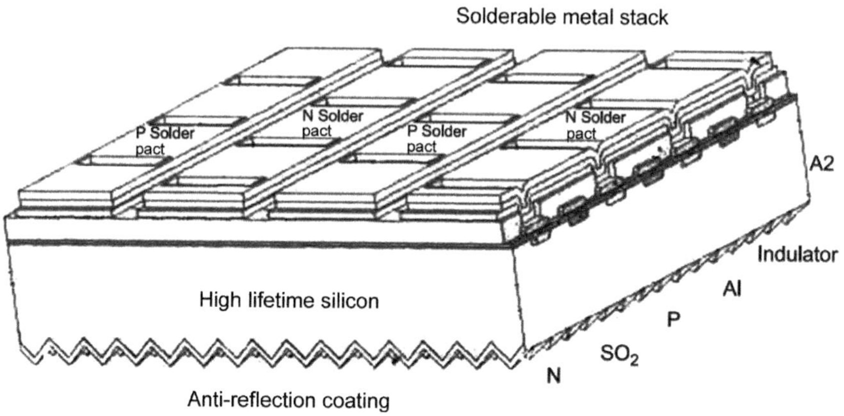

FIGURE 3.7 Cross-section image of interdigital back contact (*IBC*) cell (with permission from Elsevier) (McEvoy et al., 2012).

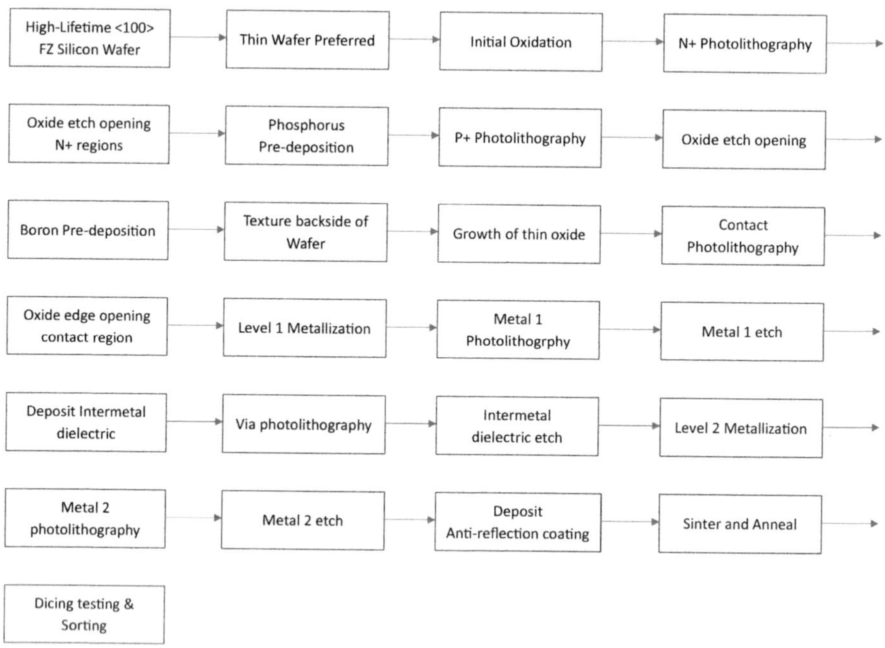

FIGURE 3.8 The manufacturing flow process of the IBC solar cells (3). *IBC*, interdigital back contact.

The degradation of the interface is due to the loss of hydrogen atoms passivating the dangling bonds at the silicon-silicon dioxide interface or the creation of new interface states. The degradation of the interface is manifested by either of the followings:

1. An increase of surface recombination velocity, which in turn results in a reduction of short circuit current over the entire range of concentration ratio.
2. An increase in the front emitter current to which results in an increase of sub-linearity (cell responsivity decreases as the concentration ratio increases) if the front surface is lightly doped or inverted due to changes in the dielectric layer.

Both mechanisms will cause a significant decrease in efficiency. The degradation has been observed so far due to the following individual or combined conditions:

1. Ultraviolet (UV) light, which creates excited electrons that overcome the potential barrier at the silicon-silicon interface and creates new interface states.
2. Elevated temperature (above 100°C), which allows hydrogen atoms to escape from the interface.
3. Mechanical stress.

It is also assumed that the following sunlight conditions enhance the degradation:

1. Highly concentrated sunlight which creates a large concentration of carriers at the surface of the solar cell.
2. The presence of moisture.
3. The presence of atomic hydrogen could break an existing Si-H bond to form an H_2 molecule.

More research still needs to be done to understand the degradation mechanism and find a definite solution to this issue.

Some of the considerations to ensure the stability of IBC solar cells are:

1. It is important that during the solar cell fabrication, the front passivation oxide layer is grown with the smallest interface state density. 1.1.1–Trichloroethane (TCA), dichloroethane (DCA), trichloroethylene (TCE), or HCL oxide growth, as well as an aluminum anneal, should be avoided as they have been recognized as giving a very unstable Si-SiO_2 interface.
2. UV light with a wavelength below 400 nm should be filtered from the solar spectrum, either at the solar cell level or at the concentrator module level. The ARC can include a UV filter such as TiO_2 or a very thin polysilicon layer.
3. The front surface of the solar cell should be coated with a thin layer of silicon nitride (as part of the ARC), or another material, hopefully, compatible with the ARC, forming great hydrogen and moisture diffusion barrier.

In recent years, substantial advancements have been achieved the development of solar cells. The conversion efficiencies have gone from 17% to 19.60% or even higher. As to the IBC cells, the conversion efficiencies have reached above 23%.

Sunpower has a recent announcement of Maxeon Gen 5 solar cells. The specifications and physcial characteristics of the Maxeon Gen 5 solar cell are shown in Fig. 3.9 and Fig. 3.10.

The dogbone designed by Sunpower to connect the back contact cell to improve the contact performance as well as the longevity of the connection is shown in Fig. 3.11, which has been widely used to connects the 125 mm Sunpower cells shown in Fig. 3.12 with 3-point connection. For the Maxeon Generation 5 of Sunpower Cells, Fig. 3.13 shows the front and back of the 6 inch (161 mm) back contact cell, which requires a 4-point connection due to the larger size.

3.4 Latest silicon cell technology and the advancements

In the past few years, aside from adding more busbars onto the cell, there has been quite a lot of research and development works done on the silicon solar cells. Major countries in the world in solar technologies are putting a sizeable amount of resources on the development of mono-like Si passivated emitter and rear cells (PERCs) through production lines to optimize the efficiency of these cells, some of which are:

1. PERC (passivated emitter and rear contact) cells.
2. PERC (passivated emitter and rear contact) bifacial cells.
3. PERT (passivated emitter and rear totally defused) cells.
4. PERT (passivated emitter and rear totally defused) bifacial cells.
5. PERL (passivated emitter and rear locally-diffused) cells.

The standard PERC cell structure and the corresponding fabrication processes are shown in Fig. 3.14. As can be seen, the solar cell is based on a p-type Si wafer. The front surface (emitter) is passivated by a dielectric such as SiNx, which also acts as an antireflection layer. The rear surface of the silicon wafer, however, is not passivated by any dielectric. Aluminum (Al) is doped into the Si to form a back surface (DSF) during the metal co-firing process to act as a high-low junction that prevents minority carriers from recombining to the rear surface.

As for the PERC solar cell, both the front and the rear surfaces of the device are passivated by dielectrics. Small pockets of the rear dielectric layer are opened (etched away) with a laser so that metal can be contacted to the rear surface device.

In comparison with the conventional solar cells, PERC is able to improve the efficiency of the silicon solar cells due to the additional passivation

Maxeon® Gen 5 High efficiency solar cell

Electrical characteristics of typical maxeon gen 5 cell												
Cell bins	Pmpp (Wp)		Eff (%)		Vmpp (V)		Impp (A)		Voc (V)		Isc (A)	
	Min	Max	Min	Max	Min	Max	Min	Max	Min	Max	Min	Max
Mn1	6.25		24.2		0.608	0.638	9.950	10.195	0.724	0.735	10.65	10.87
Ln	5.8	6.24	22.5	24.1	0.591	0.628	9.660	10.145	0.716	0.734	10.43	10.82
Kn	5.23	5.80	20.27	22.4	0.542	0.625	8.520	10.130	0.702	0.734	9.41	10.81

Temp coefficients in sunpower panels: Voltage: -1.74mV/°C, Current: 2.9mA/°C, Power: -0.29%/°C

FIGURE 3.9 Sunpower 6 inch cells specification (with permission from Maxeon).

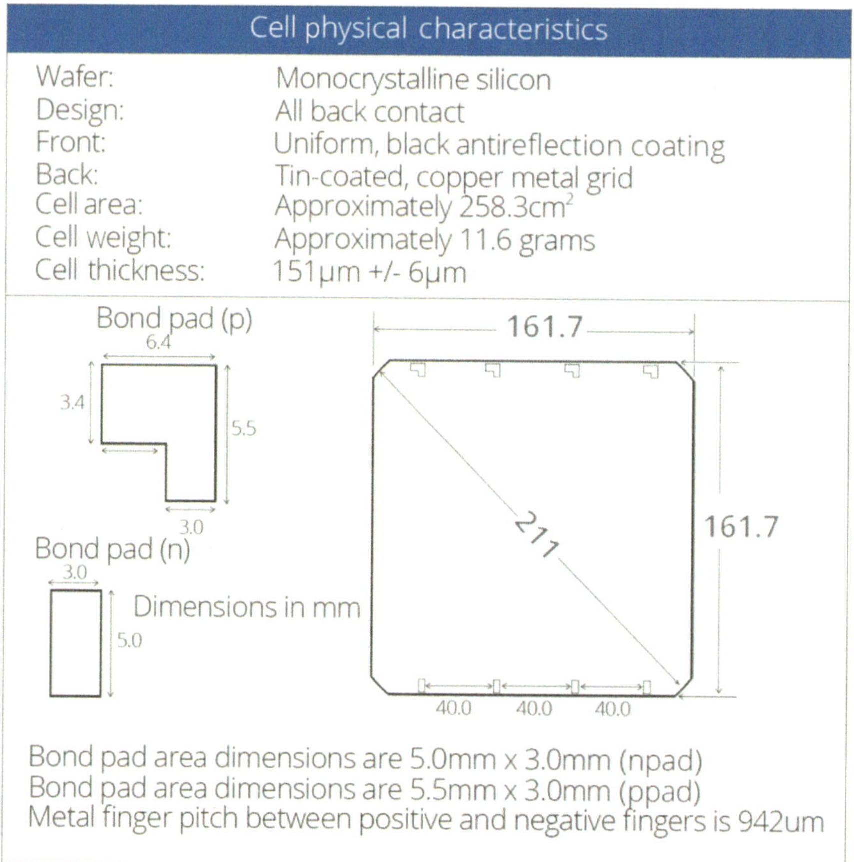

FIGURE 3.10 Sunpower 6 inch cell physical characteristics (with permission from Maxeon).

of the dielectric layer on the rear side, which prevents minority carriers (currently generated from light absorption) from recombining that is eliminated at the rear surface.

The extra rear dielectric layer reflects long-wavelength photons from the rear surface back to the device for more light absorption and current generation.

The manufacturing process of the PERC is as follows:

1. Wafer cleaning.
2. Texturing.
3. Diffusion.
4. Rear side morphology formation.
5. SE formation and PSG removal.

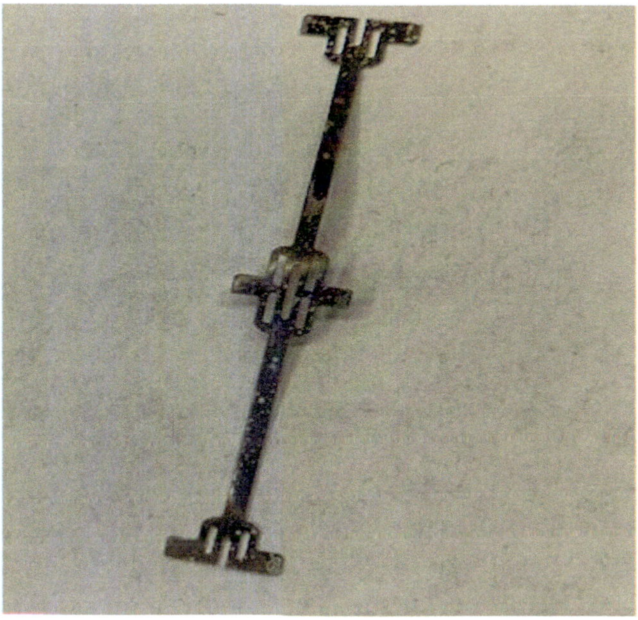

FIGURE 3.11 Dog bone to be used on tabbing of Sunpower cells (SPD Lab).

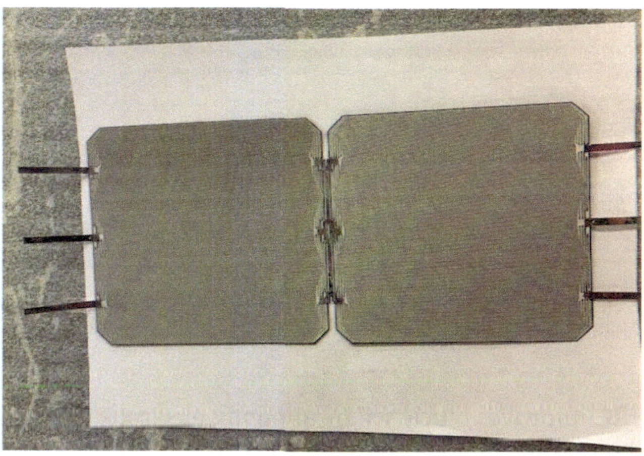

FIGURE 3.12 Dog bone Tabbing on Sunpower cells (SPD Lab).

6. Rear passivation film deposition.
7. Front ARC.
8. Rear film opening.
9. Screenprint of front and rear contacts.
10. Copfiring.

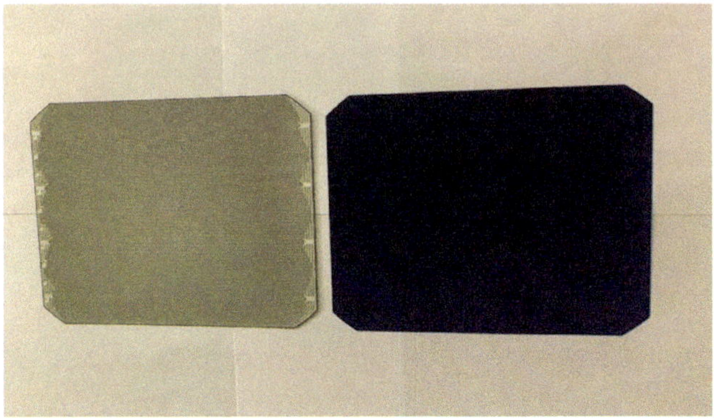

FIGURE 3.13 Sunpower Maxeon GEN 5, 6 inch Back & Front Back Contact Cell (SPD Lab).

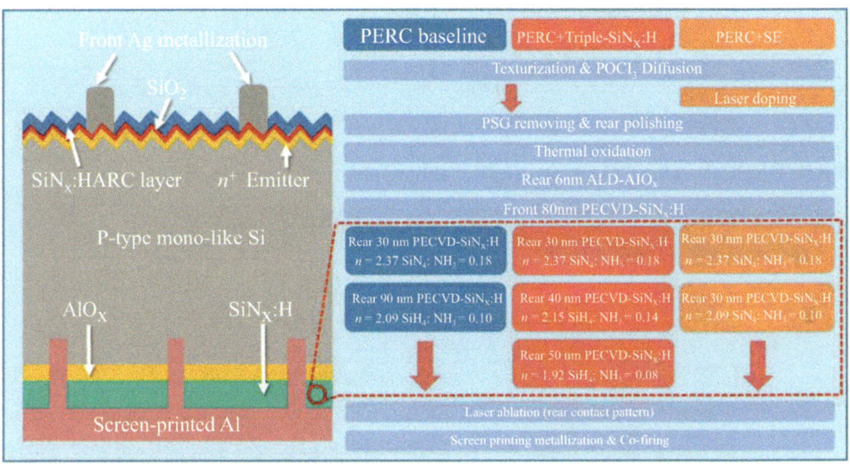

FIGURE 3.14 Cross-section of the PERC and process steps (Lv et al., 2020). *PERC*, passivated emitter and rear contact.

The PERC improves efficiency by allowing electrons to flow freely and increase the reflectiveness of the solar cells. Recent successes of the PERC solar cells have attracted the attention of the PV industry and research institutions especially due to the fact that the PERC structure seems to be reaching the bottleneck of its practical power conversion efficiency limit and thus, it has persuaded the PV researchers to look for other cell architectures to continue increasing the conversion efficiency of the silicon solar cells. The passivated emitter rear totally (PERT) diffused solar cell has attracted much attentions in the industry. Fig. 3.15 compares the PERC and PERT cell technologies.

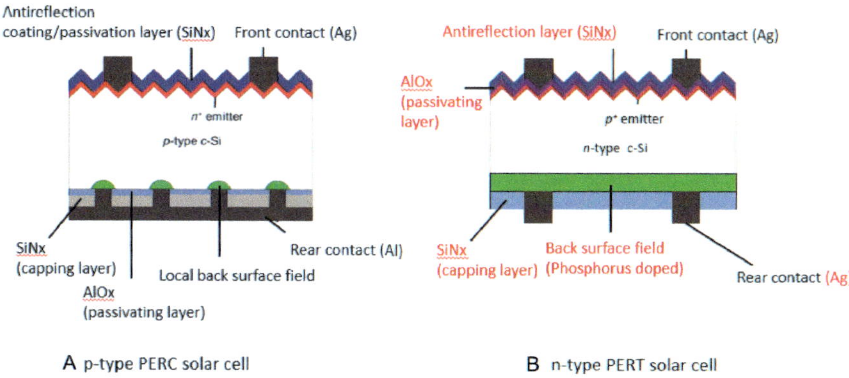

FIGURE 3.15 A p-type PERC versus an n-type PERT structure. *PERC*, passivated emitter and rear contact; *PERT*, passivated emitter and rear totally

PERC structure has a localized BSF. The BSF was created from the doping of Al into Si during metal cofiring processes. BSF helps to improve the solar cell efficiency by forming a high-low junction with a p-type Si base wafer. This junction repels minority carriers and prevents them from recombing at the rear surface of the Si wafer.

As for the PERT structure, the rear surface is totally diffused with boron (p-type) or phosphorous (n-type). Usually, PERT technology is implemented on n-type Si solar cells. This is to take full advantage of n-type Si wafers' higher tolerance to metallic impurities, lower temperature coefficient, and lower light-induced degradation (LID) than P-type Si wafers. The LID is lower in n-type Si possibly due to lower boron-oxygen pairs, as the bulk in the n-type wafer is doped with phosphorus.

However, the totally diffused BSF requires additional processes, such as high temperature and BBr diffusion which result in being more expensive to manufacture than PERC (12).

Nonetheless, the full area BSF in PERT solar cells may provide a more effective high-low junction passivation effect than localized coarser Al-based BSF in PERC. In addition, PERT also allows the integration of the so-called tunnel oxide passivated contact (TOPCon) structure which has the potential to improve the efficiency of the device.

The manufacturing process of PERT on n-type silicon is summarized as follows.

1. Saw damage removal and cleaning.
2. Boron diffusion (emitter).
3. Thermal oxidation.
4. Front-oxide removal.
5. Texturing.

6. POCl3 diffusion (FSF).
7. Thermal oxidation.
8. PECVD SiN3 deposition.
9. Laser ablation rear.
10. Metallization rear.
    1% AlSi sputtering + anneal.
11. Laser ablation front.
12. Metallization front.
    Ni/Cu/Ag plating + Sintering
13. Edge Isolation.

PERL solar cell is another high-efficiency technology that uses microelectronic techniques to produce solar cells, reaching an efficiency of 25% under the AM 1.5 spectrum. The high-quality oxide at the front surface significantly lowers the number of carriers that recombine at the surface being the passivated emitter. The rear emitter is locally diffused only at the metal contacts to minimize the recombination of the rear while maintaining good electrical contact.

The two PERL cell design variations that have been employed are: (1) altered boron emitter (deep/shallow); and (2) different dielectric layers for rear side passivation.

PERL Bifacial Cell Technology has many benefits such as:

1. Higher efficiency of the cells achieved by passivation on the n-type mono-material.
2. Bifacial architecture can feature up to a 30% higher yield of energy.
3. Having only one temperature step for deposition of both phosphorous and boron deposition minimizes the stress of the wafers.
4. The manufacturing time of the PERL cells has a low process time.
5. A different configuration like a transparent backsheet or glass/glass modules would be quite effective on performances.
6. Same ownership cost as the PERC cells.

PERL cells have a long-life expectancy based on n-type material and achieve an efficiency as high as 25%.

In the past few years, the growth of PERC solar cell production has been leading the PV industry. The growth has exponentially increased from 1 GW in 2014 to above 60 GW in 2019 according to the PV Tech. The growth has been matched by continuous cell efficiency and module power improvements. The emphasis has mainly been on the p-type PERC in the past years. Chinese companies are also looking into the n-type in the future.

The use of rear passivation layers had originally been confined to highly advanced cell concepts, such as heterojunction and IBC cells. Although depositing passivation layers on p-type cells has been well established in the research community, it has not fully justified in the industry before based on the equipment cost and additional complexity in cell lines.

Moving away from screen-printing was recognized as a key in opening the rear surface for additional benefits such as bifacial capabilities on cell performance.

In the period of 2012–2014, the paste improved significantly with the increase of busbars formation that drives the efficiency higher with minimal capital expenditures. Once the front cell processing had gone through the improvement phase, it was time for the PV industry to focus on the rear surface, and bring the rear passivation layers more to the mainstream status.

The massive cell production has been matched by substantial drive by the PV industry on R&D to further increase the cell efficiency performance, which resulted in reaching an efficiency of 19%–21% at the end of 2018. The continued capacity expansions and efficiency increases were seen from the p-mono PERC-based cell producers coming at a time when investments were also being made in the n-type alternatives, in particular the n-PERT and HJT.

Recently, there was an announcement by Sunpower Corporation on the production of a 6-inch back contact cell, the fifth generation (Maxeon Gen 5) solar cell with 6.2 W power per cell. Sunpower claims that by applying this cell on a module, it would deliver 60% more energy on a similar roof space. Sunpower's panels use a factory-integrated microinverter.

Also, the two companies of Imec and Jolywood announced a 23.2% front conversion efficiency on n-type PERT solar cells with over 80% bifaciality, while aiming at an efficiency of over 23.5% and above 90% bifaciality in the near future (Fig. 3.16). According to Imec, these cells have the potential for higher efficiencies because of their lower sensitivity to metal impurities

FIGURE 3.16 Bifacial n-PERT solar cell with an efficiency of 23.2% by Imec and Jolywood (imec, 2021). *PERT*, passivated emitter and rear totally.

and the absence of LID, which makes them competitive with the p-type PERC cells.

The research team optimized fully the screen-printed bifacial n-PERT cell process and adopted a design of 12 busbars. The cells also used a screen-printed front grid and rear soldering pads that use less silver. In particular, Imec has emphasized the limited use of silver along with the potential for multiwire interconnection schemes as factors that could lead to a low-cost production process.

They were able to achieve busbar-free PERC cells with an efficiency of 21.9% that was developed totally in-house at PV Tech. It is a monocrystalline silicon cell using a homogenous emitter on the front side and aluminum oxide passivation on the backside. They are bifacial pSPEER shingle cells with a power density of 235 W/m^2. The front side irradiation 1000 W/m^2. While the rear side irradiation 100 W/m^2. It is a patented passivated edge technology that minimizes shading on both sides of the cell (Innovations Report, 2019).

In addition, the TOPCon cells have a conversion efficiency of 22.5%. It has been manufactured through the Fraunhofer ISE process. The TOPCon technology is a charge carrier contact that is also developed at the Institute and has an ultra-thin tunnel oxide together with a thin silicon layer that would provide great charge selectivity.

In 2019, the ITRPV (International Technology Roadmap of Photovoltaic) projected the growth of different cell technologies in Fig. 3.17. We should all

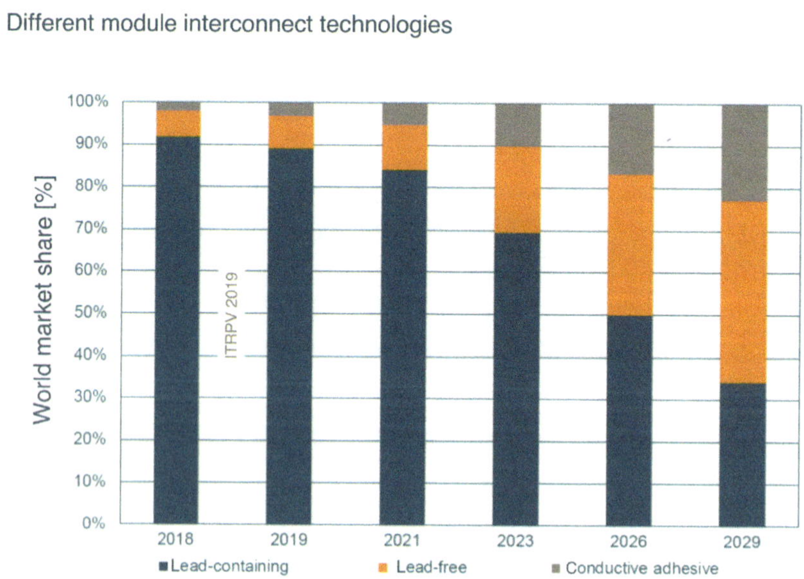

FIGURE 3.17 International Technology Roadmap of Photovoltaic (*ITRPV*) projection of the growth of different cell technologies (International Technology Roadmap for Photovoltaic (*ITRPV*) March 2019).

keep looking at all these announcements and look into the signs of progress that are being made. Meanwhile, we should all evaluate whether these new technologies in manufacturing products be sustainable in the long term.

3.5 Solar module production

3.5.1 Interconnection materials

The interconnection ribbons are basically the veins of the solar modules, and it is highly recommended that they be handled with caution. The deposited silver circuitry on the cell creates shadows on the surface of the cell. Though attaching the ribbon onto the busbars does not result in a significant loss of efficiency but it does represent an inactive portion of the module. In recent years, scientists at the Massachusetts Institute of Technology have come up with a solution by enabling the ribbons to reflect the majority of the incident sunlight back to cells within the module which results in converting the busbar area into a passive power generating side. The idea behind the concept is that the top surface of the ribbons is coated in silver embossed with parallel triangle grooves that have mirrorlike side surfaces. When struck by sunlight, the grooves reflect light to the air/glass interface at a shallow angle. The reflected light undergoes total internal reflection, making its way back to the cell surface to provide a second chance for the photons to be absorbed.

In addition, Solar Zentrum, Stuttgart, in partnership with Institute for Photovoltaic (IPV), University of Stuttgart, and Steinbeis Center Photovoltaics looked into relying on total internal reflection principle by using a white paint that scatters the incident light, instead of using silver which is much more expensive (Fig. 3.18). Bruker-Spaleck and Luvata Pori OY from Finland introduced a white-coated ribbon that performs with the same condition. On top of the white coating offered by these two companies, they introduced black and blue. The colors are not intended for performance but were mainly purposed for aesthetics since their main target is the building integrated market.

As for interconnection materials, copper will probably remain the dominating material. Copper wires are expected to gain over 30% of the market share in the coming decade according to the International Roadmap Technology of PhotoVoltaic (IRTPV) projection in Fig. 3.19. As to structured foil ribbon, it will be mainly used for back-contact cell applications, which would reach around 8 % of the market share, and the shingled and overlapping interconnection technology might gain a market share of around 10 % by 2029.

Although the innovation in the product advancements has been great thus far, the cost is always a critical issue since the PV industry has been operating at a very thin margin. The ribbon is a very small part of the solar panel though it is a very important one. The trend on the ribbon has been

FIGURE 3.18 Ulbrich White scatters the incident light that is reabsorbed by the module, thus generating more power.

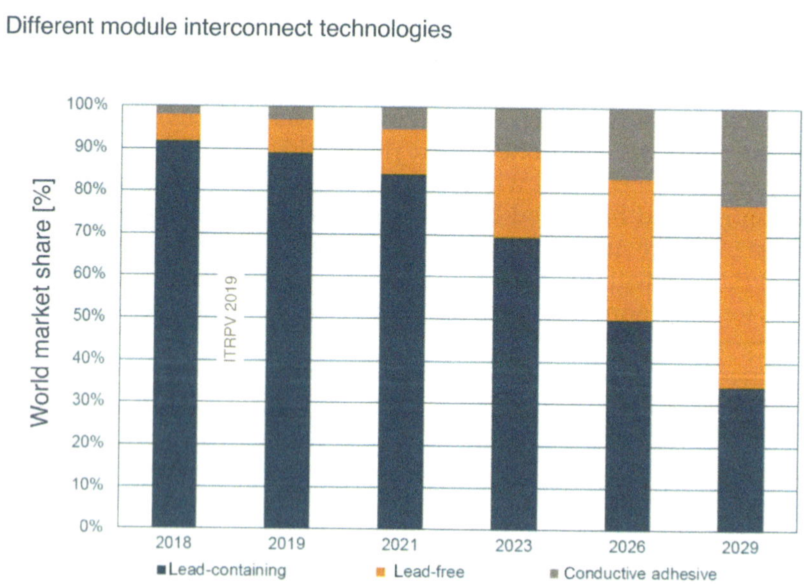

FIGURE 3.19 ITRPV projection on interconnection growth (International Technology Roadmap for Photovoltaic (ITRPV) March 2019).

to reduce the silver content as well as the lead content or to go completely lead-free in the coming years.

The major supplies of the ribbons are Alpha PV, and Ulbrich based in the United States, Bruker-Spaleck, and Schlenk Metallfolfolien based in Germany, KOS based in South Korea, Luvata Pori Oy based in Finland, and Sukriti Vidyut Udyog based in India.

It is important to note that the upcoming interconnection technologies will need to be compatible with the ever-thinner wafers that will be used in the future. In this respect, a low-temperature approach using conductive adhesives or wire-based connections has inherent advantages due to lower stress on the wafers.

Similar to the cell interconnection, there is a trend toward lead-free module interconnection covering all connections between strings and Junction box. Conductive adhesives and lead-free interconnects are expected to become the dominating alternative to the currently used lead-containing technologies.

3.5.2 Solar cell tabbing and stringing

A couple of decades ago, assembling the solar modules was manually done by initial tabbing and stringing the solar cell into a string according to the design of the module (Figs. 3.20 and 3.21).

On tabbing, the initial step was to put some flux on the busbar of the solar cell. The next step would be to put the ribbon on top of the busbar and use a soldering iron and apply it onto the ribbon to connect the ribbon to the busbar (Figs. 3.22 and 3.23).

The stringing process of the solar cells is very similar to connecting a series of batteries from positive to negative of the battery. All the cells have to be aligned to the guard shown on the stringing table.

FIGURE 3.20 Manual tabbing and stringing station (SPD Lab).

FIGURE 3.21 Latest design of manual tabbing and stringing station (SPD Lab).

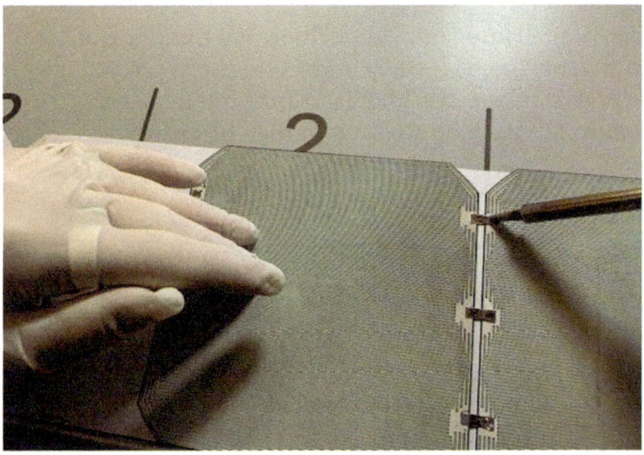

FIGURE 3.22 Manual tabbing process on back contact cells (SPD Lab).

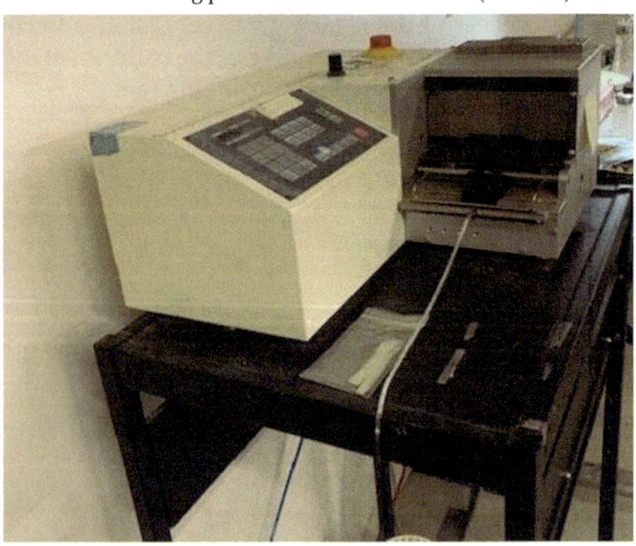

FIGURE 3.23 Ribbon cutter (SPD Lab).

Looking at the tabbing and stringing process, it looks quite simple and mainly done manually. The person handling the cells must be very cautious as to the soldering iron as well as the alignment of the string. The whole process must be handled with care and precision. Any deviation from the baseline could be time-consuming and costly that can affect the manufacturing line dearly. Especially in an industry such as solar that operates on a very thin margin, any yield issue is very critical to the bottom line. Nowadays, as the solar cell prices have decreased, and the conversion efficiency rates have been increasing, the solar cells have become a lot thinner and more difficult to handle. Thus, caution in operation is important.

Over the past two decades, the PV industry has made substantial advancements in the tabbing and stringing manufacturing processes, especially in Europe. Automatic and semiautomatic tabbing and stringing machines have been invented and developed to meet the needs of the PV Industry. However, though it has reduced the manufacturing time, the production results and the yields have not been very impressive. Although, there has been some improvement in recent years.

One of these systems is called combined tabbing and stringing (CTS). These automatic machines have several software and controlled systems built into them to successfully making them a reality to the industry. A basic software flowchart is shown in Fig. 3.24. Software and hardware interface could be quite a challenge at times especially in the precision control areas of the systems (Atlantis Energy Systems manufacturing operation).

After the strings are tested, they will be assembled with encapsulant layer and glass sheets for the lamination process (Fig. 3.25). The above is a general idea of the CTS system and module manufacturing operation. Different machine manufacturers could have some minor variations, but the basic principles are the same. As it can be seen, there are a series of tests throughout the whole module assembly process. Especially in the case of custom BIPV modules, it is strongly recommended to test prior to lamination which could also reduce the number of reject modules, as the BIPV modules could be very expensive.

$T = R\text{t(ce)}AL\text{(ns)(nd)}Y$

T = annual throughput (W/year).

R = process rate in cells/minute × 60 minutes/hour × 8 hour/shift (cell/shift).

t = system up-time, the ratio of (operating time) /total time).

ce = cell efficiency average in production. Air Mass 1.5 global conditions.

A = cell area (cm^2).

L = solar isolation = 0.100 (W/cm^2).

ns = number of operating shifts per day (shifts/day).

nd = number of operating days per week × 52 week per year (days/year).

Y = process yield, the ratio of (good cells out)/(cells in).

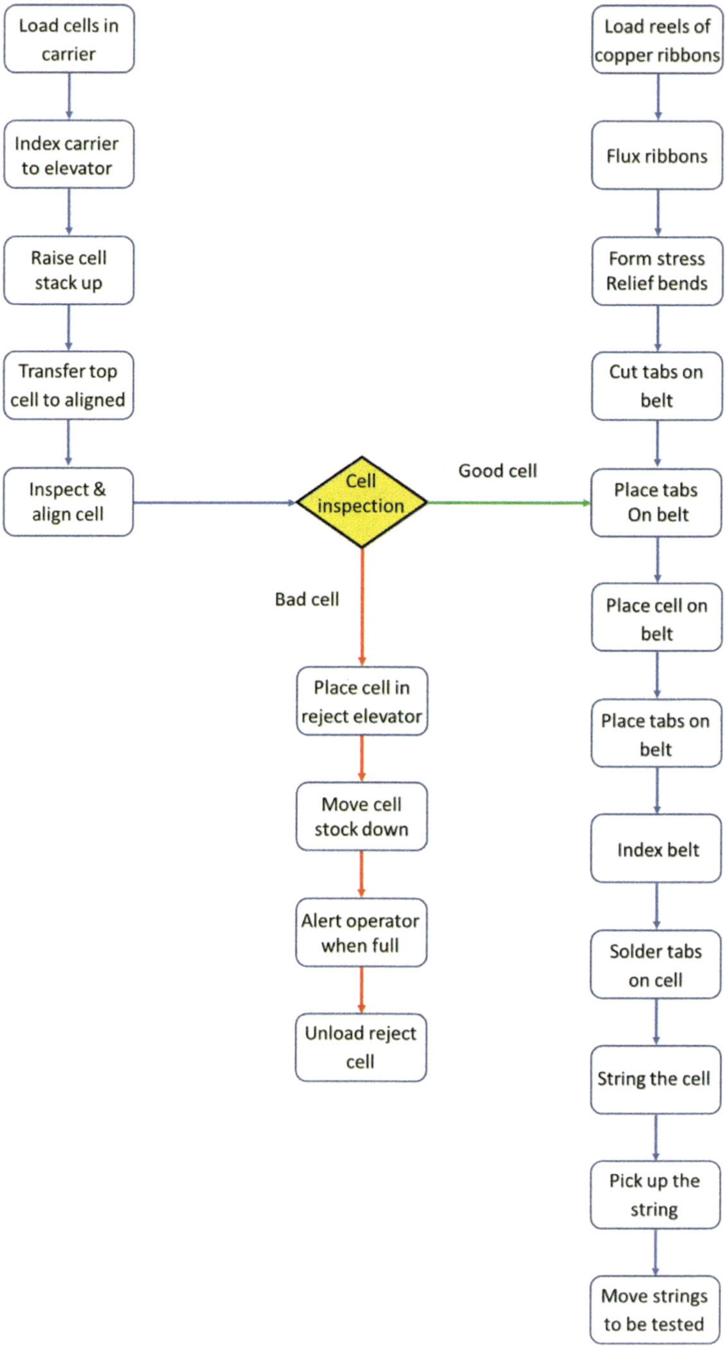

FIGURE 3.24 Flowchart of the tabbing and stringing operation.

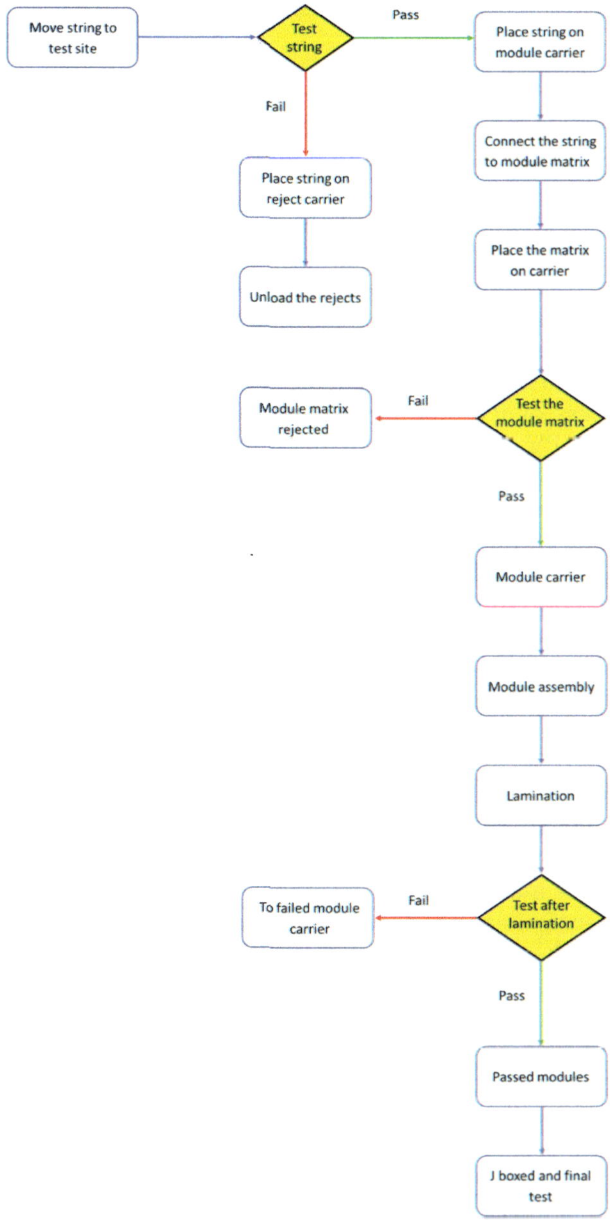

FIGURE 3.25 Flowchart of stringing to the module.

3.5.3 Module encapsulant

The encapsulant is also a critical part of the module. It is applied between the string of cells and the glass, and the backsheet or glass. Actually, the solar cells are being sandwiched between the encapsulant. The most common encapsulant used in the solar industry is ethylene-vinyl acetate (EVA).

The steps on the standard module consist of washing the glass, drying it, tabbing of the cells' ribbons, and soldering into a matrix. Module lay-up, including soldering of the cross-connections, consists of embedding, edge sealing and framing, attachment of the Junction box (J-box), and measurement of the power.

Fig. 3.26 illustrates the origin of the bubble problem in the lamination process of PV module manufacturing. The assembly of glass—encapsulant—PV cell—encapsulant—substrate layers in Fig. 3.26A is transported into a laminator in Fig. 3.26B, in which the temperature and pressure are well controlled to allow for crosslinking and curing of the polymeric encapsulant. Fig. 3.26C shows a representative lamination recipe we used recently. The laminator is heated to the curing temperature under a vacuum, following by encapsulant curing under pressure from the upper chamber to ensure strong adhesion between different layers of the PV module (Poulek et al., 2012). Meanwhile, the lower processing chamber of the laminator is evacuated to remove air bubbles in the module. However, very often bubbles may not be completely removed from a PV module Fig. 3.26D, which causes hotspots on the module under sun irradiation, PV efficiency loss, and delamination-induced lifetime reduction (Perret Aebi et al., 2010). Generally, a trial-and-error method is used with temperature and pressure as variables to develop the lamination recipe, which requires making a number of mockup modules and often leads to overdosed encapsulant and longer laminating process to get bubble-free modules. EVA has been a dominant encapsulant in the PV industry (Agroui and Collins, 2014; Agroui et al., 2006; de Oliveira et al., 2018; Schlothauer et al., 2012). However, UV cut-off and degradation can cause loss of efficiency and reduction of a lifetime. Furthermore, EVA is very "sticky" creating difficulties in panel recycling at the end of life. New encapsulating systems may enable the recovery of clean glass, improving, therefore, its recovery value (Fthenakis, 2012; Burrows and Fthenakis, 2015).

Aside from EVA, some other types of encapsulants are polyvinyl butyral (PVB), Ionomer, thermoplastic polyolefin elastomer (TPO), and silicones. The main goal of the equipment manufacturers is to decrease the processing time by developing laminators that can process more modules at the same time. Another option is to modify the encapsulant itself by adding optimized peroxide cross-link agents to achieve a faster cross-linking, or by using thermoplastic encapsulant.

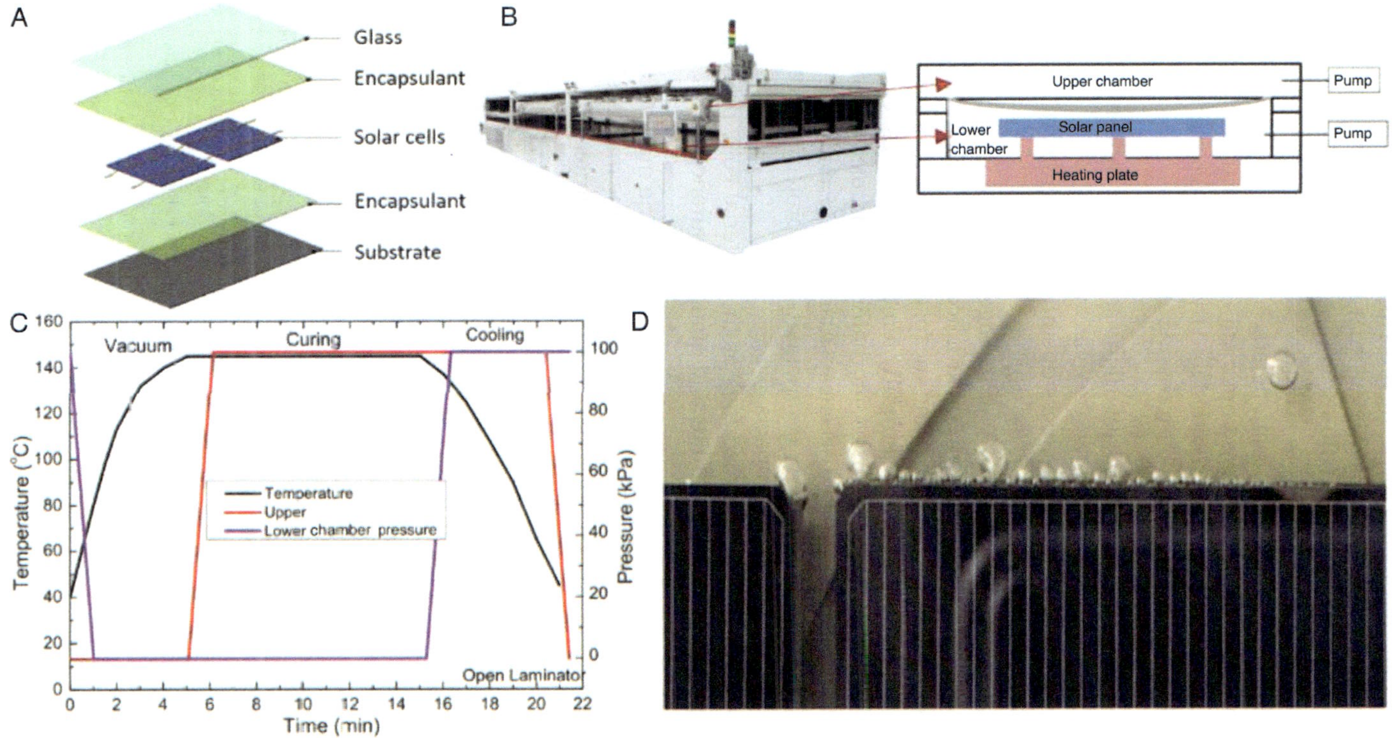

FIGURE 3.26 The lamination process and bubble formation in solar panel manufacturing. (A) The assembly of glass—encapsulant—solar cell—encapsulant—substrate layers before lamination; (B) a laminator for solar panel manufacturing; (C) the lamination recipe with temperature and pressure control; (D) the bubble formation and motion in lamination.

The main challenge in all embedding process is to achieve a uniform and sufficient curing, and cross-linking levels that results in strong adhesion and stable lamination. Consequently, the laminator must provide excellent heat and pressure uniformity, high accuracy in temperature control, and long-term stability of the process parameters.

Important requirements on the encapsulants that are critical in optimizing the module efficiency can be divided into five categories: electric yield, electrical safety, reliability, module processing, and cost.

1. The encapsulant must provide low light absorption and an adapted reflective index to minimize interface reflectance.
2. A high thermal conductivity reduces operating temperatures and thus improves electric yield.
3. For electrical safety, only very low leakage currents are allowed by standard type-approval testing in accordance with IEC 61215 and UL.
4. In terms of PV module reliability, the encapsulant properties are critical in respect of UV irradiation, humidity, temperature cycles, extremely low or high ambient temperatures, mechanical loads, electric potential relative to the ground, etc. The encapsulant has to maintain strong adhesion to the other module components and protect the cell and metallization from external impacts.
5. Module manufacturers will also look at the material cost, processing cost and processing time, shelf life, and quality assurance issues.

The need to reduce the cost of PV modules opens the markets to new encapsulant materials. Fig. 3.27 shows a breakdown of the market share of the encapsulant according to PV-tech of Fraunhofer Institute.

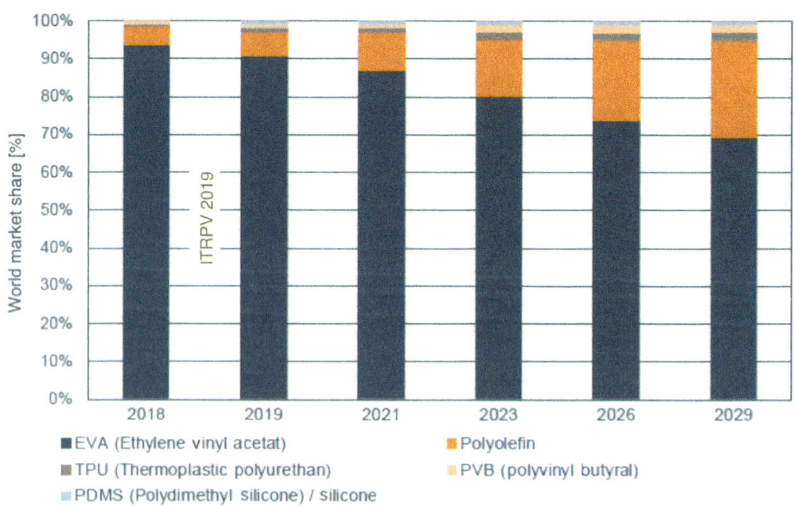

FIGURE 3.27 Encapsulant market share (Peike et al., 2013).

The major encapsulants and their processing temperature and duration are shown in Tables 3.1 and 3.2 and will be introduced below. As to the processing time, in the case of custom BIPV situations, the time could be longer than shown on the chart, especially with much thicker glass, it could take longer. It would also be subject to the design of the module.

3.5.4 Ethylene-vinyl acetate (EVA)

This copolymer is the most popular PV module encapsulant worldwide and has been used in the industry for more than 20 years. The durability of the PV grade EVA, which is highly influenced by the additive formulation used, has been improved tremendously. This is important with regard to the degradation problem of decolorization that is mostly caused by photothermal degradation such as UV light stabilizers, UV absorbers, and antioxidants.

3.5.5 Polyvinyl butyral (PVB)

PVB is probably the second widely used encapsulant material and is a thermoplastic polymer that has been used since the early 1980s as a PV module encapsulant. As compared to other encapsulant materials, PVB is very sensitive to hydrolysis because of higher water uptake. Therefore, it has to be combined with a low WVTR backsheet. While cost-wise, it is similar to EVA, it is proven by experience to the machine manufacturers that PVB does not hold up as well as EVA.

3.5.6 Thermoplastic silicone elastomer

Thermoplastic silicone elastomer represents a relatively new encapsulant class combining superior silicone performance and thermoplastic processability, but, because of price concerns, it is being used on special applications only.

3.5.7 Thermoplastic polyolefin elastomer

TPO is a polymer blend consisting of thermoplastic polyolefin (polyethylene and polypropylene), and olefinic elastomers (ethylene-propylene rubber and ethylene-octene rubber). TPO is a suitable candidate for the PV encapsulant because of its low price. The material has high electrical resistivity, does not degrade under acetic acid formation, and is resistant to hydrolysis, although the water permeation of TPO is significantly higher than that of EVA.

TABLE 3.1 Encapsulants processing conditions (Omazic et al., 2019).

Polymer	Polymer type	Parameter				
		T_g [°C]	*E* [MPa]	Refractive index (n) [•]	Volume resistivity @23 °C [Ω cm]	Moisture ingression [g/d^{-1}]
Ethylenevinyl acetate	Elastomer	− 40 to − 34	≤ 68	1.48– 1.49	10^{14}	115
Silicone		− 50	≤ 10	1.38–1.58	10^{14}–10^{15}	310
Polyvinyl butyral	Thermoplastic	+ 12 to + 20	≤ 11	1.48	10^{10}–10^{12}	310
Ionometer		+ 40 to + 50	≤ 300	1.49	10^{16}	55
Thermoplastic silicone elastomer	Thermoplastic elastomer	− 100	≤ 280	1.42	10^{16}	–[a]
TPO		− 60 to − 40	≤ 32	1.48	10^{14}–10^{18}	–[a]

[a] No data.

TABLE 3.2 Processing temperature and time based on the encapsulation material (SPDL).

Polymer	Equipment	Processing Temp (°C)	Processing time (minimum)
Ethylene-vinyl acetate	Laminator	140–165	10–35
Polyvinyl butyral	Laminator	140–160	10–30
Ionomer	Laminator	140–170	25–35

3.5.8 Ionomers

Ionomers more specifically ethylene ionomers, belong to the category of thermoplastic encapsulant materials and are produced from ethylene and unsaturated carboxylic acid comonomers (ethylene-methacrylic acid copolymer or EMAA). In the solar industry, ionomers represent a different class of encapsulation material duc to high production costs. It is stable on UV and has been demonstrated in the architectural application.

According to ITRPV projection in the coming years, EVA will continue to be the dominant encapsulant in the PV Industry.

3.5.9 Backsheet

Fig. 3.28 illustrates different types of photovoltaic (PV) modules on the encapsulation processes. The backsheet is the last layer of polymer that is used to protect the solar module. It has to withstand a severe condition. It must meet all the properties listed below:

1. Mechanical protection.
2. Electrical insulation.
3. Water/moisture vapor resistance.
4. Solvent resistance.
5. UV stability.
6. Adherence to encapsulant.

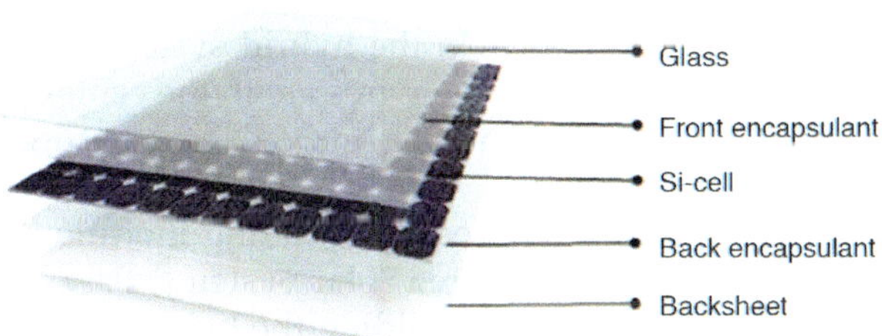

FIGURE 3.28 Different types of photovoltaic (PV) modules on the encapsulation processes (Omazic et al., 2019).

7. Adherence to sealants, labels, etc.
8. Thermal stability.
9. Dimensional stability.
10. Flame retardancy.
11. Compatible with module components.

Backsheets play a significant role in protecting the PV module and delivering the needed module lifetime. They protect the solar panel against environmental damage and ensure the panel remains electrically insulated.

Backsheet failure can result in module failures, unacceptable power, and degradation. The impact can be very significant, ranging from brand, and reputation damage to outright bodily harm. The solar backsheet is not just the simple layer of plastic film on the back surface of the module, but rather is the only layer of protection from dangerous DC voltage and any foreign material to the module.

Nowadays, the glass used in place of the backsheet has attracted interest. The glass-glass modules are often frameless, which brings the concern that the edges and corners are more prone to breakage especially on tempered glass.

As to the encapsulants especially on custom modules that are relatively large with heavy and thick glasses, sometimes more layers of encapsulant would be required to be successful in lamination. BIPV custom modules could sometimes be quite tricky. The layout and design of the module are also quite critical. A quality laminator also plays a very important role. 3S (Swiss Solar Systems AG) in Bern, Switzerland, in the early 2000s, invented and patented the first hybrid heating plate with only a 2°C difference throughout the whole lamination plate. In a thick glass-glass module, it is imperative to have this type of laminator. The long-term quality of the module would also be subject to the quality of the material, the types of machinery, and the process design on the lamination.

3.6 The hybrid lamination technology

While most laminators either use electrical wires or oil as heating elements on the lamination plates, in the hybrid system, both the heating wire and the oil is used. The heating wires are in the center of the heating tube with oil in it. This results in having an even temperature all across the lamination plate.

The 3S Industries Group has been the leading module machine manufacturer. The group consists of three Solar Systems, Pasan Test Systems, and Somont. 3S lead the world in building the hybrid lamination system and partnered with Pasan Test Systems which is a leading test system company and Somont a leading company in CTS. The Group managed to come with a very advanced module system in the later part of 2006 that drew a lot of attention to module manufacturers. Due to the cost of the Euro and Swiss Francs and the worldwide economic concerns throughout that period, the system did not sell as well as they anticipated. In later years, 3S merged with Meyer Burger and they developed the heterojunction smart wire cutting technology (SWCT). Apart from technological advancement, the Smart wire system is also aesthetically very pleasing.

When Meyer Burger merged with 3S, they developed the smart wire connection technology (SWCT) shown in Fig. 3.29. Apart from advance-

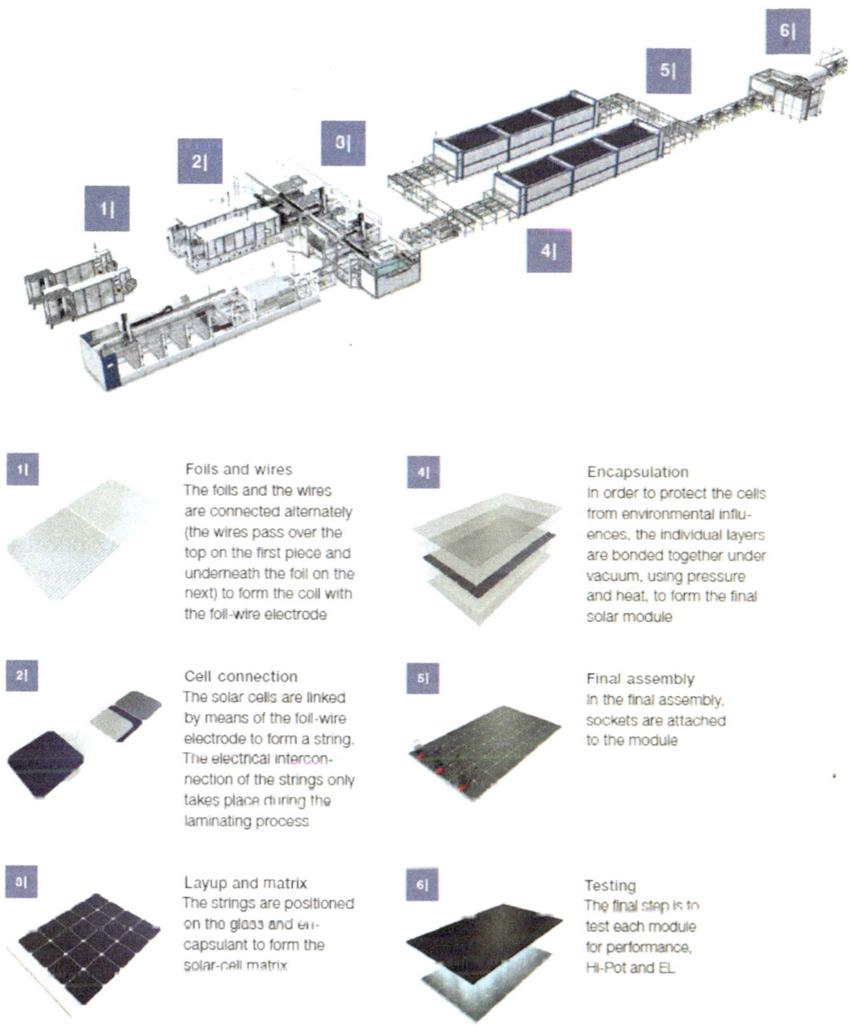

FIGURE 3.29 Meyer Burger Smartwire module manufacturing line (Burger, 2021).

ment in the technology, it is aesthetically placing as well on a completed module. In addition, instead of using the current ribbon wire process, applies the foil-wire electrode which resulted in substantially improves the efficiency of the energy conversion of the product together with reducing the chances of micro-cracks of the solar cells (Fig. 3.30).

FIGURE 3.30 Solar cell connection technology SWCT (Burger, 2021). *SWCT*, smart wire cutting technology.

The smart wire connection a totally new and patented process on module manufacturing on a massive way to produce the best quality solar products (Burger, 2021).

The technology is also versatile, flexible that could accommodate different cell types and module designs, like bifacial glass/glass modules or bifacial Heterojunction modules.

Meyer Burger has led the solar industry to come a long way because the smart wire connection technology not only substantially improves the conversion efficiency to go above 20% but decreases the price per watt as well. However, despite all these advancements in the solar module and cell technologies, to make a house or building net-zero energy, it could not achieve without including passive solar systems or hybrid solar roofing technology like the BITERS or storage and generation systems. More discussions will be presented in the later chapters.

3.7 Quality control and assurance

Quality is vital in solar panel production because solar modules are electrical energy generators, which could also be a building product if it is a BIPV system. As discussed earlier, it takes to go through many processes and involves heavy and sophisticated types of machinery just to produce the silicon wafers from sand. Afterward, the wafers have to through another set of high-tech manufacturing systems to be turned into solar cells before they are ready for module assembly.

Although the cost is important in this manufacturing process, quality plays a more significant role. The quality should not be sacrificed over the cost, since otherwise, it is the sustainability that is being crusaded which questions the whole concept of renewable energy. A quality product would last longer and in return save cost, eventually making it cost-effective in the long run. In addition, though silicon is an abundant element, we should not take it for granted. Quite contrary, if the solar module is disposed of every few years, we would face a shortage of landfills. If the solar panels could last for twenty to thirty years, that would bring a status of peace of mind to the end-user for that period instead of just a few years.

As one can see in Fig. 3.31, there is a substantial number of steps to turn silicon to silicon wafers, and from there to solar cells and modules. Every single process requires a significant amount of energy, investments, and engineering. However, it is a regret to see some companies sacrificing the quality of the product for the sake of cost by choosing cheaper materials with less quality (cheaper encapsulant, backsheet, or glass) or the types of machinery. Not only it will be a waste of material but also this practice will hurt the reputation of the organization and its brand. We are making a change in paradigm and business model of products with a lifetime of 5–10 years is not going the be sustainable. Especially with growing concerns over rapid depletion of resources, our environment, the increasing population growth, and our landfills are problems as well. Why not make a product that could last a much longer period.

Potential preventive actions on quality control include:

1. Randomly check solar products on imports by taking samples and testing them to see whether they meet the UL or IEC specified standards.
2. Randomly also check domestic manufacturers on their quality and see whether they meet the UL and IEC standards as well.
3. On the consumer side, check out the manufacturer track records on their products.
4. As to the installation contractors, good prices on modules are very tempting but performance and your reputations totally outweigh it.
5. To all solar installation contractors, getting into installing quality BIPV or Active Building Products is highly recommended.
   (a) Though BIPV systems require more details, they are more pleasing aesthetically, and usually, the modules and systems are in a well-protected environment.
   (b) Most important of all, check out the manufacturer.
6. Check the contractors on their track records.

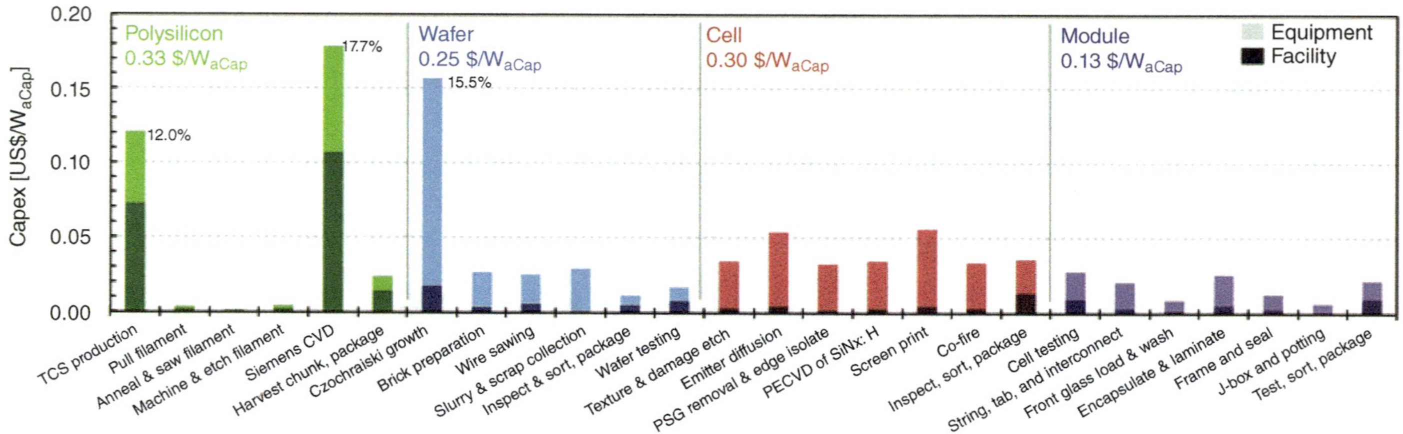

FIGURE 3.31 Initial capex for 2H-2014 c-Si photovoltaic (*PV*) manufacturing, normalized by annual production capacity. Each step includes the cost of equipment (light bar) and the facility that houses it (dark bar). Polysilicon and cell production comprise the majority of capex, with the three process steps of Siemens CVD, Czochralski growth, and TCS production comprising 45% of total capex. Reproduced from Powell et al. (2015) with permission from the Royal Society of Chemistry. *CVD*, chemical vapor deposition; *TCS*, trichlorosilane.

3.8 Vision and challenges toward future circular manufacture

Solar energy is a major renewable energy source with the potential to meet many of the challenges facing the world. The US PV module market size is expected to reach $23 billion by 2025 as a result of increasing demand due to decreased system cost and increasing environmental awareness among customers (Nuortimo et al., 2018). Significant research has been performed in the PV arena (Khudhair et al., 2018; Tonui et al., 2018; Vincent et al., 2018; Yang et al., 2018). However, the focus has been mainly on the development and improvement of PV cells. The lamination process in which PV cells are encapsulated in between a number of substrate layers to ensure efficient passing of the sunlight and protect cells from the harsh environment is often taken for granted. Better encapsulation would reduce the defects and degradation in the field, increase the life-expectancy of PV modules, therefore decreasing the levelized cost of solar electricity (LCOE) of generating solar electricity, and will improve the recyclability and thus resource circularity of PV systems. For example, Fig. 3.32 shows that by extending the lifetime via new encapsulation systems of the current 30-year life of PV modules to 50 years, we can take the LCOE from PV power plants down to $0.03/kWh, if this can be done without increasing the manufacturing cost.

There is an opportunity to achieve a 50-year module warranty with the same 90% or better output rating by switching to a thermoplastic. Not only

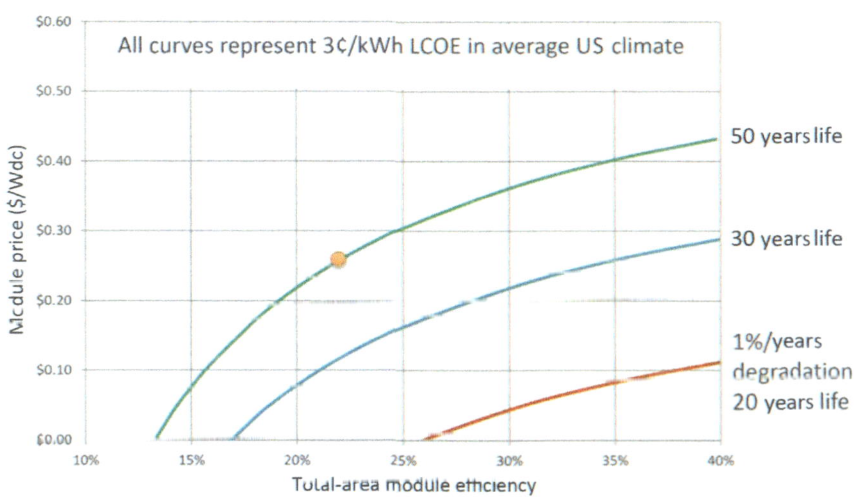

FIGURE 3.32 Pathways for accomplishing a 3 cents per kWh LCOE from 21% efficient photovoltaic (*PV*) modules. Scenarios assume: 7% weighted average cost of capital (*WACC*), 2.5% inflation, $4/kW-year O&M, 21% capacity factor (Gay et al., 2020).

can module durability improve but polymer innovations in the laminated glass and automotive sector can be integrated into the PV supply chain, increasing reliability, and lowering cost. What better way to reduce the need for recycling and drop LCOE?

Our previous research in PV module manufacturing, feedback from our industry partners, and numerous studies (Vedrtnam, 2018; Li et al., 2018; Park et al., 2013; Zaraket et al., 2019; Quansah and Adaramola, 2018) in the literature show that the lifetime of PV modules is mainly caused by encapsulation related bubbling problems that result in delamination and interconnect issues (Li et al., 2018; Kumar and Kumar, 2017; Oliveira et al., 2018) as well as the degradation of encapsulation. Proper lamination and packaging processes are of great importance to ensure the long-term reliability and lifetime of PV modules. For crystalline silicon PV modules, the cells are predominantly protected by polymeric encapsulant material (mostly EVA) that is sandwiched between a front cover (mostly glass) and a back cover (mostly glass and polymeric backsheet) (Khudhair et al., 2018; Vedrtnam and Pawar, 2017; Kumar et al., 2018). During the lamination process, the encapsulant material is crosslinked at high temperatures and different layers are bonded together. Inefficient evacuation of air during lamination tends to generate tiny air bubbles of submillimeter scale in the viscous melted EVA (Ashter, 2014; Niaounakis, 2015). Significant bubble formation is mainly caused by the volatile chemicals released from the module under the effect of heat at the beginning of the curing process.

In most cases, we found that air voids are mechanically entrapped between different layers during the lay-up of modules Fig. 3.33A, and they can be eliminated by performing sufficient air debunking during

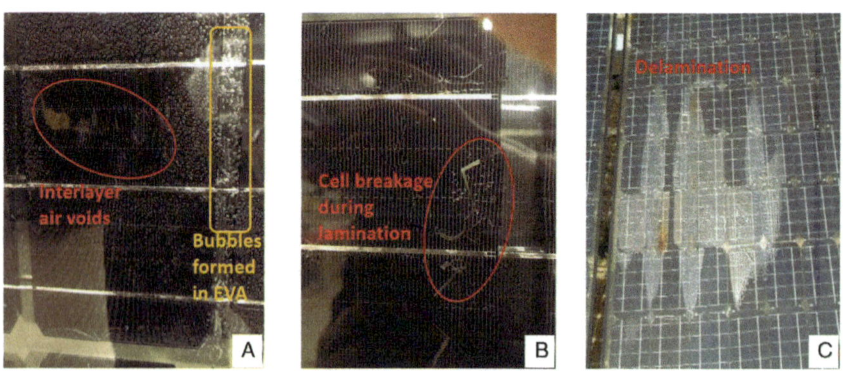

FIGURE 3.33 The problems caused by air bubbles in solar panels: (A) remaining bubbles on a cell blocking sunlight and causing hot spots; (B) bubbles caused solar cell breakage during lamination process; (c) bubbles induced delamination of modules in service.

the preheating step of the encapsulation process in a vacuum. In some cases, however, air bubbles are embedded in the EVA encapsulant with a circular geometry. Similar voids are also observed in manufacture processes such as autoclave curing (Agius et al., 2013; Torres et al., 2019), liquid compression molding (Park et al., 2011; Tatara, 2017; Lee et al., 2019), resin transfer molding (Lee et al., 2006; Yang et al., 2015; Yun et al., 2017), and packaging of polymeric composites (Kumar et al., 2018; Niaounakis, 2015; Wagner, 2016), which have been reported to result in detrimental effects on the mechanical performance and durability of the composites and increase moisture absorption in the laminates. When air bubbles form during the lamination process of PV modules, additional internal stresses could build up around bubbles in contact with the PV cells, which results in cracking of the thin crystalline silicon cell due to its brittle nature Fig. 3.33B. When the modules are in service, air bubbles and voids would allow for moisture transport in the encapsulant, which could lead to hydrolysis of EVA, metallic contact corrosion of electronics, breakage of cells, and eventually interfacial delamination (Kumar and Kumar, 2017; Oliveira et al., 2018; Peike et al., 2013; Gagliardi et al., 2017) in Fig. 3.33C. Various experiments and material characterization techniques, such as gas chromatography–mass spectrometry, differential scanning calorimetry, and thermogravimetry, have been used in the literature to identify and quantify the release of gases, such as carbon dioxide and volatile organic compounds (VOCs) from EVA during PV module lamination (Agroui and Collins, 2014; Agroui and Collins, 2003; Agroui et al., 2012; Stark et al., 2012; Schulze et al., 2015; Jaunich et al., 2016). It has been shown that only a very small amount of volatiles is needed to create a bubble trapped in EVA in the normal lamination condition (Perret Aebi et al., 2010; Song et al., 2018; Centelles et al., 2019). Quantitative understanding of bubble formation and motion during the lamination process is critical to optimize the lamination recipe for bubble-free PV modules.

The formation of bubbles in the lamination process may result from a series of factors such as chemical reactions within the encapsulant and of additives (e.g., adhesion promoters, crosslinking initiators), the degree of thermal stabilization, as well as the process conditions (e.g., temperature, pressure, degree of vacuum) (Drabczyk and Panek, 2012; Couderc et al., 2017). This project will first characterize the chemical reactivity and thermophysical and mechanical properties of the EVA at different levels of temperature and pressure to identify the creation and composition of gas entrapped in the viscous EVA fluid.

New encapsulation systems may exhibit great potentials that will improve photovoltaic (PV) module longevity and recyclability while reducing the LCOE, thus creating opportunities for new US-based PV manufacturing, labor force development, and large scales of solar system deployment. For example, new encapsulating methods have been proposed for thermoplastic polyolefin (TPO) and perovskite solar cells, and PVB, and crystalline silicon modules by spray or casting methods.

Energy and the environment have been becoming the most pressing issue in the past decades with the global population explosion and global warming. It requires sustainable development of the infrastructure and production system. Solar energy has attracted significant attention as a renewable clean energy source and has become one of the fastest-growing industries in the past decade. As more PV modules are approaching the end of their lifetime, how do we deal with them, and whether can we improve them?

The circular eco-manufacture of solar modules will be a solution, which will reduce the use of materials and energy in manufacturing products and services and then recover as much of the end products as possible through reuse and recycling. For example, for an EVA dual glass module, the glass may last ~100 years; the solar cells may still be useful after 25–30 years, but the encapsulant may exhibit yellowness or brawniness. If we can delaminate the PV module and separate the components, we can easily replace the encapsulant and recover the function with minimal cost and environmental impact.

The key to achieving this revolutionary benefit in future PV manufacturing lies in the encapsulating process. The thermoplastic encapsulant may lose adhesive strength at a certain temperature, which meets the need for circular eco-manufacturing.

3.9 Case study 1: The Water and Life Museums, Hemet, California

As the population grows, Food, Energy, and Water (FEW) have become more vital to our livelihood. This project started due to the building of the largest man-made lake in North America, Diamond Valley Lake (DVL), which is to store water for Southern California in an emergency. The lake is located between two mountain ranges. Two dams of 300 feet were constructed to hold the water of the lake, of which the East Dam is 2.5 miles long and rocks from the mountain ranges were used in constructing the dams. The museums are located relatively close to the East Dam and halfway between Los Angeles and Hoover Dam (Fig. 3.34).

The solar energy installed in the museums is 550 kW and covers around 50% of the energy requirement. About 50 kW is the BIPV module system at the entrance of the museums. The water from the Diamond Valley Lake would allow sustaining Southern California for a six-month period and will buy time in critical situations until other sources are available.

The module was relatively challenging to build partly due to size and design. Consequently, Atlantis and 3S decided to work together to make it

FIGURE 3.34 Water and Life Museum front entrance.

a reality. Two types of modules were designed and manufactured as shown in Figs. 3.35 and 3.36 with the same specification in Table 3.3. An actual module is shown in Fig. 3.37 as well.

This product is listed applicable to Underwriters Laboratories (UL) Standards and requirements by Underwriters Laboratories (UL) Inc.

The objective of the center is to showcase that design aesthetics and sustainability do not necessarily need to be separable. The Campus is about 15 acres with the two museums with 70,000 sq. ft floor space.

Water – Provides education on water and water conservation as well as stewardship.

Life – Provides education on Geography, Archaeology and Paleontology.

In addition, there are laboratories and classrooms together with the administration office to support the operation. The design of the Museums is not only attractive, but it is also low in maintenance and energy cost.

Architect: Lehrer Architects, created by Michael B. Lehrer, FAIA, Mark Gangi AIA and Frank P. Gangi, MBA.

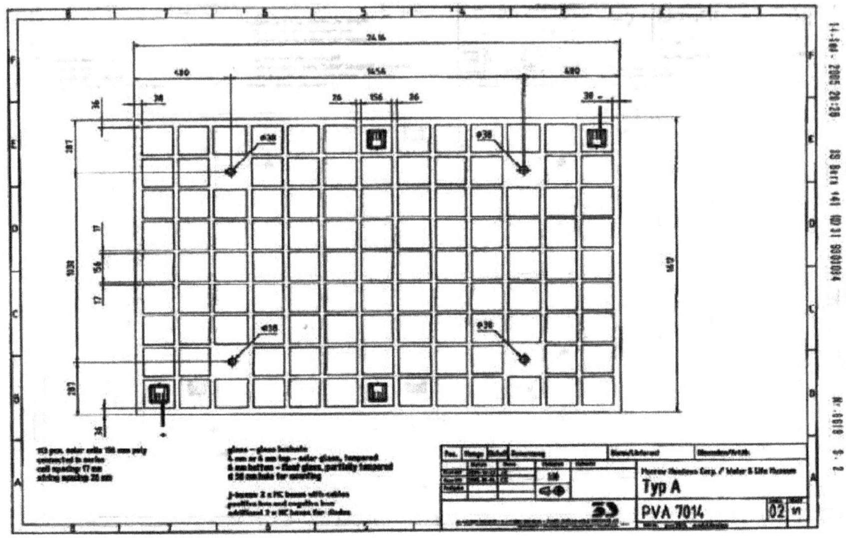

FIGURE 3.35 Water and Life "A" module.

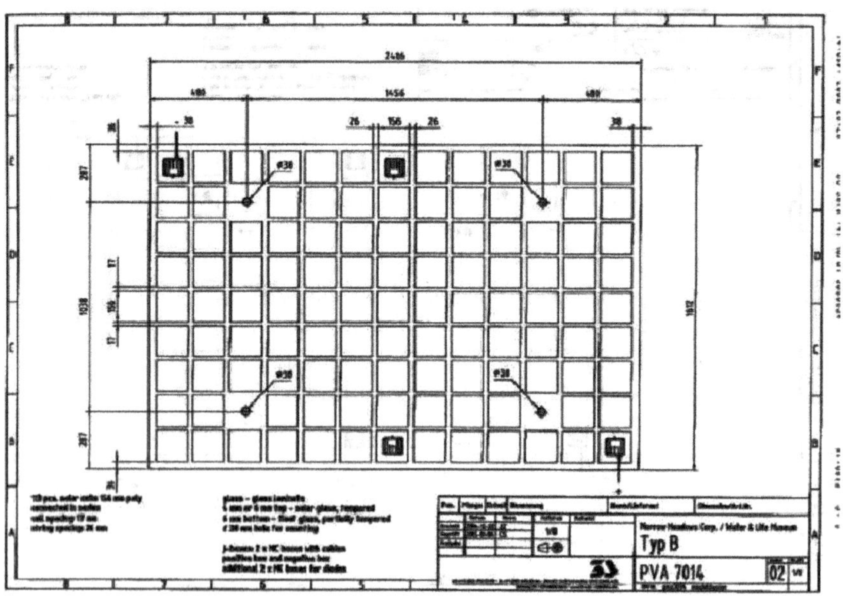

FIGURE 3.36 Water and Life "B" Module.

TABLE 3.3 Specifications of the modules used in the project.

Power	362	Watts
Vmpp	54.8	VDC
Voc	67.0	VDC
Impp	6.60	ADC
ISc	7.10	ADC
Estimated power at PTC*	330	Watts
Series fuse rating	15	Amp

** PVUSA Test ConditionsNote: All characteristics are at STC (Standard Test Conditions) unless noted*

FIGURE 3.37 Water and Life Museums BIPV module.

Awards:

- Merit Award for Best/Public Special Use Facility – Pacific Coast Conference 2007
- Beyond Green High Performance Building Award – Sustainable Building Industry Council 2007
- Commercial Project – Pacific Coast Builders Conference 2007
- Gold Nugget Grand Award for Excellence and Value for Best Sustainable Commercial Project – Pacific Coast Builder Conference 2007
- Honor Award for Excellence in Design American Institute of Architects – Pasadena Foothill Council 2007

- Honor Award – United States Green Building Council 2008
- LEED Platinum Certification – United States Green Building Council 2008
- American Architecture Award – Chicago Athenaeum 2008
- Honor Award for Excellence in Design – American Institute of Architects – Los Angeles 2008
- Merit Award for Savings by Design – American Institute of Architects – California Council 2008
- Green Good Design Award – Chicago Athenaeum 2009
- Honor Award for Excellence in Design – American Institute of Architects – California Council 2009
- Iconic Awards, Public Architecture with Distinction – German Design Council 2016
- The project is the first museum in the world to receive LEED Platinum Certification.

3.10 Case study 2: Metlife Solar Ring

The Solar Ring at Metlife (Fig. 3.38) is an iconic project as well as one of the largest if not the largest BIPV installations in the country. It consists of 1350 modules as shown below with 47 frames, generating 350 kW of clean energy to support 916 LED fixtures around the ring and other areas in the stadium. The Metlife stadium has won the Venue of the Year 2017 by Stadium Business Summit award which assesses internationally all the stadiums around the world.

FIGURE 3.38 Metlife Stadium at night with LED shining to the solar ring.

3.10.1 Module electrical characteristics

The electrical characteristics are within ± 5% of the indicated values of *I*sc, *V*oc, and *P*max under standard test conditions (STCs) (1000 W/m^2 irradiance, 25°C (77°F) cell temperature, and AM 1:5 spectrum). Under normal conditions, the module is likely to experience conditions that produce more current and/or voltage than reported at STCs (output may vary depending on the time of day, time of year, ambient conditions, ambient temperature, and shading). Accordingly, the value of *I*sc and *V*oc marked on the module should be multiplied by a factor of 1.25 when determining component voltage ratings, conductor ampacities, fuse size, and the size of controls connected to the PV output (Fiore Skylight Inc, New Jersey, http://www.fioreskylight.com/). Fig. 3.39 shows the dimension and cell layout of the solar module. The electrical characteristics and the bill of materials are shown in Table 3.4 and 3.5 respectively.

This product adheres to UL Standard 1703. All module components rated for 600 VDC. Do not exceed module maximum voltage rating of 600 VDC. Table 3.6 shows the bill of materials for each module.

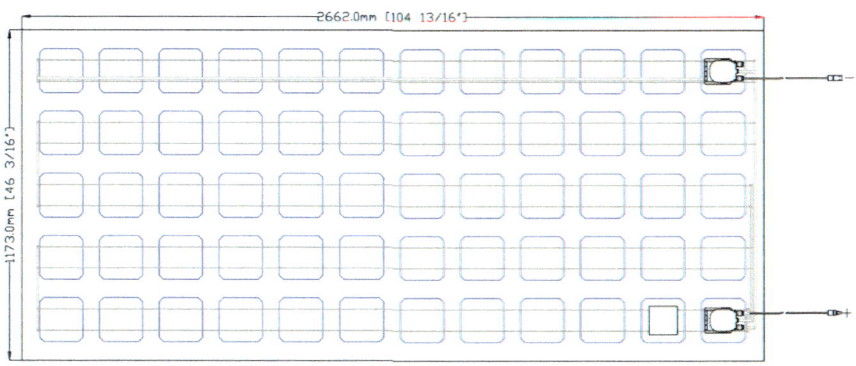

FIGURE 3.39 Module dimensions used in the Metlife stadium. Module Weight: 300 Lbs, Thickness: 16.2 mm [0.64″].

TABLE 3.4 Characteristics of the modules.

Power	236	Watts
Isc	8.48	ADC
Voc	38.0	VDC
IMPP	8.00	ADC
MMPP	29.5	VDC
Series fuse rating	15	Amp

Note: All characteristics are at STC (Standard Test Conditions)
Temperature coefficients: Isc 0.045%/°C, Voc -0.319%/°C, Pmp -0.396%/°C

TABLE 3.5 Bill of material of the module.

Item	QTY. required for assembly	Vendor	Model #	Dim. width (mm)	Dim. length (mm)	Area (m^2)
CELLS	**60**	Gintech	GIN156M. 6″ Poly 4.04W	156	156	0 024
Front glass	**1**	Oldcastle	1 /4″ Starphire Tempered	1173	2662	3.12
Rear glass	**1**	Oldcastle	3/8″ Clear Tempered w/Velour Finish	1173	2662	3.12
Encapsulant	**3**	STR	Photocap 15295P/UF	1198	2687	322
Tabbing/bussing material	**N/A**	Ulbnch	7746-9464 7746-9574	2.3 8	N/A	N/A
Solder	**N/A**	Kester	#14-6337-0062 SN63PB37 SLD	1.5	N/A	N/A
Insulation strips	**N/A**	Madico	InsulPalch (Transparent)	N/A	N/A	N/A
Tape	**N/A**	Coieme	dyMatE	19	30	N/A
Blue tape	**N/A**	3M	3M Polyester Tape 8901	38	N/A	N/A
J-box	**2**	Tyco	1740220-5	90	135	001
Diodes (included in J-box)	**3**	Tyco	SL1010	N/A	N/A	N/A
Insulation sleeves	**N/A**	Daburn	D200-2	N/A	N/A	N/A
Glue	**N/A**	Shinetsu	KE-45-T	N/A	N/A	N/A

3.10.2 Electrical installation

The modules use the MC4 connector system from multicontact (Fig. 3.40). Each module is supplied with a set of pigtail cables composed of two MC4 connectors and 12 AWG type USE-2 wire.

- WARNING –

Unplugging under load: PV plug connections must NOT be unplugged while under load. They can be placed in a no-load state by switching off the DC/AC converter or breaking the AC circuit. Plugging and unplugging

TABLE 3.6 Electrical characteristics of the module.

Rev. 1.0			
# Mods (per String)	**12**	**Power (per String)**	**2520**
	*V*mp (low)	*V*mp (high)	*V*oc
Cell temp (°C)	85	−25	−11
Temp dev from STC (°C)	60	−50	−36
% Correction	−27.00%	22.50%	12.96%
Voltage @ STC	336	336	442.8
Voltage @ cell temp	245	412	500
		Inverter type	
Vmp @ STC (V)	28.9	Temp Coeff. *V*mp (%/°C)	−0.45%
Voc @ STC (V)	37.4	Temp Coeff. *V*oc (%/°C)	−0.36%
Isc @ STC (A)	8.79	Temp Coeff. *I*sc (%/°C)	0.03%
Rev. 3.0			
# Mods (per String)	**12**	**Power (per String)**	**2748**
	*V*mp (low)	*V*mp (high)	*V*oc
Cell Temp (°C)	85	−25	−11
Temp Dev from STC (°C)	60	−50	−36
% Correction	−27.00%	22.50%	12.96%
Voltage @ STC	346.8	346.8	448.8
Voltage @ Cell Temp	253	425	507
		Inverter type	

Note: all characteristics are at STCs.
Temperature coefficients: Isc 0.045%/°C, Voc −0.319%/°C, *Pmp* −0.396%/°C.

FIGURE 3.40 MC4 connectors.

while under voltage is permitted. Make sure connectors are fully engaged without a gap between the insulators.

3.10.3 Series connections

For series connections, connect modules are shown in Fig. 3.41.

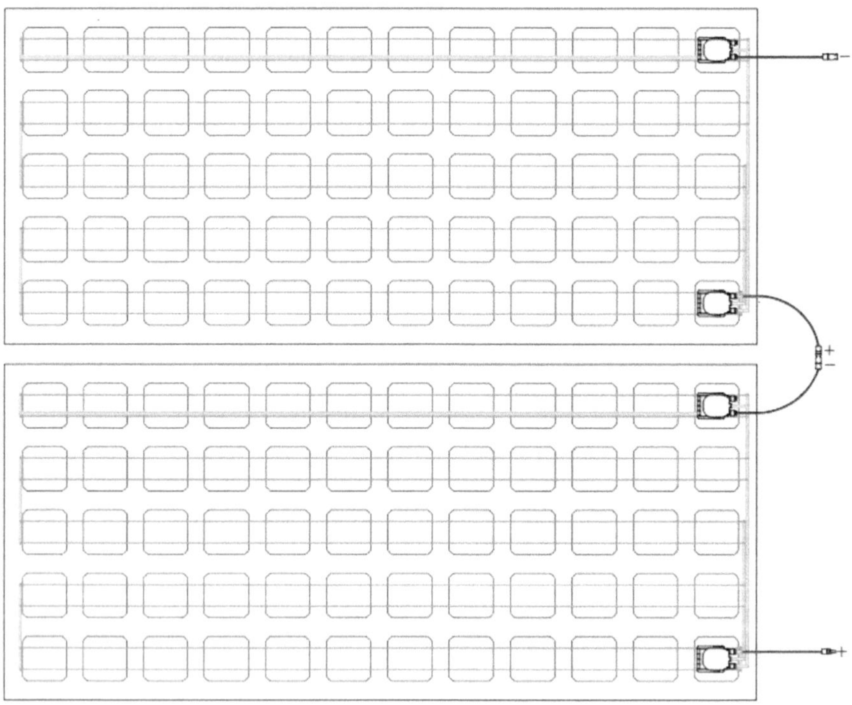

FIGURE 3.41 Outlay for series connection of modules.

3.10.4 String sizing

When determining the string sizes, be sure to follow the inverter manufacturer's instructions. In general, ensure that the open-circuit voltage

(V_{oc}) of the string on the record coldest day at the installation location does not exceed the max DC input of the inverter. Furthermore, ensure that the minimum V_{mp} at the highest monthly average temperature and the maximum V_{mp} at the lowest monthly average temperature falls within the Maximum Power Point Tracking (MPPT) Input Voltage Range of the inverter. These values can be calculated using the module electrical characteristics @ STC and voltage temperature coefficient. The following equations can be used to perform these calculations.

Max DC voltage of string = [(# of modules in string) × 38.0V] + [(# of modules in string) × 38.0V × (record lowest temperature − 25°C) × (−0.00319/°C)]

Min V_{mp} = [(# of modules in string) × 29.5V] + [(# of modules in string) × 29.5V × (Highest monthly average temperature + 30°C[1] − 25°C) × (−0.00319/°C)]

Max V_{mp} = [(# of modules in string) × 29.5V] + [(# of modules in string) × 29.5V × (lowest monthly average temperature + 30°C[1] − 25°C) x (−0.00319/°C)]

These are general guidelines only; the instructions of the inverter manufacturer shall supersede them.

3.10.5 Grounding

The AES-ST60 module will be integrated into a framed skylight system. All metal parts used in the installation of the module must be adequately grounded. Refer to NEC Sections 690 and 250.

3.10.6 Mechanical installation

The modules are installed as part of a framed skylight system (Fig. 3.42). When installing the modules, be sure to:

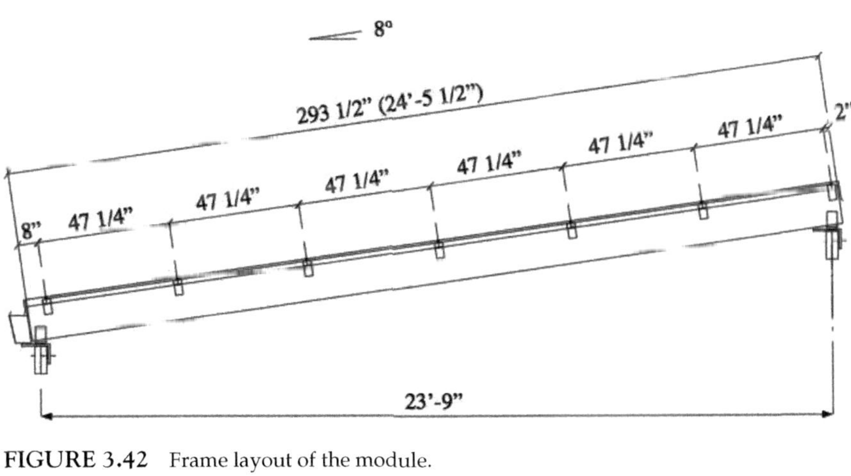

FIGURE 3.42 Frame layout of the module.

1. Use only the provided cable for module interconnection. For all other connections, use stranded 12 AWG type USE-2 wire, rated sunlight resistant. Refer to the National Electric Code and local requirements for proper wiring.
2. Observe the requirements described by the note under the Electrical Characteristics section of this Guide.
3. All metallic parts used in the installation have to be properly grounded.
4. All modules should be handled with care using proper safety equipment. Due to the weight of the module, one should never try to move a module without safety precautions and glass lifting equipment. Any glass lifting/handling equipment can be used that is designed for the proper module weight and size.
5. When handling glass always wears safety glasses and follow all safety instructions in this manual.
6. Follow recommendations made by the licensed structural engineer or architect providing the project and installation details.
7. Use only approved hardware. All hardware used to support the module has to be approved by a licensed structural engineer for the specific project.
8. Make sure that two modules will never physically touch each other.

Note: when calculating min/max Vmp, it must be taken into account that the operating cell temperature is approximately 30°C higher than ambient temperature.

Basic mechanical installation instructions are as follows:

1. The modules will be installed into the skylight framing system on the ground in sections of 30 modules (Fig. 3.43).
2. Visually inspect the purlins and rafters for proper gasket installation.
3. For each module bay, install (2) setting blocks onto the lower purlin. Install (1) 7″ from the lower-left corner and (1) 7″ from the lower right corner of the bay (Fig. 3.44).
4. Use a vacuum power cup system (capable of handling at least 500 lbs.) to pick up the module. Lower the module at a 20-degree incline until the bottom edge rests against the setting blocks. Continue to lower the module until it is fully seated.
5. Connect the modules in series according to the appropriate string sizing and array layout. The framing system is notched to allow cables to pass through.
6. Install backer rods and silicone sealant along horizontal joints (purlins) (Fig. 3.45).
7. Install pressure plates and snap caps along vertical joints (rafters) (Fig. 3.46).
8. Test each string for proper voltage reading.
9. Lift the section of 30 modules and install them into place.

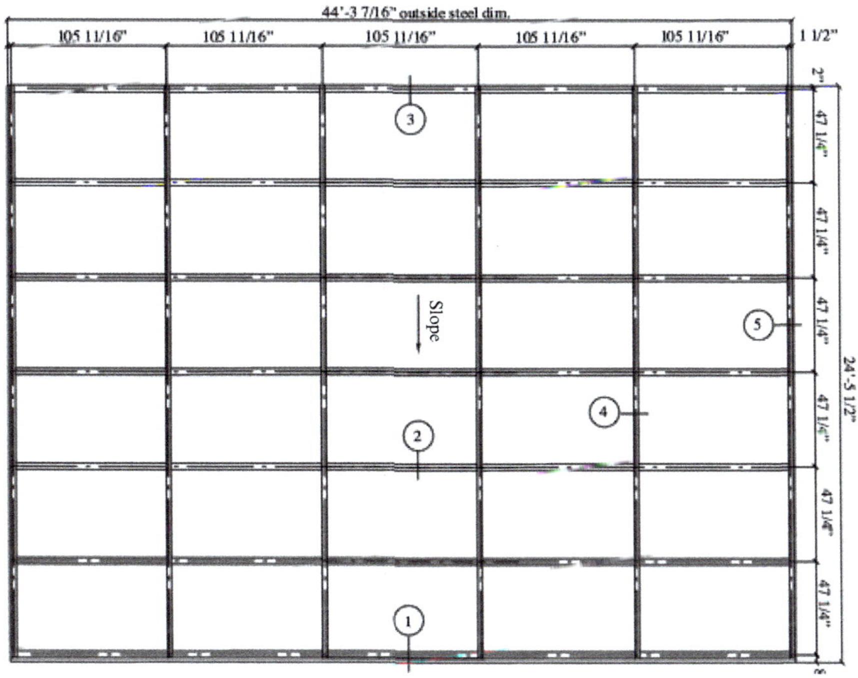

FIGURE 3.43 Layout of the frame.

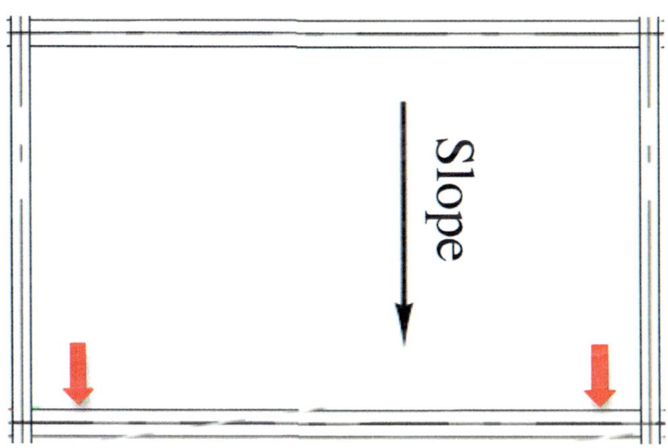

FIGURE 3.44 Layout of the module's frame.

3.10.7 Maintenance

Minimal maintenance is required to maintain the optimal performance of your solar electric system. If the front of the module becomes dirty, a reduction in energy output may result. We recommend the use of water

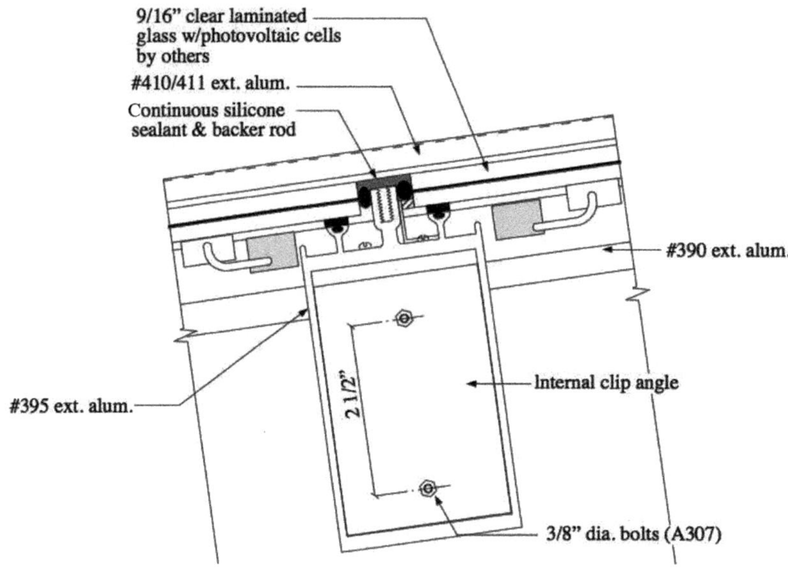

FIGURE 3.45 Backer rods and silicone sealant along horizontal joints (purlins).

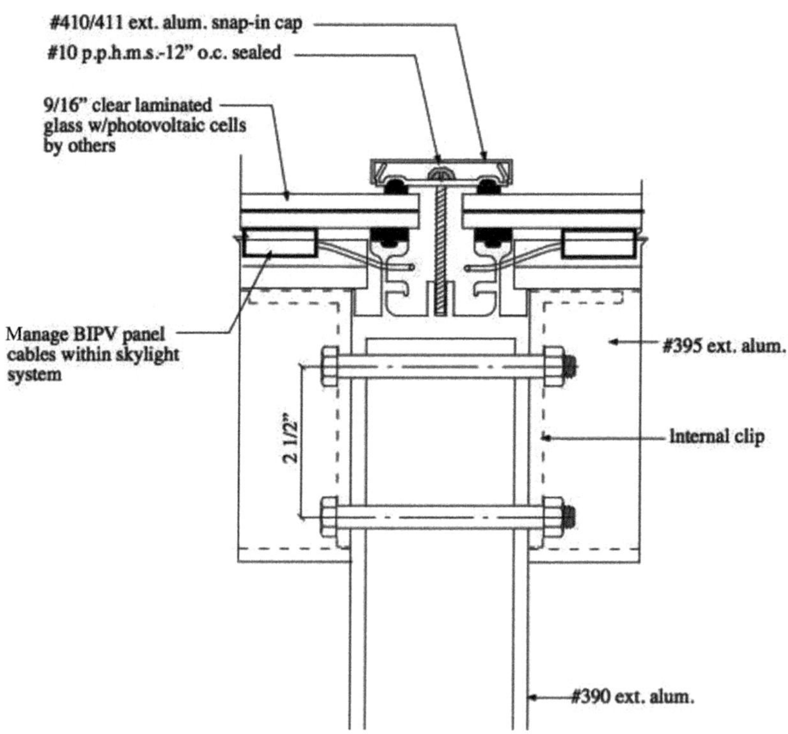

FIGURE 3.46 Pressure plates and snap caps along vertical joints (rafters).

TABLE 3.7 Part numbers for quality control.

SN 0208	SN 0212	SN 0211	SN 0210	SN 0206
SN 0209	SN 0203	SN 0204	SN 0205	SN 0207
SN 0223	SN 0224	SN 0217	SN 0218	SN 0219
SN 0202	SN 0213	SN 0214	SN 0215	SN 0216
SN 0196	SN 0192	SN 0198	SN 0220	SN 0201
SN 0199	SN 0193	SN 0194	SN 0195	SN 0197
FRAME # 38		BIPV TESTED: 5-3-12		
BIPV CRATES: 16, 17, 18		FRAME INSTALLED ON STADIUM: 5-8-12		
BIPV INSTALLED: 4-25-12		COMMENTS:		

and a soft cloth or sponge to clean the glass module surface. The mild nonabrasive detergent may be used if required (Table 3.7).

The module installation frame parts must match with the module serial number so that in case there are defects or breakages, one could easily track the location of the frame and the module.

3.10.8 System structure

The structure of the system is another critical parameter that has to be considered especially in the Northeastern part of the country where the Metlife Stadium is located. Wind load and snow load shall be taken into account. The wind and snow load design characteristics are shown in Table 3.8. Climate change is a very serious issue which causes

TABLE 3.8 Wind and snow load parameters.

Design criteria			
Dead Load:	Glass Modules + Aluminum		10 psf
Live/Snow Load:			27 psf
Wind - 110 mph:	I = 1.15	Exp. C	30 psf
Concentrated load:			250 lbs
Deflection limit:	L/175		
Materials			
Aluminum Tubes	- 6063-T5	Aluminum Alloy and Temper	
Aluminum Extrusions	- 6063-T6	Aluminum Alloy and Temper	
Sheet and Plates	- ASTM B209	(ASTM B209M)	
Bars, Rods, Wise	- ASTM B211	(ASTM B211M)	
Fasteners	- ASTM A316I	Stainless Steel E = 10,000 ksi	

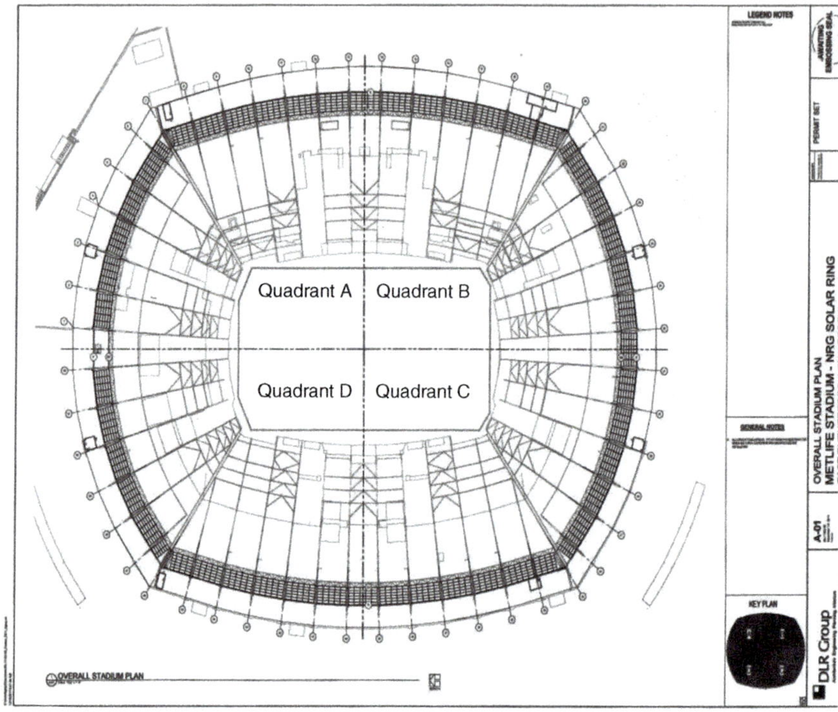

FIGURE 3.47 Metlife Solar Ring system on four quadrants.

FIGURE 3.48 Metlife module frame being lifted by crane.

extreme weathers and higher environmental loads in heavy thunderstorms and hurricanes. Though the solar ring system managed to survive the tropical storm Sandy, it is anticipated of more severe storms in the coming years, and it is vital that the system could withstand them.

FIGURE 3.49 Installation of the solar module frame onto the Metlife Solar Ring.

The solar ring system electrical system has four quadrants of operation, as shown in Fig. 3.47. Each quadrant could handle independently or coordinate with each other. They can change colors as well (Figs. 3.48 and 3.49).

References

Agius, S.L., Magniez, K.J.C., Fox, B.L., 2013. Cure behaviour and void development within rapidly cured out-of-autoclave composites. Composites Part B: Eng. 47, 230–237.

Agroui, K., Collins, G., 2003. Characterisation of EVA encapsulant material by thermally stimulated current technique. Sol. Energy Mater. Sol. Cells 80 (1), 33–45.

Agroui, K., Collins, G., Farenc, J., 2012. Measurement of glass transition temperature of crosslinked EVA encapsulant by thermal analysis for photovoltaic application. Ren. Energy 43, 218–223.

Agroui, K., Collins, G., 2014. Determination of thermal properties of crosslinked EVA encapsulant material in outdoor exposure by TSC and DSC methods. Ren. Energy 63, 741–746.

Agroui, K., et al., 2006. Thermal stability of slow and fast cure EVA encapsulant material for photovoltaic module manufacturing process. Sol. Energy Mater. Sol. Cells 90 (15), 2509–2514.

Ashter, S.A., 2014. 4 - Lamination. In: Ashter, S.A. (Ed.), Thermoforming of Single and Multilayer Laminates. William Andrew Publishing, Oxford, pp. 65–81.

Burger, M., 2021. https://www.meyerburger.com/ (accessed on 2021).

Burrows, K., Fthenakis, V., 2015. Glass needs for a growing photovoltaics industry. Sol. Energy Mater. Sol. Cells 132, 455–459.

Bye, G., Ceccaroli, B., 2014. Solar grade silicon: technology status and industrial trends. Sol. Energy Mater. Sol. Cells 130, 634–646.

Centelles, X., Castro, J.R., Cabeza, L.F., 2019. Experimental results of mechanical, adhesive, and laminated connections for laminated glass elements—a review. Eng. Struct. 180, 192–204.

Couderc, R., et al., 2017. Encapsulant for glass-glass PV modules for minimum optical losses: gas or EVA? Energy Procedia 124, 470–477.

Chapin, D.M., Fuller, C., Pearson, G., 1954. A new silicon p-n junction photocell for converting solar radiation into electrical power. J. Appl. Phys. 25 (5), 676–677.

de Oliveira, M.C.C., et al., 2018. The causes and effects of degradation of encapsulant ethylene vinyl acetate copolymer (EVA) in crystalline silicon photovoltaic modules: a review. Ren. Sustain. Energy Rev. 81, 2299–2317.
Drabczyk, K., Panek, P., 2012. A comparative study of EVA with and without thermal history for different lamination process parameters. Mater. Sci. Eng. B 177 (15), 1378–1383.
Fthenakis, V., 2012. Sustainability metrics for extending thin-film photovoltaics to terawatt levels. MRS Bull. 37 (4), 425–430.
Gagliardi, M., Lenarda, P., Paggi, M., 2017. A reaction-diffusion formulation to simulate EVA polymer degradation in environmental and accelerated ageing conditions. Sol. Energy Mater. Sol. Cells 164, 93–106.
Gay, C., Fthenakis, V., Yin, H.M., 2020. Private communication.
imec. 2019. https://www.imec-int.com/en/articles/imec-and-jolywood-achieve-a-record-of-23-2-percent-with-bifacial-n-pert-solar-cells (accessed on 2021).
Innovations Report. 2019. https://www.innovations-report.com/power-and-electrical-engineering/fraunhofer-ise-s-pv-tec-pioneers-new-technologies-and-improves-solar-cell-efficiency/ (accessed on 2021).
International Technology Roadmap for Photovoltaic (ITRPV), 2019, 10th Edition. https://pv-manufacturing.org/wp-content/uploads/2019/03/ITRPV-2019.pdf (accessed on 2019).
Jaunich, M., et al., 2016. Investigation of the curing state of ethylene/vinyl acetate copolymer (EVA) for photovoltaic applications by gel content determination, rheology, DSC and FTIR. Polym. Test. 52, 133–140.
Khudhair, A.M., Ajeel, F.N., Mohammed, M.H., 2018. Engineering and design of simple models from dye-sensitive of solar cells and photovoltaic cells applications: Theoretical study. Chem. Phys. Lett. 713, 166–171.
Kumar, M., Kumar, A., 2017. Performance assessment and degradation analysis of solar photovoltaic technologies: A review. Ren. Sustain. Energy Rev. 78, 554–587.
Kumar, A., Melkote, S.N., 2018. Diamond wire sawing of solar silicon wafers: a sustainable manufacturing alternative to loose abrasive slurry sawing. Procedia Manuf. 21, 549–566.
Kumar, N., et al., 2018. The effect of process parameters on tensile behavior of 3D printed flexible parts of ethylene vinyl acetate (EVA). J. Manuf. Processes 35, 317–326.
Lee, J.-M., Kim, B.-M., Ko, D.-C., 2019. Development of vacuum-assisted prepreg compression molding for production of automotive roof panels. Compos. Struct. 213, 144–152.
Lee, D.H., Lee, W.I., Kang, M.K., 2006. Analysis and minimization of void formation during resin transfer molding process. Compos. Sci. Technol. 66 (16), 3281–3289.
Li, J., et al., 2018. Electrochemical mechanisms of leakage-current-enhanced delamination and corrosion in Si photovoltaic modules. Sol. Energy Mater. Sol. Cells 188, 273–279.
Lv, Y., et al., 2020. Towards high-efficiency industrial p-type mono-like Si PERC solar cells. Sol. Energy Mater. Sol. Cells 204, 110202.
MaxeonCells. 2020. https://www.maxeon.com/static-files/c5e8e15e-b191-4bab-96ff-cec504040019 (accessed on 2021).
McEvoy, A., Castaner, L., Markvart, T., 2012. Solar Cells: Materials, Manufacture and Operation. Elsevier.
Niaounakis, M., 2015. Chapter 10—Manufacture of Films/Laminates. In: Niaounakis, M. (Ed.), Biopolymers: Processing and Products. William Andrew Publishing, Oxford, pp. 361–386.
Nuortimo, K., Härkönen, J., Karvonen, E., 2018. Exploring the global media image of solar power. Ren. Sustain. Energy Rev. 81, 2806–2811.
Ohl, R.S., 1946. Light-sensitive electric device, U.S. Patent No. 2,402,662.
Oliveira, M.C.C.d., et al., 2018. The causes and effects of degradation of encapsulant ethylene vinyl acetate copolymer (EVA) in crystalline silicon photovoltaic modules: A review. Ren. Sustain. Energy Rev. 81, 2299–2317.

Omazic, A., et al., 2019. Relation between degradation of polymeric components in crystalline silicon PV module and climatic conditions: a literature review. Sol. Energy Mater. Sol. Cells 192, 123–133.

Park, C.H., et al., 2011. Modeling and simulation of voids and saturation in liquid composite molding processes. Composites, Part A 42 (6), 658–668.

Park, N.C., et al., 2013. The effect of encapsulant discoloration and delamination on the electrical characteristics of photovoltaic module. Microelectron. Reliab. 53 (9), 1818–1822.

Peike, C., et al., 2013. Overview of PV module encapsulation materials. Photovolt. Int. 19, 85–92.

Peike, C., et al., 2013. The influence of laminate design on cell degradation. Energy Procedia 38, 516–522.

Perret Aebi, L., et al., 2010. Insights on EVA lamination process: Where do the bubbles come from? In: 25th European Photovoltaic Solar Energy Conference and Exhibition/5th World Conference on Photovoltaic Energy Conversion. Valencia, Valencia.

Poulek, V., et al., 2012. Towards 50years lifetime of PV panels laminated with silicone gel technology. Sol. Energy 86 (10), 3103–3108.

Powell, D.M., et al., 2015. The capital intensity of photovoltaics manufacturing: barrier to scale and opportunity for innovation. Energy Environ. Sci. 8 (12), 3395–3408.

PV Tech. 2021. https://www.pv-tech.org/news/sunpower-goes-large-with-new-a-series-ibc-module-using-ngt-technology (accessed on 2021).

Quansah, D.A., Adaramola, M.S., 2018. Ageing and degradation in solar photovoltaic modules installed in northern Ghana. Sol. Energy 173, 834–847.

Safarian, J., Tranell, G., Tangstad, M., 2012. Processes for upgrading metallurgical grade silicon to solar grade silicon. Energy Procedia 20, 88–97.

Schlothauer, J., et al., 2012. Degradation of the encapsulant polymer in outdoor weathered photovoltaic modules: spatially resolved inspection of EVA ageing by fluorescence and correlation to electroluminescence. Sol. Energy Mater. Sol. Cells 102, 75–85.

Schmid Silicon Technology (SST), 2020. https://schmid-group.com/en/business-units/photovoltaics/silicon/ (accessed on 2020).

Schulze, S.-H., et al., 2015. Cure state assessment of EVA-copolymers for PV-applications comparing dynamic-mechanical, dielectric and calorimetric properties. Sol. Energy Mater. Sol. Cells 143, 411–417.

Song, W.J.R., et al., 2018. Effect of interconnect geometry on the evolution of stresses in a solar photovoltaic laminate during and after lamination. Sol. Energy Mater. Sol. Cells 187, 241–248.

Stark, W., et al., 2012. Investigation of the crosslinking behaviour of ethylene vinyl acetate (EVA) for solar cell encapsulation by rheology and ultrasound. Polym. Test. 31 (7), 904–908.

Tatara, R.A., 2017. Chapter 14—Compression Molding. In: Kutz, M. (Ed.), Applied Plastics Engineering Handbook. Elsevier, pp. 291–320.

Tonui, P., et al., 2018. Perovskites photovoltaic solar cells: An overview of current status. Ren. Sustain. Energy Rev. 91, 1025–1044.

Torres, J.J., et al., 2019. An analysis of void formation mechanisms in out-of-autoclave prepregs by means of X-ray computed tomography. Composites, Part A 117, 230–242.

Vanguard I - the World's Oldest Satellite Still in Orbit. 2008. https://web.archive.org/web/20080919152144/http://code8200.nrl.navy.mil/vanguard.html (accessed on 2021).

Vedrtnam, A., 2018. Experimental and simulation studies on delamination strength of laminated glass composites having polyvinyl butyral and ethyl vinyl acetate inter-layers of different critical thicknesses. Defence Technol. 14 (4), 313–317.

Vedrtnam, A., Pawar, S.J., 2017. Laminated plate theories and fracture of laminated glass plate—a review. Eng. Fract. Mech. 186, 316–330.

Vincent, P., et al., 2018. Indoor-type photovoltaics with organic solar cells through optimal design. Dyes Pigm. 159, 306–313.
Wagner, J.R., 2016. Chapter 10—Blown film, cast film, and lamination processes. In: Wagner, J.R. (Ed.), Multilayer Flexible Packaging. Elsevier, pp. 137–145 Editor.
Wikipedia, 2021. Solar Cell. https://en.wikipedia.org/wiki/Solar_cell#Price_reductions (accessed on 2021).
Yang, B., et al., 2015. Influence of fabric shear and flow direction on void formation during resin transfer molding. Composites, Part A 68, 10–18.
Yang, Z., et al., 2018. Maximum efficiency and parametric optimum selection of a concentrated solar spectrum splitting photovoltaic cell-thermoelectric generator system. Energy Convers. Manage. 174, 65–71.
Yun, M., et al., 2017. Stochastic modeling of through the thickness permeability variation in a fabric and its effect on void formation during Vacuum Assisted Resin Transfer Molding. Compos. Sci. Technol. 149, 100–107.
Zaraket, J., et al., 2019. Output voltage changes in PV solar modules after electrical and thermal stresses. Experimental Analysis. Energy Procedia 157, 1404–1411.

In this long corridor of roof history, people started using materials from nature such as cold thatch and animal hire/skin to separate their living environment from the outside and provide basic protection from rain and heat. Gradually, durability, fire protection, and aesthetic considerations motivated the use of clay tile, stone slate, wood shingle, concrete roof, and metal roof, etc. With growing demand and material depletion, asphalt shingle roofs became an economic and popular option in the 20th century. Considering the growing energy crisis, a glass roof not only provides high energy efficiency with thermal coatings but also exhibits superior architectural benefits to integrating natural light and scene into the indoor environment. In the 21st century, the solar roof has been a rapidly growing industry. Solar panels are directly installed onto the roof for energy harvesting; building integrated photovoltaic (BIPV) roof directly uses solar panels as a roof structural material.

CHAPTER

4

Production and acceptance of BIPV panels

4.1 Introduction

What is BIPV and why is it important? It is basically a new field of incorporating a solar module as part of the skin of the buildings or houses. These products could be roof tiles, facades, skylights, shades, or curtain walls. When a thermal management component is added to a BIPV system, it is called a BIPVT system; so a BIPV system is a necessary part of a BIPVT system although it can be independently used as well. Another way of calling these components is Active Building Materials (ABUM) and Active Building Products (ABUP) because they can harvest and store energy by the building materials themselves instead of relying on the external energy feeding passively. It is different from the traditional building applied photovoltaic (BAPV) system as it is an integrated part of the building instead of an attachment. When the paradigm of energy generation keeps changing, it is important to investigate the variety of materials and products available. BIPV systems exhibit the following advantages over standard solar panels or modules that are commonly attached to existing roofs directly:

1. Aesthetically, BIPV panels are more uniform and pleasing compared with standard modules on existing roof materials.
2. BIPV modules are usually built with higher quality since they are custom products with a higher grade of specifications.
3. Normally, BIPV installation can withstand adverse or extreme weather conditions much better, since they are secured and protected by the building structure.
4. Although initially, they could be a bit more costly and have a high payback time, in the long run, they prove to be less expensive based on their life cycle cost.
5. The integration of PV into the building envelope avoids the leakage problem of the standard modules because of the roof penetrations.

Building Integrated Photovoltaic Thermal Systems: Fundamentals, Designs and Applications
DOI: https://doi.org/10.1016/B978-0-12-821064-2.00005-5

6. The maintenance of BIPV is less than standard modules by avoiding any moving/loose parts on the roof.
7. The chances of getting mold or trapped leaves in the fall are basically nonexistent.
8. Because the BIPV systems exhibit a smooth surface on the whole roof, they will trap less wind, water, ice, and snow loads during stormy weather for disaster mitigation and system safety.

Considering all the above-mentioned advantages, it would be wise to take building integrated photovoltaic (BIPV) installations as a superior alternative for the solar panels used in buildings.

Fig. 4.1 provides a striking view of a BIPV system installed at Hiroshima Harbor. However, this system has not been operational for some years for some reasons. This could be due to the solar panels no longer being functional, or the electrical system is not working properly, or other reasons. Because different parties were involved in this project but none of them claimed to be responsible to identify and fix problems, while some organizations and professionals have spent significant time and resources on this project, and the system could continue to yield power with the BIPV system, when a problem that disabled the system could not be identified or fixed timely, this type of early failure is not a surprise. As a BIPVT system requires teamwork from architects, construction, plumbing, electrical, mechanical, and fire protection professionals, the failure of one party will lead to malfunction of the whole system, which makes the

FIGURE 4.1 A striking building integrated photovoltaic project at Hiroshima Harbor that has been turned to not operational immaturely due to poor maintenance and support services.

diagnosis of a problem very challenging if there is no good documentation record. To establish a new technology and avoid the above failure, a clear guideline and documentation for design specifications, quality control and acceptance standard tests, installation and inspection procedures, and performance monitoring record, needs to be standardized. When a problem occurs, the potential reasons can easily be traced back to the performance data. Therefore, the cost, method, and service to fix the problem can be estimated.

The quality of products, integrated system components, and the working relationships of different parties are all critical to the success of a project. Working together and taking responsibility for selecting the best quality products, and qualified installation technicians are imperative. Although, price is a critical factor, reaching satisfactory results is even more important. Just because of a few dollars difference, one could end up with a failure that will have catastrophic consequences. By considering all these factors and paying close attention to the details, the outcomes can significantly alter toward the better side. The final work could even be a highlight in a company's portfolio and a win-win situation for both the customer and manufacturer.

The success of BIPV/BIPVT cannot solely rely on workers because there are many uncertainties in human factors, such as experience, healthy situation, training levels, etc. Standardized quality control, quality assurance, certification, and inspection can be a professional way to ensure the soundness of the system. As will be further elaborated in Chapter 5, Sunslate and TallSlate are two types of established BIPV panels. Fig. 4.2 shows an early project of Sunslate built at the Theodore Roosevelt Recreation Center at Jones Beach, NY, in parallel with a regular tile roof. After running for

FIGURE 4.2 Sunslate installation on the Theodore Roosevelt Recreation Center at Jones Beach, NY.

20 years, one can see that the regular roof has deteriorated significantly, while the Sunslate is still in great condition.

The roof on the left side had to be covered to prevent leakage while the Sunslate has endured long-term aging. Again, quality and sustainability are important criteria in a BIPV. If the Sunslate had been installed at the beginning of construction on the left side of the roof as well, the rooftop would have not needed any further maintenance at this point. Thus, not only electricity would have been generated by the panels but also the structural integrity of the roof could be preserved and leakage problems would have never happened. Leakage not only can be a headache, but it can also cause interior damages to the building. In addition, replacing the roof and dumping the material from the old roof is an additional expense that can be avoided. In light of all these costs, perhaps, if the owner or the project manager had assessed the situation more wisely at the beginning of the project 20 years ago, they might have done it differently by installing Sunslate for the whole roof and had eliminated all of these problems and concerns. The field performance and lifetime of a BIPV system directly rely on the quality of solar panels or modules, which should be controlled and monitored in the whole manufacturing process.

4.2 Material preparation and quality control

Today the manufacturing systems are highly automated and this reduces our need to learn all the detailed steps of the module assembly operation. However, it is still important to have the basic knowledge and be able to handle the module assembly manually, in the case where the automatic system is down or there are some defective modules, to fulfill the downtime and do a diagnostic, a manual or semiautomatic system can be used. The hands-on experience can be good training for workers or operators of automated systems. Particularly, for some specially designed BIPV systems, customized BIPV panels are required with certain specifications. A detailed quality control procedure should be established.

4.2.1 Inspection and acceptance of the incoming goods

Prior to manufacturing operation, one must check the preparations for all the supplies and materials especially fragile products, such as solar cells and glass. Listed below are tools that are required to handle the module manufacturing:

1. Solar cells: in transporting solar cells on-site, lightly cushioned containers of suitable size and carrying them in a vertical position are highly recommended. Prior to storing them, it is important to consider: (a) check the condition of the packages; if there are any damages, the freight forwarder should be notified of the condition of the packages. (b) In

checking the solar cells, protective gloves should be worn. (c) Check the front and back of the cells visually for any cracks and irregularities. (d) Make sure the cells meet all the criteria of the manufacturer. (e) Position the cells in a vertical orientation for all the acceptable ones and separate the rejected ones. If there are too many rejected cells, report to the supplier.

After going through all the above procedures, store the cells in a secure cupboard or a vault so that they are well-protected.

2. Glass: handling glass pieces should be done very cautiously especially for tempered ones since they are very brittle on the sides and corners. Meanwhile, make sure to look at the safety regulations in handling glass and how to protect the workers. In transportation, a mobile glass transport wagon is required, and the glass should be located on a glass rack for the intermediate storage of glass in a vertical position, leaning slightly backward. Make sure the glasses are leaned on each other with thin wooden support as a cushion in between so that it will prevent them from being scratched. Prior to bringing the glasses to the storage, check all the pieces for any defects or nonacceptable scratched condition, and if any is observed, it is vital to contact the suppliers. The checklist for glass inspection is as follows:
   - Quality of the edge and corner finishing.
   - Surface damage or scratches.
   - Quality and position of the slit.
   - Cleanliness: check whether there are any watermarks.
   - Position of the tempering stamp.
   - Stress lines at the corners by means of two optically polarized foils.
   - Maintaining a logbook on all the glass.
3. Encapsulant: Ethylene-vinyl acetate (EVA).

   Note that EVA is sensitive to light, heat, moisture, and dirt especially when it is not in the operation process. Its shelf life should be noted according to the date shown on the package.
   - Use the EVA foils within 6 months of their production date.
   - Store the EVA in a dark room at a temperature of about 30°C and low humidity.
   - Once the EVA is being cut, it should be wrapped in an airtight, black PE foil.
   - After the EVA is cut, it should not be left in the open space for more than 8 hours.

   The EVA is usually packed in rolls of 25–50 Kg using airtight packaging material. Subject to the size of the order, the rolls are packed using cardboard boxes or wooden containers. It is very important to check the shipment as soon as it is delivered. In case of any defects, notify the freight forwarder immediately. Meanwhile, also check the dimensions and volume to verify there are not any discrepancies. After verifying the order, seal all the packages and store them in a safe place. The next step after the physical inspection of the EVA is to check the gel content which is vital for the performance of the sheets. More details will be covered later.

4. Backsheet: on standard solar modules, the backsheet protects the backside of the module. Depending on the project, they are usually black or sometimes white. They should be checked and handled the same way as EVA. Although the packaging is slightly larger and heavier, each roll can weigh up to 100 Kg. Store them horizontally in a clean and safe environment, and avoid applying any pressure points.
5. Smaller items like tabbing materials, hand tools, solders, ribbons, and wires should be stored in a cupboard.
   Larger items such as foil rolls, soldering gauges, large spare parts, or larger pieces of equipment should be stored in a pallet storage rack.
   Hand pallet truck—a sturdy one is highly recommended to transport heavy loads like glass pallets and foil rolls.
6. Wire cutter—it will be used to cut wires for modules or solar roof tiles.
7. Hand tools:
   - Paper scissor and a carpet knife.
   - Nail puller, hammer, and plate.
   - Flat pliers, cutting pliers, and screwdrivers.
8. Measuring instruments:
   - Optically polarized foil and magnifying glass.
   - Multimeter.
   - Metallic ruler with millimeter and inch division, length 2000 mm.
   - Metallic ruler with millimeter and inch division, length 150–300 mm.
   - Measurement tape, sliding caliper, and set square.
9. Miscellaneous items:
   - Disposable gloves of thin latex or vinyl to protect solar cells and glass panes against corrosive materials.
   - Soiling caused by sweat and protect the staff against possible allergic reactions.
   - Additional manufacturing materials such as soldering flux, soldering wire, glass fleece/glass fabrics, junction boxes, connectors, diodes, etc., can be handled and stored without any problems. Pay attention to the instructions of the manufacturers.
   - Supporting materials such as separation foils, cushioning foils, cleaning material, etc., can be handled and stored without any issues. Be attentive to safety regulations and the instructions given by the manufacturers.

4.2.2 Encapsulant preparation

The machinery and tools are listed as follows.

A roller frame is recommended to be installed so that the EVA rolls can be opened onto the cutting table. Make sure the roller is securely positioned next to the cutting table in Fig. 4.3 as cutting the EVA needs high precision.

FIGURE 4.3 Cutting table with EVA and roller. *EVA*, ethylene-vinyl acetate.

The cutting table must be sturdy and securely attached to the foil cutting device. The working table would be subject to the product range sizes. The table should consist of an adjustable limit-stop and measuring rulers. To achieve precise dimensions, cutting the foil should be distortion-free.

Prior to starting the operation, make sure to wear gloves and safety goggles. Check all the machinery and devices. Make sure that they are all operational.

In handling an encapsulant like the EVA, take the necessary precautions as it is sensitive to light, heat, dirt, and has a limited shelf life. Prior to opening the package, make sure to check the expiration date and the exact type and specifications. As to cutting the EVA, on a standard module that is usually glass/Tedlar, two sheets are needed, and on glass/glass, three or more. In addition, a strip of EVA would be needed to fill out some gap spaces. Make sure to cut the EVA according to the requirements of the shift or the day. It is not advised to cut them in advance, even one day prior to use.

In cutting the EVA, always allow a couple of inches extra and cut a size larger than the size of the module on all sides.

Secure the EVA roll on the roll-frame. Adjust the transverse limit-stop to predefined measurement and switch on the foil cutting device. Roll out the EVA foil with caution together with the separation of the foil and let it through the cutting machine onto the cutting table without stretching the foil. The separation of the foil must be underneath, and the matte side of the EVA foil must be at the top. Hold and pull the foil precisely to the transverse limit-stop and check the position of the foil at limit-stop length. Hold the foil in position and push the cutter. After making the cut, verify the cut by measuring it and make sure that it is at the cut is at the right angle. In case it does not meet the requirement and subject to the situation, it may need to be deposed. After cutting, it is highly recommended to cover the EVA roll with a black protective foil and close it tightly until later when the material is needed again.

Abide strictly to the regulations provided by the EVA manufacturer and follow the expiration date.

- Store it in a dark room at a temperature below 30°C.
- Use the EVA within six months of purchase.
- Wrap the rolls and the cut foils airtight in a black PE foil.
- Do not leave the cut EVA open for more than 8 hours.

In case the cut foils need further adjustments: move the roll-frame and the rack for foils away from the cutting machine. Then cut the foil according to the specification, put it back into the rack and wrap it in a light-impermeable cover.

4.2.3 Backsheet (backing foil) preparation

Determine the cutting size and number first in the following steps:

- On manufacturing of a glass/foil (Tedlar) laminate, a backing foil is required that is about 15–20 mm larger than the glass on all sides. The backing foil must be in one piece.
- The backing foils can be prepared for the complete order because the storage period is unlimited.

Check the backing foil roll and measurements, then span and secure it on the roll frame. Remove the protective foil and keep it. Adjust the transverse limit-stop to confirm the exact measurement and switch on the foil cutting machine.

Roll out the cut backing foil with caution and lead it through the cutter onto the cutting table. The inner side must face downwards and the outer side upwards. The inner of the foil is usually pretreated mechanically or chemically in order to improve the bonding with the EVA.

Lead the foil precisely to the transverse limit-stop and check the position of the foil at the length limit-stop. Hold the foil in position and switch on the cutter.

Check the first cut foil for the measurement precision (±1 mm) and that it is at perfect angles. Then store it in the rack for foils with the inner side facing downwards. Cut the required number, and then store the cut ones in the rack with foils with the inner side facing downwards. Wrap the backing foil roll in the protective cover and fasten it with adhesive tape.

In case the backsheet requires correction: move the roll-frame away and the rack from the foil to the cutting machine. Then cut the foil to the exact measurements and put them back into the rack with the inner side facing downwards. Cut the exit slit according to the drawing (the point where the ribbon leads out of backing foil to connect the J-box). Apply a carpet knife with a metal ruler and cut the slid precisely. Store the backsheet onto the foil rack with the inner side facing down.

4.2.4 Covers and insulators

These materials are used in the laminates as an optical cover for the string interconnections and as insulators between the connection ribbons and the solar cells. They are usually made of the same material as the backsheet foil of the relevant laminate.

The cutting procedures are similar to EVA and backsheet as mentioned in the previous section, however, more precise adherence to the measurements is required.

The measurements and number of covers and insulators are according to the product-specific. In some situations, mats of glass fleece (35 g/m^2) or glass fabric (50 g/m^2) are required for the mechanical durability of the cell matrix during the manufacturing process and for the longevity of the product especially in glass/glass modules. Normally the glass mats are of a similar size to the cell matrix or maybe slightly smaller. The procedure in cutting them is similar to the backsheet and EVA presented above. Make sure to handle the mats gently and not stretch them. Cutting the glass fibers requires a large force on the cutting knife due to the hardness of the glass fibers.

4.2.5 Ribbon connection preparation

The machinery and tools for ribbon connection are listed as follows:

- Level and sturdy working table-top and adjustable height on working chair.
- Ribbon cutting machine—adjustable on width, length with number control on the ribbons.
- Ribbon transport containers that should be suitable for the application, e.g., plastic binds.
- Soldering plate, estimate size: 100 cm × 50 cm.
- Soldering station with a smoke extractor is highly recommended. A power of 80 W or higher is required to handle the high heat loss of the ribbon connection and the heating plate. The soldering heat guarantees the necessary transfer of heat and effective work performance. The smoke station extractor protects the workers from health risks such as gases.
- Cutting pliers—precision diagonal cutting pliers are highly recommended.

- Tweezers—angulated steel model, antimagnetic, e.g. Knipex.
- Measuring instruments—metallic ruler: millimeter or inch division—2000 mm length, metallic ruler: millimeter or inch division—150–300 mm length set square

The ribbon connection will need the following materials:

- Tabbing materials: tin-plated soft copper in various widths and thickness ribbons.
- Soldering wire: with integrated soldering flux and lead-free.
- Adhesive tape with holder: 3M Magic 810 wide.
- Protective gloves: disposable gloves made of thin latex or vinyl that can protect the connection ribbons against sweat and protect the workers from a possible allergic reaction to soldering flux and cleaning solutions.

The ribbon connection typically follows these working steps:

1. Preparation and control of machineries and materials.
   Prior to starting operation, it is important to check all the machineries, tools, and materials. Make sure each is in place and operational.
   - Prepare the machineries and materials in the way they are operated.
   - Wear protective gloves.
2. Ribbon connection on front tabbing.

Pull the ribbon out of the ribbon roll according to the required length. The dimensions of the ribbon on 125 mm cells are 2.00 mm × 0.10 mm, and on 156 mm cells, 2.40 × 0.15 mm. It is important to mark the required length and make a sizeable allowance so that there is some room to adjust. While rolling the ribbon out, do not bend over the mark and cut at the right angle with pliers. Lay the cut connection ribbons onto the transport container without bending them. An automatic cutting machine (Fig. 4.4)

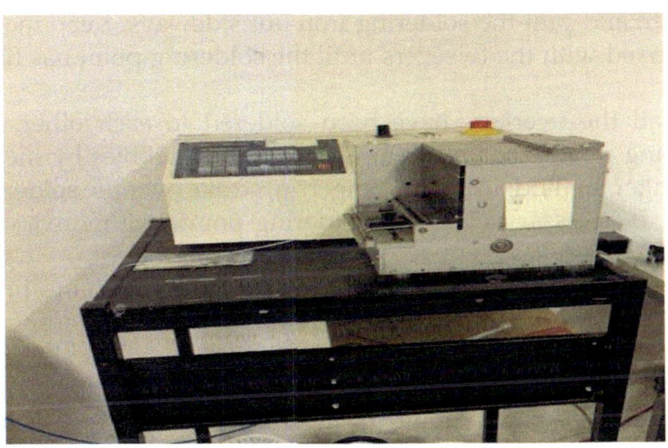

FIGURE 4.4 Ribbon cutting machine.

is highly recommended in a manufacturing operation but is optional in a laboratory setup.

On connecting the strings to each other using ribbons, as a rule: ribbon size 4.0 to 8.0 mm × 0.2 to 0.5 mm depending on the cell size and the module design. Mark the required length on the worktable and set the mechanical stopper. Roll the connection ribbon without bending it over the marking to the stopper and cut at the right angle using a plier with a tolerance of ±1 mm. When connecting ribbons to the J-box, in case the contacts are comprised of one piece, it works the same way to the string-to-string connection as mentioned above.

In case, the connection ribbons for the J-box contacts are comprised of more than one part, operation direction has to be changed as listed below:

- Assemble the pieces as mentioned in the previous steps.
- Mark the joints on the soldering plate with a metallic ruler and a set square.
- Switch on the soldering station and set it to a temperature of about 400°C.
- Place the pieces so that they overlap each other according to the markings on the soldering plate and fix them temporarily with adhesive tape.
- Hold the tweezer and the soldering wire in one hand and the soldering iron in the other. Clean the soldering tip on a moist sponge.
- Set the soldering iron at a flat angle and with light pressure on the upper connection ribbon on the first overlap and add a small amount of soldering wire.
- As soon as the tinning of the lower connection ribbon becomes liquid next to the soldering tip, fix the upper connection ribbon with the tweezer, and pull the soldering iron out sideways. Keep the soldering point fixed with the tweezers until the soldering point has fully solidified.
- Once all the overlaps have been soldered to each other, check the soldering points for unevenness, and solder bubbles by means of the one-finger method: using one fingertip, stroke over the soldering points gently. If necessary, clean the soldering points with ethyl-acetate and cleaning paper.
- Remove the adhesive tapes and place the soldered connection ribbons for J-box contacts in the transport container without bending them.

It should be noted that unevenness and solder bubbles can lead to cell breakage and short-circuit during further processing. Thus, further attention is required when soldering the ribbon pieces.

4.3 Soldering of solar cells

4.3.1 Front tabbing

Front tabbing is making electrical contact on the front side of the solar cells (minus polarity). For this purpose, two tinned connection ribbons are soldered to the conductor paths of the solar cells foreseen for them. The overhanging part of the tabs serves for the electrical connection with the reverse side (plus polarity) of the neighboring cell.

The following equipment, devices, and tools are required:

- Heated aluminum plate with necessary materials for solar cells: the adjustable temperature (50–60°C) reduces the thermal stress in the solar cells during soldering and accelerates the soldering process. The soldering plate must be adjusted to the size of the solar cells.
- Working table: a sturdy working table with an even tabletop to accommodate the transport of the solar cells in the workspace.
- Mobile side: table with rollers to accommodate the soldering station, tools, and tabs.
- Mobile work chair: the length of the workplace requires a height-adjustable chair with rollers.
- Soldering station with smoke extractor: soldering station with regulated soldering tip temperature. A minimum power of 80 W is required to compensate for the high heat loss of solar cells and the soldering plate. The soldering tip guarantees the necessary transfer of heat and efficient work. For example, the soldering station WELLER is comprised of PUD 81, FE 80, AK 20, and soldering tip HT1.
- Solar cells container: lightly cushioned container of appropriate sizes for storing the cells vertically.
- Tools: stainless steel angulated tweezers, e.g., KNIPEX 92 34 28 or 92 32 29, a metallic ruler with millimeter division, a 150 mm long fine paintbrush for additional flux application.
- Spring scale and down holder:
  - Precision spring scale on peel test—Presola 40010 (10 newtons).
  - Two down holders of suitable size, e.g., 200 × 40 × 10 mm of plastic for fixing of the solar cell during a peel test.

The soldering materials include:

- Soldering flux and container—soldering flux Stannol X33 F-071 or Kester 959T, filled into a sealable plastic container of appropriate size.
- Front-tabs—connection ribbons on front tabbing (minus polarity): the cut length would be according to the cell size and the cell spacing of the module design, especially in a BIPV case.

Prior to getting started, check all the machineries, tools, and materials that are required. The operation steps are as follows:

- Prepare the machines and materials in the order in which they are used.
- Switch on the tabbing table and set it to 50–60°C, then switch on the soldering station and set it to the appropriate temperature about 400°C.
- Wear protective gloves.
- Pretreatment of the tabs with soldering flux:
  - The soldering flux guarantees the necessary flow of the tin and the required transfer of the heat during the soldering process.
  - Moisten the tabs in the soldering flux container. Keep the tabs in bundles temporarily on the absorbent paper. Close the container afterward to avoid evaporation.
  - The pretreated tabs must be dry and be processed on the same day. If they should be processed on the next day, the pretreatment will need to be repeated.
- Lay in the cells and lay tabs onto them:
  - Hold the tweezer in one hand and the soldering iron in the other with high caution. Clean the soldering tip on a moist sponge.
  - Place the tag precisely on the front conductor line of the first cell to the left and fix it outside the cell with the tweezer. Check the position of the tab for:

    (a) Overlapping part of the left cell.
    (b) The end of the tab to be approximately 5 mm inside the right-hand edge of the cell.
    (c) Position the soldering iron approximately 10–15 mm inside the left-hand edge of the cell onto the tab at a flat angle and without applying any pressure.
    (d) As soon as the tinning of the tab becomes liquid to the left and of the soldering tip and it turns glossy, lead the soldering iron evenly to the right over the tab and flatly out over the right-hand edge of the cell. At the same time, check the position of the tab and correct it if necessary, with the tweezer.
    (e) Additional flux applied to the cell using a fine paintbrush may be helpful for beginners.

The duration of the soldering process is of the decisive importance of the quality of the soldering joints. A soldering process that goes on for too long reduces the quality and results in "burnt joints," while a soldering process that is too short reduces the quality and results in "cold joints." Clean the soldering tip and then solder the rear tab the same way.

Final check and temporary storage: once all the solar cells have been fitted with two tabs on each, check the soldering joints for unevenness and solder bubbles using the two-finger method in Fig. 4.5: hold the overhanging tabs with the left hand and stroke gently over the soldered

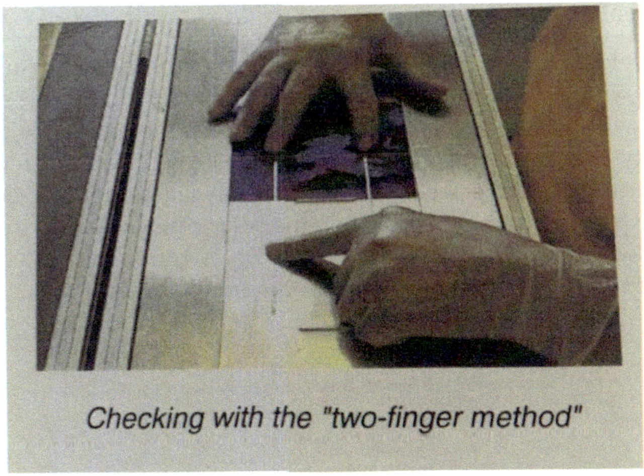

FIGURE 4.5 Two-finger method.

tabs with two fingertips of the right hand. Unevenness and solder bubbies can lead to cell breakage and short-circuits.

4.3.2 Visual check of the soldering joints and cell surface cleanliness

Remove the solar cells carefully from the soldering plate and store them temporarily in an upright position with overhanging tabs upwards, in the transport container in Fig. 4.6.

FIGURE 4.6 Tapped solar cell container.

4.3.3 Peel test

The peel test (Fig. 4.7) measures the adhesion of the encapsulant to the substrate which plays an important role in the durability of the solar module and is vital in the quality and performance of the product. Considering the quality assurance reasons, the bonding between the soldered tabs and the solar cells must be regularly checked, at least once per employees' shift. The peel strength of the tabs on the solar cells measures the bonding strength with a spring scale of about 10 Newtons. The tested cells are chosen at random from the current production batch.

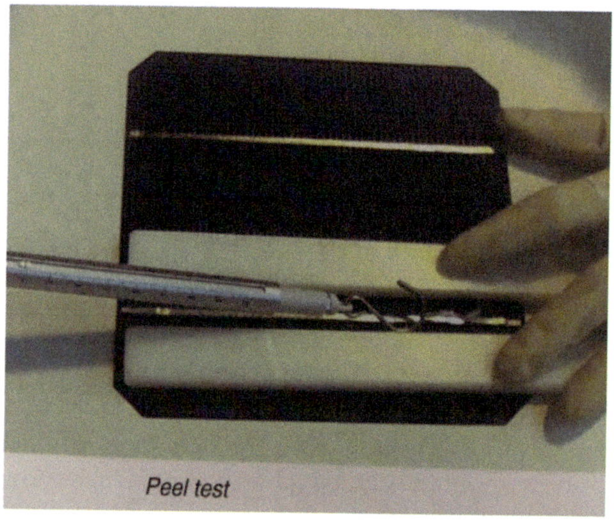

FIGURE 4.7 Peel test operation.

To perform the test, solder the overhanging tabs onto a loop. Lay the solar cell onto a level underlay with the front side facing upwards. Lay the plastic down holder to the left and right of the tab which is to be tested and hang the spring into the tab loop. Fix the solar cell with the down holder with one hand and grip the spring scale with the other hand. At a 180 degrees angle, pull the spring scale until the tabs peel off the entire length of the solar cell and simultaneously read the value shown on the spring scale. The peel angle and the strength are defined by the manufacturer of the cells. The peel strength usually measures between 1 and 3 Newton. The visual inspection of the surface of the peeled cell and tab shows an important appraisal of the quality regarding the moistening intensity and uniformity of the bonding.

4.3.4 Soldering string

String soldering is the electrical connection of solar cells into a string (Fig. 4.8). For this purpose, the overhanging tabs of the solar cell's front

FIGURE 4.8 Front side and backside of a string.

side (minus polarity) are soldered onto the relevant conductor lines on the reverse side (plus polarity) of the neighboring cell. The number of cells per string is subject to the design of the module.

The cell stringing table includes a heated aluminum plate with pockets of solar cells. The adjustable temperature ranging between 50 and 60°C results in reduced thermal stress on the solar cells during the soldering operation while also accelerating the soldering process. The soldering plate must be adjusted to the size of the solar cells. A turning and pushing device, halogen lamps, and a measuring system enable reliable functional checking of the finished string.

Sunpower back contact cell string soldering: the back-contact cells do not have any contact on the front side of the cell. All the stringing operations would be done at the back (Fig. 4.9). On the backside,

FIGURE 4.9 Back contact cell string.

one side has positive polarity while the other side is negative. In stringing (Fig. 4.10), the cell just connects the tabbing from the negative polarity to the positive polarity. In the beginning and with high caution, put a small drop of flux on each of the three small pads on the negative side and carefully put three tiny drops of solder. Immediately, place the ribbon onto the pad and connect to the positive side.

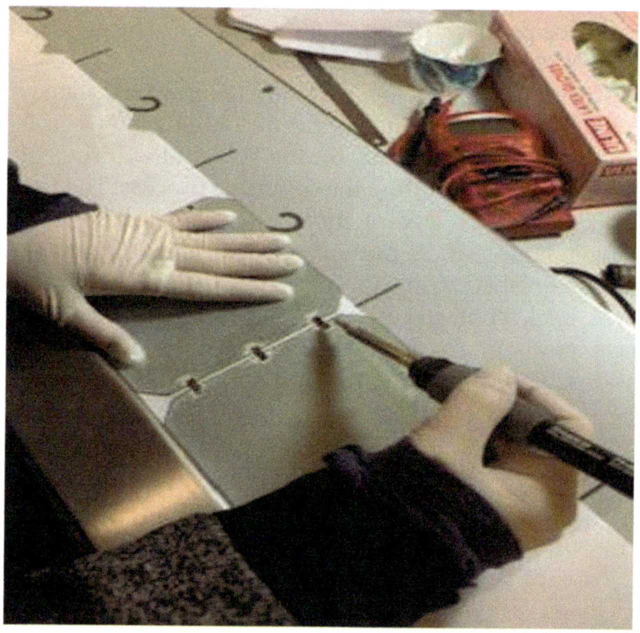

FIGURE 4.10 Stringing back contact cell operation.

In stringing back contact cells, there is another process which is using dog bones (Fig. 4.11), which are applicable to the latest Maxeon Gen 5 of the 161 mm back contact cells with the 4 point connection.

After stringing is completed, carefully remove the strings from the string soldering table using both hands, and store them temporarily with the cell front facing downwards on a cushion-mat in the transport container for strings. Transport the strings in a sturdy container on wheels for intermediate storage and transport of the finished strings. The size: 230 × 75 × 80 cm (L × W × H). Make sure to observe the polarity since the plus pole of the strings must always be on the same side of the transport container.

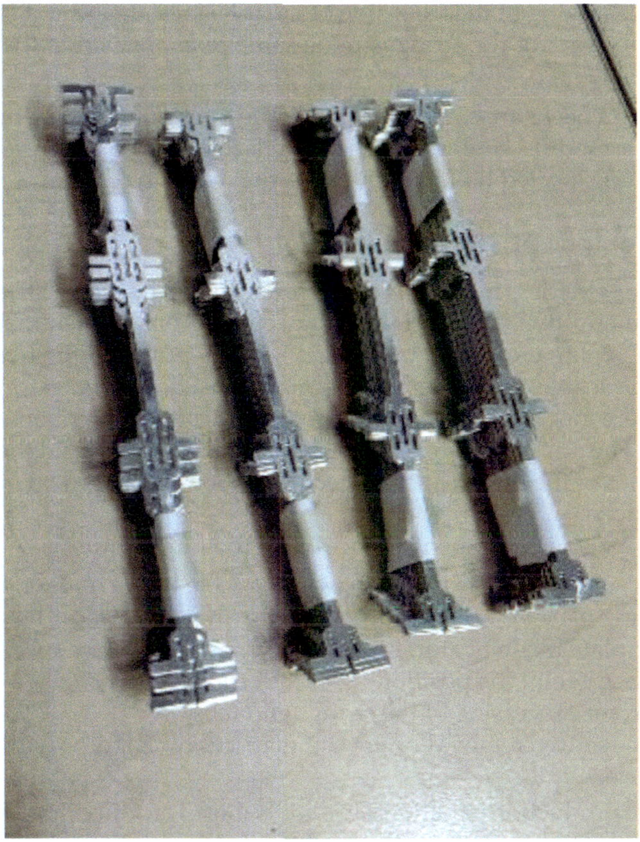

FIGURE 4.11 Dog bones use on back contact cell stringing (with permission from Maxeon).

4.3.5 String soldering: quality control

Quality control should be performed constantly by the workers at the operation site, before, during, and after each step. The steps are:

- Control of the exact specifications of the solar cell and tabs.
- Visual checking of the soldering plate for cleanliness.
- Visual checking of the soldering tip for cleanliness.
- Visual checking of the soldering joints for precise geometry.
- Mechanical checking of the soldering joints for unevenness and solder bubbles.
- Visual checking of the reverse side of the cells for damage and cleanliness.
- Electrical function control of the strings.

- Visual checking of the front side of the cells for damage and cleanliness.
- Visual control of the strings on the same pole position in the transport container.

All these checks are extremely vital while the layout is being built. When building a module, any issue can waste sizeable time and resources and cause an unexpected halt in the process that could be crucial.

4.3.6 Equipment, devices, and tools

Matrix soldering is the electrical connection of several strings into a matrix. In addition, the electrical connections from the cell matrix to the junction box are created at this stage. For this purpose, the overhanging tabs of the strings are soldered to larger connection ribbons for string-to-string interconnects. These are soldered to the connection ribbons for J-box contacts. The number of strings, their position, geometry, and the electrical connections of the matrix and connection ribbons are based on the product specifications.

The following equipment, devices, and tools are required:

- Matrix soldering table: a sturdy working table with a horizontal rotating soldering plate to carry and process the strings is needed. The matrix soldering plate made of aluminum should have the below specifications:
  - Precise geometric positioning of the strings and the tabs.
  - Efficient processing of the strings and tabs.
  - Damage-free removal of the cell matrices by hand or with a suction device.

In large production situations, it is advisable to produce product-specific plates similar to the plate for the strings.

- Intermediate storage on cell matrices: a sturdy and mobile working table with an even tabletop for intermediate storage transport of the finished cell matrices (Fig. 4.12).

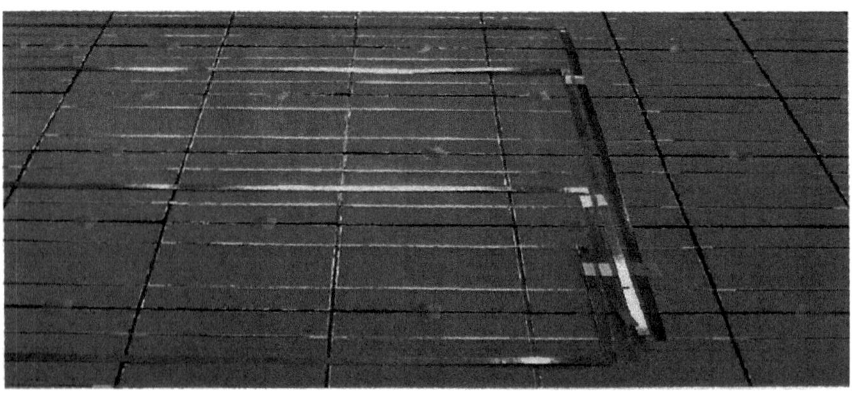

FIGURE 4.12 Cell matrix.

Some additional materials are required for cell matrix assembly as:

- Lead-free soldering wire with integrated flux, e.g. FELDER ISO-Core ELR 1 mm Sn95.5AG3.8Cu0.7 or BALVER LF 2200NC Sn95.95.5 Ag.3.8Cu0.7.
- Tape for cell fixation: Holder with adhesive tape Scotch 3M Magic. 19mm wide.
- Soldering underlay: Strips of Polytetrafluoroethylene impregnated fiberglass as an underlay for matrix soldering, e.g., MAAGTECHNIC, type 7510, thickness 0.25 mm, measurements as requested.

The matrix operation steps are as follows:

1. Control of machinery and materials.
   Prior to starting, check that all the equipment, tools, and materials to make sure they are operational and in the right order.
   - Turn on the module tester and carry out the reference measurements. Make sure to comply with the operation instruction of the module tester.
   - Prepare the machines and the materials in the order in which they will be used.
   - Switch on the soldering station and set the temperature to about 400°C.
   - Wear protective gloves and goggles.
2. Setting up the matrix soldering plate.
   Set up the matrix soldering plate in accordance with the production drawing:
   - Set the mechanical stoppers for the beginning and the end of the string.
   - Set the stoppers for the side limitation and space of the strings.
   - Fix the soldering underlay and add the markings for string-to-string connections.
   - Mark the points for plus and minus polarity of the strings.
   - Mark the points for connections.
3. Laying the strings and the string-to-string interconnects.
   Lay the string to string interconnects as per the markings and if necessary, fix them with adhesive tape on the soldering plate. Lay the strings cautiously, with the front of cells facing downwards, between the mechanical stoppers of the soldering plate, and with the plus and minus polarity as per the marking. Bring the strings into zero position, i.e., all the strings to the stopper at the front right.
4. Shortening and soldering tabs.
   Shorten the overhanging tabs with the cutting pliers to the following size: the outer edge of the string to string interconnects +0/-1 mm. Hold the tweezer and the soldering wire in one hand and the soldering iron in the other. Clean the soldering tip on a moist sponge. Take high caution

as the tip is very hot. Position the soldering iron at a flat angle and with a little pressure on the first tab, add a small amount of soldering wire. As soon as the tinning of the string to string interconnect of the left and right of the soldering tip becomes liquid (it becomes glossier), fix the tab with tweezers and put the soldering iron flatly out to the side. The duration of the soldering operation is of decisive importance for the quality of the soldering joints. Avoid burnt joins and cold joints

5. Laying and soldering the J-box contacts.
   Lay the prepared connection ribbons for J-box contacts carefully on the cell matrix as per the drawing and the marking, and then fix them temporarily with adhesive tape. Hold the tweezer and the soldering wire in one hand and the soldering iron in the other. Clean the soldering tip on a moist sponge. Solder the J-box contacts on the string-to-string interconnect using the procedure described above. Mark the exit line (the place at which the J-box contacts lead out of the laminate into the junction box) on the connection ribbons and mark the plus/minus polarity.
6. Checking the cell matrix and securing the transport.
   Once all the tabs, string-to-string interconnects and J-box contacts have been soldered to each other, check the soldering joints for unevenness and solder bubbles using the "one finger" method: stroke gently over the soldering joints with one fingertip. Unevenness and solder bubbles can lead to cell breakage and short-circuit in further processing. Visual check of the electrical connections, geometry, and measurements as per the production drawing is required. A visual check on the soldering joints and cell surfaces for cleanliness should be done. If necessary, clean the soldering joints with cleaning paper and ethyl acetate. Temporarily, secure the strings with each other by means of the adhesive tape to avoid any shifting during transportation.
7. Electrical function control of the matrix.
   Two persons are required to transport the cell matrix. Hold the cell matrix carefully at the string-to-string interconnects and lift it from the soldering plate. Then lay it on the center of the module tester. Insulate the connection ribbons electrically from the cell matrix and connect the measurement cables with the correct polarity. Measure the cell matrix and register the results according to the operating instructions of the module tester. Check whether the measurement matches the actual designed value.
8. Final check and temporary storage.
   Remove the measurement cables and electrical insulation from the cell matrix. Visually inspect the cell matrix on damage and cleanliness. It is recommended to have two persons carefully pick up the cell matrix from the module tester and lay it on a cushioned mat on the intermediate storage table. Lay a cushioned mat for the next cell matrix.

In a manufacturing environment, a semiautomated lay-up station would be highly recommended in case the fully automated line is interrupter by technical issues. It is quite time-consuming in going through all the operating procedures as presented above. A system of this type is not a costly investment with the following advantages:

- Safe time on handling the matrix transport handling.
- Safe time on reference measurement of the module tester.
- Setting up of the matrix soldering plate.
- Temporary securing the cell matrix transport.
- Safe time on module preparation before lamination.

4.4 Laying up of glass/foil laminates

In this part of the operation, although it can be done with the semiautomatic layup station (Fig. 4.13), the manual procedure will be introduced.

The prepared materials, including glass pane, EVA encapsulant, solar cell matrix, and backing foil (blacksheet), are layered on each other and prepared in such a way that in the laminating process, they would be able to firmly bound together into a final laminate. The glass/foil (Tedlar) laminate is usually comprised of a front cover of toughened, low iron white glass and a back cover of Tedlar/polyester, and between them, the solar cell

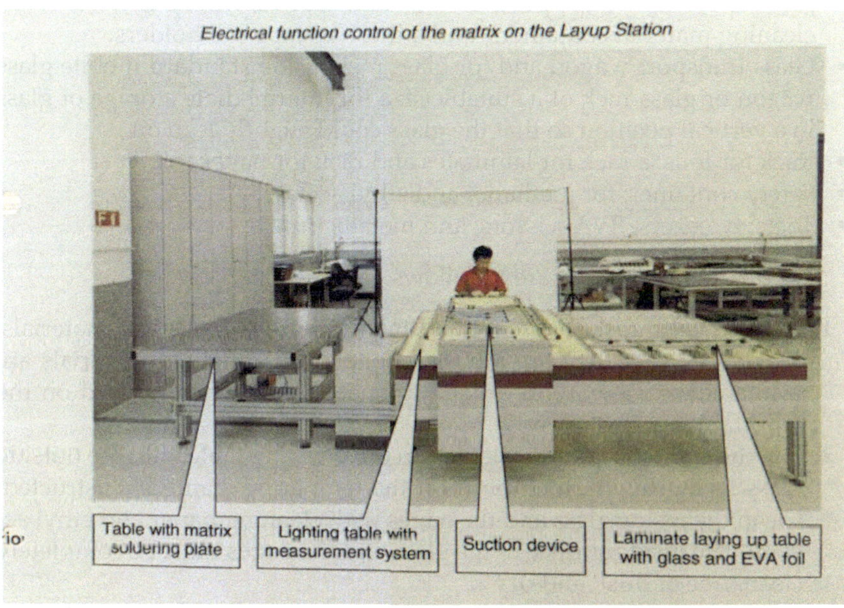

FIGURE 4.13 The semiautomatic layup station.

matrix is embedded in the EVA. Due to the weight of the glass panes and the handling of cell matrices, EVA foils, and backing foils (Fig. 4.14), two persons are usually required to lay up the laminate.

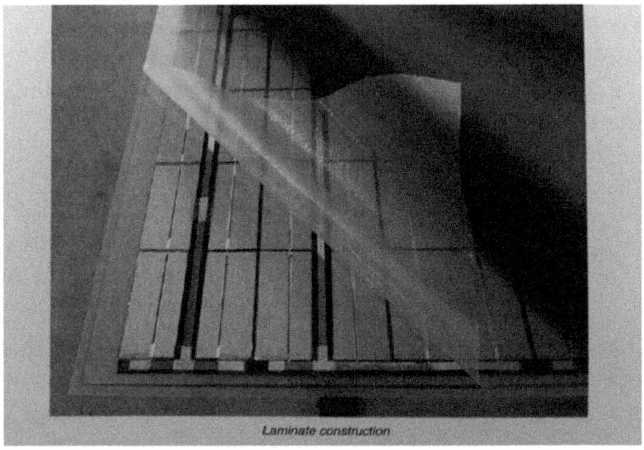

FIGURE 4.14 Glass/foil laminate layup.

The equipment, devices, and tools include:

- Sturdy mobile working table: for laying up of the laminates with adjustable cushioned supports. Mobile side table: it will be used for tools, cleaning material containers, and adhesives tape roll holders.
- Glass transport wagon and/or glass rack: trade standard mobile glass wagon or glass rack of a suitable size for intermediate storage of glass in a vertical position so that the glass could slightly lean on.
- Rack for foils: a rack for laminates and rack for paper roll.
- Safety container: for methanol and ethyl acetate.
- Tools: tweezers, EVA scissors, and metallic rulers.

Layup operation steps are as follows:

1. Prior to getting started, check all the machinery, tools, and materials. Make sure that they are all operational, and sufficient materials are available, as they are in order. Laminates are usually layered on the worktable for this purpose.
2. Lay the front glass pane releases on the working table with the outside glass facing down, then position the tempering stamp as instructed. Clean the inside glass and the edges with cleaning paper and ethyl acetate. The inside surface of the glass and the edges must be completely clean, free of dust, and dry.
3. Hold the cut EVA foil and remove the protective foil. Check the foil for cleanliness and lay it with the matt side facing down onto the glass. The EVA foil should overhang the glass edges by 5–10 mm.

4. In laying the cell matrix, two persons are required. Hold the functionally tested matrix at the string-to-string interconnects and lay it on the EVA foil. Make sure the front side of the solar cells is facing downwards. Check the position of the cell matrix if it turned 180 degrees. Double-check the position of the cell matrix in relation to the glass edges on all sides and correct if necessary.
5. Lay the covers: the covers hide the string-to-string interconnects and mechanically stabilize the cell matrix against any flow movements during lamination, although they are not compulsory for electrical purposes. In accordance with the production drawing, lay the covers between the string-to-string interconnects and the front EVA foil as shown in Fig. 4.15A. Then adjust the inner length edge of the covers to match the edges of the cells. Stick the covers and the string-to-string interconnects together with small pieces of Scotch 3M Magic. Carefully remove the adhesive tapes for temporary transportation safety of the cell matrix.
6. Lay the insulators—the insulators separate electrically the connection ribbons for J-box contacts from the solar cells and optically hide the connection ribbons. As shown in Fig. 4.15B, carefully lay the insulators in accordance with the drawing between the connection ribbons and the cell matrix with the outer transverse edges of the insulators tabs overhanging the outer edges of the cell matrix by at least 2 mm; all the other edges of the insulators overlap the connection ribbons by 5 mm. In order to guarantee a secure bond between the insulators and the cell matrix, additional EVA strips are now inserted which are at least 2 mm bigger than the insulators on all sides.
7. Check, in accordance with the drawing, the position of the insulators and the connection ribbons, particularly in the area of the exit line, and secure it with Scotch 3M Magic on the cell matrix to avoid any shifting during the lamination process.

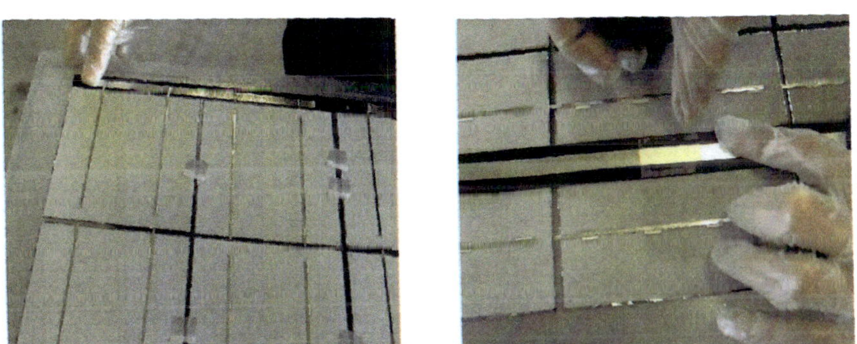

A B

FIGURE 4.15 (A) Laying the covers. (B) Laying and fixing the insulation.

8. Additional stabilization of the cell matrix: flow movements of the liquid EVA during the lamination process can lead to shifting of the solar cells, and that would have a significantly negative effect on the quality and function of the product. The factors that could increase the risk of cells shifting during lamination are listed below:
   - Cell distance of more than 2 mm within the strings.
   - Distance of more than 2 mm between the strings.
   - Long strings.
   - A Large EVA surface and/or a low cell surface area in relation to the laminate surface.

   If necessary, using the 3M Magic adhesive tape connections between cells of neighboring strings will reduce the risk of the cell shifting during lamination:
9. Lay the rear EVA encapsulant: hold down the cut EVA, remove the protective foil for cleanliness, and lay it on to the cell matrix with the matt side facing upwards, with coverage exactly the same as that of the front EVA.

   Cut the exit slit for the connecting ribbons (Fig. 4.16): using the EVA scissors, cut the rear EVA from the edge over the marking of the exit line. Open the EVA foil. Prepare the connection ribbons (Fig. 4.17) and thread the connection ribbon out (Fig. 4.18) and close the EVA again. Gently bend the connection ribbon.
10. Lay the backing foil (backsheet): hold the cut backing foil. Check the inner side of the foil (lying underneath) for cleanliness and if required, clean it with methyl alcohol and paper. Lay the completely clean, dust-free, and dry foil onto the rear EVA foil with the inner side facing downwards. Thread the connection ribbons through the exit slit (Fig. 4.18) and check the position of the foil. It should have an equal overlap on all sides. Stick the connection ribbons to the rear foil with Scotch Blue adhesive tape (Fig. 4.19).

FIGURE 4.16 Exit slit for connections.

FIGURE 4.17 Prepared connection ribbons.

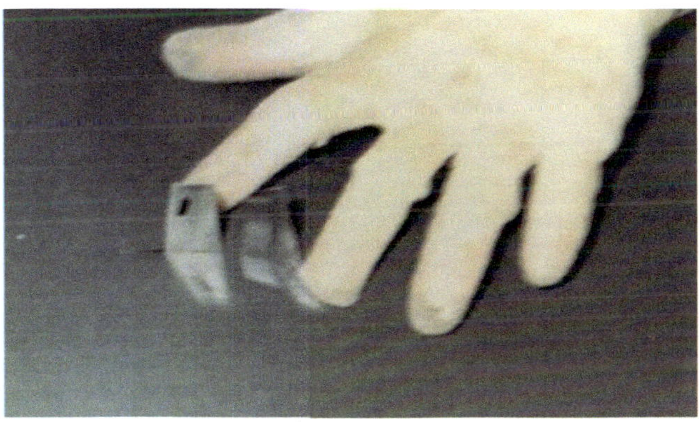

FIGURE 4.18 Thread the connection ribbons.

Technically, in laying up of glass/glass laminates, the layout is quite similar to the glass/foil laminates (Fig. 4.20) but there are some differences. The glass/glass laminate is usually comprised of a front cover of tempered, low-iron white glass, and a tempered float glass (window glass) back cover. Between them, the solar cell matrix is embedded in the EVA.

1. Additional stabilization of the cell matrix.

   In addition to what was mentioned in the glass/foil process on stabilization, follow the procedures outlined below:

   - Connect the cells of the neighboring string together with 3M Magic adhesive tape, which is usually insufficient in the case of glass/glass laminates.
   - Ensure precisely measured spacers of transparent 0.35 mm backing foil are laid between the cells of neighboring strings and stuck with the cells using stripes of 3M Magic.

FIGURE 4.19 Sticking the connection to the backsheet.

FIGURE 4.20 Glass/backsheet laminate composition.

- Mats of glass fleece (30 g/m^2) or fiberglass (50 g/m^2) effectively stabilize the matrices but have a negative effect on the optical transparency of the laminates. Refer to work procedures before using them.

2. Lay on the glass fleece or glass fabrics.
   A layer of glass fleece or glass fabrics on the rear side of the cell matrix effectively reduces the shifting of the solar cells during the laminating process because the soldering points become entangled with glass fibers. However, glass fleece and glass fabrics also reduce the optical transparency of the laminate.
   Hold the cut glass fleece or glass fabrics without stretching it, check it for cleanliness, and lay it with precise measurement onto the cell matrix. Lay the connection tabs behind the glass fleece or glass fabrics.

Then carefully wipe the glass fleece or glass fabrics that is flat. After laying the glass fleece or glass fabrics, lay the first rear EVA foil. The connection ribbons must be freely accessible wherever they must be insulated against other conducting components. Cut the first rear EVA foil with the EVA scissors in such a way that the connection ribbons can be laid behind the first rear EVA foil.

3. Insert the insulators.
   The insulators separate electrically the connection ribbons from the solar cells and other conducting components (Fig. 4.21). Insert the insulators in accordance with the production drawing between the connection ribbons and the first real EVA foil. The outer cross edges of the insulators overlap the outer edges of the cell matrix by at least 2 mm: all the other edges of the insulators overlap the ribbons by at least 5 mm.

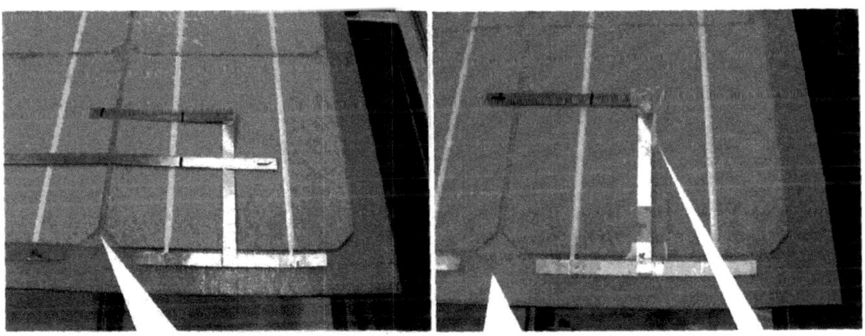

FIGURE 4.21 The configuration of glass fleece and insulator on the laminate.

   Check the position of the insulators and the connection ribbons in accordance with the production drawing, particularly in the area of the exit line, and secure them among each other against shifting during lamination with Scotch 3M Magic.

4. Laying the second rear EVA foil.
   As mentioned before, check the cleanliness of the EVA foil and lay it with the matt side facing upwards, onto the first rear EVA, with coverage exactly the same as the front glass panel and other EVA foils.
   Cut the exit slit for the connection ribbons and cut out the second EVA foil from the edge over the marking of the exit line. Then fold the EVA foil open, thread the ribbon through the slit and then close the EVA foil again. Bend the ribbon at the exit line to 90 degrees. Check the measurements, presence, and position of all laminate components.

5. Lay the backing glass.
   Prior to laying the backing glass on, make sure it is clean. The inner side and the edges must be cleaned with ethyl acetate and cleaning

paper. Before laying it on, check again for being completely clean and dust-free.

As shown in Fig. 4.22, lay the backing glass onto the second rear EVA foil (glass/glass sometimes requires a third layer) with the inside of the glass facing upwards and the tempering stamp as instructed. During the process of laying the glass, thread the connection ribbons through the predrilled slot of the backing glass. Check that the backing glass is overlaid at exactly the same coverage as the front glass.

FIGURE 4.22 Layup of glass/glass laminate.

6. Edge sealing of the laminate.
   Flowing out EVA foil through the glass/glass laminate slit (Fig. 4.23) during the lamination process causes the development of bubbles in the glass/glass laminate. The edge sealing of the laminates avoids this incident. Seal the edge of the glass/glass laminates on all sides with adhesive tape Scotch Blue as Fig. 4.24. Apertures in the adhesive tape

FIGURE 4.23 Glass/glass laminate slit.

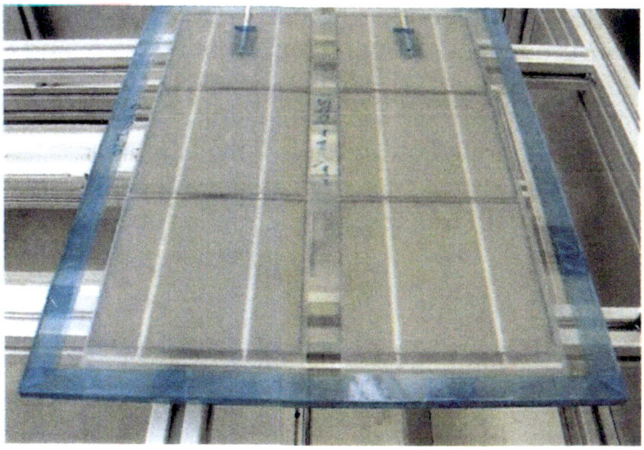

FIGURE 4.24 Edge sealing on glass/glass laminates.

every 30–50 cm all trouble-free outgassing (air volatile matters of EVA) of laminates during the lamination process.

4.5 Lamination of glass/foil laminates

After laying up the laminates by the above procedure or through a layup station (Fig. 4.25), the lamination process can start. The machinery, devices, and tools are as follows:

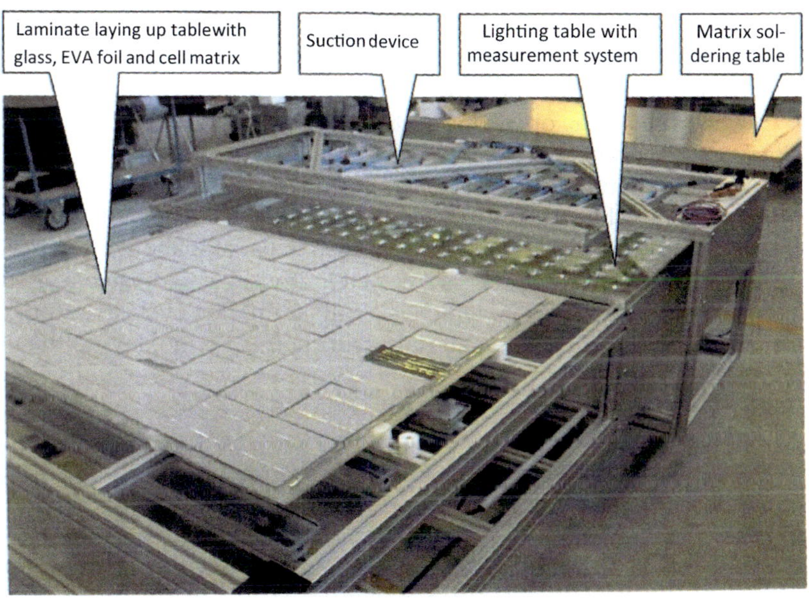

FIGURE 4.25 Layup station.

- A laminating line is required: the fully automatic laminating line consists of an insertion table, a laminator, a cooling press, and an extraction table.
- Working table: a sturdy working table with a level tabletop for quality control and the intermediate storage of the laminates.
- Glass transport wagon and/or glass rack, overhead laminates transport with suctions cups to transport the laminates to and from, before and after lamination.

The materials are as follows:

- Completed layup laminates, separation foil, e.g., PTFE impregnated fiberglass, quality standard-smooth, Type 5311, thickness 0.08 mm. Width and length are subject to the size of the laminator.
- Taicum powder: to increase the separation characteristics of the silicone membrane of the laminator.

The operation steps are as follows:

1. System control and material.
   Prior to starting the operation, check all the systems and materials.
   - Turn on the compressed air supply and cooling water circulation of the laminating line.
   - Prepare the equipment and materials in the order in which they will be used.
   - Wear protective gloves.

   To prevent accidents and defects, the laminating line may be operated by people who are:
   - Trained to operate the machine.
   - Aware of the regulations and safety.
   - Strictly follow the operating instructions for the laminating line.
2. Starting the laminating line.
   Turn on the laminating line according to the instructions. Before the first lamination begins, the machine and its components must reach the appropriate operating temperature which is usually about 30 minutes. In order to enhance the process safety, it is advisable to do an empty run before the first lamination, which is a full laminating cycle in automatic mode without lamination materials. During this period, lay up the first laminate accordingly.
3. Prepare the laminate.
   Though the overhanging transport with suction cups can be used, it is advisable to have two persons to handle the operation to put the laminate onto the insertion table. In case, there is no integrated separation foil system, lay and center a clean separation foil without moving the backing foil of the laminate. The separation foil is one piece and overlaps the backing foil on all sides by a maximum of 10 mm.

4. Begin the lamination process.
   Check the laminating line to make sure that all the preconditions for the successful lamination are met:
   - The correct recipe loaded, corresponding heating plate temperature reached, and control of the laminator is in automatic mode.
   - Extraction table is free and confirmed, laminate is correctly prepared on the insertion table.

     Press the push button "Module Ready" on the insertion table. From now on, the lamination process is fully automatic. Adjustments and corrections of the laid-up laminate are no longer possible. During the lamination process, remove the finished laminates from the extraction table and press the button "Table Free." Lay up the next laminate and place it on the insertion table.
5. Extract the laminate and perform the quality control.
   To ensure the quality standards are met and to reduce the rejection rate, the laminates of the first three laminations of a production series should be intensively checked for quality before the next laid-up laminate is released for the lamination process. In case there are any concerns about the quality issues, the potential problems must be evaluated and resolved before the next laminate is released for the lamination process. As soon as the laminate has been transported out of the cooling press and lied on the extraction table, carefully remove the separation foil, check it for any EVA residue. Then roll it up and lay it aside. If the laminating line is not fitted with a cooling press, wait until the laminate has reached room temperature before applying the mentioned step.

To remove the laminate, two people would be required. Carefully take the laminate from the extraction table either by hand or with the assistance of an overhanging suction cups system without pulling the overlapping backing foil. Lay the laminate on the cushioned worktable with the glass side facing upwards.

Check the laminate according to the predefined quality and standards:

- Check any breakages, damage, soiling, and irregularities of the cells.
- Verify the geometry of the cell matrix, the edges, and the spacing
- The position of the cover, insulators, and connection ribbons.
- The size and the location of the bubbles, foreign bodies, and soiling of the EVA
- Damage and soiling of the glass.

Allow three tries on the lamination process if needed to perfect the process. Once again, check out the electrical function and make sure they verify according to the predefined parameters.

Fig. 4.26 shows the laminating line, which can be used for different types of laminates. For glass/glass laminates, basically, it is quite similar

FIGURE 4.26 Laminating line.

to glass/foil laminate, but there are a couple of extra safety checks of the following items:

- The rear side of the laminate according to the specified quality standard.
- The breaks, damage, soiling, and irregularities of the cells.
- The geometry of the cell matrix, the edges, and the spaces.
- Position of glass fleece or fiberglass.
- Presence and position of the insulators, spacers, connection ribbons, and glass panes in relation to each other.
- Bubbles, foreign bodies, and soiling in the EVA.
- Damage and soiling of the backing glass pane.

By applying the overhanging crane with suction cups and with the help of two workers, carefully lay the laminate onto the cushioned worktable with the glass facing upwards. Check according to the procedures as given in glass/foil laminates. Additional device and materials required are:

- Frame during glass/glass module lamination.
- Fiberglass material for glass/glass module lamination.

Quality control for lamination of "glass/foil" and "glass/glass" laminates includes:

- Strict observance of the operating instruction for the laminating line.
- Visual checking of the laminating line for cleanliness.
- Checking of the correct operating temperature after startup.
- Checking of the lamination program for product specification.
- Supervision of the laminating line for correct operating conditions.
- Visual checking of the prepared laminates on the insertion table: position of the laminate, laminate frame, and separation foil.
- Visual checking of finished laminates (Fig. 4.27) for:
  - Damage and cleanliness of glass panes.

FIGURE 4.27 Quality control tools.

- Damage and cleanliness of cells.
- Geometry of cell matrix, edges, and spaces.
- Position of the glass fleece or fiberglass.
- Position of covers, insulators, spacers, connection ribbons, and glass panes.
- Bubbles, foreign bodies, and soiling in the EVA.

- Electrical function control of the laminates from the initial three laminates as development cycles of the production review.
- Checking of laminates for careful and secure intermediate storage.

4.6 Assembly of modules

After the laminating process (Fig. 4.28) finishes, make sure the laminates are cooled and clean prior to trimming since the EVA has the necessary

FIGURE 4.28 Laminating line operation display.

firmness when it is cold and makes it easy to cut, then conducts the edge trimming of laminates.

The operation procedure is as follows. Recommend two workers to cool the laminate and place it on the suction table with the glass side facing upwards. Turn on the vacuum-suction system and make sure the laminate is firmly placed without any slipping.

- Preliminary cut (Fig. 4.29): using the carpet knife with a new blade, cut into the emanated EVA along the upper edge of the glass.
- Main cut: drag the carpet knife from below, parallel to the edge of the glass, and cut the overhanging backing foil and EVA in a single movement, flush the edge of the glass.
- Fifteen-degree cut (Fig. 4.30): using the carpet knife with a new knife, drag it from below at a 15 degrees angle vertically across the lower edge of the glass and separate the overhanging backing foil with a smooth fine cut.

Note: only use sharp blades.

Optional: apply using a Hot Knife as an alternative.

Caution: hot blades could lead to the stripping of the backsheet.

Finally, clean the glass edges and surface with cleaning paper and ethyl acetate, completely remove all the EVA residue and any other soiling.

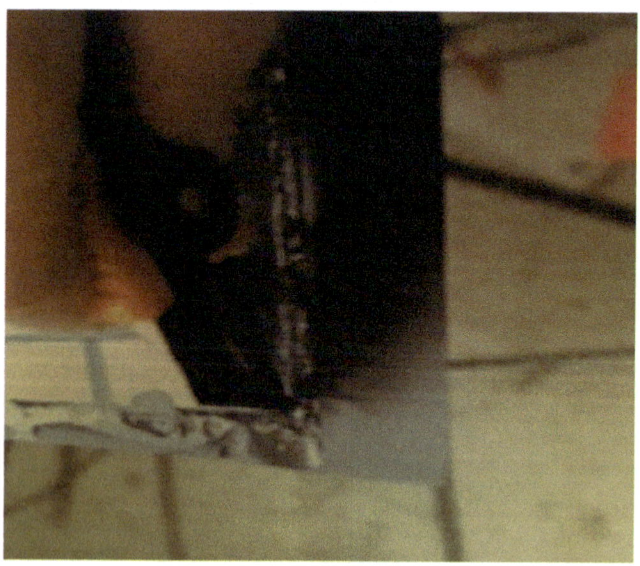

FIGURE 4.29 Preliminary trim of EVA. *EVA*, ethylene-vinyl acetate.

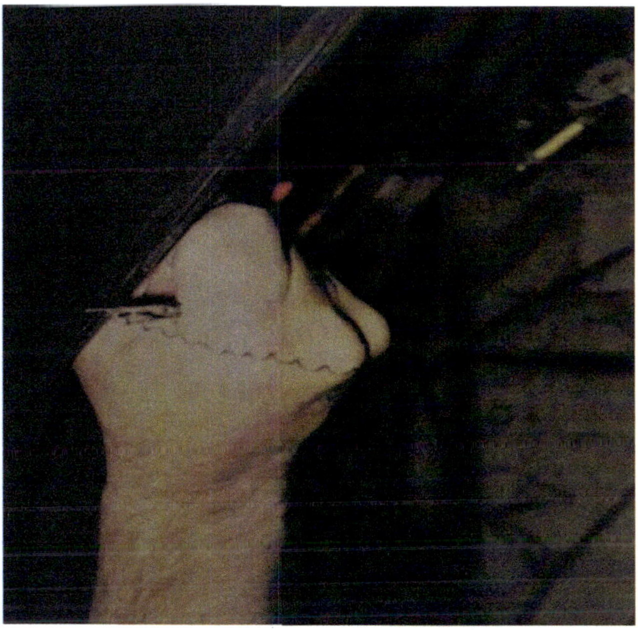

FIGURE 4.30 The 15 degrees cut.

4.7 Junction boxes assembly

The multicontact method is used for junction box assembly. The equipment, devices, tools, and materials are as follows (Figs. 4.31 and 4.32):

- Pressing device: two pieces of hand-press devices are needed that guarantee a pressure of 600 N while the junction box is being glued

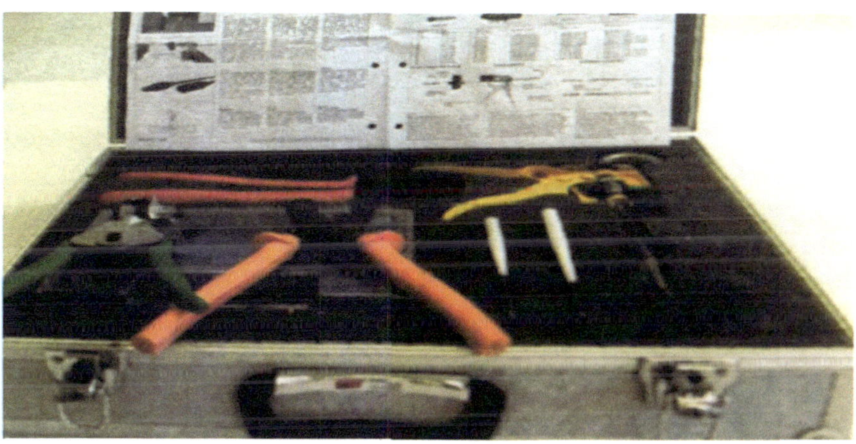

FIGURE 4.31 J-box tool kit.

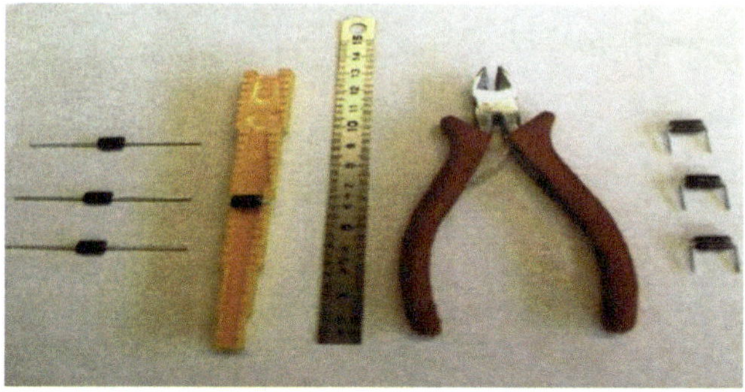

FIGURE 4.32 Diode and tools.

on, without causing any damage to the junction box or the laminate (Fig. 4.33).

- Positioning plate: product-specific plate for precise position of the junction box on laminate.
- Diode plate: plate for cutting and bending of the diode connections.
- Ring spanner: for tightening the screwed cable.
- Junction boxes: junction boxes with all the necessary components such as double-sided adhesive foil, terminals, screwed cable gland, and diodes.
- Connection cables and connectors: double-insulated connection cables and connectors.
- Insulating sleeves: flexible electrical and thermal insulating sleeves of fiberglass substrate coated with silicone rubber for insulation of the connection ribbons.

After checking all equipment, devices, tool, and materials, the operation procedure is as follows:

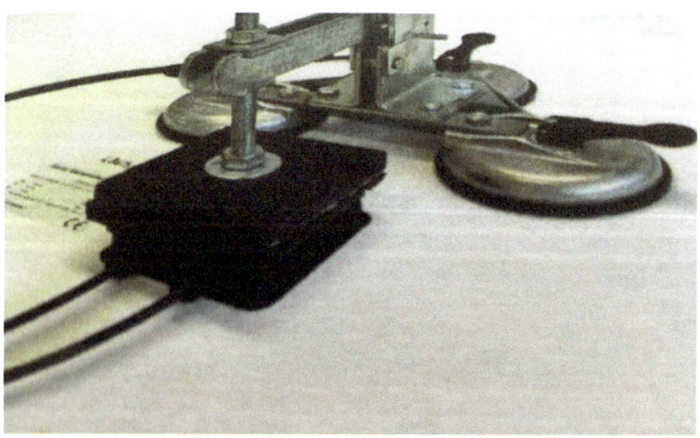

FIGURE 4.33 J-box press.

1. Prepare the connection cable with connectors (Figs. 4.34 and 4.35). The connection cables are preassembled by the manufacturer and can be prepared in self-assembly. In this case, the assembly regulations of the manufacturer in terms of handling and tools must be strictly observed.

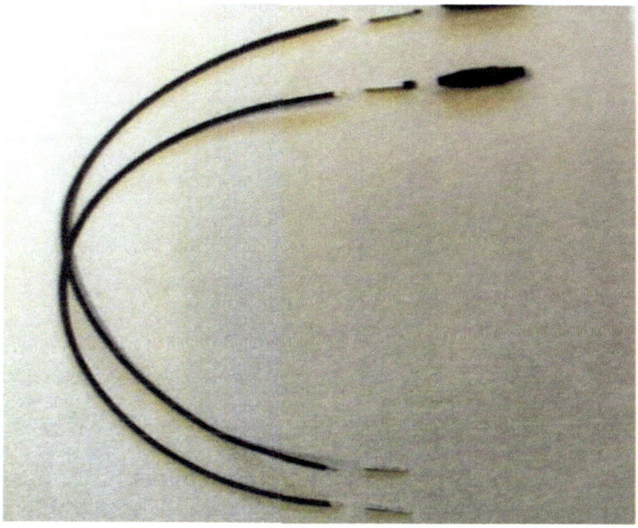

FIGURE 4.34 Cable components.

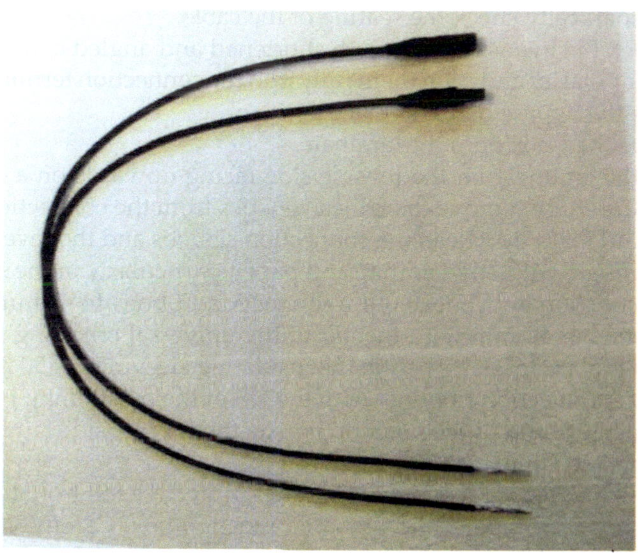

FIGURE 4.35 J-box cables.

2. Prepare the J-box.
Clean the reverse side of the box with cleaning paper and methyl alcohol. Attach the double-sided adhesive foil to the dry reverse side of the J-box without removing the outer protective foil. Add all the prescribed components in accordance with the production drawing:
- Lock in place the necessary number of connection terminals with the correct polarity identification (Fig. 4.36).

FIGURE 4.36 Terminals—connecting to J-box.

- Fit the screwed cable gland with dummy plugs.
- Lead the connection cables through the screwed cable gland and plug them fully into the connection terminals. Check the polarity and tighten the screwed cable gland with the ring spanner (ranch). Mechanically check the seating of the cable.
- Check the bypass diodes with shortened and angled connections for their polarity and plug them fully into the connection terminals. Make sure the bypass diode in place

3. Prepare the J-box onto the laminate.
Place the laminate on the reverse side facing upwards on a cushioned table. Carefully remove the adhesive tapes from the connection ribbons and bend them up. Clean the connection ribbons and the reverse side of the laminate with ethyl acetate and paper, particularly in the sticky part of the junction box. Shorten the connection ribbons to 65 mm and fold their ends of 10 mm with the aid of the universal crimping pliers into a rectangular cross-cut. Guide the insulating sleevings of 50 mm length over the connection ribbons. With the aid of the positioning plate, mark the exact position of the junction box on the reverse side laminate.

4. Stick on and connect the junction box.
As shown in Fig. 4.37, remove the outer protective foil from the double-sided adhesive tape of the prepared junction box. Carefully guide the open junction box over the connection ribbons and stick it to the reverse of the laminate as per the marking, then press it on by hand. Using the screwdriver, open the spring-type terminal and stick the relevant

FIGURE 4.37 J-box installation

connection ribbon fully into the spring-type terminal with angulated flat pliers. Hold the connection ribbon in a fixed position and close the spring-type terminal (Fig. 4.38). Mechanically check the seating of the connection ribbon. As soon as all the connection ribbons have been connected, check the cover seal, close the J-box cover and click into its final position. Position the pressing device and press the J-box on for 2 minutes at 600 N.

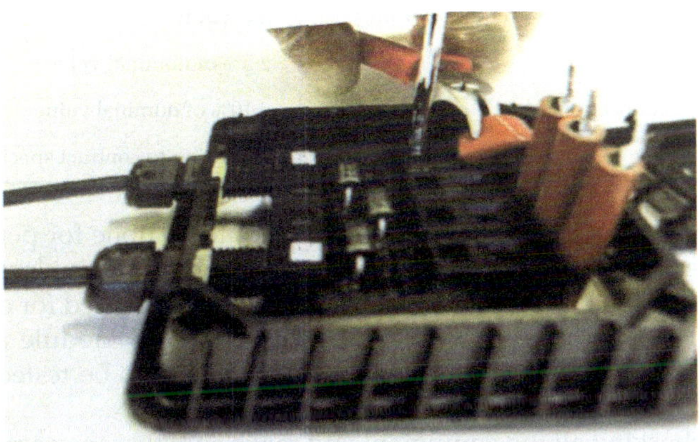

FIGURE 4.38 J-box installation procedure

4.8 Quality control standards

The modules are examined in accordance with the following specifications of cosmetic tolerance for glass/foil and glass/glass PV modules (Table 4.1) unless otherwise specified by the customer.

TABLE 4.1 The specification of building integrated photovoltaic modules for quality control after the laminating process.

Items	Tolerance
Measurements of modules	up to 500 mm. ±1.0 mm
	up to 1000 mm ±1.5 mm
	up to 1500 mm ±2.0 mm
Scratches on the glass pane	two pieces with $L = 20\ mm/m^2$
Foreign particles in the EVA	$< 20\ mm^2/m^2$
Glass edge chips	$< 50\ mm^2/m^2$
Cell to cell spacing standard module: 3 mm	±2.0 mm (no cell touching)
String to string spacing standard module: 3 mm	±2.0 mm (no cell touching)
Custom modules: subject to the design specified	
Connection ribbon to cell spacing	±2.0 mm/m
Gas bubbles in the EVA^2/m	a) any summation. $< 50\ mm^2/m$
	b) no bubbles from edge 5 mm inward
Color of cells and backsheet	Subject to cell manufacturers and module specified design
Position of tabbing ribbons on the cell front	±0.5 mm
Position of Junction Box	±5% of nominal value
Length of the connection cable	±10% of nominal value
Electrical values	according to contract specifications

Power measurement and final testing should be done for performance assurance. A reference PV module measured by independent institutes (for example, National Renewable Energy Laboratories) is used for calibration of module tester at regular intervals. The reference module should be identical or at least similar to the module which is to be tested in terms of cell-type, layup, connection, and size.

After checking all the equipment and make sure that they are clean and operational. The testing steps are as follows:

1. Prepare the module tester.
   Abide by the operation instructions, switch on the tester, the regulation transformer, the power measurement unit, and the computer. Run the computer up to the measurement program and enter all the product specification data. Check the glass surface of the light table for cleanliness and set the module carrier rails to the size of the reference

module. Check the reference modules for cleanliness and lay them centrally on the rails of the light table. Connect the reference module to the correct measurement cables (pay attention to the polarity). Perform the reference measurement, save it. and print it if necessary. Remove the module afterward and store it in a safe place. Reference module: at times when the module tester has been idle for a while, it is advisable to recalibrate the tester using the reference again to verify again.

2. Power Measurement.
   Measurement conditions: 25°C, 1000 W/m^2, AM 1.5.
   Set the module carrier rails of the light table to the length of the module that is being tested. On the final measurement, two persons should hold the released module carefully on the carrier rails and center it there. Connect the module to the power measurement cables (paying attention to the polarity!) and perform the power measurement. Check the data of the power measurement and save it under the correct series number. If a second measurement is necessary, remover the module and let it cool down for at least 30 minutes (module's temperature 25°C) since the heating of the cells can negatively influence the result of the measurement.
   The successful power measurement guarantees the following product qualities:
   - Correct connection of the solar cell matrix, connection ribbons, and connection cables.
   - Correct polarity of the cell matrix, bypass diodes, connection cables, and plugs.

   The successful power measurement guarantees the absence of a bypass diode. Stick on the type label with a relevant series number at the position foreseen for it. Remove the measurement cables from the module and then lay the module on to the cushioned worktable for the final quality control.
3. Final quality inspection.
   The final quality inspection is the summary of all quality controls from the previous fields of work and power measurement. Execute the final quality control in accordance with the specified standards. Fix the connection cable to the module with adhesive tape.

4.9 Determining the gel content of the encapsulant (EVA)

4.9.1 Purpose

The gel content (insoluble fraction) produced in EVA plastics by cross-linking can be determined by extracting with toluene. This procedure describes the method used to determine the gel content of the EVA.

4.9.2 Safety precautions

1. Toluene is a flammable substance. Do not use near-open flames.
2. Use this chemical in a fume hood.
3. Wear safety glasses.
4. Wear chemical-resistant gloves.
5. Read the Material Safety Data Sheet for toluene before using.

4.9.3 Principle

The purpose of the method is to determine the gel content of the EVA. Gel content provides a means to both controlling the process and rating finished product quality. Gel content levels of 70% or higher are recommended for the longevity of the module. It is a critical part of the quality control of the module.

Encapsulant gel content testing according to the ASTM D2765.

For more information, check with ASTM or the Encapsulant, or the EVA manufacturer.

4.10 BIPVT products

4.10.1 Sunslate

The Sunslate is the first BIPV solar roof tile in the world. In the past 20 or so years since it was introduced, the product has achieved significant success in terms of performance. Though just as for most new products in the market, there were some problems initially, later after it was perfected, the Sunslate has carried itself very well throughout all these years in terms of quality and durability through extreme weather conditions in places such as Montana, Aspen, Colorado, Lake Tahoe, California, Wisconsin, Toronto, Canada, and Joes Beach, New York, or in areas with extreme hurricanes like Florida and Jamaica.

Apart from having high-quality components used in this product, the most critical part of it is the junction box. It took numerous years for the company to develop the junction box (Fig. 4.39) and the connection system. There were quite a few iterations before they managed to come up with an appropriate design that could be able to satisfy the weather conditions. Water is very tenacious and when the temperature drops below freezing, there could be problems. Thus, a product with high-quality design components is vital.

The image shown in Fig. 4.40 is one of the iterations on the development of the J-box. Wago has been well known for its turn and lock connector which has a wide range of different applications including submarines. Though the contacts in this design have been reasonably successful, they

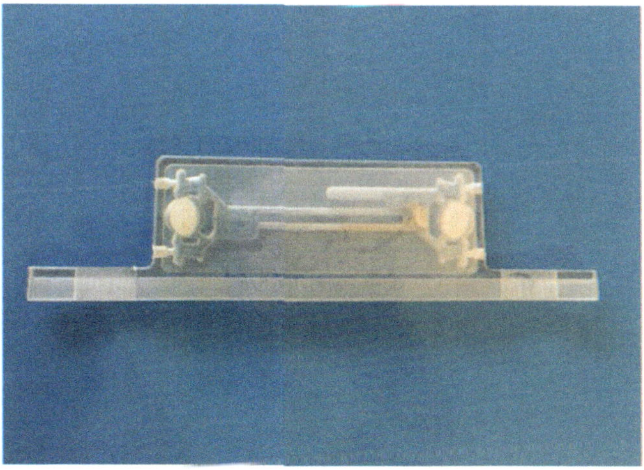

FIGURE 4.39 Initial Sunslate J-box development

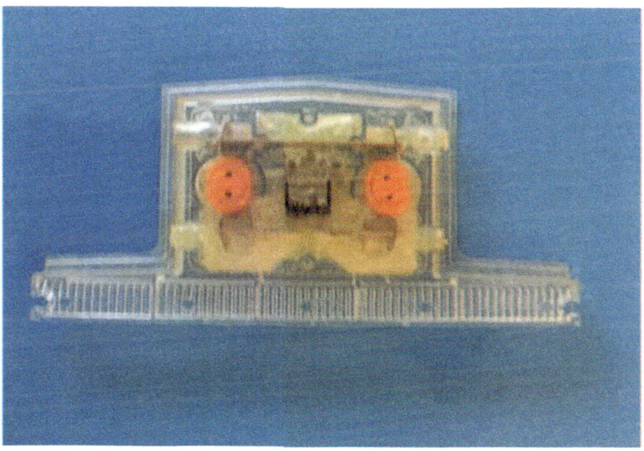

FIGURE 4.40 Sunslate J-box with Wago turn and lock connector with a by-pass diode

still have some minor problems. After making some small changes, the connector was perfected. Since then, there was not a single problem with the Sunslate connection system. In addition, there has not been a single roof leakage (Fig. 4.41).

Fig. 4.42 shows the final version of the Sunslate J-box together with the shield that is used to cover the front of the J-box. The shield is latched on the J-box and seal off by applying RTV silicone which waterproofs the whole system. In case, should there be any problems with the Sunslate, the shield could be taken off by a screwdriver so that one can check the product.

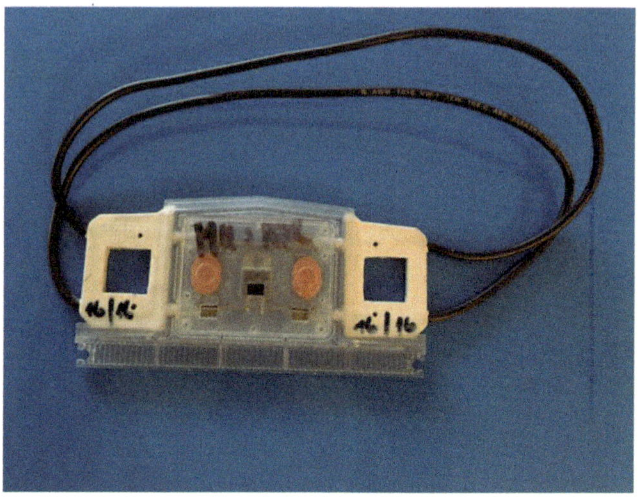

FIGURE 4.41 Sunslate J-box development.

FIGURE 4.42 The final version of the Sunslate J-box together with the shield that covers the front.

The basic principle of assembling the Sunslate is quite similar to the regular module. There are six solar cells in series. Due to the tight space situation, it requires extra tabbing and stringing in order to be accommodated into it. Aside from such an operation, the remaining parts are just the same as the regular glass/Tedlar module. Again, the quality of material and process is vital. The image shown in Fig. 4.43 is the Sunslate layout.

FIGURE 4.43 Sunslate cell layout.

The tabbing goes right to the back of the J-box and is soldered onto the metal contact. After the contacts have been connected, the encapsulant will be covered, and following it would be the backsheet before going to lamination. The power will be tested before and after lamination. Provided that everything goes well after all the testing, the Sunslate modules will be put on a secure container and ready to be glued to the substrate which is the Swiss Pearl Eternit roof tile as shown in Figs. 4.44 and 4.45[1].

As the solar cell standard transformed from 125 mm to 156 mm, most manufacturers discontinued the 125 mm or the 5-inch cell (Fig. 4.46). Respectively, module and BIPV manufacturers had to follow suit in order to be competitive. Especially for BIPV manufacturers, it was a costly shift because they had to redesign the product and recertify. The problem is that the sale volume of the product is relatively small as compared to the regular standard. In manufacturing, we always look at the economy of scale and that is how to make a product cost-competitive. BIPV manufacturers are niche players in the solar energy market until more people recognize and appreciate their product. As more people get to learn about all the advantages of active building materials, although it could be a bit costly in the beginning since it requires paying attention to more details in the manufacturing process, and this per se might result in dissuading the manufacturers, however, in the long run, it could worth every penny and

[1] Frank C. Pao, US Patent No. US 10,958,212 B2, Date of Patent: 3/23/2021, Electrical connection support assembly and method of use.

FIGURE 4.44 The Sunslate roofing as an active building material.

FIGURE 4.45 New Sunslate with the latest connection system and back contact solar cells.

FIGURE 4.46 Sunslate 5, Sunslate 6, and A-100 solar skylight.

time they spent after a successful installation with the following appealing features.

1. Aesthetic and integrity of the building.
2. A quality product could withstand extreme weather conditions.
3. Make sure the product is UL fire A certified.
4. Product longevity is vital: (a) does not require frequent replacement, (b) replacement end up in the landfill which affects sustainability.
5. Standard module on the roof especially slope ones, roofers would not guarantee the roof due to roof penetration.
6. Most cheap import modules, though they are certified, are likely to have performance issues and it would be hard or impossible for installers to go after the manufacturers.

The Sunslate module has passed the Underwriters Laboratories (UL) and International Electrotechnical Commission (IEC) certification. UL is required mainly in the United States while IEC is required worldwide. Recently, the two organizations have combined the testing operations into one through some of the tests are done separately and some tests do not require repetition which saves time and costs. There are numerous tests that are required but the most critical ones are:

1. Humidity freeze test: to test the module whether it can withstand extreme weather conditions of −40–85°C with relative humidity at 85%.
2. Thermal cycling test: to test the module on longevity.
3. Fire test: on the UL 790, for fire test on roofing products there are three classifications: A, B, and C.
   (a) On Class A roof, the material can withstand severe fire conditions exposure effectively.
   (b) On Class B roof, the material can withstand moderate fire condition exposure effectively.
   (c) On Class C roof, the material can withstand light fire condition exposure effectively.

In California, all roofs are required to be Class A but on standard modules are normally Class C. However, if the roof to which the modules are mounted is Class A, it would be acceptable. The Sunslate in this case stands out because it is the roof and it has Class A certification.

The UL and IEC certifications do not cover the wind load situation. To certify the wind load, there are organizations in Florida that provide tests on Miami Dade County wind load tests. An area of caution to watch for is that in case the hurricane wind speed goes beyond the certification, put some 3M foam glue at the bottom edge of the Sunslate.

Prior to submitting the modules for certification, the bill of materials of the module must also be included since it contains all the components of the module. All the components together with their specifications have to be approved by the test lab.

In addition, prior to submitting the samples, it is quite important to have the hipot test or dielectric voltage-withstand test.

1. Using the appropriate cable as shown in Fig. 4.47, short the positive and negative terminals of the module.
2. If the module is frameless, wrap the edge of the module in foil. Apply the positive lead of the hipot tester to the shorted terminals, and the negative lead to the edge of the module.
3. A voltage of 2200 V direct current (DC)* will be applied with the hipot tester between the shorted terminals of the module and the edge of the module.
   WARNING: When the high voltages are being applied, the voltage must be applied gradually and smoothly. The voltage is to be increased from

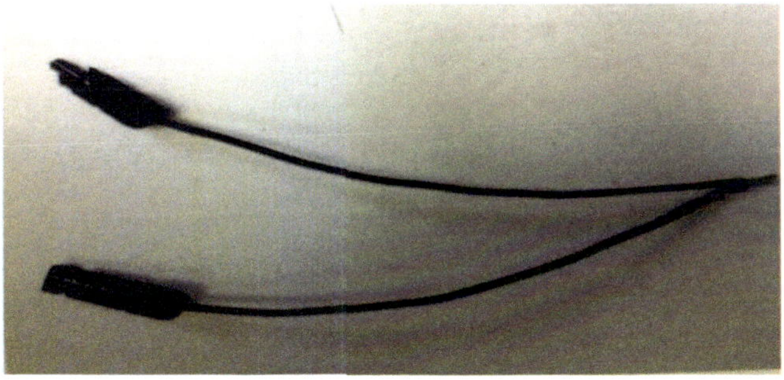

FIGURE 4.47 Hipot testing cable.

0 to 2200 V DC with 5 seconds and then maintained at 2200V DC for 1 minute.

4. Measure the leakage current. The leakage current must not exceed 50 μA.
5. Repeat the measurement for each edge of the module.
6. Reverse the polarity of the hipot tester and repeat steps 1–3.

*In general, the applied voltage is twice the maximum system voltage plus 1000 V DC. However, if the maximum voltage rating is less than 30 V, only 500 V DC will be applied.

4.10.2 TallSlate

TallSlate is another BIPV module applicable to different type of solar cells (Figs. 4.48 and 4.49). It is different from Sunslate in the way that it

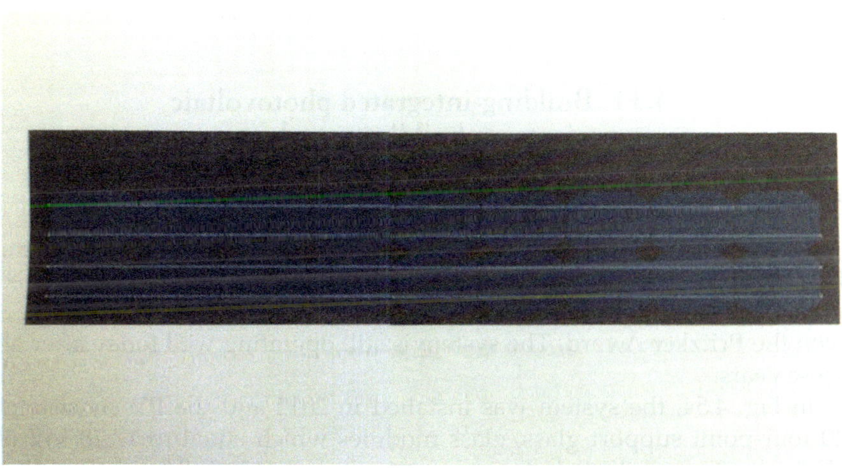

FIGURE 4.48 TallSlate module.

FIGURE 4.49 TallSlate back contact cell module.

does not have a substrate as the Sunslate. Aesthetically, it is pleasing as well and relatively simple to install. The bill of materials for all the TallSlate products is shown Fig. 4.50. Currently, there are three types of TallSlates: the TallSlate Grandee, the TallSlate Economy, and the regular TallSlate for different sizes and customer needs.

There are times that some of these tests are demanding, therefore specimen preparation is of utmost importance including doing visual appearance that there are no defects or bubbles are in a very bare minimum, preferably flawless. Upon successful completion of the certification, the product is ready to be promoted to potential customers. Specifications of the product should have been prepared to illustrate the appearance and performance capabilities of the product in Figs. 4.51 and 4.52. In addition, an application note, installation manual, and a press release would be helpful to the customers as well as the success of the product.

4.11 Building-integrated photovoltaic projects on building envelopes

4.11.1 Building-integrated photovoltaic glass projects

This is a 100 kW BIPV system in Fig. 4.53 built in 2003 when it was the largest BIPV project in the country. Installation of the solar system started in late 2003 and was completed in early 2004. In 2005, the whole project won the Pritzker Award. The system is still operating well today after all these years.

In Fig. 4.54, the system was installed in 2012 and the PV consists of 40 four-point support glass/glass modules which equal to 17.28 kW of electricity. The whole project won numerous awards including outstanding

MATERIAL COMPARISON FOR NEW TALLSLATE

Version 1
02/24/15

ITEM	Grandee (Report #3125558CRT-001c)	Economy TallSlate (To be added)	Old TallSlate (Report #3125558CRT-001c)	New TallSlate (To be tested full UL1703, IEC61730, IEC 61215)
Size	51-13/16" x 38-3/8"	51-13/16"x15-3/4"	47-1/4"x15-3/4"	47-1/4"x15-3/4"
Number/type of cells	(40) 156mm cells	(8) 156mm cells	(18) 125mm cells	(18) 125mm cells
Cells	Qcell poly Q6LTT Qcell mono Q6LM LOF mono 156mm Gintech mono G156S	Qcell poly Q6LTT Qcell mono Q6LM LOF mono 156mm Gintech mono G156S	Qcell mono Q5M150 LOF mono 125mm Gintech mono G125S	Sunpower 125mm mono back contact cells
Glass (Superstrate)	PPG 1/4" Starphire tempered (Starphire is low iron glass. Can I change this to 1/4" low iron tempered, various manufaturers?)	PPG 1/4" Starphire tempered (Starphire is low iron glass. Can I change this to 1/4" low iron tempered, various manufaturers?)	PPG 1/4" Starphire tempered (Starphire is low iron glass. Can I change this to 1/4" low iron tempered, various manufaturers?)	1/4" Low Iron Tempered, various manufacturers
Backsheet (Substrate)	Madico TPE HD	Madico TPE HD	Madico TPE HD	Madico Protekt HD
Encapsulant	STR Photocap 15295P/UF	STR Photocap 15295P/UF	STR Photocap 15295P/UF	STR Photocap 15580P/UF
Jbox	Tyco 1740220-1, 1740220-5 (Add 1740220-4, see note two columns over)	Tyco 1740220-1, 1740220-5 (Add 1740220-4, see note two columns over)	Tyco 1740220-4 (Add 1740220-5, see note one column over)	Tyco 1740220-5, 1740220-4 (As discussed with Andrew, I will send a loose sample of the 1740220-4 to evaluate it is equivalent to the 1740220-5)
Diode	*Dioetec SB1540* Tyco SL1110, SL1110B	*Dioetec SB1540* Tyco SL1110, SL1110B	*Dioetec F1200D* International Rectifier 90SQ045	Dioetec F1200D
Cable	USE-2, 12 AWG, Sunlight Resistant, 90°C	USE-2, 12 AWG, Sunlight Resistant, 90°C	USE-2, 12 AWG, Sunlight Resistant, 90°C	Huber+Suhner Solar Cable 12 AWG (4.0 mm^2) P/N: 12583780
Connectors	Tyco 6-1394462-3, 6-1394462-4, 7-1394461-3, 7-1394461-4 and 7-1394461-5 Tyco 4-1394462-8, 4-1394462-9, 7-1394461-0, 7-1394461-1 and 7-1394461-2	Tyco 6-1394462-3, 6-1394462-4, 7-1394461-3, 7-1394461-4 and 7-1394461-5 Tyco 4-1394462-8, 4-1394462-9, 7-1394461-0, 7-1394461-1 and 7-1394461-2	Tyco 6-1394462-3, 6-1394462-4, 7-1394461-3, 7-1394461-4 and 7-1394461-5 Tyco 4-1394462-8, 4-1394462-9, 7-1394461-0, 7-1394461-1 and 7-1394461-2	Multi-Contact MC4 Connectors, 12 AWG Male P/N: PV-KST4/6I-UR Female P/N: PV-KBT4/6I-UR
Cell Interconnects	Ulbrich 7746-9464 2.3mm x 0.15mm Copper dimension 60% Sn / 40% Pb 10-15 Micron solder thickness	Ulbrich 7746-9464 2.3mm x 0.15mm Copper dimension 60% Sn / 40% Pb 10-15 Micron solder thickness	Ulbrich 7746-9464 2.3mm x 0.15mm Cooper dimension 60% Sn / 40% Pb 10-15 Micron solder thickness	Luvata 54180 4.0mm x 0.1mm Copper dimension 62% Sn / 36% Pb / 2% Ag 15 Micron solder thickness
String Interconnects	Ulbrich 7746-9574 8.0mm x 0.203mm Copper dimension 100% Sn 0.5-2.5 Micron solder thickness	Ulbrich 7746-9574 8.0mm x 0.203mm Copper dimension 100% Sn 0.5-2.5 Micron solder thickness	Ulbrich 7746-9574 8.0mm x 0.203mm Copper dimension 100% Sn 0.5-2.5 Micron solder thickness	Ulbrich 7746-9574 8.0mm x 0.203mm Copper dimension 100% Sn 0.5-2.5 Micron solder thickness
Bus Sleeving	Daburn D200-2	Daburn D200-2	Daburn D200-2	Daburn D200-2
Jbox Glue	Shin-Etsu KE-45-T	Shin-Etsu KE-45-T	Shin-Etsu KE-45-T	Dow Corning PV-804
Solder	Kester #14-6337-0062 SN63PB37 SLD	Kester #14-6337-0062 SN63PB37 SLD	Kester #14-6337-0062 SN63PB37 SLD	Kester #14-6337-0062 SN63PB37 SLD
Label	Tech Tag and Label Thermlfilm Select 21940 (listed as Rtherm in UL file)	Tech Tag and Label Thermlfilm Select 21940 (listed as Rtherm in UL file)	Tech Tag and Label Thermlfilm Select 21940 (listed as Rtherm in UL file)	Tech Tag and Label Thermlfilm Select 21940 (listed as Rtherm in UL file)

FIGURE 4.50 Bill of materials of the TallSlate products which includes the manufacturer or supplier's names and part numbers.

TALLSLATE™ PHOTOVOLTAIC ROOFING SLATE 01

The beauty of a genuine slate roof integrated with our glass photovoltaic modules. The TallSlate is designed to meet your aesthetic and economic needs.

We have partnered with Greenstone Slate Company to provide a product that is both attractive and easy to install. TallSlates are mounted using the Nu-lok roofing system. Nu-lok offers an alternative to traditional slate roof installation that still creates a genuine Vermont slate roof but uses up to 40% less slate, reducing material cost and roof weight. For more information, please visit greenstoneslate.com.

TALLSLATE™ MATERIAL COMPOSITION 02

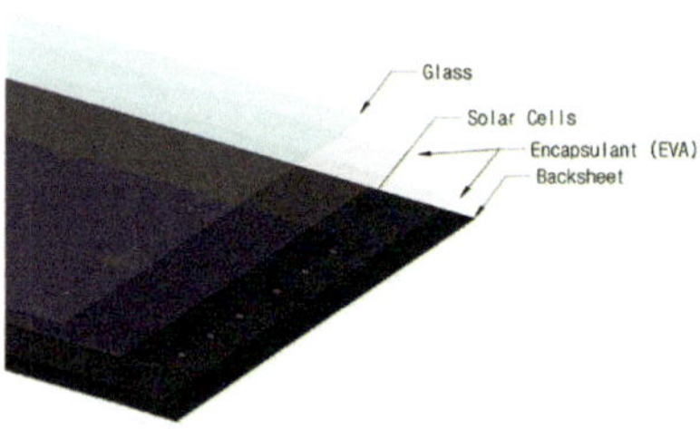

The TallSlate is constructed of solar cells laminated between tempered glass and photovoltaic backsheet.

Our next generation TallSlate will feature back contact mono-crystalline silicon cells for a uniform roof appearance.

atlantisenergy.com | REV. 3.0 06-26-14

FIGURE 4.51 TallSlate back contact cell specification p1.

ATLANTIS ENERGY SYSTEMS

TALLSLATE™ SPECIFICATIONS 03

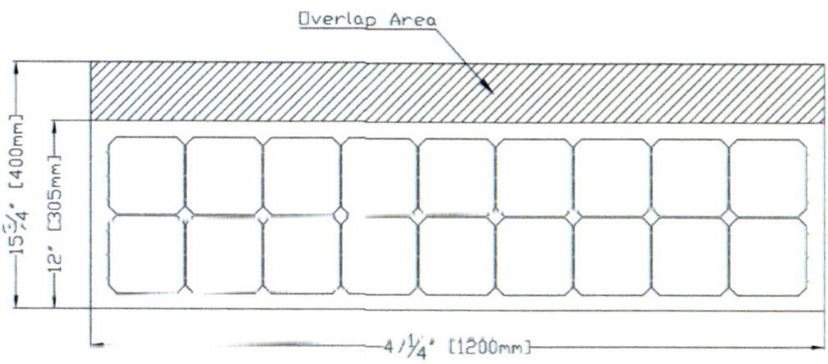

ELECTRICAL SPECIFICATIONS	
MODEL	TS125SM
POWER	54 Watts
I_{SC}	6.10 ADC
V_{OC}	12.4 VDC
I_{MPP}	5.50 ADC
V_{MPP}	9.8 VDC
SERIES FUSE RATING	10 Amp
RATED SYSTEM VOLTAGE	600 VDC

All electrical specifications are at Standard Test Conditions (STC): 1000 W/m^2 irradiance, 25°C temperature, AM 1.5

MECHANICAL SPECIFICATIONS	
DIMENSIONS	15.75" x 47.25" (400 mm x 1200 mm)
EXPOSED DIMENSIONS	12" x 47.25" (305 mm x 1200mm)
EXPOSED AREA	3.94 SQFT (0.366 SQM)
WEIGHT	17 lbs. (7.71 kg)
GLASS	0.25" (6mm) low iron tempered
ENCAPSULANT	EVA
CELLS	125 mm back contact mono-crystalline silicon
BACKSHEET	Madico Protekt HD / TPE HD

This product will undergo testing according to UL Standard 1703 and will be CEC listed.

atlantisenergy.com | REV. 3.0 06-26-14

FIGURE 4.52 Specification page 2.

FIGURE 4.53 Caltrans District 7 Headquarter in Los Angeles.

FIGURE 4.54 San Diego County Operation Center Dining Area.

FIGURE 4.55 BIPV panels are used for both electricity production and shades.

government building of the year, LEED Platinum Award on the Dining Area, and AIA California Council Merit Award.

Figs. 4.55 and 4.56 show a BIPV glass project at Utah State University, Logan, Utah. The solar part consists of two systems. Located on the rooftop are the regular standard PV modules while the BIPV modules mounted on the south facing side of the building act as shades. The system consists of 108 translucent modules at 310 W each, making in total about 33.5 kW. It was the first type of such system as well as the largest system installed in Cache Valley at that time. It was supported as part of the American Recovery and Reinvestment Act. Additionally, it was expected to impress the students that the University is introducing cutting-edge energy-saving technologies. The BIPV solar panels not only produce electrical energy but also act as shades to the building which helps improve the efficiency of the building in the summer months even though the building is already quite efficient. As a whole, it is said to be the state of art of building technologies and a division of the US Department of Agriculture.

FIGURE 4.56 Another view of the BIPV panels at Utah State University.

In addition to the advanced solar system, the building design optimizes energy and water usage. This optimized efficiency and reduction of water plus the latest building materials that were used resulted in the system being awarded the Gold LEED Certification.

4.11.2 Sunslate installations

Back in 1999, Atlantis was asked to install the barn with Sunslate at St. Ignacio, Montana, that was an interesting surprise. The Sunslates barn roof supplies 12.7 kilowatts of building-integrated solar for a two-home complex as shown in Fig. 4.57. It was originally an off-grid complex. The roof and its 48 Volt battery backup that is made up of 2 Volt marine batteries drives a 450 foot deep well to supply two 1500 gallon polypropylene water storage tanks as well as energies for the two homes which are around 150 feet away. The roof is a 55 degrees slope. The head of the contractor praised the Sunslate that it is the best component in the whole building throughout the years.

Another project is a 20.0 kW Sunslate system that was installed in 2002 in Aspen Colorado (Fig. 4.58). Initially, the contractor tried to convince the

FIGURE 4.57 Barn in Montana was built in 1999.

FIGURE 4.58 Sunslate Barn in Aspen, Colorado.

owner to use a standard module. However, not only did he refused, but, he took off the copper roof and replaced it with the Sunslate.

Domestically, the Sunslate projects have won numerous awards throughout the years. From the solar conference in the Moscone Center, San Francisco, as the "Coolest Product" to LA Times and Readers' Digest.

Sunetic, the company that was praised for the marketing of photovoltaic solar panels won the "Stars et Metiers 2013 department" prize awarded by the Banque Populaire Atlantique (Fig. 4.59) in partnership with Chamber of Trades and Crafts in the category "Global Innovation Strategy" to Thierry Auguste who, with his team provides solutions of renewable energy solutions and solar projects.

As it is obvious from the photos (Figs. 4.60–4.68), the houses that are built are quite solid as compared to most houses that have been built in the United States. We should really take a good look at how and why we have not been doing the construction the right way.

As one can imagine, the roof is a relatively complicated part of the house, but the Sunslate managed to be accommodated into it. The installation was

FIGURE 4.59 Award winning installation in at Pays de Loire, France.

FIGURE 4.60 Schneider House installed Sunslate as façade in Bern, Switzerland.

FIGURE 4.61 Sunslate installation as facades in an apartment building in Bern, Switzerland.

FIGURE 4.62 Sunslate installation on Swiss Chalet.

FIGURE 4.63 Sunslate installation on a Swiss Residence.

FIGURE 4.64 Sunslate installation in Walkringen, Switzerland.

FIGURE 4.65 Sunslate installation in Roehgi, Switzerland.

FIGURE 4.66 Installing Sunslate in Jamaica.

FIGURE 4.67 Installing the Sunslate.

FIGURE 4.68 US Homes Sunslate installation in Sacramento, California in late 1999–early 2000s in the subdivision gate community. The system size ranges from 2 to 4 kW of Sunslate.

done more than 10 years ago, but it has survived numerous hurricanes and to this day, there has not been a complaint.

4.11.3 TallSlate installations

These panels are designed for ultra-high winds and are Dade Country certified as well as designed for natural slates surround. They can either be grid-connected or their products can be stored on-site with a battery solution of your choice.

FIGURE 4.69 TallSlate panels on the roof of a house in Nova Scotia, Canada.

The house in Fig. 4.69 is located on the Nova Scotia coast of Canada. It is relatively simple to install and in addition, one could pick matching slates to make it more aesthetically pleasing.

TallSlate is relatively easy to install and can be seen as very suitable on strait slope roofs. Figs. 4.70–4.74 shows the installation projects in the

FIGURE 4.70 TallSlate installation in Jefferson, New York.

FIGURE 4.71 TallSlate back contact cells installation.

FIGURE 4.72 TallSlate back contact cell installation.

FIGURE 4.73 TallSlate installation in Jefferson NY.

FIGURE 4.74 TallSlate installation in Colorado.

united states. Price-wise, it is less costly than the Sunslate. Although the TallSlate is UL 790 Class C fire certificated, it can achieve class A by adding an asphalt layer beneath the slate. Both solar roof slates do not require penetration on the roof which can be harmful to the integrity that is very critical in preventing leakage.

Sunslate is more agile as compared with TallSlate. It can be applicable to a large variety of shapes of sloped roofs including hips and valleys. The product has been around for more than 20 years and has great success in all the installations. All the previous projects are still standing and without a single leakage after all these years. Though they are a bit pricey, it is well worth it due to the durability, being able to withstand extreme weather conditions, in addition to aesthetically being pleasing. Most customers managed to have already got their return on the investment.

4.12 Case study: Natomas parking lot

This is a building-integrated cover car park in a modern office building located in Sacramento, California. The total energy production for this car park is about 200 kW which is quite sizeable considering the size of the building that is around 300,000 square feet. Due to the hot climate zone of this region, there is a high demand for air conditioning usage. However, the solar energy generated by this system can only cover all the lighting of the whole building and the energy usage in the lobby. This project was completed in 2009. Looking at the current solar energy and storage technologies, one could suggest a system that can handle the total energy requirement of the whole building using a micro-grid system with a hybrid storage generator. The module for this project was a custom module as requested by the designer and was manufactured by Atlantis Energy Systems Inc.

The modules are partially translucent so that they could provide some sunlight onto the parking area. The majority of the parking cover used the Sunpower standard modules (Fig. 4.75).

In Natomas, there are eight car park covers of different sizes. Tables 4.2–4.5 list the bill of materials for the modules. Here is the breakdown of the modules used in this project:

206 Atlantis translucent modules (120 W) = 24,720 W
564 Sunpower modules (305 W) = 172,020 W
Total = 196,740 W $\cong$ 200 kW

Fig. 4.76 shows the structural design of supporting frame and the project layout is provided in Fig. 4.77. Figs. 4.78–4.80 show different views of

FIGURE 4.75 Natomas custom-made module.

TABLE 4.2 Natomas modules bill.

Bill of Materials		Premier Power/Natomas							
Series Item#	Identifier	Quantity for assembly	Vendor contact info.	Model or part number	Certification file attached	Width Dim (Inches)	Length Dim (Inches)	Area (Inches^2)	Additional data
100	Glass				↓				
101	FRONT	1	SYRACUSE GLASS ONE GENERAL MOTORS DRIVE SYRACUSE, NY 13206 1-800-962-3211	as per customer ID#		41.18	61.38	2558.32	1 FRONT LOW IRON TEMPERED GLASS
102	BACK	1	SYRACUSE GLASS ONE GENERAL MOTORS DRIVE SYRACUSE, NY 13206 1-800-962-3211	as per customer ID#		41.18	61.38	2558.32	1 BACK LOW IRON PARTIALLY TEMPERED
103	FRONT	1	PILKINGTON 811 MADISON AVE. TOLEDO, OH 43697 1-419-247-4926	as per customer ID#		41.18	61.38	2558.32	1 FRONT LOW IRON TEMPERED GLASS
104	BACK	1	PILKINGTON 811 MADISON AVE. TOLEDO, OH 43697 1-419-247-4926	as per customer ID#		41.18	61.38	2558.32	1 BACK LOW IRON PARTIALLY TEMPERED

(continued on next page)

TABLE 4.2 Natomas modules bill—cont'd

Bill of Materials		Premier Power/Natomas							
Series Item#	Identifier	Quantity for assembly	Vendor contact info.	Model or part number	Certification file attached	Width Dim (Inches)	Length Dim (Inches)	Area (Inches^2)	Additional data
105	FRONT	1	QUIGDAO YUESHUI CHUAN INT TRADE COMPANY KAIXUAN MOUNTAIN HONGKONG EAST ROAD QUINDAO CHINA	as per customer ID#		41.18	61.38	2558.32	1 FRONT LOW IRON TEMPERED GLASS
106	BACK	1	QUIGDAO YUESHUI CHUAN INT TRADE COMPANY KAIXUAN MOUNTAIN HONGKONG EAST ROAD QUINDAO CHINA	as per customer ID#		41.18	61.38	2558.32	1 BACK LOW IRON PARTIALLY TEMPERED
200	Encapsulant				↓				
210		2	SPECIALIZED TECHNOLOGY RESOURCES 10 WATER STREET ENFIELD, CT 06082 1-860-749-8371	15295/UF		41.18	61.38	5116.63	2 complete layers of EVA (one on each side of the cells between the top & bottom of the cells and the glass)

TABLE 4.3 Natomas modules bill of material.

Series item#	Identifier	Quantity for assembly	Vendor/contact #	Model or part number	Certification file attached	Width dim (inches)	Length dim (inches)	Area (inches^2)
300	tabs/buss				↓			
301		see calculated values	ULBRICH PRECISION FLAT WIRE 692 PLANT ROAD WESTMINSTER.SC 29693 1-864-647-9524	2MM 6MM		0.08 .24	750 115	60 27.5
400	cells				↓			
401		53	Q-CELL GUARDIANSTRABE 16 06766 THALHEIM GERMANY 011-49(0)3494 668 60	QSM-15O		4.92	4.92	24.21 PER CELL
402		53	MOTECH SOUTH TAIWAN SCIENCE PARK No. 2 DA-SHUN 9TH ROAD HSIN-SHI.TAINAN 74145 TAIWAN 011-886-6 5050789	IM125-S3		4.92	4.92	24.21 PER CELL

(continued on next page)

TABLE 4.3 Natomas modules bill of material—cont'd

Series item#	Identifier	Quantity for assembly	Vendor/contact #	Model or part number	Certification file attached	Width dim (inches)	Length dim (inches)	Area (inches^2)
500	backsheet		N/A		↓	N/A	N/A	N/A
600	j-box				↓			
601		1	MULTI-CONTACT USA 5560 SKYLANE BLVD, SANTA ROSA, CA 95403 1-707-575-7575	PV-JB/2-2 BOX	YES	TBD	TBD	TBD
602		1	TYCO ELECTRONICS USA HARRISBURG, PA 1-717-564-0100	2 1394723-7 BOX WITH DIODES AND CUPS	YES	TBD	TBD	TBD
700	diodes							
701		3	DIOTEC SEMICONDUCTOR AG KREUZMATTENSTR, 4 79423 HEITERSHEIM GERMANY 011 49-76-345-2660	P1200 SERIES	YES	N/A	N/A	N/A

TABLE 4.4 Natomas modules bill of material.

Series item#	Identifier	Quantity for assembly	Vendor/contact #	Model or part number	Certification file attached	Width dim (inches)	Length dim (inches)	Area ($inches^2$)
800	connector				↓			
801		2	MULTI-CONTACT USA 5560 SKYLANE BLVD. SANTA ROSA, CA, 95403 1-707-575-7575	PV-AK-3 TERMINAL CUPS	YES	N/A	N/A	N/A
802		2	TYCO ELECTRONICS USA HARRISBURG, PA 1-717-564-0100	1740699 WITH CONTACT RAILS	YES	N/A	N/A	N/A
900	cable / conductor							
901		2	US WIRE & CABLE 6301 E. Stassney Ln Expo 6, Suite 100 Austin, Texas 78744 1-800-880-9473	#12 awg		TBD	TBD	TBD

(continued on next page)

TABLE 4.4 Natomas modules bill of material—cont'd

Series item#	Identifier	Quantity for assembly	Vendor/contact #	Model or part number	Certification file attached	Width dim (inches)	Length dim (inches)	Area ($inches^2$)
1000	cable ends							
1001		2	MULTI-CONTACT USA 5560 SKYLANE BLVD. SANTA ROSA, CA 95403 1-707575-7575	PV A KST3(M)/PV A KBT3(F)	YES	N/A	N/A	N/A
1002		2	TYCO ELECTRONICS USA HARRISBURG. PA 1-717-564-0100	1394462 3(+)/1394462-4()	YES	N/A	N/A	N/A
1100	frame	1SET	TBD	TBD	↓	TBD	TBD	TBD
1200	edge sealant		TBD			TBD	TBD	TBD
1300	hardware/ hooks		N/A		↓	N/A	N/A	N/A

TABLE 4.5 Natomas modules bill of material.

Bill of Materials	4 page document		Premier Power/Natomas					
Series ITEM	Identifier	Quantity for assemblyr	Vendor/contact	Model or part number	Certification file attached	Width dim [inches]	Length dim [inches]	Areas (inches)
1400	label							
1401		1	TECHTAG & LABEL, 1901 North Penn Road, Hatfield, PA 19440 1-800-982-8224	TBD	N/A	102	102	0.0104
1500	manuals							
1501	Installation		ASSEMBELED BY MANUFACTURE		N/A	N/A	N/A	N/A

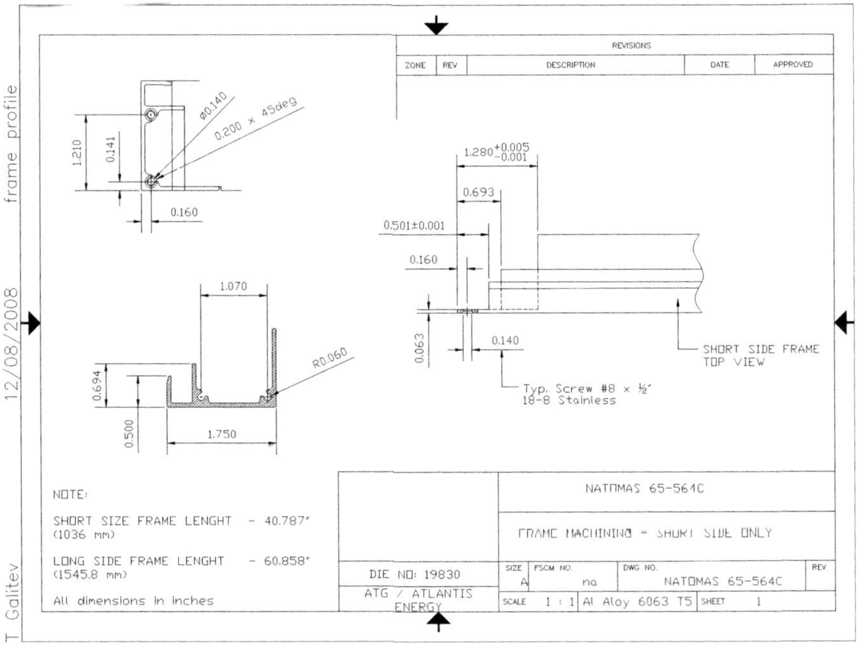

FIGURE 4.76 Supporting frame design structure of the Natomas modules.

the BIPV system in the project. It is quite a large solar system in an office building of this size but could not get close to the required energy demand of the building. To meet the demand, it would probably require an additional 1–2 Mw of energy. However, space and the cost prevent it from happening. That is why the property can only do net metering. This project was built back in the 2008–2009 period. In recent years, there are storage technologies and generation systems that could make a building completely self-sustaining.

It is vital to have a well and structurally sound support in mounting the solar panels, especially in severe weather locations. Though initially, it could be a bit more costly, in the long term, it will return for the initial investment. In case anything falls in the storms or modules being blown away, it could cost damages to the items in the surrounding areas. In addition, in case there is anyone around, any parts coming off from the frame could be extremely dangerous and harmful. To prevent these types of incidents from happening, there should be some regulations set by the local town board and state government in order to

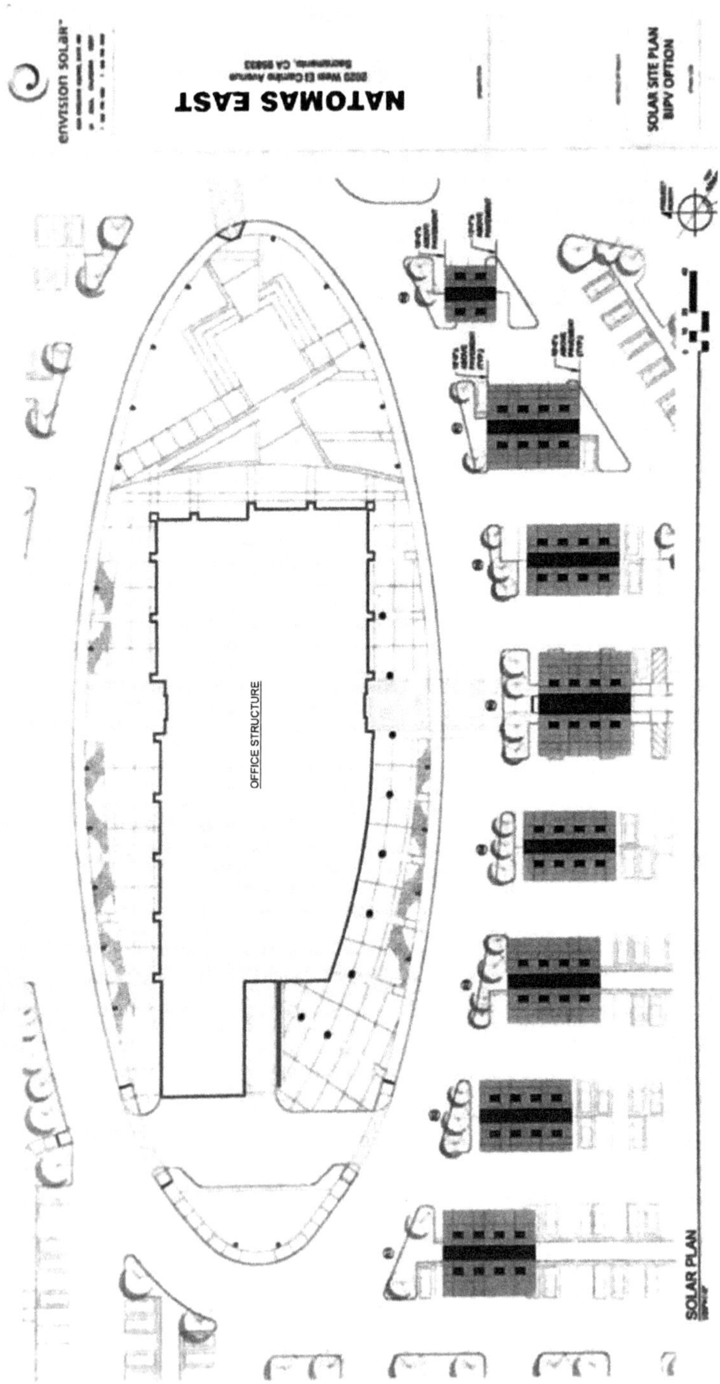

FIGURE 4.77 Natomas project layout.

FIGURE 4.78 The parking cover in Natomas with translucent modules that allow sunlight to come through.

FIGURE 4.79 The parking covers in front of the office building in Natomas.

FIGURE 4.80 The frame and the support structure of the covers.

meet the required building code standards. Too often, in many situations especially in heavy hurricane locations like Florida, the Caribbean, and the Gulf states or heavy snowstorms areas like the Northeastern states, Midwestern states, and Canada, that should be of serious concern. Therefore, special attention is required regarding the projects in these regions.

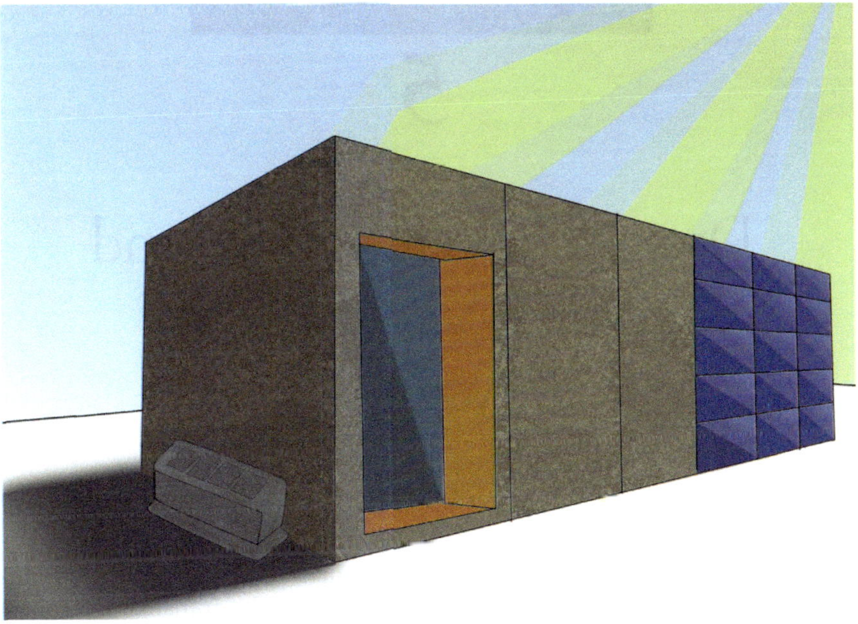

The right half of this concrete building is covered with a building integrated photovoltaic thermal (BIPVT) system. The sunlight strikes the building envelope. In the left half, it turns into heat, and extra electricity is required for air conditioning, and accordingly, more heat is released into the air. On the right side, the solar energy is intercepted by the building integrated photovoltaic thermal system and converted to electricity and heat to power the building's operation. The scale-up of the two parts leads to two distinct directions toward heat island and sustainable cities, respectively. Quantitative understanding of heat and energy flow through the building envelope will be critical for the design, construction, operation, maintenance, and management of the building. A BIPVT system not only harvests photovoltaic and thermal energy, but also addresses all other needs on serviceability, durability, sensing, and control, with a modernized appearance as well.

CHAPTER

5

Design, development, and applications of BIPVT systems

5.1 Introduction

When a thermal management component is integrated into a building integrated photovoltaic (BIPV) system, it can be transformed into a building integrated photovoltaic thermal (BIPVT) system, which enables simultaneous conversion of sunlight into both PV electricity and thermal energy at higher PV efficiency. In parallel to BIPV technology, a building-integrated solar thermal system has been developed with a long history for air and water heating. The early work to use sunlight to heat buildings can be traced back to Morse's invention in 1881 (Morse, 1881). The invention used a blackened surface of metal or other suitable material with a flat or corrugated surface protected by glass in front. Behind the blackened surface is an airway. The sun-heated air, as it ascends, may be directed into the room or building, so as to warm the building, or it may serve to draw the air from the room for ventilation. The idea led to the design and application of the Tromble wall (Trombe and Michel, 1974) and later air-based BIPVT technologies.

Solar water heating can be universally found in nature. The engineered application can be traced back to Mouchot's work (Wikipedia, 2009) to concentrate sunlight to generate steam power at high temperatures, which led to another application of concentrated solar power, but it is challenging to integrate it into building envelope. Flat plate collector systems have been developed practically for building applications. It has no moving parts and is a robust and simple system to heat a working fluid over the building roof although the leakage risks have to been carefully addressed. This system appeared in Florida and Southern California in the 1920s and grew more popular in North America after 1960, particularly after the 1973 oil crisis.

Building Integrated Photovoltaic Thermal Systems: Fundamentals, Designs and Applications

DOI: https://doi.org/10.1016/B978-0-12-821064-2.00002-X

The concept of a BIPVT system can be traced back to Wolf's work in the 1970s. The flat plate thermal collector was used in coupling with the PV array for a feasibility study. The analysis results showed the combination system to be a viable approach for the maximization of solar energy harvesting. However, the concept is only materialized in the 1990s after BIPV was established in the market. Because of the seamless packaging of BIPV systems, the heat accumulated in the solar panel will not dissipate as fast as standard solar cells with space to the roof, the efficiency reduction by temperature will be more severe, which accelerated the development of BIPVT systems.

The coauthor Frank Pao was a pioneer in the design and development of BIPVT in the United States and led the development of the building integrated thermal electric roofing system (BITERS) (Pao and Gottlieb, 2012; Pao, 2013) system in his previous company—the Atlantis Energy System (AES). Shortly after AES deployed the building-integrated solar roof tile around the early part of 2000, the company sought to increase the solar conversion efficiencies in every potential way. In those days, the solar cell efficiency was around 13%–14% max and the actual module efficiency was around 12%–14% without consideration of the efficiency loss caused by temperature. After looking at the rate of the solar cell advancements, which were relatively incremental, the priority to reduce the temperature loss rose, and the company decided to think of other alternatives.

Throughout the next few years, the company started looking for ways to capture more solar energy on the rooftop especially on slope roofs, the attics are usually quite hot in the summer. Using air or liquid for temperature control and heat collection was carefully reviewed and investigated. There are some uses of heated air for drying or temperature conditioning in some locations at specific seasons, but it is hard to generalize it to regular houses or buildings. It means that adding an air cooling system can only achieve a sole goal for PV efficiency recovery from the temperature-induced loss. However, using a liquid system can store heat energy with a much higher energy density and control the PV temperature in a much easier way. Eventually, the company decided to use liquids.

In the summer of 2007, with the help of a thermal company, Dawn Solar, the first experiment was set up in Brentwood, New Hampshire to evaluate the thermal energy performance with the Sunslate. As shown in Fig. 5.1, two 1 kW Sunslate systems were installed on the roof, where one has a thermal system beneath the Sunslate while the other does not.

The energy and water flow within the testing configuration is shown in Fig. 5.2. Two 1 kW inverters were used to feed the electricity to the grid and cold water pumped from the well and heated by the solar system was stored in a 115-gallon water tank. The temperature and flow rate of water and the electricity generated was carefully monitored for 21 days.

FIGURE 5.1 One Kilowatt of Sunslate on each side installed on a barn in new hampshire.

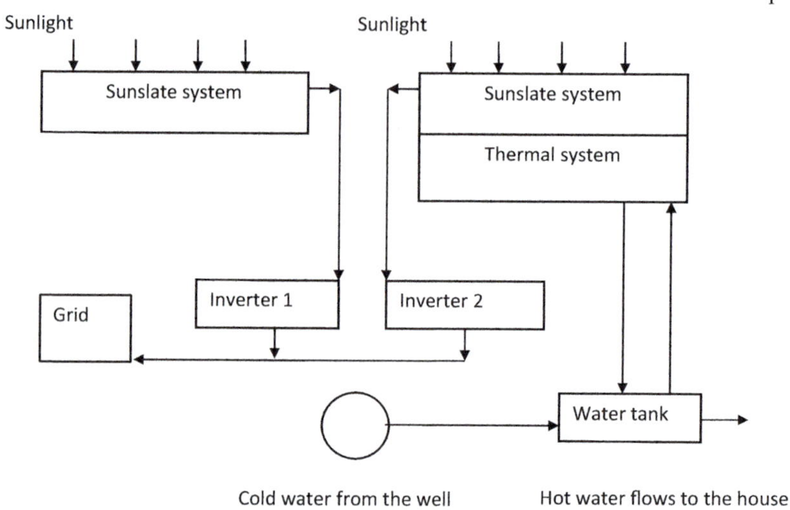

FIGURE 5.2 The solar energy flow in two Sunslate systems for energy harvesting comparison.

The energy efficiency analysis showed as follows:
Three weeks in Brentwood, New Hampshire

Total sum of degrees	665.9
Volume	120 gallons tank
Density	8.33 pounds/gallon
Energy	665,633.6 BTU (3410 BTU/kWh)

Power	195.15 kWh
Total Power generated by Sunslate	54.6 kWh
Total Power generated by Thermal	195.15 kWh
Total Power by the BITERS	249.75
Approx:	250 kWh
Temperature range:	55–102°F

Apart from the above results, as the temperature reaches 85°F, the conversion efficiency of the Sunslate without the thermal system started to decrease and as the temperature goes above 100°F or higher, it reaches around 10% of the efficiency reduction. The thermal extract of the heat from the tubing could also reduce air conditioning costs in the summer. Based on these results, the BITERS system was started for development. Though the tests only covered a three weeks range, the data can be generalized to a sizeable part of the country in this range of temperature.

In addition, recent advancements in solar cell technologies have reached the 23%–24% range efficiency. Therefore, the module conversion efficiencies would be in the order of 20%–21%. Once the thermal system is included, it would easily reach or go above 50% conversion efficiency in a place like California which could make the house or building totally self-sustaining or net-zero energy. In case, some houses have swimming pools can use the excess energy to extend a couple of months use or whirlpool. In addition, considering the temperature situation in parts of California, Arizona, New Mexico, Texas, Nevada, and Southern and Gulf states, the summer months could get very hot. The roof temperature could go up to 70–80°C. The thermal system could be used for water desalination as well.

Although there are many other BIPVT technologies in the research and demonstration phase, which have been introduced in Chapter 1 and will be revisited in Chapter 10, only a few BIPVT technologies are available on the market. The BITERS has been developed with the Sunslate and TallSlate products, which will be elaborated in the following section.

In parallel, Dr. Yin also established and led a team to design, develop and fabricate the BIPVT system at his research laboratory at Columbia University since 2008. The research progress of BIPVT at Columbia University has been introduced and some new technologies will be reviewed. Future applications to energy-independent houses have been envisioned.

5.2 Building integrated thermal electric roofing system

The BITERS is a solar hybrid system that could convert the sun's energy into thermal energy and electricity. The system comprises several metal battens having a longitudinal channel mounted horizontally onto a number of wooden battens, a thermal tubing containing liquid mounted on the longitudinal channels, PV modules mounted on the metal battens and connected in series to form a string, an inverter connected to each of the strings, a heat exchanger connected to the thermal tubing, and a pump connected between the thermal tubing and the heat exchanger. It is totally building integrated with the controlled roof temperature and the improved PV efficiency. AES has used two types of PV modules in BITERS products: the Sunslate module has a small size to mimic the classic roof tile, whereas the TallSlate module is a regular solar panel for modernized roof applications.

5.2.1 Components and functions of Sunslate building integrated thermal electric roofing system

The Sunslate is a BIPV roof tile with a solar module glued to the Eternit tile, which is a brand of fiber cement board with a flexural high strength. Fig. 5.3 shows the design and actual product of the Sunslate roof tile for BITERS, which divides the surface into two areas: the PV module covers the lower part and the upper part is overlapped with the PV module part in installation for leakage proof and additional structural support. The Sunslate tile is mounted on a metal batten which also holds the thermal tubing. Sunslate is the photovoltaic (PV) solar roof tile that converts solar energy into DC electricity. The metal batten holds the thermal tubing that collects solar energy and converts it into thermal energy by running the liquid in the thermal tubing.

The thermal part of the system shall be installed first. The metal batten will be mounted horizontally across the roof onto the battens that go vertically up the roof. Once the metal battens are installed, the thermal tubing would be mounted on the gutter of the batten horizontally across as well. The spacing between the metal battens could vary subject to the size of the Sunslate and the thermal required specification. As the metal battens are more tightly spaced, more thermal tubing could be installed which results in higher thermal energy conversion efficiency. The thermal tubing is connected to the pump and a tank with a heat exchanger. The metal batten performs three functions: (1) mounts the thermal tubing, (2) transmit the heat, and (3) supports the Sunslates. When the metal battens are used to mount the Sunslates, hooks are used to secure the mounting of the Sunslate. One hook holds down the Sunslate at the top and the bottom hook supports it as it overlaps the other Sunslate.

Fig. 5.4 shows the Sunslate BITERS system installed on the roof. To fix the Sunslate tile to the roof, holes are drilled in advance according to specified positions on the metal batten which will save time and also simplifies the installation procedures. The specified position would be based on the size of the Sunslate. The Sunslates would be connected in series horizontally across the roof through the custom-designed j-box. The series connection of the Sunslates could form a string and the number of

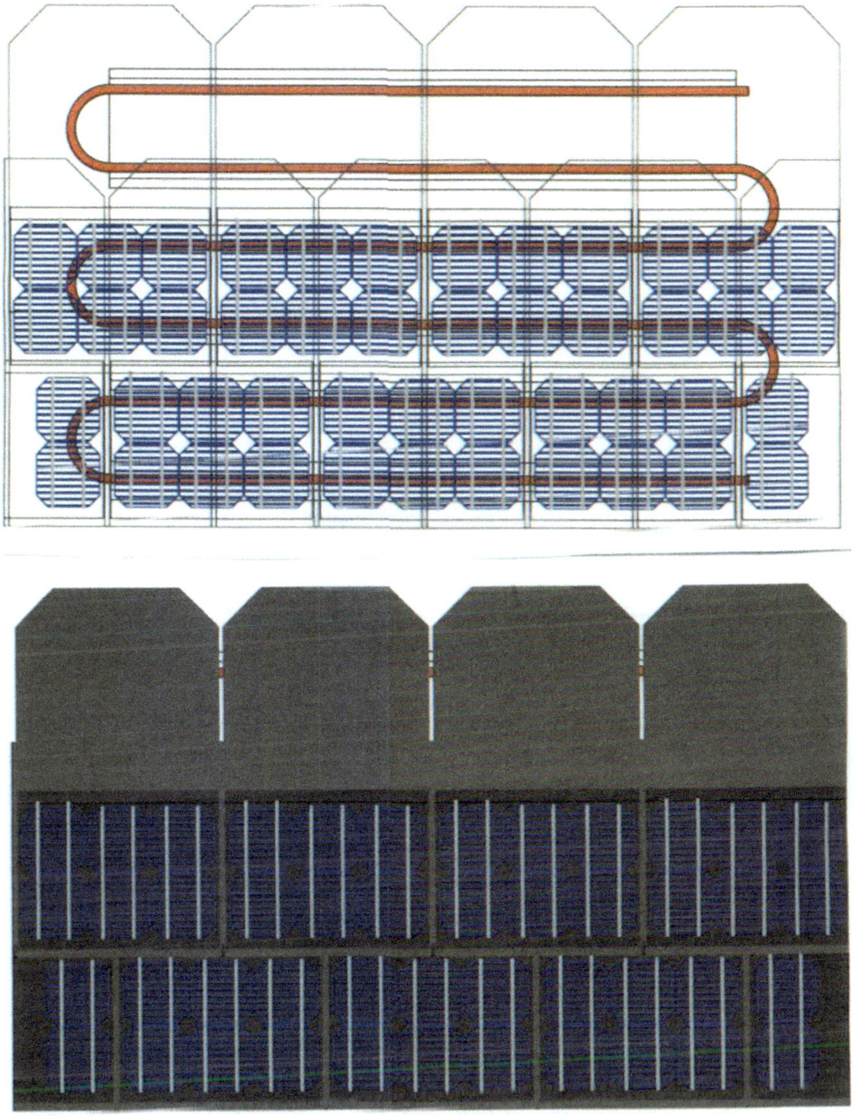

FIGURE 5.3 The design and product of the Sunslate roof tile for BITERS, *BITERS*, building integrated thermal electric roofing system.

connections in the series would be subject to the design. Each string could be connected to a microinverter or numerous strings could be connected in parallel to a regular inverter or numerous Sunslates could be connected in series feeding to a regular inverter which would be subject to the size of the inverter and the system specifications. The design would depend

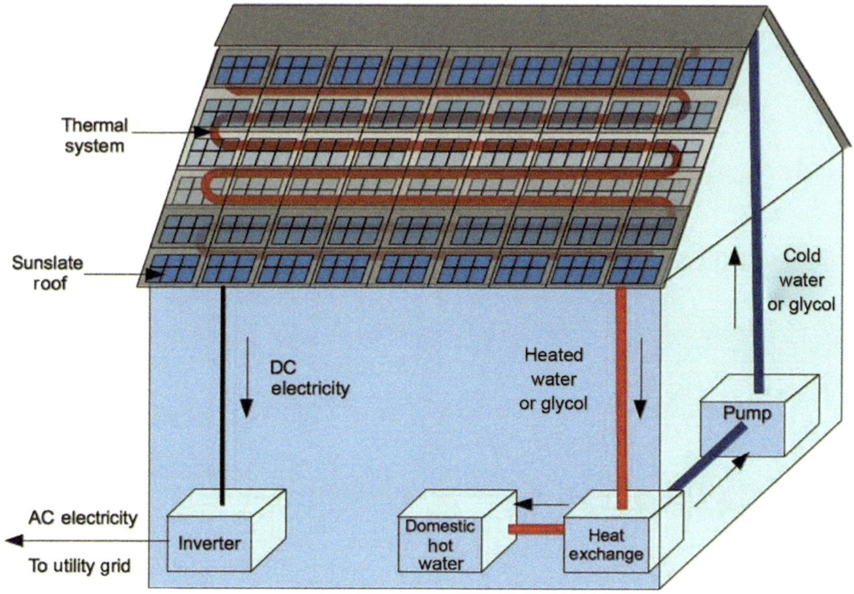

FIGURE 5.4 A Sunslate panel containing glass, solar cells, back sheet, and eternit tile is attached to a metal batten together with the tubing.

on the specification of the system. The inverter could convert the DC electricity to AC and feed it into the grid. However, if one would like to have a standalone system, one could use the Sunslate system to charge up batteries by configuring the Sunslate at 12, 24, & 48 Volt strings.

5.2.2 Operation of the Sunslate building integrated thermal electric roofing system

Fig. 5.5 shows the operation of the Sunslate BITERS system. Once all the strings of Sunslate tiles are checked and if they reached the acceptable power, they will be connected to the inverter. Then the PV system would be turned on and start supplying AC power to the grid. In a standalone system without connection to the grid, the strings could be connected to the battery via a charge controller and start storing electricity that could be used to power the dc appliances and lighting application. Meanwhile, on the thermal system, the circulation pump could get the heat exchanger fluid to circulate through the thermal tubing and extract the heat from the roof or façade through the metal batten down to the tank with the heat exchanger where it would heat the domestic water supply. The pump could be connected to a separate standard PV module that could meet the wattage of the pump. Based on the roof temperature, cold water or warm

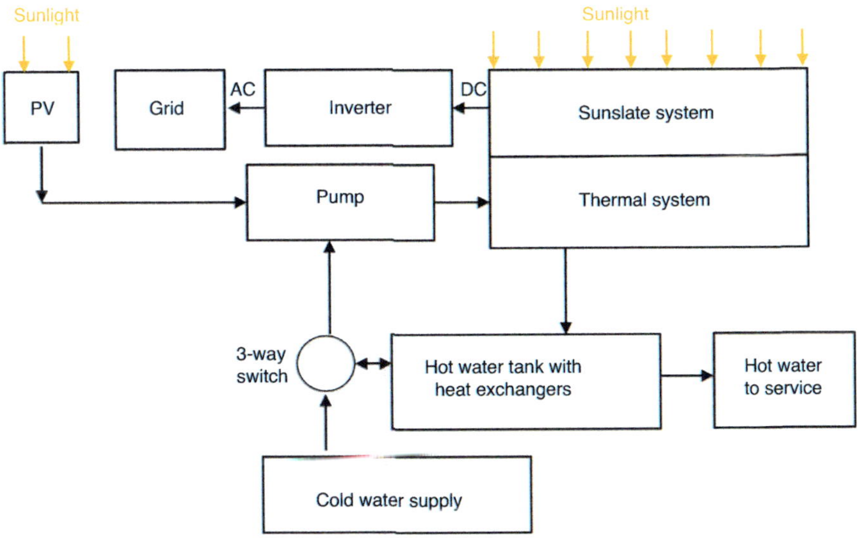

FIGURE 5.5 Block diagram of the Sunslate BITERS system. *BITERS*, building integrated thermal electric roofing system.

water can be pumped to the roof through a 3-way switch for PV cooling or excessive heat dissipation. In winter with snow or icy weather, warm water will be used for snow/ice removal.

As the intensity of the solar power increases so would be the throughput of the liquid in the thermal tubing resulting in generating more thermal energy. In addition, while heat is being extracted through the thermal tubing from the Sunslates that is being cooled thereby prevent the degradation of efficiency in converting sunlight to DC and ultimately AC electricity. Since a fairly amount of heat energy had been taken out from the attic, the house would be cooler which results in a reduction in the air conditioning energy consumption. In case, there is an excess of required energy on standard domestic hot water application, it could be stored for later use or it could be used for hot tub or heater pools which would also be subject to locations.

When more heat energy is harvested than the demand, it should be dissipated to the environment through a radiator or circulated back to the roof at night or early morning. Actually, this brings some uncertainties for the design and control of the BITERS system, for example, under extreme weather with continuous high temperature, the thermal fluid may reach a stagnation temperature when the heat cannot be used or dissipated to the environment, so that the cooling function may be jeopardized. More research is underway. Using geothermal energy for seasonal energy storage is a practical way in some regions. A cascade heat pump can be another potential solution.

5.3 Installation projects of the Sunslate building integrated thermal electric roofing system

The main applications of the Sunslate BITERS system would be on the slope roofs whether they are commercial or residential buildings. However, it could also be applied on curtain walls or facades on buildings. Sunslate BITERS is commercially available on the market. In the following, some actual Sunslate BITERS projects will be introduced.

5.3.1 Sullivan estate project—aesthetical roof in a resort

Sullivan Estate is a holiday resort place when families rent this location for a certain period to enjoy their stay in Hawaii. In Fig. 5.6, the installed Sunslate BITERS system including 50.5 kW of Sunslate panels and 100 m^2 of solar thermal beneath is shown in Fig. 5.6, which is fairly large and produce more power than what would be required. Since it is in a remote area, it is always wise to have some redundancy in case anything unanticipated happens. The system can be totally self-sustaining for electricity in the resort. A sizeable portion of the Sunslate system has a thermal system beneath that generates thermal energy for domestic applications as well as Jacuzzis and outdoor heated pool. In addition, shown in the above photo is the Solar Tree which is being used to pump a water fountain.

FIGURE 5.6 Sunslate BITERS at the Sullivan Estate with Solar Tree in Oahu, Hawaii. *BITERS*, building integrated thermal electric roofing system.

This roof is very difficult for any solar panels or roof tiles to be mounted to generate sufficient energy of the building's demand, but the Sunslate mounting system has achieved excellent performance. In total, a roof area of 9,250 sq. ft was covered by 3,700 Sunslate tiles. In addition, the thermal systems are right beneath the Sunslate that will obtain more energy for the building especially in a place like Hawaii. This project was completed around 2006 when AES initially started installing the BITERS. Throughout all these years, despite the island had been hit by some heavy wind, the Sunslate BITERS is still operational. Aesthetically, it is quite pleasing (Fig. 5.7).

FIGURE 5.7 The other angle of the Sullivan Estate.

5.3.2 Virginia project–building integrated thermal electric roofing system and skylight on a residential home

The customer on this project put a new extension on this existing home and uses the Sunslate BITERS in Fig. 5.8 It is a relatively small system which is 2.7 kW and the owner has made a substantial saving on the utility bill. Although the total size of the building is around 4000 sq. ft because the family is a heavy consumer of electricity with a Jacuzzi and an outdoor swimming pool which requires water heating. Their monthly electric bill is well over $500.00 and the Sunslate BITERS cut it almost half. Consequently, the family has been a very happy customer after the installation. Virginia is considered a mid Atlantic state where the climate is somewhat milder than the northeast. However, it could get down to quite cold in the winter months which would be a little below freezing. A similar region in Europe would be London, Paris, or Geneva.

FIGURE 5.8 Sunslate BITERS and skylight installation at Virginia. *BITERS*, building integrated thermal electric roofing system.

5.3.3 The future house USA project

The Sunslate BITERS could be ideal in a relatively tropical or Mediterranean climates places like Southern France, Italy, Greece or Florida, Bahamas & Singapore to achieve total self-sustaining. On all these locations, winters are relatively mild or do not have winter at all. In situations, when the temperature goes down to below high 30°F, a reasonably good size storage tank would be sufficient to maintain the thermal energy throughout the winter months. One customer installed a Sunslate BITERS in Northern California, throughout the last couple of years, he managed to store more than enough energy than he needed and plans to release some on other applications. Around early 2007, China invited 10 countries around the world to participate to demonstrate their latest house technologies, especially, self-sustaining types or net-zero energy types at the 2008 Olympic Village in Beijing. In the case of the Future House in Beijing, since the sun hours is around 4 hours which is quite low and the climate in the winter is quite cold which would be somewhat similar to cities in Germany and Northern Europe, a 7.5 kW Sunslate BITERS was installed together with a geothermal system so that it could be totally self-sustaining in Fig. 5.9. Among the 10 countries that were invited by China to display their future

FIGURE 5.9 Sunslate BITERS Future House USA at the Olympic Village in Beijing, China. *BITERS*, building integrated thermal electric roofing system.

homes at the Olympic Village before the Game, Future House USA was the only one that managed to achieve net-zero energy status. Since it is a high-profile project, there is some redundancy as well to make sure that it would always maintain the total self-sustaining condition. Apart from the high solar energy conversion efficiency ratio, the Sunslate BITERS is aesthetically pleasing, agile in application on a wide variety of roof shapes, the only BIPV roofing product that has a roof leakage warranty.

The Sunslate BITERS system exhibits unique features in this FHUSA project as follows:

1. FHUSA was engineered to a commercial specification. It looks like a house but designed for classroom/meeting applications, sustaining higher occupancy rates than a typical house. Its' designs took local seismic conditions into consideration and are engineered to be earthquake resistant. All designs were to International Building Code standards. The house was pre-engineered to EnergyStar specifications.
2. FHUSA is a structurally insulated panel (SIP) house. It features R26 walls and an R44 roof. It exceeds infiltration standards and is essentially an air tight structure. The SIPs are manufactured in the United States.
3. FHUSA uses a top-of-the-line geothermal heat pump system. The house was equipped with a 56 tons geo system, and additional tons were

installed to cover the high occupancy rate. Three energy recovery ventilators were used to provide compliant indoor air quality with any occupancy load. The loop system was placed in the backyard utilizing 9–70 meter vertical loops. The DC-super heater option to create domestic hot water is installed but currently not activated. All ducts are in conditioned space, and air sealed as well. The geo system components were manufactured in the United States.

4. FHUSA employs a 7.2 kW Sunslate roofing system, coupled hybrid system creating 65,000 BTUs of water beating capacity. Additionally, eight Atlantis Solar Trees are used to produce up to –4 kW bringing the total Atlantis system capacity to –11 kW. The inverter system is a grid-tie unit. For electric storage, 48 gel batteries are used to create storage for 1 day of autonomous run time, which can be subdivided for up to 10 days of emergency power, running the house on 10% per day. Hot water is stored in three heavily insulated solar storage that holds for 5–6 days. Last in the hot water chain is an on-demand electric tankless water heater that acts as a backup. All electric and space conditioning systems are coupled with an energy management system to control and monitor. There are 130 monitor sensors stalled throughout the house and systems to provide data for Research and Development. The PV solar roof tile were all built in the United States.

The Future House USA energy system (Fig. 5.10) consists of three main components:

1) The Solar Electric system comes from the Sunslate and the solar tree. Sunslate is the solar roof tile installed on the south-facing roof of the

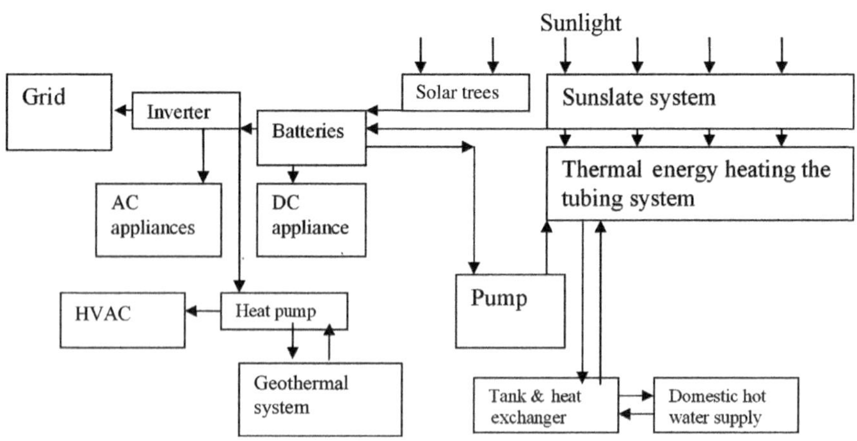

FIGURE 5.10 The future house USA—system operational diagram.

house and is an integral part of the roofing system while the solar trees are layout in the backyard. Both systems generate DC electric power and charged up the batteries. The DC power supports some appliances and lighting. In addition, the batteries are connected to the inverters which support the AC appliances and feed electricity to the grid as well.

2) The solar thermal system is directly beneath the Sunslate system. It consists of thermal tubing with glycol being pumped up and circles throughout the roof extracting the solar thermal energy to heat the domestic hot water through the heat exchanger. Considering the size of the system installed, there could be a sufficient amount that could power the swimming pool and Jacuzzi in addition to regular daily requirements.
3) The geothermal system has a heat pump with a geothermal loop buried beneath the ground. It handles the heat ventilation and air conditioning requirements of the house. In the summer months, the system extracts the heat from the house and stores it in the ground which results in cooling the building. In the winter months, it pulls the heat stored from the loop in the ground by circling the glycol and warms the house. Consequently, reduces the size of the system to meet the heat and cooling needs.

All the three systems basically compensate each other. The Sunslate PV system generates electricity and the installation of the thermal tubing beneath it would allow the system to maintain the performance on the solar energy conversion efficiencies in the summer months. In addition, the heat extracted through the thermal tubing from the attic together with the geothermal system could cool the house substantially which results in reducing the air conditioning requirement. On top of all these systems, the house is constructed with SIP that has very high R values which are vital to energy conservation.

The Future House USA project in Beijing, China, employs an advanced design that integrates the architectural style and innovation of Frank Lloyd Wright with the Chinese practice of Feng-Shui. This approach integrated the home into the rhythms of the surrounding landscape, thus creating a refreshingly smooth flow of air and light. The US Future House design represents the latest in sustainable energy and nontoxic technologies, clothed in tried and true, award-winning, design style. Interior space fluidity is enhanced by the minimization of interior walls. Open and multitiered interior rooms feature much natural light, regally high ceilings, and are all finished with Chinese indigenous material and labors (Fig. 5.11) following the established standard.

The BIPVT system in FHUSA achieved the following distinguished performance:

FIGURE 5.11 Thermal system installation in Beijing, China.

- High energy performance—a completely balanced space conditioning system within a high-performance structure, including maximized indoor comfort, minimized energy bills, and accentuate environmental stewardship!
- Nontoxic building—only user-friendly, nontoxic materials, adhesives, and laminates are used at the FHUSA. No formaldehyde or volatile organic compounds are needed. Additional healthy features include kitchen cabinets being made with wheat straw, foot pedal control for faucets in the kitchen and bathrooms, completely filtered indoor air. The nonpolluting roofing system, combined with the filtering action of the porous pavers, filters all stormwater runoff before it returns to the watershed.
- Zero carbon footprint—the FHUSA creates, on-site, all the energy that it uses annually so that its operational carbon footprint is zero.
- Zero energy home—using geothermal and solar energy, both renewable, FHUSA represents the state of the art as a self-sustainable, healthy home system. Officially known as a Zero Net Energy Home.
- Modernized electricity accessibility—designed with connections for electric or hybrid cars in the three-car garage thus creating a complete "sustainable home system."

Though some of the components like the batteries system or storage systems or the geothermal system have made substantial advancement in recent years, the design and performance of the FHUSA is still the state of the art even after it was built more than 10 years ago. The whole system had been performing very well. When US Commerce Secretary Gary Locke and Energy Secretary Steven Chu visited the project site on July 16, 2009 (China.org.cn, 2021), it was a very hot summer day. The initial plan was to limit the number of visitors and reporters to be allowed into the house because the heat in July was above 90°F. However, with the excellent performance of Sunslate BITERS, nearly everyone was able to get on the site with desirable comfort. All the data looked well, despite so many people flooded into the building. Unfortunately, the afterward performance data of the house cannot be released from Beijing, China, yet. It will be exciting to know the long-term performance data of this FHUSA project in the future. The model home of a Sunslate BITERS system has a 9.7 kW Sunslate system and 1000 sq. ft of the thermal system. Considering the size of the system, the amount of hot water it could generate in the summer months could more than handle our daily usage including Jacuzzis and swimming pools. On the excess amount of heat generated in the summer months, one should look into storing it in the ground and extracting it in the winter by cycling cold water from the attic through the ground. Through the case studies, the hybrid system not only cools the attics and generate a sizeable amount of hot water of above 100 F, but it also maintains the efficiencies of the electric power generation as the outdoor temperature goes above 80 F. Normally, as the outdoor temperature goes above 80 F, the solar cell efficiencies decrease. In addition, the amount of electricity saved on air condition should also be considered. The payback time without subsidies could range of less than 10–15 years subject to location and the system size and design. However, with incentive, the return could range from 3 to 6 years, subject to the state and location as well.

Another project with the BITERS was done in a residential house in Gloucester, Massachusetts in Fig. 5.12. The customer has been very pleased

FIGURE 5.12 Sunslate BITERS installation in Gloucester Massachusetts.

with his Sunslates BITERS combination for the past 10 years since the installation. A special seasonal system in this house allows the owner to cut his winter fuel oil usage by 50%. This is done by creating a copper coil for the BITERS thermal to connect to the forced air HVAC, thus, minimizing the demand for fossil fuels. In the warmer seasons, the swimming pool is used as a heat sink. In 2019, after switching the BITERS thermal output from HVAC to pool heat, the water temperature in the pool rose 15°F in just two weeks. Based on the owner's testimonial, he had more thermal energy generated than he needed. In the past years, the pool's temperature has increased to 90°F due to this transition.

5.4 Development of the TallSlate building integrated thermal electric roofing system

Sunslate BITERS has been a successful BIPVT product for agile installation procedure and stable performance. However, because it covers small footage, the cost for manufacturing and installation is relatively high. To reach a larger market, making larger roofing panels with higher efficiency means a shorter payback period. AES developed the TallSlate BITERS later for lower manufacturing costs and rapid installation.

In parallel to the development of the Sunslate BITERS system, the TallSlate product was developed using a larger roof panel directly without the Eternit tile. Therefore, the thermal system can be incorporated beneath the PV module without compromising the mechanical and thermal performance as well as the aesthetic of the installation. There are three different types of products for different market needs: the TallSlate, the TallSlate Grandee, and the TallSlate Economy. All of them could have a thermal system beneath them.

Fig. 5.13 shows the layout of the TallSlate. Similarly to Fig. 5.3, the metal batten with the thermal tubing is mounted beneath the TallSlate module. The TallSlate module is made of a glass-EVA-Cell-EVA-backsheet structure, so no cement tiles are used, which saves the material redundancy

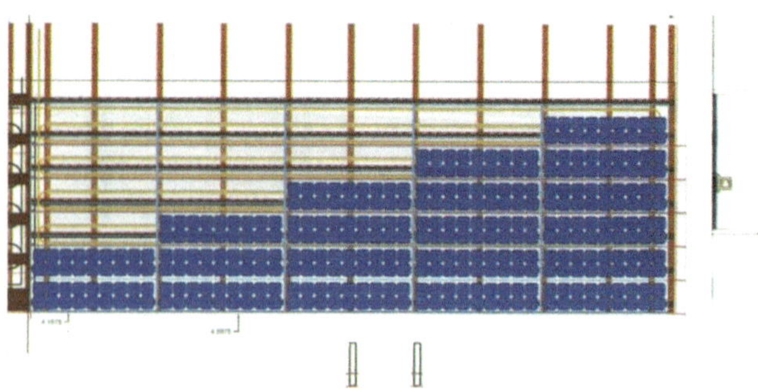

FIGURE 5.13 The TallSlate hybrid system.

as well as cuts off the heat transfer path. Solar thermal energy could be extracted in higher efficiency and the PV temperature can be controlled responsively. Although the operation of the TallSlate BITERS is similar to the Sunslate BITERS, the TallSlate BITERS can address the following challenging problems:

1. Installation cost and time.
2. Parts availabilities on site.
3. Chance on errors.

By modular manufacturing and installation, multiple TallSlate panels are assembled at the plant into a modular system, which could be 500–700 watts. The modular system could be lifted to the slope roof and could be securely mounted onto a latch. The solar roof tiles could be connected to a micro-inverter and the thermal tubing could be connected to the next modular system. The heat collected from the roof will be used for the building or dissipated in a horizontal geothermal system The system is controlled by the pump and the thermal control system (TCS). The components and relationship diagram are shown in Fig. 5.14.

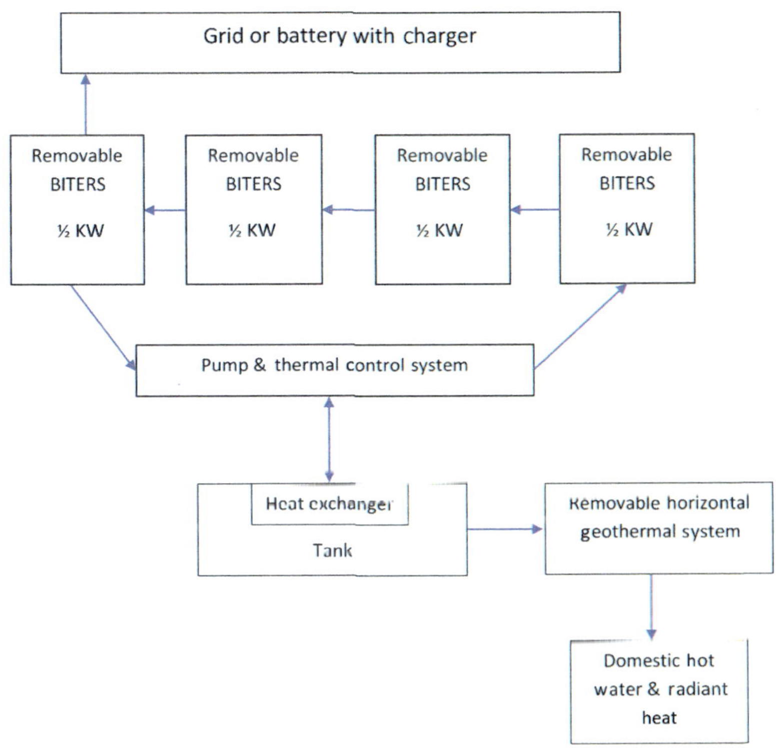

FIGURE 5.14 Modular BITERS with a horizontal geothermal. *BITERS*, building integrated thermal electric roofing system.

The PV slates, using the TallSlate, TallSlate Economy, or TallSlate Grandee modules, will be integrated as a part of the roof while generating solar electric and thermal energy on the same plane. The electrical energy collected could either be fed to the grid through the inverters or could be used to charge up the batteries or fuel cell on standalone systems or uninterrupted power systems (UPS). Right beneath the PV solar roofing system is the thermal system being used to collect the solar thermal energy by circulating liquid (glycol or water) through the thermal tubing on the system. The system will be controlled by the pump and the thermal controlled system. The pump could either be AC or DC. However, it is better to use a separate PV system operate a DC pump to circulate the liquid in the thermal tubing because as the intensity of the solar energy goes throughout the day, the pump could drive increase the flow of the liquid which results in collection more solar energy in hot summer days. The excess thermal solar energy collected apart from being able to support the domestic hot water could be stored in the horizontal geothermal for winter months in generating thermal energy on radiant heat for the building, which will be elaborated in Chapter 6 for the geothermal applications. When a borehole geothermal well is used, the geothermal energy can be stored under the ground and utilized in the winter season.

The thermal tubing and wire connection in a modular system are illustrated in Figs. 5.15 and 5.16, respectively. Because 12 regular PV panels are assembled for one modular system in the plant with hand-on tools

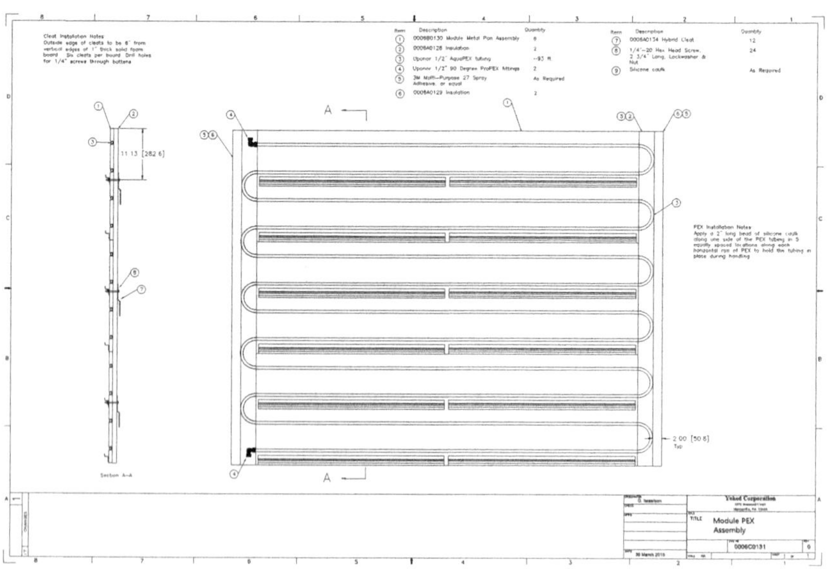

FIGURE 5.15 Thermal tubing in the modular control system.

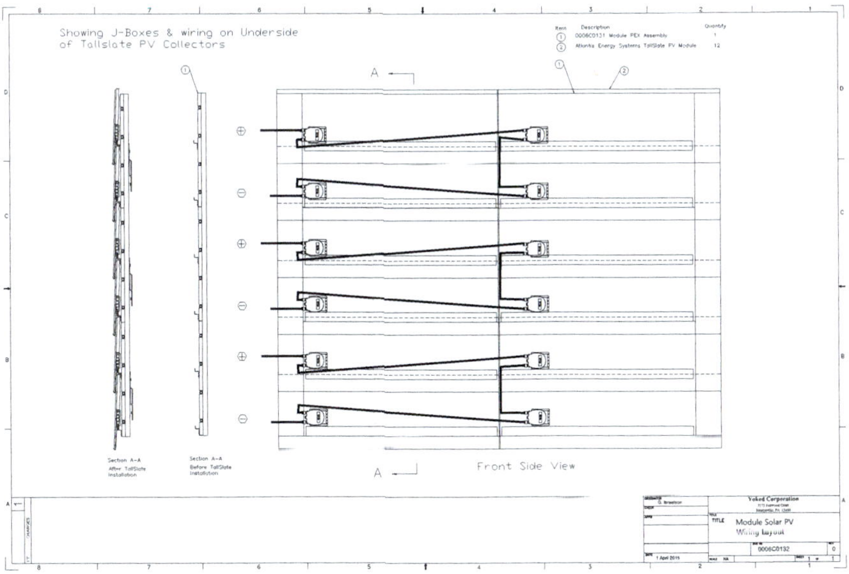

FIGURE 5.16 Wire connector on TallSlate modular system.

and standard quality control, the installation speed and quality can be controlled well, so that both the hard cost (materials) and the soft cost (labors and certifications) on installation can be significantly reduced.

To reach higher system efficiency, back contact cells are used to avoid surface shading loss (Fig. 5.17). Once the thermal system is connected and passes the pressure test, the solar panels can be installed to the modular

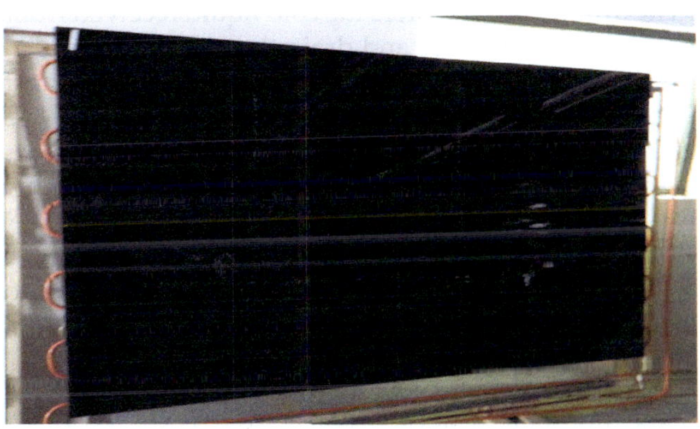

FIGURE 5.17 Modular BITERS with back contact cells. *BITERS*, building integrated thermal electric roofing system.

FIGURE 5.18 Modular BITERS with the thermal system. *BITERS*, building integrated thermal electric roofing system.

system in a "plug-and-play" fashion, as shown in Fig. 5.18. Multiple modular systems can be delivered to a construction site and installed and connected rapidly.

In Fig. 5.19, the TCS handles the flow of the liquid in the thermal tubing structure. It is connected to the input section and the output section of the thermal tubing structure and it comprises a liquid storage unit, heat exchanger, a pump, a drain valve, a check valve, a fill valve, a forward gauge assembly, a backward gauge assembly, a flow sight glass, and an air eliminator. The liquid storage unit stores the liquid received from the thermal tubing structure. The heat exchanger is connected to the liquid storage unit to extract and store the heat energy. The pump circulates the liquid received from the liquid storage unit to the thermal tubing structure. The drain valve transfers the cold liquid received from the pump to the thermal tubing structure and drain the excess liquid in a controlled manner. The check valve regulates the flow of air while transferring the cold liquid to the thermal tubing structure and the fill valve receives the cold liquid from the check valve to the check valve and regulates the filling of cold liquid into the thermal tubing structure through the input section. The forward gauge assembly includes a first temperature gauge and a first pressure gauge to check the temperature and pressure of the cold liquid flowing to the input section of the thermal tubing structure and the backward gauge assembly includes a seconde temperature gauge and a second pressure gauge to check the temperature and pressure of the hot liquid flowing through the output section of the thermal structure. The flow sight glass regulates the flow of the hot liquid from the output section to the thermal tubing structure and the air eliminator releases the air transferred from the output section to the liquid storage unit to an expansion tank via a

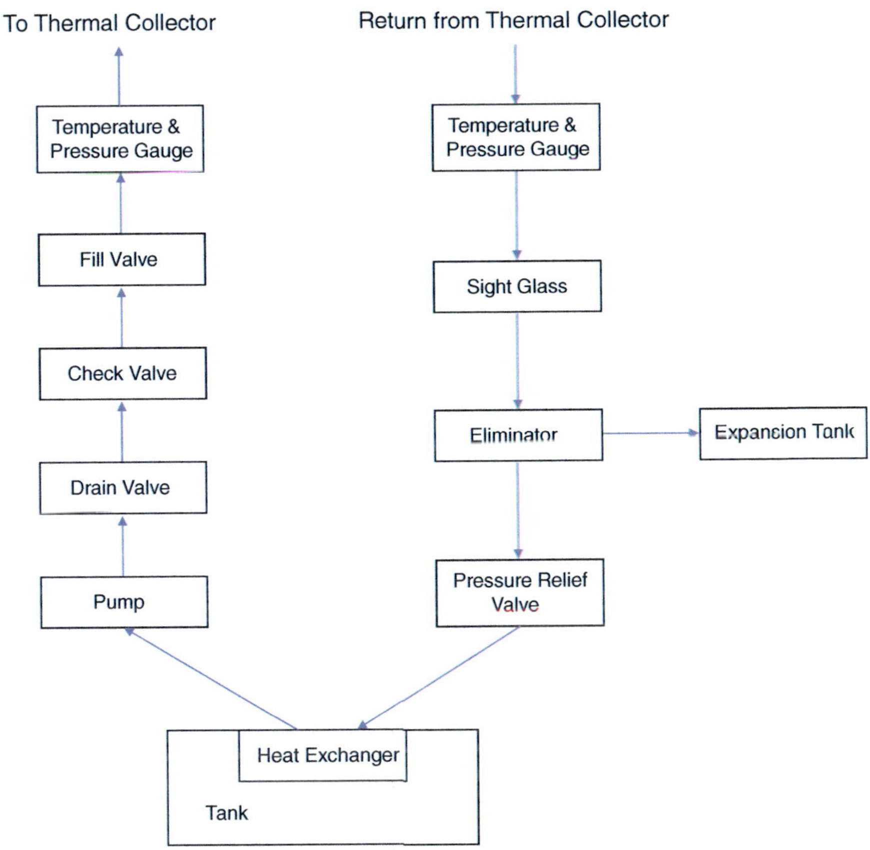

FIGURE 5.19 Thermal control system in the modular system.

pressure relief valve. The thermal tubing structure and the PV cells operate simultaneously to generate hot liquid and electricity, respectively, and at least one sheet structure integrated with the PV cells enables the building integrated power generating to be used as a skylight, windows, atriums, and facades. The TCS significantly simplifies the installation of the solar thermal system by combining all components into one package which is easy to install, including all necessary equipment needed to circulate the heat transfer fluid through the array and the heat exchanger in the storage tank. The TCS is both simple to install with only five connections and is elegant in size and shape. The pump inside the TCS is activated by two temperature sensors, one located on the array and one inside the storage tank. These temperature sensors work with the variable speed pump to control the flow and maintain an appropriate difference in temperature between the tank and the array, maximizing efficiency. The TCS can be used in BITERS systems size from 0.5 to 3 kW.

The components in the TCS box are listed below and connected per the process in Fig. 5.20:

- Pump
- Sight glass (visual flow checker)
- Fill and drain valves
- Check valve
- Two temperature/pressure gauges
- Super vent air eliminator (purges air)
- Pressure relief valve
- Expansion tank 40 psi

After the modular system is securely mounted on the roof, all connections should be examined and a pressure test will be conducted to identify and clear any potential leakage problems. In addition, the removable modular system not only makes the installation in the field much simpler but also allows homeowners to remove the system in case they decided to move to a new location or replace the part which does not function well. Basically, the TallSlate BITERS shared many features of BIPVT designs but applied the modular construction concept into the manufacture and installation procedure. Sunslate roof tile is close to the classic roof

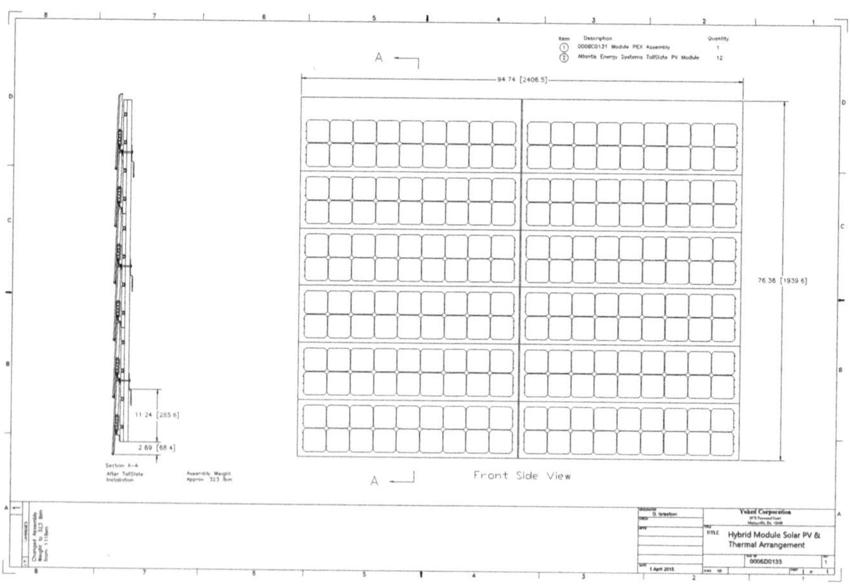

FIGURE 5.20 Modular BITERS thermal system connection process. *BITERS*, building integrated thermal electric roofing system.

tile with overlaps between tiles, whereas TallSlate panels minimize the overlap and space between solar cells for maximized PV harvesting and material efficiencies. However, the two BITERS systems are compatible and they can be linked up together if a building needs two systems installed together.

5.5 Recent development and products of building integrated thermal electric roofing system

Recently, the engineers have been extending the modular system to Sunslate BITERS as well and standardizing the assembling procedure.

Fig. 5.21 shows the latch-holding connection for the TallSlate modular system and the bridge connection for Sunslate modular system. Because the Sunslate tile is much smaller than the TallSlate panel, much more tiles can be assembled for one Sunslate modular system. For the architectural purpose, the Sunslate modular system has a tile layout with a horizontally offset pattern in Fig. 5.22.

An actual Sunslate modular system is shown in Fig. 5.23, 45–50 roof tiles are assembled in the plant and be delivered to construction sites as one unit. It can be used for small and large projects in a "plug-and-play" fashion. For many residential applications, one unit can be a good start at the garage roof for a small investment.

In small towns and remote countryside areas, a lot of times electrical and clean water could hardly be available. Portable or transportable BITERS

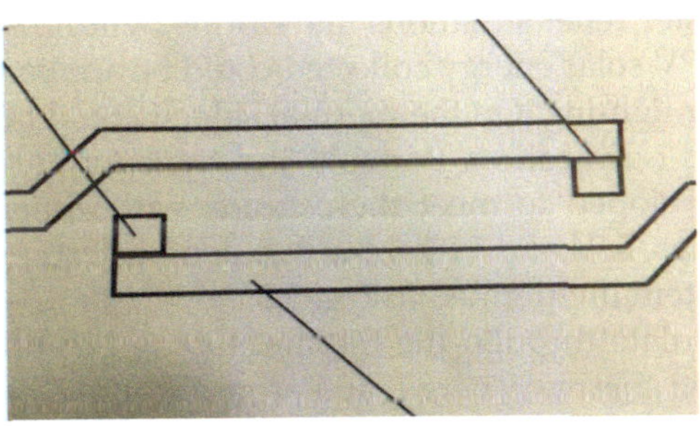

A

B

FIGURE 5.21 (A) The Latch-holding connection for the TallSlate modular system. (B) The bridge connection for the Sunslate modular system.

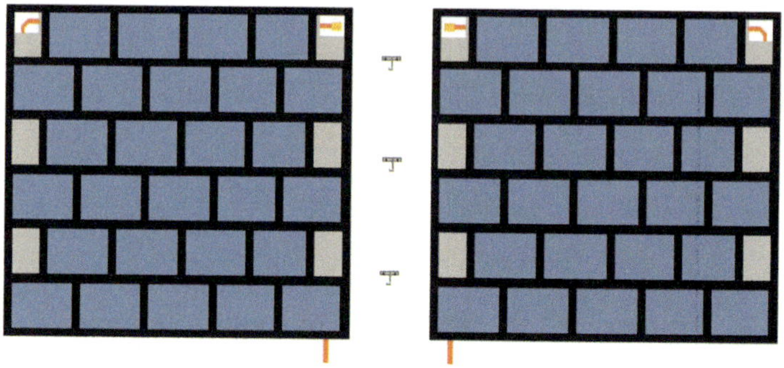

FIGURE 5.22 Sunslate modular overlap connection shown on the above diaELSST100-05-gram.

FIGURE 5.23 Sunslate modular BITERS system. *BITERS*, building integrated thermal electric roofing system.

could be handy to have in disaster relief situations like storms, wildfires, and war or a camping site. The PV solar energy collected could be applied to charging electronic devices or lighting. On the thermal side, it could be used for water desalination and purification. Portable and transportable BITERS systems have been developed to meet these needs with either Sunslate or TallSlate modules. Figs. 5.24 and 5.25 show two demonstration prototypes with the thermal system highlighted in Fig. 5.26.

The thermal system will circulate through the thermal tubing behind the solar system, and collect the thermal energy, and store through the heat exchanger to the tank that contained the saltwater or other types of water. In addition, water could also be collected from the rain. As the tank reaches a certain temperature, the liquid in the tank would be transferred to the distiller. When the thermal system started circulating

FIGURE 5.24 Portable BITERS. *BITERS*, building integrated thermal electric roofing system.

FIGURE 5.25 Portable BITERS—back contact cell module. *BITERS*, building integrated thermal electric roofing system.

FIGURE 5.26 Thermal system highlighted in a portable BITERS. *BITERS,* building integrated thermal electric roofing system.

through, the charged battery through the solar electric system so that the boiler would convert the heated water into steam. The steam would go to the condenser and distilled water will be collected. The foreign material will be transferred to a separate container. Fig. 5.27 illustrates the process of the water desalination system operation. With this technology, one can design future zero energy—zero water home by collecting energy from sunlight and water from rain and the environment. It will enable exciting mobility for people that go on vacation or travel with a mobile home.

In the water desalination system of Fig. 5.27, the steam converter is powered by the stored electricity. Due to the high latent heat of water from liquid to vapor, depending on the design and working condition, a

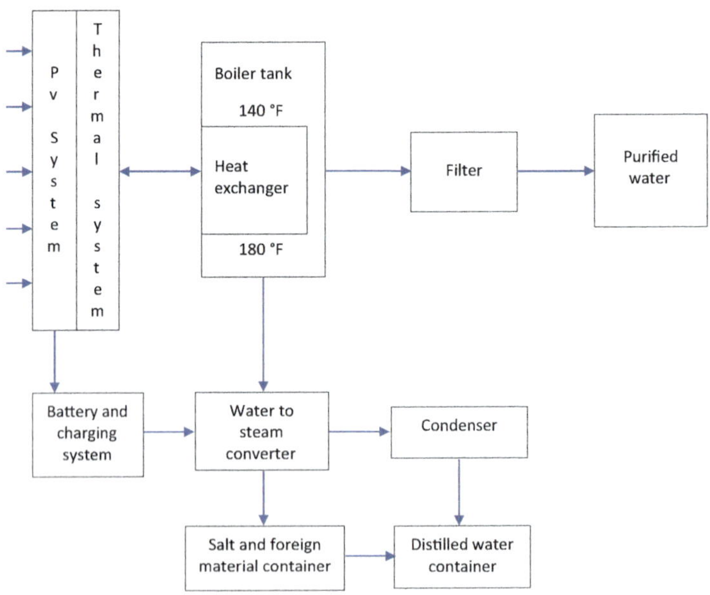

FIGURE 5.27 BITERS—water desalination system operation. *BITERS,* building integrated thermal electric roofing system.

concentrated solar power system can be added to save electricity for better use. Provided the day is sunny, the reflectors could concentrate sunlight and heat the water to boiling, the steam could go in one direction while the foreign material could go in another direction. Concentrated solar thermal collectors can be made with solar reflectors, such as parabolic trough and dish, or Fresnel lens.

As the hot water temperature level reaches a range of 140~180°F, it will be flowing or pumped into a transparent container with two reflectors, one on each side. The reflector will shine the sunlight onto the container. The solar energy would heat the water to boiling point which result will convert water to steam. Once the steam reaches a certain pressure, the valve would release the steam to the condenser.

Fresnel lens can convert hot water to steam in a similar fashion. The Fresnel lens would aim at an appropriate angle onto the water container. The hot water comes from the tank through the thermal system from the roof. As the water heats up, it will be converted to steam. Once the steam reaches a certain pressure, it will go through a valve and be released to the condenser.

The above system can be easily set up for a small water purification process in a short time window. However, to scale it up for a long-term application, solar trackers are necessary as the sun is moving and the solar thermal collectors need to move accordingly to maintain the concentration ratio. Some emerging technologies will be introduced in Chapter 10.

5.6 Design and development of building integrated thermal electric roofing system at the research laboratory

In parallel to industry's research and development, coauthor Dr. Huiming Yin started research on BIPVT when he became an Assistant Professor at Columbia University in 2008. When the gas price rocketed up to >$4 per gallon, infrastructure received grades "D" or "F," global warming was under debating those years, Dr. Yin saw the critical need to change the civil infrastructure for sustainable development and climate adaption. He invented a novel BIPVT panel integrating all components into one piece with a functionally graded material (FGM) and established the Sustainable Engineering and Materials Laboratory (SEML) at the Carleton Laboratory of Columbia University. The SEML was later grown to the Pao Laboratory in 2017 and operated at a large facility at Poughkeepsie, NY, for 2 years and became the base for National Science Foundation (NSF) Industry/University Cooperative Research Center—Center for Energy Harvesting Materials and Systems at the Columbia site, which has been directed by Dr. Yin as well. Although for some reasons, SEML has to move back to Carleton Laboratory at Manhattan of New York City, the Pao

Laboratory became a nursery for several exciting BIPVT technologies. This section will walk through the research on the design and development of BIPVT by Dr. Yin's team at Columbia University.

5.6.1 Early building integrated thermal electric roofing system prototype

As a licensed Professional Engineer in Civil Engineering, Dr. Yin's early work mainly focused on the application of solar panels in building envelopes. Because PV modules exhibit the optimal efficiency at about the room temperature (Radziemska, 2009; Ibrahim and El-Amin, 2012), shown in Fig. 5.28, which is consistent with human's indoor comfort, thermal management of the building envelope becomes a win–win solution for both solar energy harvesting and building energy efficiency. Moreover, the heat can be useful if it can be managed wisely. However, most solar projects were based on the building applied photovoltaic (BAPV) technology by attaching solar modules to the existing roof with many potential problems to civil engineers, such as additional loading, leakage, and compliance with wind/snow codes, which make the energy benefits significantly discounted if no governmental incentives.

Although solar energy spreads thin, it is renewable and clean. It is available and accessible to everyone but the right way is needed to use it. Therefore, some design objectives were scratched to change the engineering practice:

- Self-supply energy in both electricity and heat to meet building's needs.
- High performance in the separation of water, moisture, sound, and heat from outside as a building envelope.

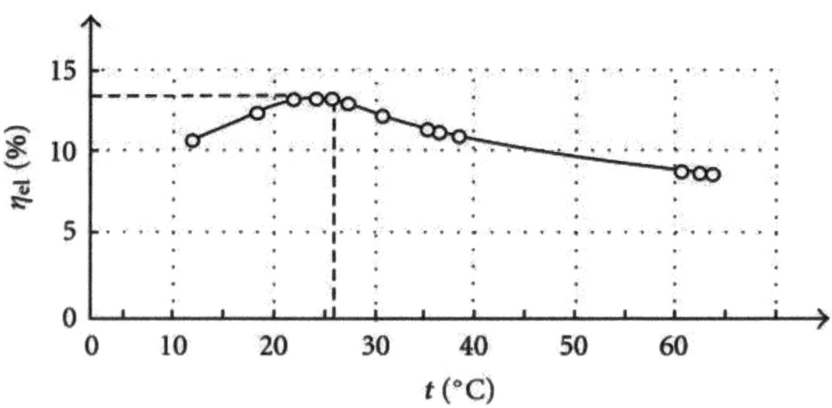

FIGURE 5.28 The photovoltaic efficiency of a silicon solar cell changing with temperature (Radziemska, 2009).

- Strong materials and structures for mechanical and temperature loads including wind, snow, and service loads.
- Aesthetic and durable building skin with desired aging resistance and fracture toughness.
- High benefit-cost ratio in its life cycle including materials recycling, energy, and environmental impacts, etc.
- Excellent workability and maintainability for installation and operation of the system Inspired by the sandwich construction method which was developed in the past century, Dr. Yin invented the early BIPVT prototype in Fig. 5.29.

The novel hybrid solar panel has been designed for PV, thermoelectric (TE), and hot water utilization through a multilayered configuration, which is schematically illustrated in Fig. 5.29. A PV surface layer and the TE layer below transfer the photoelectric and TE energy into electricity, respectively. The TE layer is bonded to a plastic lumber plate through an FGM interlayer. The FGM contains aluminum (Al) powder dispersed in high-density polyethylene (HDPE) matrix with a graded microstructure seen in the left top of Fig. 5.29. Water pipelines are cast within the FGM to control the panel's temperature. The plastic lumber, made of recycled polymeric materials, provides mechanical loading support and heat insulation of the building skin.

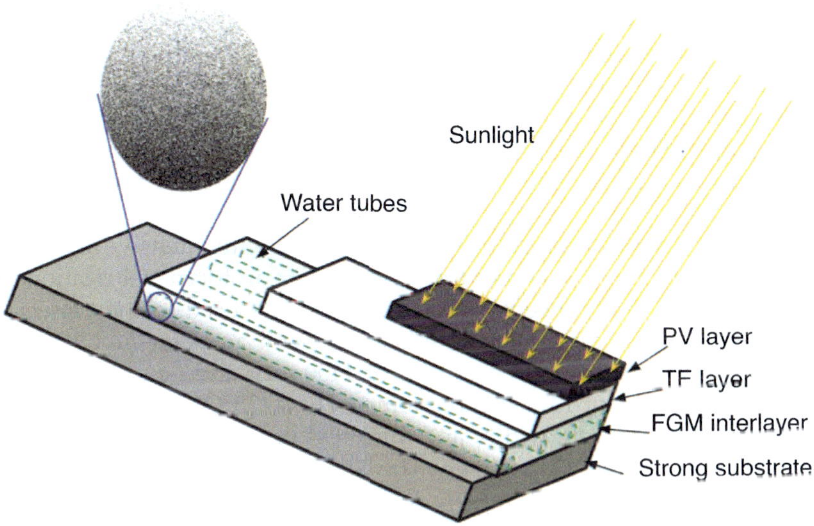

FIGURE 5.29 The photovoltaic efficiency of a silicon solar cell changing with temperature (Yang and Yin, 2011).

The innovations of the design are summarized as follows:

- The temperature difference between the PV module and the water tubes provides a considerable temperature gradient within the TE layer for higher efficiency of TE utilization.
- The hot water, whose temperature is partially controlled by the flow rate, can be directly utilized by water heating systems for domestic usage.
- Due to the temperature control on the roof, the room temperature can be significantly reduced and thermal comfort in the building can be much improved.
- The various materials and components will be chemically and physically compatible to assure a durable end product.
- A high percentage of Al powder makes rapidly transfers heat into the water tubes, but below them, the heat conduction is blocked by the HDPE and the plastic lumber substrate.
- The thin-film PV layer improves the heat conduction and structural integrity within the panel and protects the polymer materials underneath from ultraviolet radiation.
- The plastic lumber substrate provides mechanical and structural support for the upper layers and thermal insulation for indoor air as well as to prevent water condensation.

In addition, the design of this roofing panel is generalized, such that new high efficient PV and TE modules can be easily integrated into this structure. Because plastic lumber is mainly made of recycled HDPE, at the end of their lifetime, solar panels can be recycled into plastic lumber again.

This early research has been sponsored by US Department of Energy (DOE) and New York State Energy Research and Development Authority through the collaboration with a small business company—Weidlinger Associates, Inc. on the project DOE STTR DE-SC0003347. Fundamental research was later sponsored by NSF per Dr. Yin's NSF Career Award—CMMI 0954717.

Early prototypes were fabricated in the SEML with intensive labor work from the team. Fig. 5.30 shows the materials for prototype fabrication. For demonstration and concept proof, the PV layers used in the panels are commercial single-crystalline Si solar cells with an open-circuit voltage V_{oc} of 0.55 V, short circuit current I_{sc} of 4400 mA, and an energy conversion efficiency η_{pv} of 13% at room temperature (25°C) under 1000 W/m^2 irradiation. The length, width, and thickness of the solar cell are 12.5 cm, 12.5 cm, and 270 μm, respectively. The TE layer is currently made of bulk bismuth telluride (Bi2Te3) alloys. A total of 97 Bi_2Te_3 thermocouples are arranged electrically in series and thermally in parallel to form the TE module with a length of 2.9 cm, a width of 2.9 cm, and a thickness of 0.5 cm. Pure Al powder and HDPE matrix are used to fabricate the FGM

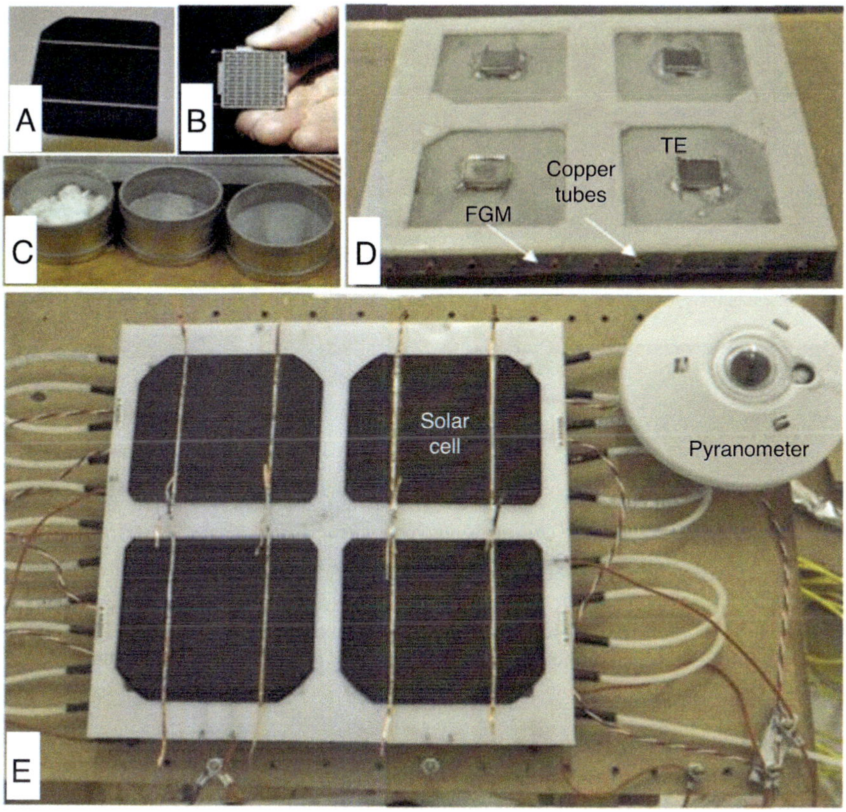

FIGURE 5.30 Prototype fabrication materials: (A) single crystalline Si solar cell, (B) TE module, (C) HDPE and Al powder, (D) solar panel before attachment of solar cell, and (E) prototype of hybrid solar panel with water tube connections (Yang and Yin, 2011). HDPE, high-density polyethylene; *TE*, thermoelectric.

layer. The thermal conductivity of Al and HDPE are 238 W/m.K and 0.26 W/m.K, respectively. The water tubes are made with copper (Cu) pipes with a diameter of 6.4 mm. The plastic lumber substrate is made of recycled polymeric materials.

One very time-consuming and challenging step is to fabricate the FGM layer. Fig. 5.31 illustrates the steps:

1. Mix the Al and HDPE powders at a certain ratio, place them on a fine sieve (#100), and sieve the powers to fill the mold uniformly. Because it is hard to disperse the powders, extra effort to press and scratch the agglomerated powders.
2. Repeat Step 1) with different Al:HDPE ratio, and fill the mold layer by layer.

FIGURE 5.31 Fabrication procedure of FGM for the early prototype of the BIPVT module. *BIPVT*, building integrated photovoltaic thermal; *FGM*, functionally graded material.

3. At a certain depth, insert the copper tubes in the mold, and continue step 1) until the mold is loaded as planned
4. Heat the mold in a vacuum oven to 200°C and ~10 mTorr.
5. When the mix is fully melted, cover the mold with a plate and press it over a universal test machine for shaping and curing.
6. Demold and obtain the FGM plate for prototype integration

Three types of solar modules were fabricated and assembled according to Fig. 5.32. Thermal conductive paste with a conductivity of 1.9 W/m.K is used as an adhesive between the PV, TE, and FGM layers. The bare PV cells are directly bonded with the plastic lumber and the FGM layers to form the reference sample and sample A, respectively. For sample B, the TE layer is bonded with the FGM layer followed by the bonding of the PV layer above the TE layer. The complete panels are degassed in a vacuum oven at 90°C for 2 hours.

The early prototypes were tested on the roof and the laboratory environment (Fig. 5.33) and the results are shown in Table 5.1.

Although a significant cost and effort were taken for integration of the TE layer into Sample B, the TE layer only provided a minor or negative contribution to the overall electric conversion efficiency. However, Sample

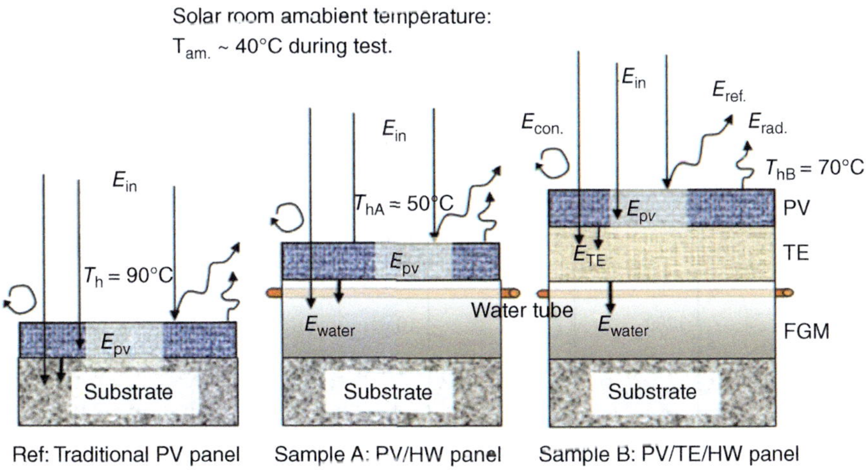

FIGURE 5.32 The solar module configuration during the performance testing (Yang and Yin, 2011).

FIGURE 5.33 The field test of the prototype on the Pupin Building of Columbia University.

A provides excellent performance in comparison with the reference sample fabricated in a traditional way without thermal management. Therefore, the original design was simplified with PV on FGM directly.

TABLE 5.1 Results of performance testing of the building integrated photovoltaic thermal (Yang and Yin, 2011).

Performance index	Irradiation (W/m^2)	Traditional photovoltaic panel	Sample A photovoltaic/ functionally graded material	Sample B photovoltaic/TEFGM
Photovoltaic cell efficiency	750	12.6%	16.6%	16.0%
	1000	8.4%	12.1%	11.0%
	1200	7.1%%	10.6%	8.5%
Total electric power (W/m^2)	750	94.5	124.5	(Photovoltaic) 120+(Bulk TE) 7.2 = 127.2
	1000	84	121	(Photovoltaic) 110 (Bulk TE) 9.4 = 119.4
	1200	85.2	127	(Photovoltaic) 102 + (Bulk TE) 9.0 = 111.0

5.6.2 A novel manufacturing method

As the PV–FGM design showed excellent performance, the team started to work on scaling up the fabrication to mass production. However, the procedure in Fig. 5.31 is not affordable or practical due to the long processing time and difficulty for quality control. Inspired by the natural FGM produced by sedimentation, the team started to experiment with the solution-based method. Fig. 5.35 shows the early experiments (Lee and Yin, 2015) to fabricate FGM by sedimentation of suspension solution. In the mixing procedure, the 30 g 1:3 of Al-HDPE power mix was introduced into 80 mm height of ethanol and well separated by ultrasonic devices for 10 minutes, then, also stirred by a mechanic mixer with mixing speed 500/s for 10 minutes. After mixing, the homogenous suspension would be placed in the glass tube and started settling. Through 24 hours of sedimentation, more than 98% of particles were completely settled; minor particles were still floating in the suspension above the deposition. The deposit body was obtained after draining the suspension off and vaporing remaining ethanol out of the deposit body by 6 hours air-dry. In this way, we can obtain the FGM mix in one step.

However, it was not very useful because it took too long time. Moreover, the results were very sensitive to particle size and a solid load of suspension (Yin et al., 2014; Liu and Yin, 2014), which has been investigated by Dr. Yin's team. To accelerate the fabrication process and increase material output, increasing solid load is necessary to shorten the falling distance and thus the sedimentation time. However, once the solid load reaches above 25%, it is very challenging to create a desirable particle gradation for FGM fabrication while our goal is to reach at least 60% and ideally 75% of the solid load. When the student Dr. Pohua Lee was struggling in failure with testing many batches of samples with different solid loads, a miracle occurred to him. Among many samples, he found one with beautiful material gradation in spite of a high solid load of 60%. To find the reason, he took the sample to the concrete lab and realized that the change could be caused by the vibration while he was walking. Then he further used a shaker to repeat the tests. Through some experiments, he managed to use 75% of the solid load to make FGM mix in 15 seconds in Fig. 5.36.

This discovery gave birth to a novel manufacture method, which is scalable for the FGM panels. The team can make BIPVT prototypes in a short time and could use them for many other designs and configurations. Different substrates using wood, foamed materials, concrete were tested for this novel design (Chen and Yin, 2016).

With the significant reduction of PV cell's price in the recent year, although we have conducted significant research, the cost weight of FGM is relatively increasing in the BIPVT module (Chen and Yin, 2016), which gave the research team a new task to further simplify the design. Some emerging work is to integrate the heat exchanger into the structural substrate directly.

5.6.3 Modernized experimental facility

Developing a new BIPVT technology, which is designed for >25 years of lifetime, requires a rigorous theoretical foundation and precise measurement for the prediction and optimization of BIPVT's performance and lifetime. In the beginning, the team conducted tests in the field but found that the performance of solar panels changes significantly with the cloud moving (Fig. 5.33). They build a sunroom facility to use a metal halide global lamp (Fig. 5.34), which can simulate the solar irradiation spectrum well but the temperature cannot be controlled. The team started to design and acquire a multifunctional environmental chamber for three needs:

- Reproduce different weather conditions for new material and structure design and performance tests.
- Provide accelerated weathering environments for life-cycle analysis with short-term tests.
- Measure the energy and environmental impacts during material production and degradation.

They started to work with three chamber manufacturers, including Weiss Technik, Envirotronics, and Atlas, to design and develop a multifunctional environmental chamber in 2009, and eventually, a multifunctional weathering system, as shown in Fig. 5.37, was designed which includes an environmental climatic chamber, calibration components, and

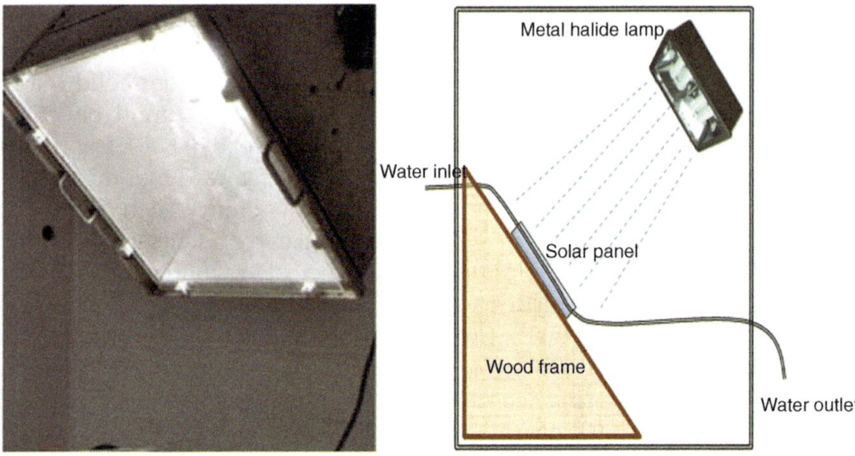

FIGURE 5.34 The laboratory test of the BIPVT module in the Sun Room of the Carleton Laboratory. *BIPVT*, building integrated photovoltaic thermal.

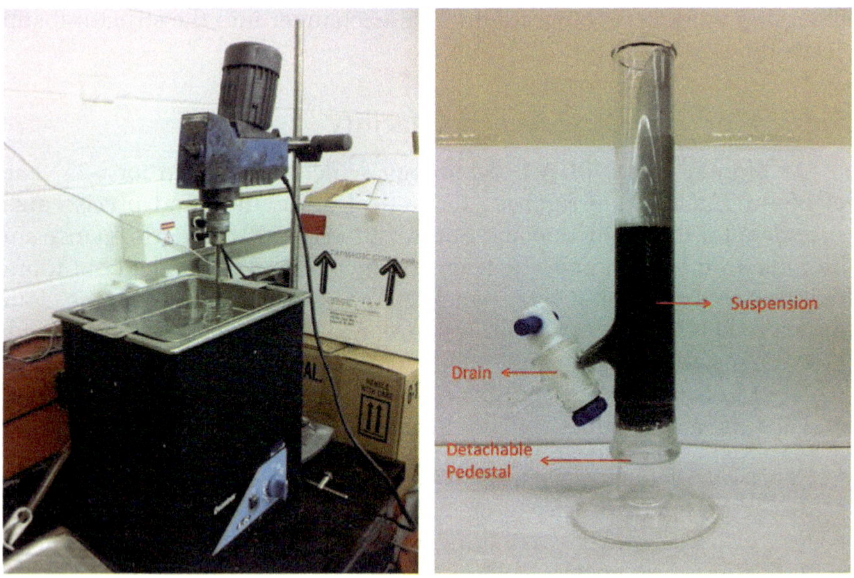

FIGURE 5.35 The solution-based method to make FGM by sedimentation. *FGM*, functionally graded material.

peripheral facilities. The chamber, which is to be airproof, waterproof, and thermally insulated, will be equipped with multiple weathering actions of temperature, moisture, ultraviolet radiation, and solar radiation at different aggressiveness levels for large scale accelerated performance tests.

FIGURE 5.36 The solution-based method to make FGM by sedimentation by vibration. *FGM*, functionally graded material.

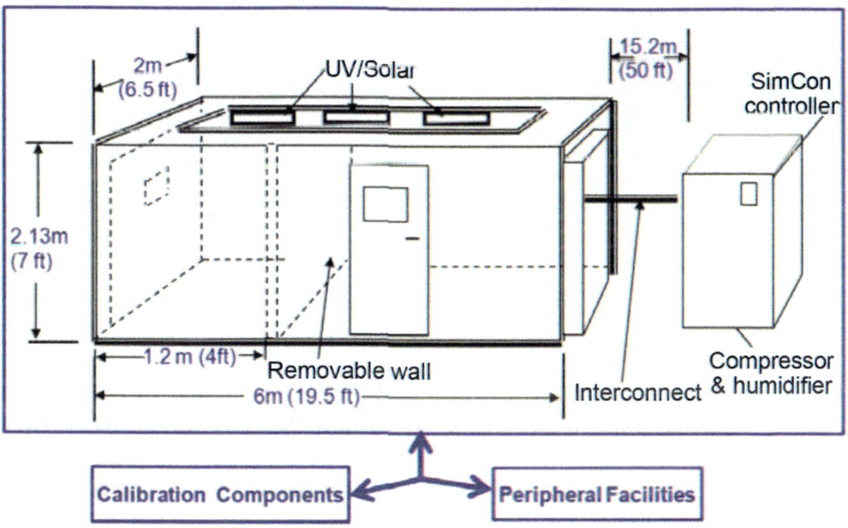

FIGURE 5.37 A multifunctional chamber system to simulate different weathers and accelerated environmental tests.

With the funds of research and equipment projects from the Department of Homeland Security CU09-1155, NSF CMMI 1301160, Office of Naval Research DURIP N00014-13-1-0795, and Henry Mitchell Weitzner Research Fund, this chamber, and measurement system have been gradually acquired and installed at the Pao Laboratory. Due to the significant support from the Weitzner family, the chamber is named "Weitzner Chamber" along with a group of other characterization instruments.

This facility is not only critical for BIPVT research and development, it can be a very useful tool for climate and environmental-related research. Many new materials are used on the outside of buildings, like roof coverings, paints, metal products, and window coatings. The Weitzner chamber will help create testing environments for life-cycle analysis. In these cases,

FIGURE 5.38 The environmental chamber installed in the Pao Laboratory.

the question is not just "will it last in the elements" but "how long will it last." Can we build a skyscraper with new "green" materials without knowing how long the materials will hold up? Durability and low impact on the environment are paramount considerations in modern construction, which will be revealed by the chamber.

Eventually, on April 20, 2018, the chamber passed the commissioning tests, as shown in Fig. 5.38. The unique aspect of this instrument is to combine multiple aging methods to reproduce some complex weathering environments and to provide the capability to conduct simultaneous performance tests under those conditions. This instrument is applicable to general materials and integrates various aging effects into one system. It was designed for both IEC standard tests and research-based experiments with easy control and programming. Specifically, this multifunctional environmental chamber provides the test capabilities as Table 5.2.

Inside the chamber, it can host a car for accelerated performance tests. It can go through a temperature cycle of −50–125°C in 90 minutes, and eight metal halide global lamps can simulate 1.8 suns uniformly and >5 suns when they are concentrating on the central area (Fig. 5.39). Basically, the chamber can reproduce almost all kinds of weather on earth in such a manner that you can experience the Phoenix's summer and Antarctic winter in

TABLE 5.2 Specifications of the environmental chamber.

Function	Capacity	Specifications
Temperature	−50–125°C	Cooling and heating rate 4°C/minute; Sensitivity 2°C
Moisture	20–95% RH	Applicable temperature range of 4.5–85°C; Sensitivity 5%
Solar irradiation	0–1800 W/m^2	Temperature −10–60°C; moisture −90 RH; Uniformity 10%
UVA irradiation	0–200 W/m^2	Temperature −10–60°C; moisture −90 RH; Uniformity 10%
UVB irradiation	0–20 W/m^2	Temperature −10–60°C; moisture −90 RH; Uniformity 10%

FIGURE 5.39 The interior space of the Weitzner Chamber for large-scale accelerated tests.

~20 minutes. Through a certain similarity of the material behavior under different environments, it enables different accelerated test methods so that one can measure the BIPVT system's life cycle performance in several weeks. This unique facility enables researchers and industry members to develop new BIPVT systems with standard reference tests and customized performance tests, which can provide reliable parameters for the life-cycle assessment and life-cycle cost analysis of new products. Therefore, we can address the economic, environmental, and social aspects of sustainability and minimize the risks of infrastructure under extreme weather conditions. This approach also provides a rational standard for federal and state agencies to make wise decisions for future infrastructure construction.

Through industry–university collaboration, university researchers can use the industry's manufacturing facility for sample fabrication, and the industry can use the university research laboratory's scientific research facility to obtain reliable data so as to optimize their manufacturing recipe. This is an intelligent way to accelerate the development of new BIPVT technologies.

5.6.4 Improved building integrated thermal electric roofing system technologies

Since the Pao Laboratory was established in July 2017 at Poughkeepsie, NY, Columbia, researchers can use the manufacturing line for rapid prototype fabrication and performance characterization. The BIPVT technologies have been improved in several ways.

First, to reduce the high-cost weight of Al-HDPE FGM in the BIPVT panel, the team has attempted to use a tube embedded in foamed Al to replace the FGM on the substrate for lower costs and better efficiency in Fig. 5.40) (Yin et al., 2021). Because Al-HDPE FGM includes dispersed Al particles in the HDPE matrix, the thermal conductivity is not as good as foamed Al with continuous Al, which exhibits higher strength with a lighter weight, so that the structural substrate can be removed for higher material efficiency. However, currently, the manufacturing cost of foamed Al is still high although it essentially uses less Al and eliminating HDPE. When a more mature, economic manufacture method becomes available,

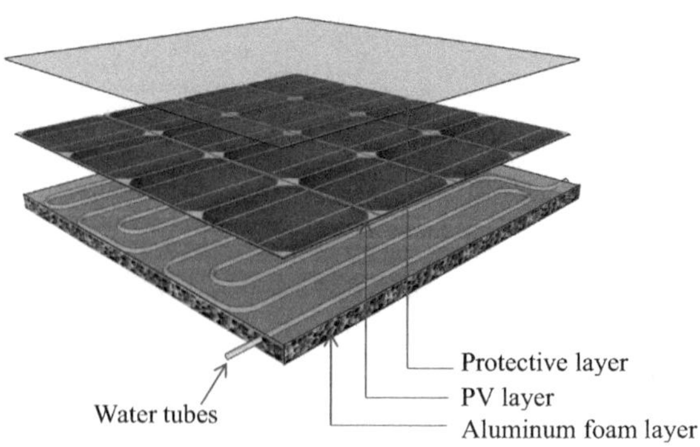

FIGURE 5.40 A BIPVT panel with a foamed aluminum including the heat exchanger. *BIPVT*, building integrated photovoltaic thermal.

it can replace the existing Al-HDPE FGM layer. It will be excellent for BIPVT systems for hot areas which require high efficient heat transfer from building skin. To resolve the economical issue, the Columbia team has used foamed concrete with metal mesh reinforcement to replace the foamed Al and obtained very promising results.

In addition, to further simplify the design, the team tried to remove the heat exchanger from the BIPVT panel and use alternative ways for thermal management, which will be introduced in Chapter 10. Here we can provide a brief preview.

Using a phase change material (PCM) filled foamed concrete for the substrate of the BIPVT panel as a modular system can use the latent heat of PCM to reduce the temperature variation of the building envelope (Yin et al., 2021). For those areas that the daily temperature variation is large but the average temperature is not high, this system can be ideal. Currently, the team is extracting PCM of fatty acid from waste oil or animal fat as a renewable, degradable material for this application. Because both foamed concrete and PCM exhibit low density and thermal conductivity, using metal mesh and polymer fiber reinforcement can tailor the thermal resistance of the panel for a high heat transfer rate to PCM but a high R-value of the overall panel. The team has developed a procedure to fabricate lightweight concrete boards for BIPVT application in Fig. 5.41. Using a surfactant mixed with water under a certain pressure to make stable bubble foam first, one can mix the foam into cement paste and fill it in a mold with metal mesh to form a lightweight concrete at the desired thickness. Incorporate liquid PCM into the foam concrete at a higher temperature, and encapsulate with PV cells to form the BIPVT. Because no air ducts or liquid tubes exist, the BIPVT system will be much simpler and affordable. However, the thermal management capacity depends on the match of PCM amount/latent heat with the temperature variation and average temperature. If continuous hot weather happens, the PCM may remain in the liquid, so that it cannot function as designed .

Another application of PCM in BIPVT application is to use heat pipes to replace the air ducts or water tubes so that one can use liquid/gas phase

FIGURE 5.41 The procedure to fabricate lightweight concrete panels for BIPVT application. *BIPVT*, building integrated photovoltaic thermal.

change for heat exchange at high efficiency while saving the metalwork for tubing. For high-temperature heat storage, PCM at a higher temperature, say 250°C, is recommended so that the stored heat can reach a practical Carnot efficiency for electric generation.

In addition, another mechanism for thermal management of BIPVT is to use a meta-surface coating to selectively absorb sunlight based on the solid-state physics, in which low-energy photons are reflected, and only high-energy photons pass the coating to trigger electron-hole generation. Therefore, the temperature increase can be significantly reduced.

5.7 Building integrated thermal electric roofing system applications to energy independent buildings

The concept of the zero-energy (ZE) building, or zero net energy (ZNE) building, has been widely discussed in the solar energy society. There are many buildings that have already reached the balance between the total amount of energy used by the building on an annual basis and the amount of renewable energy created on the site, using solar energy along with heat pumps, high-efficiency windows and insulation, and geothermal energy. This is a practical way to cut down the carbon footprint in buildings. However, it does not necessarily lead to energy independence. Today we can see many ridiculous cases that after a storm some areas were blacked out by the damage of powerline; some ZE buildings in the areas have sufficient PV harvesting but no power is available because their electricity relies on the grid instead of their solar panels. Energy-independent building requires comprehensive energy management. Fortunately, BIPVT systems can handle a certain redundancy through multiple energy resources. For example, solar energy can be harvested by BIPV and BIPVT systems; geothermal energy will be used to recover PV efficiency, heat storage and exchange, space cooling and heating, as well as water heating; building components use the state-of-the-art energy efficient materials such as PCM and foamed concrete wall panels, smart window with light sensing and control of blinds for energy efficiency. Therefore, we can envision the energy-independent buildings with BIPVT

5.7.1 From zero energy to energy independency

In light of climate change, we are anticipating more extreme weather, such as severe storms, drought, forest wildfire, which could seriously affect the reliability of our grid and cause disasters by loss of power. Many families or organizations prepare redundant electricity backup systems, such as uninterruptible power system (UPS), gas-electric generator, or batteries, in case that the grid stops work. As the existing power infrastructure is often overly loaded in response to the increasing energy demands, the centralized power supply becomes more fragile and vulnerable to all problems caused by climate change. While we try to slow climate change, we need to adapt to the new "normal lifestyle."

The most aggressive way to change is to implement energy independent house. Although not everyone needs it, it does have a decent market for remote areas, vacation purposes, or disaster mitigation. Even in areas with grid access, deployment of energy-independent building technologies can significantly improve the power infrastructure resiliency and security.

Although ZE buildings are not necessarily energy independent, an energy-independent building must be a zero energy building, or, more accurately, a negative energy building considering the redundancy for the mismatch of energy demands and supplies. To have a zero energy house or building, some of the basic areas to consider are:

1. Sound structure of the building could be durable.
2. Fireproof is a very serious consideration.
3. Insulation is very critical.
4. Heat ventilation and air conditioning system design that could optimize the use of energy especially using radiant heat.
5. Window applications: tilt and turn windows or electrochromic windows (smart windows).
6. Optimize the electricity usage by using LED lighting.
7. Optimize the heat usage by using a heat pump instead of direct heat.

All the above items could play an important role to make an energy-independent home or zero energy building on a solid foundation. Otherwise, the energy loss or low efficiency will ruin the work of energy harvesting. BIPVT is truly the solution for energy-independent building because of the universal availability and accessibility of solar energy as well as the compatibility with the building energy needs in both electricity and heat, particularly when it is coupled with geothermal energy which will be elaborated in Chapter 6. Although wind and wave are very exciting energy resources, they are limited to a certain area or season.

To make energy independent, the energy supply and consumption need to be balanced at all times. The smart window project in Chapter 10 provides an algorithm to use energy storage to achieve equilibrium. Because of the uncertainty of the weather, both energy supply and consumption will be exposed to some random factors, so certain redundancy or safety factors should be applied to make a practical estimation.

The BITERS system has proved that it could make the house totally self-sustaining even in severe weather conditions in places like northern midwestern states such as Wisconsin. A horizontal geothermal system together with the BITERS could work out quite well with thermal management.

Using a borehole geothermal well, the thermal energy collected in the warmer days could be stored and in the winter months, the thermal energy could be applied to keep the house warm (see Chapter 6). More details on the Wisconsin-located house will be presented in a case study in the next chapter.

In addition, another application on the BITERS is that on the PV system side, it could be connected to an electrolyte, which could use water to generate hydrogen. Hydrogen gas could be used on fuel cells to generate electricity in the household or could be applied to cars because the coming generation of cars will be using hydrogen as fuel because fuel cells will also be used as car engines. Recent advancements in fuel cells have made them more affordable. It could be by far more durable than a battery. This idea will be revisited in Chapter 8.

On the current geothermal system especially out in the field, they are thermal tubing just being laying openly in the ground which could be of concern. Though the soil is a great insulator, it constantly has movements due to weather which results could have problems with the tubing and integrity of the installation. Cost is always a concern but if the system could not operate after a number of years is an even greater concern. Just look at the radiant heat floor, the main reason it lasts so long is that it is relatively in a well-protected environment. We should change our thoughts in considering doing the same to the geothermal system. The geothermal system basically should not have any moving parts, similar to the PV system. Also, it is buried in the ground. One should anticipate they should be able to last in the order of forty to fifty years.

5.7.2 Microgrid and energy storage for building integrated thermal electric roofing system applications

An energy-independent building basically makes it possibly an off-grid one although it often needs to connect to the grid to feed extra electricity back to it. Therefore, the building itself can be a grid or a microgrid to manage the electricity use in the building. Actually, it well fits the philosophy of a decentralized power system for infrastructure resilience and redundancy, which may use our homes, buildings, and communities as microgrids to share the load of the central grid or macrogrid. Besides, most grids in the nation are still using fossil fuels which is a critical concern. More expansion of microgrids with BIPVT systems will cut the CO_2 emission.

A microgrid consists of PV generators, electricity storage, grid connection, and DC/AC loads. The PV generators are the DC power source with the efficiency depending on the irradiation and panel temperature. Because of the mismatch of power generation and consumption, certain energy should always be saved in the electricity storage for the electricity supply at nighttime. Although DC loads are favorable on efficiency and compatibility, many electrical loads in buildings are designed for AC, which requires each element to have identical wave shapes in order to operate. The synchronization requires an inverter in order to provide the AC power link and be compatible with the AC from the macrogrid when the microgrid is not standalone.

Because the power generated from the PV generator exhibits significant variation with irradiation and temperature, a smart power balancing strategy is needed to smooth grid peak demand, avoid undesirable grid power injection, and make full use of local PV production.

At daytime, most time PV generators can harvest more energy than the energy demand so as to save a portion in the storage. Although several energy storage media will be reviewed in Chapter 8, in building applications, today batteries are the most common and practical choice with its maturity in electric vehicles. Because both PV and batteries are DC, the microgrid can perform in a similar fashion. In the future, more applications may turn to heat storage in the form of high-temperature heat. The heat engine will be used for electricity supply, and AC–AC synchronization will be required.

Using PCM with a melting point of ~250°C, which is similar to most geothermal electrical generator's heat sources, one can save energy at a much lower cost and generate electricity at a decent efficiency as well. However, it may require a cascade heat pump to use some electricity to boost the heat from BIPVT to such a higher temperature.

5.8 BIPVT design and demonstration at the Cherokee home

The heat on solar panels could affect the solar cell efficiency and the durability of solar modules, as well as the indoor comfort. BIPVT is a valuable, practical, and viable approach to change the building and energy practice. The research and experiments in the past will be an invaluable asset for us to move forward. As we continue to work on net zero energy houses or buildings, we should also look in every way for zero-emission technologies which are vital to climate change. It is a tremendious task to coordinate the BIPVT subsystems for the optimal performance and maximal benefits under the uncertain working environment, which requires seamless design and management among architectural, structural, electrical, mechanical, water and enviromental, geotechnical, and industrial and operational engineering aspects. The BIPVT technologies are still at the beginning phase, and industry standardization and optimization are still in the process to reach more durable and affordable products. Here we use one example for BIPVT design and demonstration for residential applications.

The Cherokee home in Fig. 5.42 at Raleigh, North Carolina, was designed by William McDonough—Renowned Solar Technology Architect. This home gives a great example that a traditional home could be modernized into net-zero energy with all the latest solar electric and thermal roof technologies together with a geothermal system to make the house net zero energy which demonstrates that most traditional homes could be renovated with modern energy systems to conserve energy. It is the first residential home in the country under the NAHB (National Association of Home Builder) Green House Building Guidelines, which was certified by EPA's Energy Star Program.

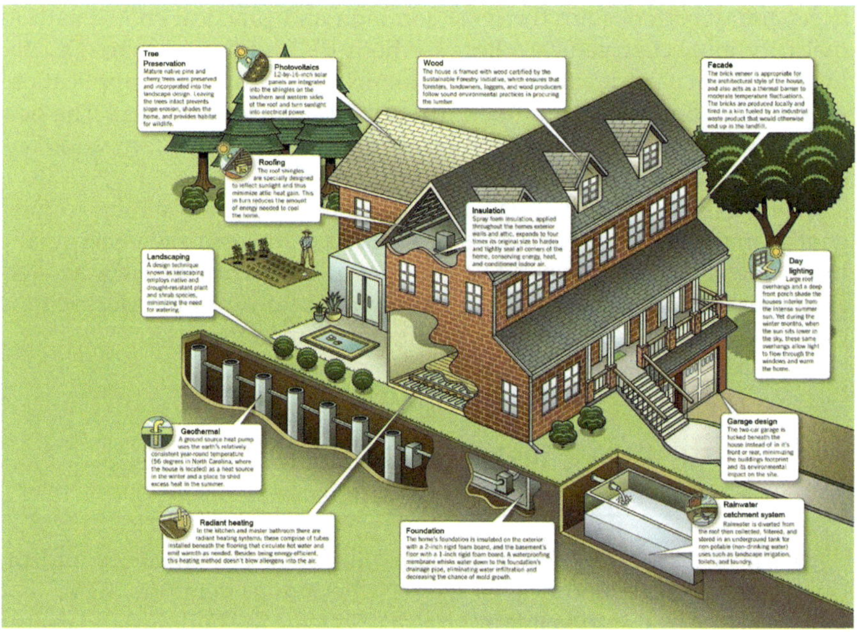

FIGURE 5.42 Cherokee Home BIPVT at Raleigh, North Carolina. *BIPVT*, building integrated photovoltaic thermal.

Fig. 5.43 shows the layout of the 385 Sunslate panels for a 5.12 kW roof system. In addition to what is shown on the side, there are some unique aspects in the design and operation of the BIPVT system (Weber, 2006):

1. Sunslate—the solar roof tiles which reduce the electricity consumption of the building and beneath it is the solar thermal system which collect the heat the assist in domestic hot water as well as cooling off space which resulted in saving air conditioning cost.

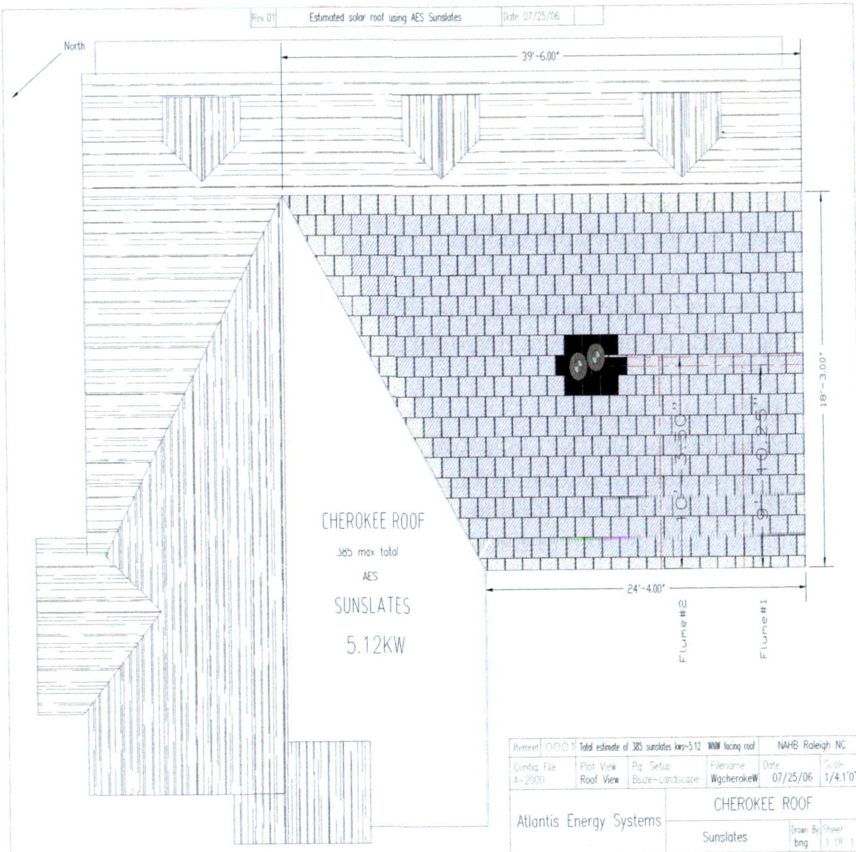

FIGURE 5.43 Layout of the 385 Sunslate panels for a 5.12 kW roof system.

2. Radiant floor heat—shown at the bottom of the building.
3. Geothermal—beneath the building is the geothermal system which collects the heat in the summer month and storages the thermal energy for the winter months.
4. Rainwater collection—the water collected from the roof will go to the tank shown at the bottom of the building. As we all well know that water is a precious commodity.

Together, the design of this house has these items in mind:

(a) On the environmental side, it has the ability to recycle 90% of organic waste.
(b) Consume 50% of less water.
(c) Retain 95% of stormwater collected and reuse them.

(d) Recycle 75% of all demolition and construction waste.
(e) Provide great indoor quality of air which includes that 95 percent of all the products that do not contain or very low amount of volatile organic compounds.
(f) On sustainability areas: (1) energy star rated windows, (2) appliances, (3) the best type of insulators, (4) ground source heat pump, (5) solar thermal water heater.
(g) It also creates a wildlife habitat

This project created an excellent example for next generation sustainable buildings and has been recognized for a long list of Awards, including:

- US DOE Gold Award.
- National Home Builder Gold Award.
- The city of Raleigh overall environmental award (all categories).
- LEED platinum award (first in South East US).
- National Wildlife Federation: Certified Wildlife Habitat for backyard.
- Gold-certified under the North Carolina Center of solar healthy homes.
- First Home in the nation built in the typical subdivision under NAHB Model Green Home Guidelines.
- 2008 International Builders Show (IBS), Gold Energy Value Housing Award.
- National Association of Home Builder Research Center, Highest Honor Award.
- NREL: Gold Award for successfully integrating cutting-edge energy features.
- Active Energy Building: Energy Globe Award.

5.9 Case study 1: Turn an old house into the Beauty at the Beach

The original house was built in 1932 overlooking Truesdale Lake, Westchester County, NY (Figs. 5.44 and 5.45). In late 2011, its renovation began with planning a design to provide maximum comfort as well as a minimal carbon footprint. The renovation was done by Sylvain Cote, an architect and builder, and was finished in 2014 (Fig. 5.46). For optimizing the solar energy performance on the tight south-facing roof, Atlantis Energy Systems combined electric and thermal roof systems and used the Sunslate BITERS (Building Integrated Thermal Electric Roofing System). In addition, the Sunslate system satisfies the historic and zoning code of this location (Fig. 5.47).

FIGURE 5.44 The 83 year old Beach House in 2011 (front view).

FIGURE 5.45 The Beach House facing Truesdale Lake (back view).

FIGURE 5.46 The Beach House in Truesdale Lake, South Salem, New York in 2014.

FIGURE 5.47 The finished renovated Beach House in 2014.

Installed components [15]:

1) 3.30 kW Sunslate PV with thermal system beneath provides electricity and domestic hot water.
2) Unico high velocity small duct system that effectively removes humidity and delivers conditioned air throughout the walls and ceiling with hardly any energy loss.
3) Viega radiant heat system which is very healthy for space heating without the need for air circulation.
4) A Fantech energy recovery ventilator that constantly brings fresh air from the outside which improves the indoor air quality.
5) All the above systems can be controlled remotely to optimize energy efficiency.

The house project has been recognized by the following awards:

1) Best Renovation Home of the Year by Green Builder 2015.
2) LEED Platinum Certification of 90.
3) Energy Star Certified at HERS (Home Energy Rating System) rating 30.

5.10 Case study 2: Dover building integrated thermal electric roofing system installation

The project is located in Dover which is a suburb of Boston, Massachusetts, and the client's main interest in the BITERS was for his indoor swimming pool.

BITERS system size: PV system 390 Sunslate modules at 22.1 W each about 8.6 kW

Thermal System 800 sq. ft.
Tilt angle – 45 degree
A closed-loop system
1 A/F Storage Tank with a capacity of 40 gallons
Taco 009 Pump in Collector Loop
RESOL DeltaSol MX Controller
A-Thermal-Grid DL 2to send data to servers
39 Enphase inverters, each whole a string of 10 Sunslates
Rated Power Output: 15.3 kW
RESOL, V40-15 flowmeter for DHW flow
RESOL, V40-15 flowmeter for Pool flow

The schematic in Fig. 5.48 shows the layout of the sensors in the whole system. To the client's request, due to the high cost of the utility bill, the thermal system is a sizeable compensation to the cost according to readings. More detailed information has not been able to obtain.

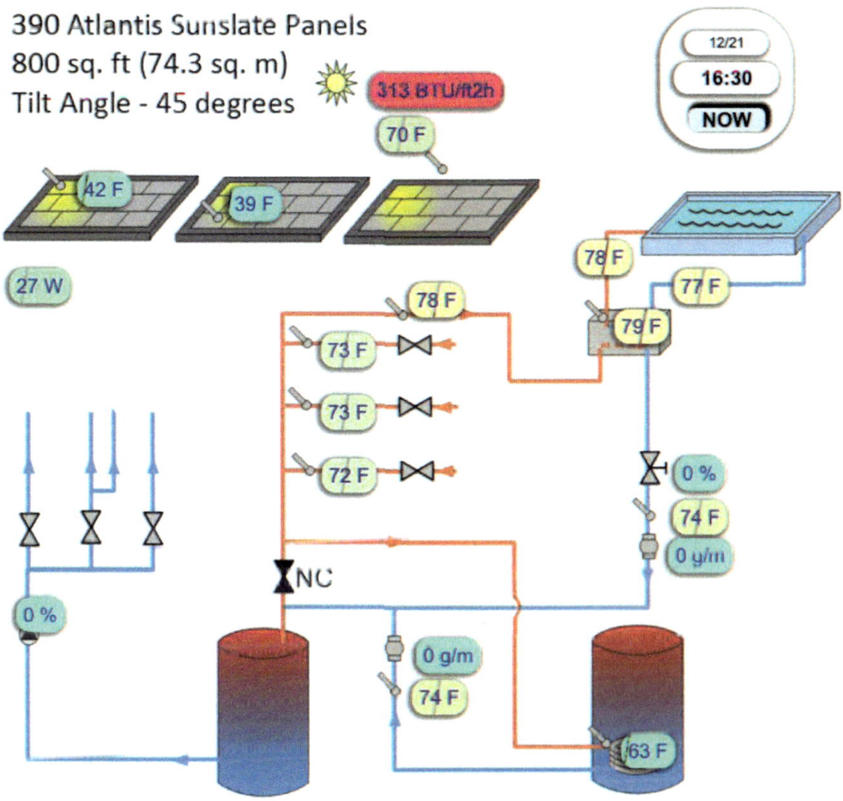

FIGURE 5.48 Dover system layout and performance.

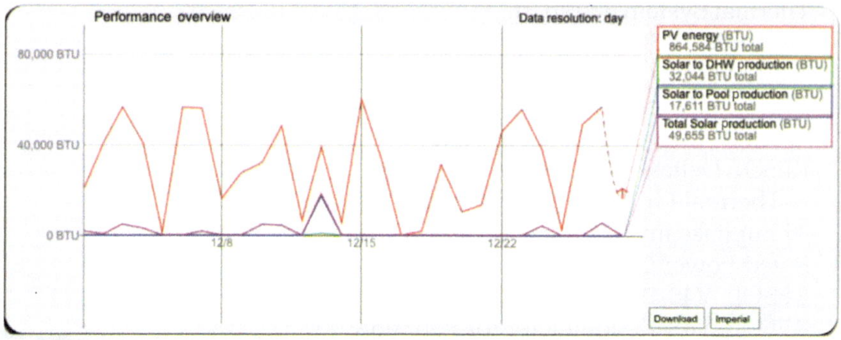

FIGURE 5.49 System performance on May 24, 2020.

The temperature near Boston and the neighboring towns was about low 60°F to high 50°F which is quite normal for a spring day (Fig. 5.49). In the following tables, the collected data on the performance of the project throughout each month on the average temperature in degree Fahrenheit are shown in Tables 5.3–5.7.

The pool data maintained quite steadily throughout the year and the tank data had some fluctuation but not too much. The variation of the solar collector is observed quite a bit throughout the year as it was anticipated

TABLE 5.3 Average temperature of the system in each month of 2015.

Month (2015)	Pool temperature (°F)	Bottom tank temperature (°F)	Collector temperature (°F)
Jan.	76	62	30
Feb.	76	60	27
Mar.	76	64	41
Apr.	77	66	59
May	81	81	80
Jun.	82	77	79
Jul.	79	92	90
Aug.	81	86	88
Sep.	83	81	81
Oct.	81	70	62
Nov.	80	66	53
Dec.	74	67	48

TABLE 5.4 Average temperature of the system in each month of 2016.

Month (2016)	Pool temperature (°F)	Bottom tank temperature (°F)	Collector temperature (°F)
Jan.	74	63	36
Feb.	76	64	42
Mar.	73	67	41
Apr.	70	70	59
May	80	74	71
Jun.	83	79	87
Jul.	85	87	87
Aug.	85	88	87
Sep.	83	77	77
Oct.	82	72	64
Nov.	79	66	52
Dec.	78	65	37

TABLE 5.5 Average temperature of the system in each month of 2017.

Month (2017)	Pool temperature (°F)	Bottom tank temperature (°F)	Collector temperature (°F)
Jan.	68	63	39
Feb.	76	64	43
Mar.	74	64	41
Apr.	77	69	62
May	84	85	85
Jun.	82	79	82
Jul.	84	85	85
Aug.	83	85	85
Sep.	81	75	78
Oct.	82	71	70
Nov.	80	64	49
Dec.	80	63	34

TABLE 5.6 Average temperature of the system in each month of 2018.

Month (2018)	Pool temperature (°F)	Bottom tank temperature (°F)	Collector temperature (°F)
Jan.	73	60	33
Feb.	68	61	44
Mar.	73	62	43
Apr.	73	64	53
May	82	72	74
Jun.	83	76	79
Jul.	86	88	88
Aug.	86	83	86
Sep.	85	72	75
Oct.	81	66	59
Nov.	79	62	46
Dec.	78	62	42

TABLE 5.7 Average temperature of the system in each month of 2019.

Month (2019)	Pool temperature (°F)	Bottom tank temperature (°F)	Collector temperature (°F)
Jan.	78	60	36
Feb.	79	61	40
Mar.	79	66	48
Apr.	82	72	75
May	81	67	70
Jun.	85	76	85
Jul.	86	84	95
Aug.	85	84	95
Sep.	86	84	95
Oct.	82	72	64
Nov.	79	66	52
Dec.	78	65	37

due to being located in the New UK region. During the winter, spring and fall, the collector temperature could go up as high as 80°F and drop down to as low as 20°F at night. Based on the average temperature in the tank, it is about low 70°F. It should not take too much more energy from the utility to get up to the high 70°F or low 80°F. Normally, the water from the tap is about the low 50°F in the summer and low 40°F in the winter which is a substantial difference to warm the pool especially, it is a heat sink.

In going through the past five years of operation in Tables 5.3–5.7, the thermal system operated and performed quite consistently. The Sunslate system since the installation time has generated 40,609.8 kWh energy. Normally, in the northeast region, the thermal energy required is at least one to one and half times the electrical energy. Considering the size of the pool which is about 40 × 18 ft being quite large in a single-family home, the thermal system has done a reasonable performance.

Looking at the BIPVT installation shown above, it seems to be one of the major solutions to bring houses and buildings to net-zero energy. In some locations, the design of the building may require additional systems to reach this goal, while in warmer regions of the country such as California and Southeastern states, the insolation throughout the year is strong and consistent enough to achieve net-zero energy. The initial cost of BIPVT could be high but in the long run, it will pay off for itself since the building has become energy independent.

References

Chen, F., Yin, H., 2016. Fabrication and laboratory-based performance testing of a building-integrated photovoltaic-thermal roofing panel. Appl. Energy 177, 271–284.

China.org.cn, 'Future House USA' good demo of Sino-US green cooperation. 2009 http://www.china.org.cn/environment/news/2009-07/16/content_18150544_3.htm (accessed on 2021).

Ibrahim, A., El-Amin, A., 2012. Temperature effect on the performance of n-type μc-Si film grown by linear facing target sputtering for thin film silicon photovoltaic devices. Optoelectron. Adv. Mater.-Rapid Commun. 6, 73–77.

Lee, P.-H., Yin, H., 2015. Size effect on functionally graded material fabrication by sedimentation. J. Nanomech. Micromech. 5 (2), A4014008.

Liu, Y., Yin, H., 2014. Equivalent inclusion method-based simulation of particle sedimentation toward functionally graded material manufacturing. Acta Mech. 225 (4), 1429–1445.

Morse, E.S., Warming and ventilating apartments by the sun's rays. U.S. Patent 246,626, issued September 6, 1881

Pao, F., Optimized building integrated hybrid roofing system. U.S. Patent 8,196,369, issued June 12, 2013.

Pao, F., Gottlieb, R. Building integrated thermal electric hybrid roofing system. U.S. Patent 8,196,369, issued June 12, 2012.

Radziemska, E., 2009. Performance analysis of a photovoltaic-thermal integrated system. Int. J. Photoenergy 2009. doi:10.1155/2009/732093

Trombe, F., Michel, J. Naturally air-conditioned dwellings. U.S. Patent 3,832,992, issued September 3, 1974. Harvard.

Weber, C., Home magazine, 2006.
Wikipedia, Augustin mouchot. 2009 https://en.wikipedia.org/wiki/Augustin_Mouchot.
Yang, D., Yin, H., 2011. Energy conversion efficiency of a novel hybrid solar system for photovoltaic, thermoelectric, and heat utilization. IEEE Trans. Energy Convers. 26 (2), 662–670.
Yin, H., Lee, P.-H., Liu, Y., 2014. Equivalent inclusion method for the Stokes flow of drops moving in a viscous fluid. J. Appl. Mech. 81 (7), 071010-1–071010-12.
Yin, H., Pao, F., Zadshir, M., Lou, J., Wei, Z., 2021. Building Integrated Photovoltaic Thermal Systems. U.S. Patent Application No. 63/248,502.

The winter and summer scenes of a residential house are shown in this figure. In summer, solar energy is harvested by the BIPVT system. The excess part of the heat is stored under the ground. In winter, the heat and geothermal energy under the ground can be pumped to the roof through a passive heat exchange system to warm the building. On a heavy snow day, the roof can be heated for snow removal, which will recover the BIPVT energy harvesting rapidly. This seasonal heat storage by a geothermal system with a passive heat exchange enables the building to operate as if it were embedded in the earth.

CHAPTER

6

BIPVT coupling with geothermal systems

6.1 Introduction

Geothermal energy is derived from the thermal energy generated and stored in the earth. The energy is accessible by heat transfer from rocks to the surface either through boreholes or natural cracks and faults (Dickson and Fanelli, 2013; Fridleifsson and reviews, 2001). Geothermal energy is a renewable resource because there is a constant heat flow to the earth's surface and the atmosphere from the immense heat stored within the earth. Geothermal energy is a commercially proven renewable form of energy that can generate power and provide the heat (Verda et al., 2016; Hepbasli, 2003; Boyle, 2004; Vieira et al., 2017; Glassley, 2014) without additional consumption of fuel or other resources. A high-temperature geothermal plant can directly use steam to generate electricity when the heat reservoir temperature is higher than 150°C. However, it requires a sizeable geothermal reservoir at a certain depth and high temperature. Fig. 6.1 illustrates the geothermal resource distribution in the United States, which is concentrated in the west and the south areas and is often located in the deeper earth (>1000 ft).

For the most area, the ground temperature in the depth of 40–400 ft is relatively stable, which can be commonly seen from the small basement temperature change over a year. Depending on the location, the stable ground temperature is a constant from 45 to 75°F, which is typically at the mid-range of the yearly ambient air temperature variation. Direct use of the earth as a stable heat sink/source for heat exchange may provide high-performance heating and cooling services to space and district heating, greenhouse heating, snow melting, bathing, and swimming. In recent years, district heating systems incorporated with shallow geothermal energy have attracted increasing research interests (Ozgener et al., 2005; Hepbasli, 2010; Yildirim et al., 2010). Ground source heat pumps (GSHP) use shallow geothermal energy which is available almost everywhere. They convert the low-temperature geothermal energy to thermal energy

Building Integrated Photovoltaic Thermal Systems: Fundamentals, Designs and Applications
DOI: https://doi.org/10.1016/B978-0-12-821064-2.00001-8

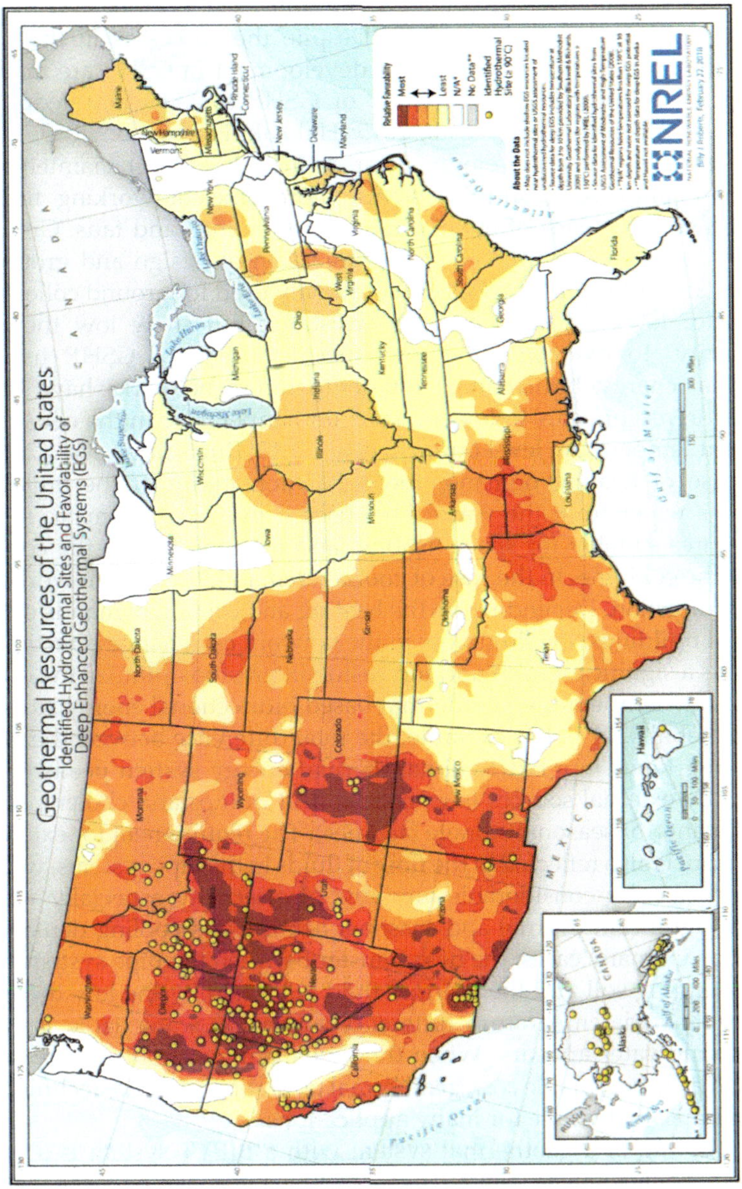

FIGURE 6.1 The geothermal resources of the United States for high-temperature utilization (NREL) (NREL, 2018).

at a higher temperature which can be used for space or water heating (Spitler and Gehlin, 2015; Luo et al., 2015; Emmi et al., 2015). Usually, a refrigerant is used as the working fluid in a closed cycle, and an antifreeze solution is circulated inside a closed coil and exchanges heat with the heat source/sink through the GHE. Despite the successes in the past and continuous growth, research and development in GSHP are mainly focusing on increasing the efficiency of GSHP systems and reducing costs (Rees, 2016; Carvalho et al., 2015; Hakkaki-Fard et al., 2015; Esen et al., 2017). The main development areas include ease of maintenance and repair, improved control systems, more efficient working fluids, and increased efficiency of auxiliaries, such as pumps and fans. Ground collectors need to be improved by optimization of design and grouting material. Currently, plastic tubes are commonly used for ground collectors which offer low cost and corrosion resistance but show low thermal conductivity. To overcome the low-efficiency issue of GSHP use of geothermal energy, bidirectional low-temperature networks have been proposed in recent years to lower heat fluid temperature in the networks to ambient temperature level (kräuchi et al., 2015; Ruiz-Calvo et al., 2017). In the case of heating demand, the circulation pump of the building withdraws water from the warm line, uses it in a heat pump to reach temperatures suitable for space heating, and then discharges the cooled water to the cold line. In the case of cooling demand, the system works in the other direction. Depending on the heating and cooling demands of the connected buildings, the fluid flow in the network can change direction.

There is a significant need and increasing interest to optimize the operations of low-temperature geothermal-based bidirectional energy storage (Prasanna et al., 2017). For example, the total difference between heating and cooling flows in a bidirectional thermal energy system needs to be balanced by external sources (e.g., central heat pumps and chillers, solar thermal plants, or seasonal storage facilities). The requirement for external energy input is also reflected in the flow of fluids in the pipe systems, which is often driven by considerable amounts of electricity especially when there are significant diurnal and seasonal heating and cooling demands. In addition, in many cases, the heat transfer between the pipe system and the surrounding soil and the thermal fluid is limited by the low thermal conductivity and temperature difference when the water table is low, or no groundwater appears. When those challenging problems are well addressed at an acceptable cost, geothermal technologies can be technically and economically feasible for many applications.

The coupling of a geothermal system with a BIPVT system is a very promising application to energy independent buildings or zero-energy buildings. Although it is not an established technology yet, a similar idea has been explored to use photovoltaic system to power geothermal systems, such as the solar assisted ground source heat pump (SAGSHP) systems (Sakellariou et al., 2021; Nouri et al., 2019), which can be traced back to the earlier work in the 1970s by Metz (Metz, 1979). SAGSHP may share similar components as BIPVT coupled geothermal system, but the solar system is often independent of buildings. SAGSHP takes advantage of ground source heat pump systems (GSHPS) for their high coefficient of performance (COP) and their low operational costs and of solar harvesting systems for their wide availability. Ground coupled heat pump absorbs heat from the earth by a GHE with pipes installed horizontally that are in the depth of 6–10 ft m below ground or pipes with different shapes that can be put into vertical boreholes with the depth of 150–1200 ft. Due to the high thermal mass of the earth and its relatively steady temperature, a higher COP can be obtained compared with air-based heat pumps.

Although geothermal energy is considered a reliable, renewable clean energy source with high efficiency and low maintenance, the design and scale of geothermal energy applications are subject to some constraints. First, geothermal heat is generated and transferred slowly, rapid heat utilization may deplete the geothermal source, which will disable the system or at least reduce the efficiency. Secondly, drilling and installation of the geothermal system require high initial costs although the investment payback period is not too long. Thirdly, because each geothermal well has a certain interference radius for heat transfer, a fairly large land area is required for the installation and operation of a sizable geothermal system. Although the high-temperature geothermal utilization can supply both electricity and heat, it relies on the heat reservoir location, whereas a bidirectional geothermal coupling with BIPVT uses the thermal mass at a relatively stable temperature of the shallow earth only, so that it can be installed on-site of most buildings with much less initial cost.

As building consumes both electricity and heat with cooling and heating needs, bidirectional use of geothermal energy in BIPVT can avoid or relieve the above problems and provide extra benefits on energy storage, transmission, and distribution. As the geothermal-BIPVT coupled system mainly use the earth to reduce the daily or yearly temperature variation, the average heat utilization is self-balanced with charging and discharging in the long term, so that the depletion of geothermal source can be avoided. Moreover, because in hot summer, it is very costly to dissipate the excess heat harvested by BIPVT to the environment at a high ambient temperature, which is necessary to keep the BIPVT system operational. Using a geothermal system can automatically resolve the problem by storing the heat underground while achieving seasonal heat storage for heat utilization in winter. A recent research project from the Center of Energy Harvesting Materials and Systems developed a method to use the heat to pump water as a passive water circulation system, which can operate a building as if it were embedded in the earth such as a basement without any energy inputs. Therefore the investment payback period can be significantly reduced. Because bidirectional geothermal coupling with BIPVT can apply to most buildings, the potential of the technology is unlimited.

6.2 Geothermal well design and characterization

Geothermal energy harvesting relies on the temperature difference between the ambient and underground environments and requires a geothermal well to transfer heat between the two parts. Geothermal wells are not simply holes in the ground in ancient time but must be designed and constructed in an engineering sense with a flow sensing and control system as well as a certain strength and durability for a service lifetime commonly over 30 years.

6.2.1 Geothermal well construction

Based on the survey conducted by the National Renewable Energy Laboratory (NREL) (Beckers and Young, 2018), there are five categories of geothermal wells based on the size, functions, power scale, drilling effort, among other considerations, as Table 6.1.

For applications of a bidirectional geothermal-BIPVT system, the first type of temperature-gradient holes shall be used. Although the geothermal well design and drilling procedures are based on engineering principles, the practice is almost an art rather than a science because of a high degree of uncertainty in drilling with different geological conditions and environmental constraints. In general, a temperature gradient geothermal well includes the following parts, as illustrated in Fig. 6.2, as an example:

The upper part is made of three concentric casings which are screwed together:

1. Above the ground surface is a protective casing to prevent damage from surface loads.
2. The surface casing (15~30 ft) penetrates the soil or sediment layer of the ground surface, and is subjected to dynamic temperature change with the ambient temperature at the top but reaches a stable temperature at the bottom.
3. The anchor casing (30~50 ft) carries the wellhead and stabilizes the well system, and reaches the rock layers

The bottom part is an uncased open hole embedded in the rock. To prevent cross-contamination or leakage, a PVC or steel tube is installed in the hole. In some circumstances, more casings might be added to stabilize the system in the drilling process. The casings are cemented into the ground, and where they pass through an already cemented casing, the annular gap is filled with cement. Once the hole is drilled for the first

TABLE 6.1 Comparison of different geothermal well types.

Well type	1	2	3	4	5
Well name	**Temperature-gradient hole (TGH)**	**Core Hole (*into reservoir*)**	**Slim Hole (*into reservoir*)**	**Standard-completion confirmation well**	**Standard-cosregumpletion production/injection well**
Main objective	Measure shallow temperature-by-depth profile	Obtain subsurface core samples	Penetrate reservoir to assess T/P, chemistry and geology	Conduct flow and interference test	Regular production/injection
Data obtained	T gradient, geology	T/P, chemistry, limited flow test, geology	T/P, chemistry, limited flow test, geology	T/P, chemistry, limited flow test, interference test, geology	T/P profiles, flow test, interference test, chemistry, geology
Range of typical measured depth	150–500 ft (46–152 m) upto 1500 ft [457 m] in some areas	1500–5000 ft (457–1524 m)	1500–5000 ft (457–1524 m)	Up to 12,000 ft (up to 3658 m)	Up to 12,000 ft (up to 3658 m)
Conductor size	Typically N/A	Hole diameter: 15–20″ (0.38–0.51 m) Casing OD: 9–14″ (0.23–0.36 m)	Hole Diameter: 15–20″ (0.38–0.51 m) Casing OD: 9–14″ (0.23–0.36 m)	Hole Diameter: 22–24″ (0.56–0.61m) Casing OD:16–20 (0.41–0.51m)	Hole diameter: 30–40″ (0.76–1.0 m) Casing OD:20–30"(0.51–0.76 m)

(*continued on next page*)

Surface string size	Typically N/A	Hole diameter: 10–15″ (0.25–0.38 m) Casing OD: 6–10″ (0.15–0.25 m)	Hole Diameter: 10–15″ (0.25–0.4 m) Casing OD: 6–10″ (0.15–0.25 m)	Hole diameter: 16–18″ (0.41–0.46 m) Casing OD: 13–16″ (0.33–0.41 m)	Hole diameter: 20–30″ (0.51–0.76 m) Casing OD: 18–22″ (0.46–0.56 m)
Size of final cemented string	None (2″ [0.05 m]) blind tubing	4–6″ (0.10–0.15 m)	4–6″ (0.10–0.15 m)	6–10″ (0.15–0.25)	8–18″ (0.20–0.46 m); most common; 9–5/8″ (0.24 m) and 13–5/8″ (0.35 m)
Final hole size	3–6″ (0.08–0.15 m)	<6″ (<0.15 m)	<6″ (<0.15m)	6–10″ (0.15–0.25 m)	8–12″ (0.20-0.30m)
Drilling time	1–5 days	15–45 days	15–45 days	30–60days	30–70 days
Cost range ($US)	$20–150 K	$0.5–2 M	$0.5–2 M	$2–6 M	$3–10 M
MW. Potential	0	$<2MW_e$	$<2MW_e$	$<5MW_e$	$<20MW_e$
Lead time (designing and permitting)	1–3 months	4–8 months	4-8 Months	5–10 months	9–12 months

T, temperature; P, pressure; OD, outer diameter (Beckers and Young, 2018).

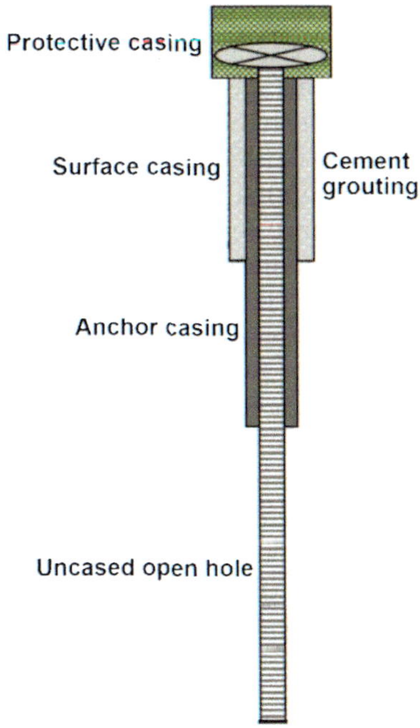

FIGURE 6.2 An example of a temperature gradient geothermal well.

(shallowest) length of casing, the casing is gradually assembled in the hole piece by piece at a certain length. The piece which will be lowest in the well is first fitted with a "shoe," to protect the bottom edge. It is lowered into the well, held while the next piece is screwed to it, the pair is lowered and held and the process is repeated until the required length is assembled and in place, hanging in the hole. A volume of cement slightly greater than the volume of the annulus formed by the outer surface of the tube and the surrounding hole is pumped into the tube, and a rubber piston is inserted after it. Water is pumped into the tube behind the piston, driving it downwards, and causing the cement to flow upward to fill the annulus, eventually overflowing at the surface. Once the overflow has been seen, the annulus is assumed to be full of cement (Watson, 2016).

Once the cement is fully cured, the thermal fluid circulation system and wellhead can be installed.

6.2.2 Ground temperature profile

The design and operation of a temperature gradient geothermal well rely on the ground temperature profile. Especially, when a bidirectional geothermal-BIPVT system is used, the heat storage and discharge efficiency will be significantly affected by the ground temperature. Ground temperature profile can be measured on-site but it takes a long period to collect the data because it changes with time and depth. For example, Fig. 6.3 uses the Central Park at New York City as an example and estimates the temperature profile, in which the analysis procedure be elaborated later in this section. The stabilized temperature is at 55°F for the depth at 40 ft or deeper. Close to the surface with a depth <30 ft, the temperature changes considerably with seasons and depths. Fig. 6.3A shows that the ground surface temperature exhibits a sinusoidal variation with time with the peak temperature in Jul., with the depth increase, the temperature variation range reduces rapidly to nearly zero. The peak temperature also delays to Sep. for the depth at 10 ft. Because the ground temperature change is mainly driven by the interaction between earth and the air through the ground surface as well as radiation. As the weather and ambient temperature data are well documented, it is possible to predict the ground temperature profile by numerical modeling.

For simplicity, assume the ground in Fig. 6.4 is homogeneous with physical properties written as diffusivity α, conductivity k, specific heat c_p at a unit mass, surface convective heat transfer coefficient h, emissivity e, and density ρ. The temperature field $T(z, t)$ changes with the depth z and time t, which satisfies the Fourier Law

$$q = -k\frac{\partial T}{\partial z} \tag{6.1}$$

Where q is the heat flux density. For a small depth of earth δz with a unit area in the dotted box in Fig. 6.4, the temperature change satisfies the energy conservation, i.e. $\delta q.1 = -\rho c_p \frac{\partial T}{\partial t}\delta z.1$, where δq is the heat flux difference between the upper and lower surfaces of the dotted box, which leads to the following partial differential equation

$$\frac{\partial q}{\partial z} = -\rho c_p \frac{\partial T}{\partial t} \tag{6.2}$$

Combining the above two equations yield the governing equation of heat conduction

$$\frac{\partial T}{\partial t} = \alpha \frac{\partial^2 T}{\partial z^2} \tag{6.3}$$

where heat diffusivity $\alpha = k/\rho c_p$. Consider the boundary conditions as follows (Badache et al., 2016):

1. Surface temperature changes periodically as

$$T(0, t) = T_0 + A\cos(wt) \tag{6.4}$$

2. Far-field temperature is stable as

$$T(+\infty, t) = T_0 \tag{6.5}$$

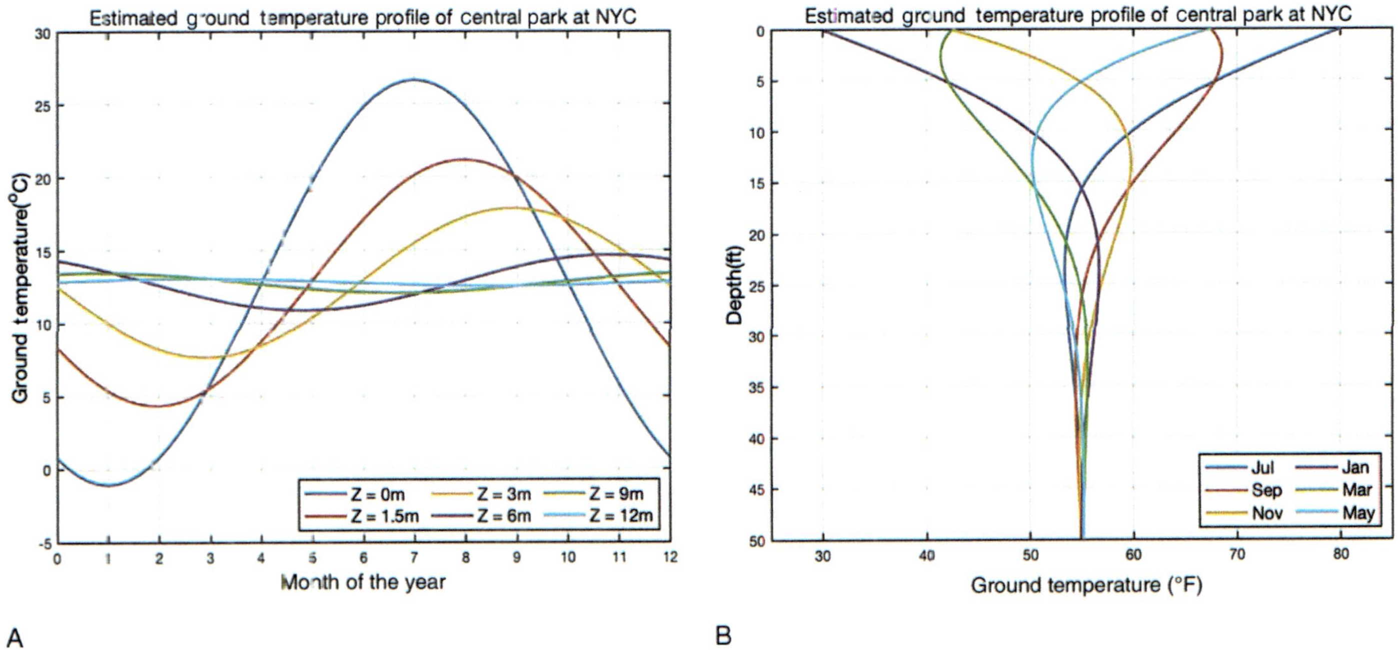

FIGURE 6.3 The ground temperature profile at Richmond, VA: (A) temperature versus time at different depths and (B) temperature versus depth in different months.

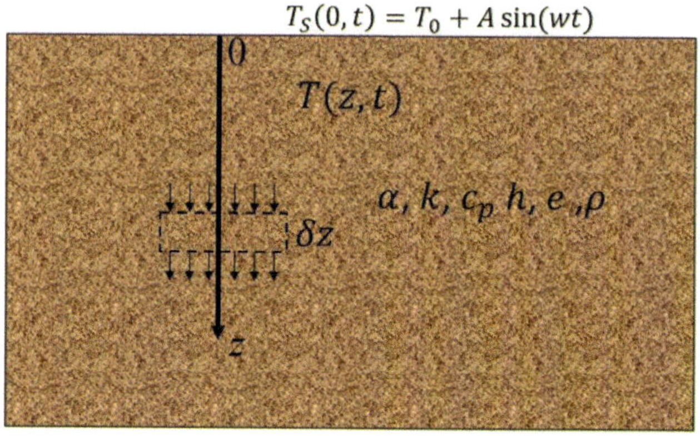

FIGURE 6.4 Heat conduction in the ground caused by the surface temperature change.

Notice that here the yearly temperature change is considered, so t is counted in days, and the angular frequency $w = 2\pi/364$. A periodic condition for temperature distribution is imposed as

$$T(z,t) = T(z, t+364) \tag{6.6}$$

For this boundary value problem of a semi-infinite solid with a periodic condition in time, the solution is periodic in time, with the angular frequency w, and decays exponentially along z with a characteristic penetration length d to be determined. To solve it, separation of variables can be conducted, say $T - T_0 = F(t).G(z)$, where $F(t)$ can be written in the complex form of e^{iwt} due to the periodic condition and the surface temperature change, i.e.

$$F = e^{iw} \tag{6.7}$$

so that Eq. 6.3 can be rewritten as

$$\alpha\frac{G''}{G} = \frac{F'}{F} = iw \tag{6.8}$$

Then the ordinary differential equation of $G(z)$ is obtained as

$$G'' - \frac{iw}{\alpha}G = 0 \tag{6.9}$$

which yield the general solution

$$G = C_1 e^{(1-i)\frac{z}{d}} + C_2 e^{-(1-i)\frac{z}{d}} \tag{6.10}$$

where $d = \sqrt{2\alpha/w}$ is called the ground damping depth. Because the temperature decreases with z and converges to a stable value at $z \to \infty$,

$C_1 = 0$. Therefore, the temperature can be rewritten as

$$T = T_0 + Re\left(C_2 e^{iw-(1-i)\frac{z}{d}}\right) \tag{6.11}$$

where *Re* ( ) means the real part of the enclosed function. Using the boundary condition (6.4), it is obtained $C_2 = A$. Therefore the temperature field is obtained as

$$T = T_0 + Ae^{-z/d}\cos\left(wt - z/d\right) \tag{6.12}$$

For example, for Central Park in New York City, approximately the ground surface temperature changes from a minimum of ~30 °F in January to a maximum of ~80 °F (Weather.gov, 2021). One can estimate $T_0 = 55°F$, $A = 25°F$. Using the thermal diffusivity of rock at ~1 mm^2/s, one can approximate the ground damping depth $d \cong 10ft$. Eq. (6.12) can be rewritten as

$$T = 55 + 25e^{-z/10}\cos\left(\frac{\pi}{6}t - \frac{z}{10}\right) \tag{6.13}$$

where time is counted by month here starting at Jul. One can draw the monthly temperature change at depths 0, 5, 10, 20, 30, and 40 ft, as shown Fig. 6.3A. The temperature at the surface changes significantly with months but the variation rapidly reduces with the depth. When $z = 20$ ft, the variation range reduces to 52–58°F; when $z = 30$ or 40 ft, the curves are almost overlapped and a stable temperature is obtained. The peak temperature shifts with depth by ~0.2z in the month, which means every 5 ft, the peak temperature delays ~1 month. The temperature-depth profiles in Jul., Sep., Nov., Jan., Mar., and May are shown in Fig. 6.3B. The temperature exponentially decays with the depth at a rate $e^{-z/10}$. At a depth of 30ft, the variation is <5% of the surface variation and can be considered as a stable temperature.

Given the ambient temperature change and geological and environmental parameters, the parameters T_0 and A can be determined by the physical models. For example, Badache et al. considered the surface heat conduction, the convective heat flux transferred between the surface and the ambient air, the short-wave global solar radiation (radiant flux) absorbed by the ground surface, the longwave radiant flux exchanged with the surroundings (sky), and the latent heat flux of evaporation, and provided a good prediction of the temperature profile (Badache et al., 2016).

Obviously T_0 and A can be determined by the experimental method as well, so that one can estimate the temperature profile by surface temperature measurement only, which avoids the costly, time-consuming measurement of temperature at the different depths. However, the formulation is based on the following simplifications and assumptions, which may lead to disparity from the actual measurements:

1. The material is assumed to be homogeneous. Although geothermal wells typically penetrate through different layers, because the temperature variation concentrates in the top 50 ft, if the material properties in that range do not change significantly, the average properties can be used.
2. Water motion in soil, especially under rain or water table change, may accelerate the heat transfer and significantly change the temperature profile.
3. The material properties also change with the moisture content of underground materials. It introduces more complexity and uncertainty of temperature profile. The periodic condition may not be accurate anymore.

Therefore, although an advanced model can provide a useful prediction of the temperature profile for design and analysis, on-site temperature sensing and monitoring are still necessary for the operation of a geothermal well as extreme weathers are more frequent and the GSHP may also change the local temperature profile as well.

Notice that although the above analysis was demonstrated for daily average temperature change in a sinusoidal curve in 1 year period, it can be extended to an arbitrary temperature change pattern with the Fourier series. It can also be used for instant temperature change with a daily period as well, except that the period shall be replaced by 24 h and the ground damping depth shall be much smaller.

6.3 Geothermal well systems

The bidirectional geothermal—BIPVT system shares the same feature with the GSHP for the aspect of geothermal wells. There are two types of geothermal well systems: an open-loop system directly uses nearby groundwater for the heat exchange process. It can directly connect with the water body of a pond or pump water under the water table. The water does not circulate but instead is directed into drainage after use or refilled to the water body. Although it is relatively economic and efficient, the operation of the system significantly relies on the water body condition and is susceptible to environmental contamination as well. It is often constrained by the availability of the quality water body and may fail due to water table or water level change. Because the operation may produce disturbance to the water body, it may be restricted in some areas.

A closed-loop system generally circulates a mixture of antifreeze (e.g., glycol) and water without any direct water exchange from the environment. It can apply to anywhere with a long lifetime, but the heat exchange rate could be much slower to transfer heat to the ground with low thermal conductivities. Therefore, it requires long pipelines, which can be horizontal or vertical.

A horizontal ground loop is generally installed over a wide area of ground and requires a large space to dig trenches hundreds of feet long at 6–10 feet deep. Horizontal ground loops can only be used when adequate yard space is available and trenches are easy to dig. A series of plastic pipes are installed in the trench for the heat exchanger. Because trenching is more cost-effective than drilling, it is widely used in farms. However, because it cannot reach the depth with the stable temperature, it is mainly for daily temperature conditioning and cannot perform seasonal heat storage. Therefore, vertical closed-loop geothermal systems should be used for bidirectional geothermal—BIPVT systems for better long-term performance.

Vertical closed-loop systems use heat exchange pipes in geothermal well with typically three configurations: (1) most borehole wells use coaxial pipes, U-bend pipes, or double U-bend pipes for heat exchange; (2) helical coils use a spiral heat exchange for a larger contact area at a unit depth; (3) energy pipes assemble a group of heat exchange pipes (U-bend or other configuration) together through a certain pattern. The last two require a larger hole to host the heat exchanger, and therefore relatively a smaller drilling depth is needed. The fluid running through the closed-loop transfers heat from the earth to the building during the heating season and moves heat into the earth from the building during the cooling season. Considering the heat transfer depending on the ground damping depth, it is favorable to have a deeper geothermal well, so that the heat dissipated in summer can be stored and used in winter. Therefore, a borehole-based closed-loop system is recommended for bidirectional geothermal wells in the following section. Notice that vertical closed-loop should be built in a very safe manner. Otherwise, they may pose a threat to aquifers and nearby drinking water wells if improperly installed or abandoned.

6.3.1 Design procedure of bidirectional geothermal wells

Bidirectional geothermal—BIPVT systems perform thermal management through two mechanisms: (1) dissipate heat or extract heat from the ground by borehole heat exchanger in a passive manner, which fully depends on the temperature difference between the ambient temperature and the stable ground temperature T_0 but does not cost extra energy except the water pump, (2) increase or decrease the temperature of geothermal heat to a desirable level by heat bumps, which require extra energy for the compressor in the heat pump cycle but enable active control of temperature subjected to the Carnot efficiency of an ideal heat cycle. Both mechanisms can be bidirectional. Although the current practice mainly focuses on the second mechanism while the first mechanism is subjected to the second one. However, from the energy efficiency point of view, it is more desirable to maximize the use of the first mechanism. The design procedure of a bidirectional geothermal system includes the following steps:

1. Feasibility study.
   A geothermal project should start at the feasibility study on how the project should be developed before drilling any well. Because the construction and operation of the system will be subjected to technical, economical, legal, and environmental constraints, although borehole geothermal systems can be applied to most areas, the construction procedure, system efficiency, and operation & maintenance costs can vary in a big range. If a designer or a driller ignores some constraints and limitations, it may lead to the failure of the project at a late stage. On the contrary, if the system can be well designed and optimized, huge life cycle benefits can be achieved as the lifetime of the system is long. A typical feasibility study includes the following parts:
   (a) Identify the peak heating and cooling demands of the building through the architectural design or the operational data of the building or other buildings in the area. Therefore the capacity of the geothermal system can be estimated and the required drilling area can be calculated and compared with the available land area.
   (b) Review the feasibility to drill wells from the local authority or agency. For environmental or groundwater protection, some areas may be restricted to drill, particularly for areas with artesian aquifers.
   (c) Collect the geological information of the area, such as rock type and hardness for well drilling; ground thermal characteristics for system operation; groundwater situation for drilling and operation, which can be obtained from geological maps or neighboring boreholes.
   (d) Conduct the predesign of the system. Based on the required load capacity, the necessary size and layout of the geothermal system can be determined. Particularly the drilling length can be calculated and geothermal well distribution can be designed.
   (e) Calculate the project costs and benefits, including both construction and installation, operation and maintenance (O&M), etc. The investment payback period can be estimated.

   The feasibility study can identify rigid hurdles for the project and provide a rough boundary for the system design.
2. Field investigation.
   Notice the feasibility study is a concept initiation. Because the actual geological situation can change in a large range, the proper design for

the installation needs to use measured data for ground parameters. Therefore, pilot wells should be drilled to collect field data. Surely they can be used for borehole heat exchanger installation later as well. The following data can be collected from field investigation with core samples and used in the simulation and design of the geothermal system:

- The type of soils and rocks underground and their distribution along with the depth.
- Thermal conductivity, density, the specific heat of the soils and rocks.
- Undisturbed ground temperature.
- Water table and porosity distribution of soil.

In addition, the temperature, moisture, climate information need to be collected for performance simulation and analysis.

3. System design and optimization.
The system design includes the well distribution, and the borehole length and number, borehole heat exchanger configuration, thermal fluid system parameters, as well as construction, installation, and operation guidelines and interfaces with the BIPVT and electric systems. The design parameters can be optimized by numerical simulation or analysis to achieve high performance, reliability, and system safety while keeping cost-effective.

6.3.2 Heat and mass flow in a bidirectional geothermal system and the energy efficiency

In a bidirectional geothermal system, the heat extracted from the ground is transferred by the thermal fluid, so that the heat and mass flows are coupled with each other. Although the heat and mass flow in a bidirectional geothermal system relies on the heat exchanger configuration, geological conditions, and operational aspects, etc., this section uses a coaxial heat exchanger as an example to show the heat and mass flow analysis, as shown in Fig. 6.5. Notice that here only the heating cycle of a bidirectional geothermal system is demonstrated. By switching the heating cycle to the refrigeration cycle, the heat storage process can be implemented in an inverse fashion. We will not discuss it. The analysis can be extended to multiple heat exchangers.

The coaxial heat exchanger is embedded in the rock and composed of an injection annulus and a production inner pipe. A thermally conductive well casing can be added between the outer conductive pipe and borehole wall for protection. The inner pipe is thermally insulated and located concentrically to the well casing, acting as the inside channel of the exchanger. A segment of thermally insulated pipe is installed along the top of the well casing to prevent heat loss from the injected fluid to the ground, which will dissipate to the ambient air. The working fluid is continuously injected through the annulus space and produced through the center tube, without direct contact with the surrounding geothermal reservoir. The cold working fluid is heated by the increasingly higher temperature of surrounding rocks as it flows down the annulus of the heat exchanger. The heated fluid eventually returns to the surface with some heat loss due to heat transfer from the hot-produced flow to the cooler injected flow. Therefore, the inner pipe should be well insulated to minimize the heat loss in the returning pathway.

6.3.3 Heat transfer in the system

Fig. 6.5 shows the heat harvesting from the ground to evaporate the refrigerant in the heat pump cycle. The cold fluid is circulated into the well through the annulus and extracts heat from the surrounding rock at a higher temperature. The length of the heat exchanger or the depth of the well should be large enough so as to heat the thermal fluid sufficiently. The transient heat transfer problem is illustrated in Fig. 6.6 as an axisymmetric problem with coordinate (r, z).

Here z-axis is along the borehole axis and the r-axis along the radial direction of the borehole with the origin at the top of the conductive casing. The fluid with an initial temperature T_0 is injected downhole with constant

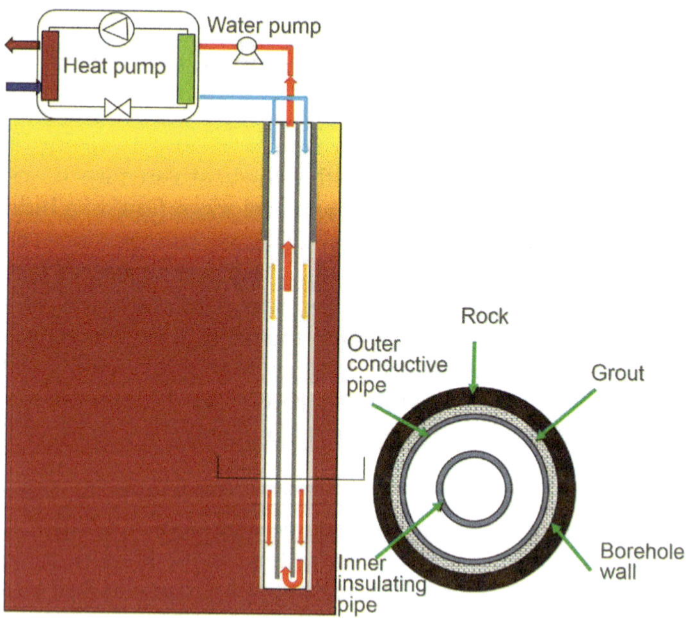

FIGURE 6.5 Schematic illustration of a bidirectional geothermal heat pump system with a coaxial heat exchanger. Here the heating cycle can be replaced by a refrigeration cycle and multiple boreholes can be connected to the heat pump.

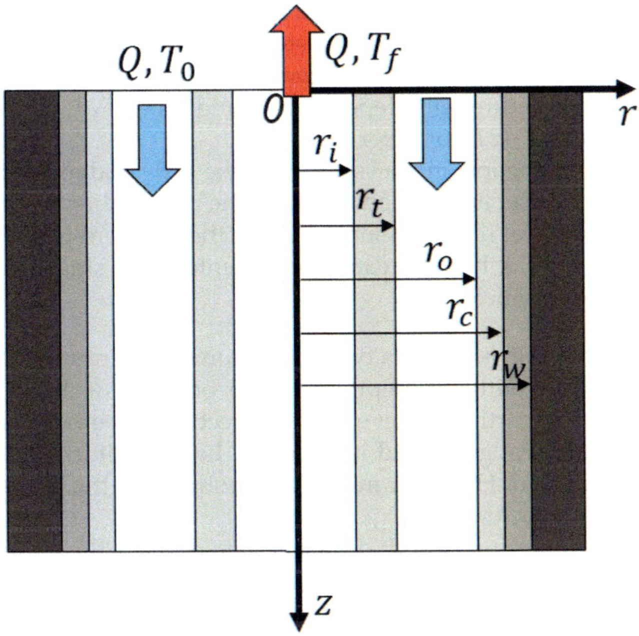

FIGURE 6.6 The axisymmetric boundary value problem for heat transfer in a borehole heat exchanger.

flow rate Q. The far-field temperature of the ground is T_∞ and the fluid temperature field depends on the coordinates and time as $T(r, z, t)$. The geometric configuration is shown:

r_i: The inner radius of the inner insulating pipe - $0 \le r \le r_i$: the discharge flow.

r_t: The outer radius of the inner insulating pipe - $r_i \le r \le r_t$: the inner pipe wall.

r_o: The inner radius of the outer conductive pipe - $r_t \le r \le r_o$: the injected flow.

r_c: The inner radius of the casing - $r_o \le r \le r_c$: the outer pipe wall

r_w: The inner radius of the well - $r_c \le r \le r_w$: the casing wall

Similarly to Eq. (6.3), the governing equation of this axisymmetric boundary value problem can be written

$$\frac{\partial T}{\partial t} = \alpha \left[\frac{1}{r} \frac{\partial}{\partial r} \left(r \frac{\partial T}{\partial r} \right) + \frac{\partial^2 T}{\partial z^2} \right] \tag{6.14}$$

for the heat conduction in the solid. However, in the injected and discharge flow, the heat convection and conduction are more complex. Ramey provided a simplified solution to this problem (Ramey Jr, 1962). This section will follow Ramey's assumptions:

1. An infinite rock domain is assumed at $r > r_w$ with isotropic and homogeneous material properties and a steady far-field temperature T_∞.
2. The fluid in the tube flows with a constant flow rate Q in the z-direction and the temperature only depends on z and t, i.e., $T_1(z,t)$ for $0 \leq r \leq r_i$ and $T(r, z, t) = T_2(z,t)$ for $r_t \leq r \leq r_o$.
3. The heat conduction in the system is the only radial, therefore the analysis can be separately conducted in the z and r directions.
4. Steady radial heat flow is assumed inside the wellbore-formation interface (- $0 \leq r \leq r_w$), hence it can be represented by a single steady-state heat transfer coefficient U.

Notice that Ramey's formula typically assumes the ground temperature linearly increases with the depth for deep borehole wells due to heat generation in the earth. However, the bidirectional geothermal wells do not focus on the heat generated in the earth, here a uniform far-field temperature is assumed. However in the neighborhood of the heat exchanger the temperature decreases.

Given an infinitesimal depth dz at a certain depth z, the temperature at $r = \infty$, r_w and r_o is written as T_∞, $T_w(z,t)$, and $T_2(z,t)$. Based on Fourier's Law, the heat transfer through the pipe wall and casing can be written

$$\delta q = -2\pi r_o U[T_w(z,t) - T_2(z,t)]dz \tag{6.15}$$

where the steady-state heat transfer coefficient U depends on the thermal conductivities of the outer pipe and casing as well as the convective heat transfer coefficient of the water pipe surface as well as their thicknesses. Here, $T_w(z,t)$ depends on the far-field temperature and the heat flux as well as written

$$\delta q = -\frac{2\pi k(T_\infty - T_w)}{f(t)}dz \tag{6.16}$$

where k is the thermal conductivity of the rock and $f(t)$ is a function to be determined by the radial heat conduction analysis. For simplicity, the inner pipe is considered to be perfectly insulating, so that $T_1(z,t) = T_1(t)$. Based on the energy conservation, the heat transferred from the rock will heat the fluid as

$$\delta q = -cQ\frac{\partial T_2}{\partial z}dz \tag{6.17}$$

where c is the specific heat of the rock. From Eqs. (6.15) and (6.16), one can obtain

$$T_w(z,t) = \frac{kT_\infty + r_o U f(t) T_2(z,t)}{k + r_o U f(t)} \tag{6.18}$$

Substituting Eq. (6.18) into Eq. (6.15) and combining it with Eq. (6.17), one can write

$$\frac{\partial T_2}{\partial z} + \frac{T_2}{A(t)} - \frac{T_\infty}{A(t)} = 0 \tag{6.19}$$

where function $A(t)$ is written as

$$A(t) = \frac{cQ\left[k + r_o U f(t)\right]}{2\pi r_o U k} \tag{6.20}$$

Given a time t, Eq. (6.19) can be treated as an ordinary differential equation in terms of the coordinate z. The general solution is

$$T_2(z,t) = T_\infty - C(t)e^{-z/A(t)} \tag{6.21}$$

Using the boundary condition,

$$T_2(0,t) = T_0 \tag{6.22}$$

one can obtain $C(t) = T_\infty - T_0$. Therefore, the temperature along the heat exchange pipe is obtained as

$$T_2(z,t) = T_\infty - (T_\infty - T_0)e^{-z/A(t)} \tag{6.23}$$

Here $A(t)$ depends on $f(t)$, which was obtained by the radial heat conduction analysis for the temperature field through the equation:

$$f(t) = -\frac{2\pi k(T_\infty - T_w)}{dq/dz} \tag{6.24}$$

where $T_w = T(r = r_w, t)$ in the radial plane. The governing equation in the solid is reduced to

$$\frac{\partial T}{\partial t} = \frac{\alpha}{r}\frac{\partial}{\partial r}\left(r\frac{\partial T}{\partial r}\right) \tag{6.25}$$

with the boundary conditions of constant far-field temperature and heat flux to well as

$$T(\infty,t) = T_\infty; \; -kT_{,r}(r_w,t) = q_0 \tag{6.26}$$

and the initial condition

$$T(r,0) = T_\infty \tag{6.27}$$

The solution can be obtained through the Laplace transformation (Carslaw and Jaeger, 1992). For $\frac{4\alpha t}{r_w^2} \gg 1$, the surface temperature can be approximated as

$$T(r_w,t) = T_\infty + \frac{q_0 r_w}{2k}\ln\frac{4\alpha t}{cr_w^2} \tag{6.28}$$

where $\ln c = 0.57722$ is the Euler–Mascheroni constant. Therefore Eq. (6.24) can be rewritten as

$$f(t) = -\frac{2\pi k(T_\infty - T_w)}{\frac{dq}{dz}} = \frac{2\pi k}{2\pi r_w q_0}\frac{q_0 r_w}{2k}\ln\frac{4\alpha t}{cr_w^2} = \ln\frac{2\sqrt{\alpha t}}{r_w} - 0.2886 \quad (6.29)$$

Therefore the water temperature distribution along z can be explicitly written as

$$T_2(z,t) = T_\infty - (T_\infty - T_0)e^{-\frac{2\pi r_0 Ukz}{cQ\left[k + r_0 U\left(\ln\frac{2\sqrt{\alpha t}}{r_w} - 0.2886\right)\right]}} \quad (6.30)$$

With the increase of depth, the water temperature increase. Because water takes heat away and the surrounding rock's temperature keeps reducing, the water temperature also exhibits a decreasing trend with time. The temperature at $z = H$ or the bottom of the well can be considered as the product temperature for use.

For example, consider a well with a water injection flow rate Q, surface water temperature $T_0 = 100°F = 37.8°C$, the casing size $r_w = 3.5$ in $= 8.8$ cm, the ground temperature $T_\infty = 55°F = 12.8°C$, water specific heat $c = 4.2 \times 10^3 J/kg/°C$, the ground thermal conductivity $k = 1.86$ W/m/°C, pipe outer radius $r_0 = 3$ in $= 7.6$ cm, the U value of the casing wall $U = 100$ W/m^2/°C, soil's thermal diffusivity $\alpha = 0.8 \times 10^{-6} m^2/s$, and the shallow borehole depth $D = 200$ m. One can use Eq. (6.30), to predict the temperature of the liquid changing with the depth and time. Notice that Eq. (6.30) is based on the assumption $t \gg \frac{r_w^2}{4\alpha} = 2.420$ s or 40 min in this case. Therefore, start to count the time at 24 h. Fig. 6.7 shows the fluid temperature change with the depth at 24, 60, 120, and 240 h at the flow rate $Q = 0.5$ kg/s. The warm water is cooling down along the depth of the well. Notice that because the water flow rate is high, the leaving water temperature (LWT) is not low compared with the ground temperature of 12.8°C. If the flow rate reduces, the LWT will be lower. When the warm water keeps running, it will heat the surrounding earth, which will reduce the heat transfer rate and cause a higher LWT. Fig. 6.7B shows the LWT at $z = 200$ m changing with time for different injection flow rates at $Q = 0.25$, 0.5, 0.75, and 1 kg/s. It clearly shows that the higher the flow rate, the higher the LWT. When the system keeps running, the LWT will increase. Notice that Eq. (6.30) is used for the long-term temperature prediction. When the BIPVT system needs to control the roof temperature, the simplification in Eq. (6.28) does not work anymore. A numerical method needs to solve the transient boundary value problem with the given governing equations for instantaneous temperature prediction with varying flow rates.

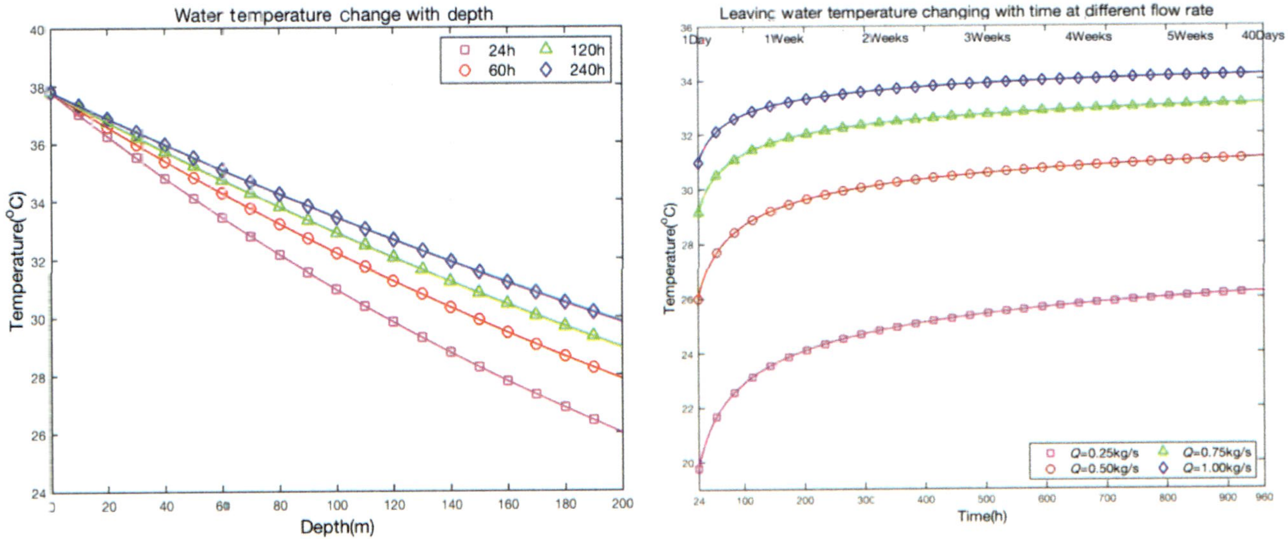

FIGURE 6.7 (A) The water temperature (°C) distribution along the depth (m) of the well (four observation times at 24, 60, 120, and 240 h after starting water flow); (B) the output water temperature (°C) changing with time (hour) at the different injection flow rate.

6.3.4 Thermal fluid circulation in the system

The heat exchange depends on the fluid flow rate Q in the heat exchanger. To drive the water circulation, extra power is required to provide enough flow rate and overcome the viscous force. Here a Newtonian fluid is considered, such that the shear stress τ is linearly related to velocity gradient or strain rate. Fig. 6.8 shows a Newtonian fluid between two parallel plates. The upper plate is moving at a constant velocity u_x. The shear stress can be written

$$\tau_{yx} = \mu \frac{\partial u_x}{\partial y} \tag{6.31}$$

where μ is the dynamic viscosity or absolute viscosity. It is also simply referred to as viscosity, which is related to kinematic viscosity by $\nu = \frac{\mu}{\rho}$ with ρ being the density.

Considering the fluid moving in a pipe with temperature or density change, the continuity of the fluid requires

$$\frac{\partial \rho}{\partial t} = \frac{\partial(\rho u_x)}{\partial x} + \frac{\partial(\rho u_y)}{\partial y} + \frac{\partial(\rho u_z)}{\partial z} \tag{6.32}$$

For simplicity, because the temperature and pressure ranges for the bidirectional geothermal system are relatively small, incompressible fluid

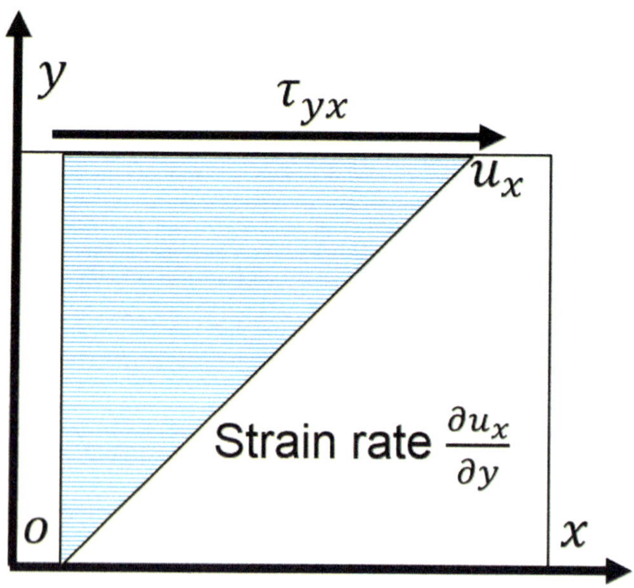

FIGURE 6.8 A linear flow of a Newtonian fluid between parallel plates.

is assumed so that

$$\nabla \cdot \mathbf{u} = 0, \text{ with } \nabla = \left(\frac{\partial}{\partial x}, \frac{\partial}{\partial y}, \frac{\partial}{\partial z}\right) \text{ and } \mathbf{u} = \left(u_x, u_y, u_z\right) \quad (6.33)$$

The conservation of momentum can be written by the Navier–Stokes equation as follows:

$$\rho \frac{D\mathbf{u}}{Dt} = -\nabla p + \nabla \cdot \tau + \rho \mathbf{g} \quad (6.34)$$

where $\frac{D}{Dt}$ is the material derivative, defined as $\frac{\partial}{\partial t} + \mathbf{u} \cdot \nabla$; p the pressure field; t time; τ the deviatoric stress tensor, which has order 2; $\mathbf{g}$ body accelerations acting on the fluid particles, for example, gravity, inertial accelerations, electrostatic accelerations, and so on.

If a laminar flow is considering in a horizontal straight pipe, say in the x direction, the above equations can be significantly reduced because all fluid particles are moving in the same direction with $u_x(x,r)$ only. Eq. (6.33) shows that $u_{x,x}(x, r) = 0$ so that it can be written as $u_x(r)$. The Navier–Stokes equation is reduced to

$$0 = -\frac{\partial p}{\partial x} + \mu \frac{1}{r} \frac{\partial}{\partial r}\left(r \frac{\partial u_x}{\partial r}\right) \quad (6.35)$$

The general solution can be written

$$u_x(r) = \frac{r^2}{4\mu}\left(\frac{\partial p}{\partial x}\right) + C \ln r + D \quad (6.36)$$

where C and D are to be determined by the boundary conditions. Due to the fixed boundary and the symmetry at the central axis, the boundary conditions are written

$$u_x(a) = 0 \text{ and } \frac{\partial u_x(0)}{\partial r} = 0 \quad (6.37)$$

where a is the radius of the pipe. Therefore, the velocity distribution is obtained as

$$u_x(r) = -\frac{\frac{\partial p}{\partial x}}{4\mu}\left(a^2 - r^2\right) \quad (6.38)$$

which is the Poiseuille solution. The flow rate can be related to the pressure gradient as follows

$$Q = \int_{r=0}^{a} 2\pi r u_x(r) dr = -\frac{\pi a^4}{8\mu} \frac{\partial p}{\partial x} \quad (6.39)$$

The pumping power for a pipe with length L can be calculated from the pressure gradient as

$$P = \Delta p Q = -\frac{\partial p}{\partial x} L \left(-\frac{\pi a^4}{8\mu} \frac{\partial p}{\partial x} \right) = \frac{\pi a^4 L}{8\mu} \left(\frac{\partial p}{\partial x} \right)^2 = \frac{8\mu L}{\pi a^4} Q^2 \tag{6.40}$$

Notice that when the velocity of the flow is high, which can be characterized by a constant of $Re = \rho \bar{u} D / \mu$, namely the Reynolds number, with $\bar{u} = \frac{Q}{\pi a^2} = \frac{a^2}{8\mu} \frac{\partial p}{\partial x}$ and $D = 2a$, the laminar flow will transit to a turbulent flow, which requires much more power to circulate the thermal fluid but provides benefits for a higher convective heat transfer coefficient. Because turbulent flow also produces noise and vibration, some applications may require to keep $Re < 2100$ to maintain the laminar flow. When $Re > 2100$ or 3000, the Darcy-Weisback equation is commonly used to calculate the frictional loss or pumping power (Brown, 2003), which is much higher than the Poiseuille flow for a high Re turbulent flow.

For a Poiseuille flow, if the fluid is moving in an annulus with an inner radius a and outer radius b, the boundary conditions are changed to

$$u_x(a) = u_x(b) = 0 \tag{6.41}$$

The solution can also be provided as

$$u_x(r) = -\frac{\frac{\partial p}{\partial x}}{4\mu} \left(\frac{a^2 \ln \frac{b}{r} + b^2 \ln \frac{r}{a}}{\ln \frac{b}{a}} - r^2 \right) \tag{6.42}$$

The pumping power can be calculated in the same fashion in Eqs. (6.38)–(6.40). The analysis of frictional pressure loss can be extended to different shapes of pipe cross-section with a hydraulic diameter, curved pipe, or fitting loss with an equivalent length, and roughness of the pipe surface.

For thermal fluid moving in the geothermal well, the actual pressure will also change with the depth by gravity. One can take the average of the velocity of the cross-section, and transform the Navier–Stokes equation to the Bernoulli equation. Although the Bernoulli equation originally considered an incompressible flow where net frictional forces are negligible, here the frictional effect is added with the above pressure gradient by the viscous force. Consider the x axis is along the vertical direction. The gravity will play a role in Eq. (6.34) but all quantities are taken as the average of the cross-section for a steady flow. Therefore, the flow velocity only exhibits u_x with $u_r = 0$ and all other quantities only depend on x. Eq. (6.34) can be rewritten as

$$\rho u_x \frac{du_x}{dx} = -\frac{dp}{dx} + f_x + \rho g \tag{6.43}$$

where $f_x = \nabla \cdot \tau = \tau_{rz,r} + \frac{\tau_{rz}}{r}$ is the frictional force at a unit length. For example, using Eq. (6.38) in Eq. (6.31) one can obtain $f_x = \frac{\partial p}{\partial x}$.

For any given two points x_0 and x_1, by integral of the above equation from x_0 to x_1, one can obtain

$$x_0 + \frac{p_0}{\rho g} + \frac{u_0^2}{2g} = x_1 + \frac{p_1}{\rho g} + \frac{u_1^2}{2g} + \Delta h_f \tag{6.44}$$

where the subscripts 0 and 1 denote the variable's value at x_0 and x_1, respectively; and Δh_f is the equivalent length of the frictional loss. In reality, the fluid entries, exits, fittings, turning points, and curvatures can also be added. When a close-loop is considered, a water pump should be provided to match the head loss or the pressure difference to power the thermal fluid circulation.

For the coaxial heat exchange system in Fig. 6.5, once the peak power demands from the building are given, one can design the well geometry for depth and radius, as well as the thermal flow rate Q. Then the pressure gradient or heat loss will be calculated with the above formulation, and the water pump power can be estimated. Notice that the heat loss depends on the total length of the pipes, although the pressure changes significantly with the depth of well.

6.3.5 Energy consumption and coefficient of performance of the system

The heat exchange with the earth shall mainly be along the thermally conductive casing range of the borehole wall and the bottom. The thermal fluid circulation system needs energy inputs to pump the fluid while supplying heat for the heat pump to provide desired heat supply. Notice that the heat flow rate is subjected to constraints of water flow rate, ground temperature, and well design; while the heat demands in the building are more dynamic and change with human activities and preferences, weathers, and ventilation conditions, etc. To meet the dynamic demand, reversible heat pumps work in either direction to provide heating or cooling to the internal space. They employ a reversing valve to reverse the flow of refrigerant from the compressor through the condenser and evaporation coils.

Fig. 6.6 shows that the bidirectional geothermal system is connected with a heat pump for the heating mode. The thermal fluid coil serves as an evaporator, while the indoor is a condenser. The refrigerant flowing from the evaporator carries the thermal energy from the ground and enters a gas compressor. Vapor temperature is augmented within the compressor by the mechanical load. The indoor coil then transfers heat including energy from the compressor to the indoor air, and the vapor is condensed into liquid again. Through integration with the indoor air circulation, the heating process can be tailored to the customized needs. The liquid finally flows back through a valve that reduces its pressure in order to cool it down so it can repeat the cycle.

In cooling mode, the cycle is similar, but the geothermal fluid coil is now the condenser and the indoor coil (which reaches a lower temperature) is the evaporator. This is the familiar mode in which air conditioners operate.

The COP of a heat pump is a ratio of useful heating or cooling provided to the mechanical work required for the compressor. Geothermal pump systems reach a fairly high coefficient of COP, 3–6, on the coldest of winter nights, compared to 1.75–2.5 for air-source heat pumps on cool days.

6.4 A novel passive bidirectional BIPVT-geothermal system

A traditional geothermal system is operated when there is a heating and cooling need so that it commonly works in an intermittent fashion to save energy for water pumping. Therefore, a good sensing and control system is required. When a geothermal system is coupling with a BIPVT system, the system needs to be constantly operated to prevent the temperature loss of PV efficiency. Therefore, it is critical to keep the system operating sustainably and efficiently. As the basements of most buildings consume much less energy than the rooms in the air, a novel passive BIPVT-geothermal system is introduced in this Section.

6.4.1 The passive BIPVT-geothermal system design

The goal of this design is to enable a sustainable operation of the BIPVT-geothermal system as the building was operated like a basement, in which no energy is required to power the system most time although a backup system could be integrated for extreme cases or active control.

Fig. 6.9 shows a schematic illustration of the novel energy-efficient bidirectional geothermal energy storage and exchange system, which can be scaled up for large-scale applications. This technology utilizes geothermal wells coupled with thermal energy storage by phase change materials (PCMs) to regulate the temperature in buildings through efficient heat exchange between a wall panel and the underground. Aluminum heat pipes are embedded into the wall panels and roof panels for heat exchange with indoor environment and temperature control of PV cells, and a copper heat pipe (60–150′ long) will be installed in a shallow geothermal well filled with PCM for heat dissipation and storage to the underground. To enable efficient heat transfer and storage, heat pipes and an elastic manifold will be designed and used to achieve auto-adaptive and self-driving thermal flow between the building and geothermal well by automatically circulating water through a novel passive pumping mechanism. The manifold will connect the bottom of the wall panel and the top of the geothermal heat well to bring up heat from underground to the room by the passive pumping mechanism. With the passive system keeps running, only

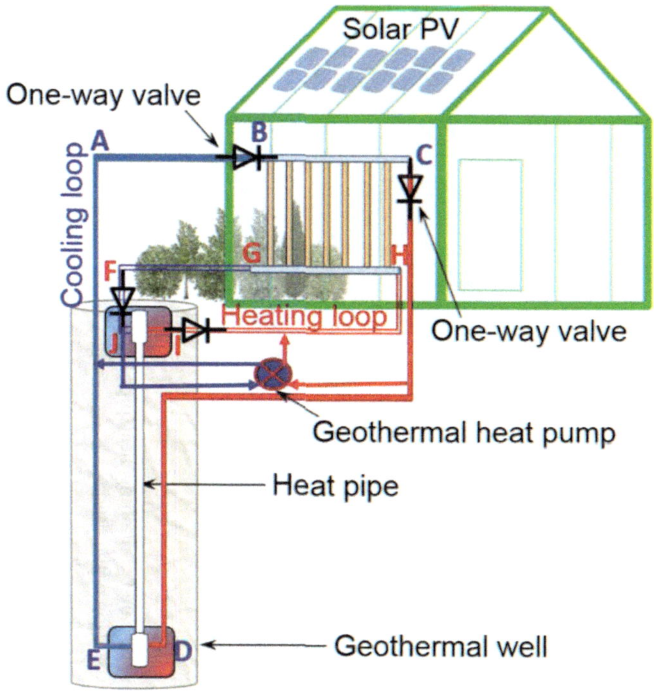

FIGURE 6.9 Schematic illustration of a passive BIPVT-geothermal system.

a small backup geothermal heat pump is required to provide customized air conditioning.

This disruptive technology can reach high system efficiency by removing or minimizing the use of the heat pump while achieving equivalent indoor comfort. Under regular operation, the O&M costs may approach zero, particularly for some geographic areas where a stand-alone passive geothermal system can sufficiently meet the cooling demand without any heat pumps. This is a game-change technology to break the industry criterion based on the COP as no energy input is required for the operation in those cases. Although a backup heat pump system, which can be easily integrated into the system, could be needed in some areas, the frequency and capacity of the heat pump use will be much lower, so that the effective COP will be significantly increased.

6.4.2 Technology innovations and merits

This technology will lead to self-powered solutions for temperature conditioning of buildings with high efficiency and improved performance. Fig. 6.10 shows the three key innovations (Yin et al., 2021).

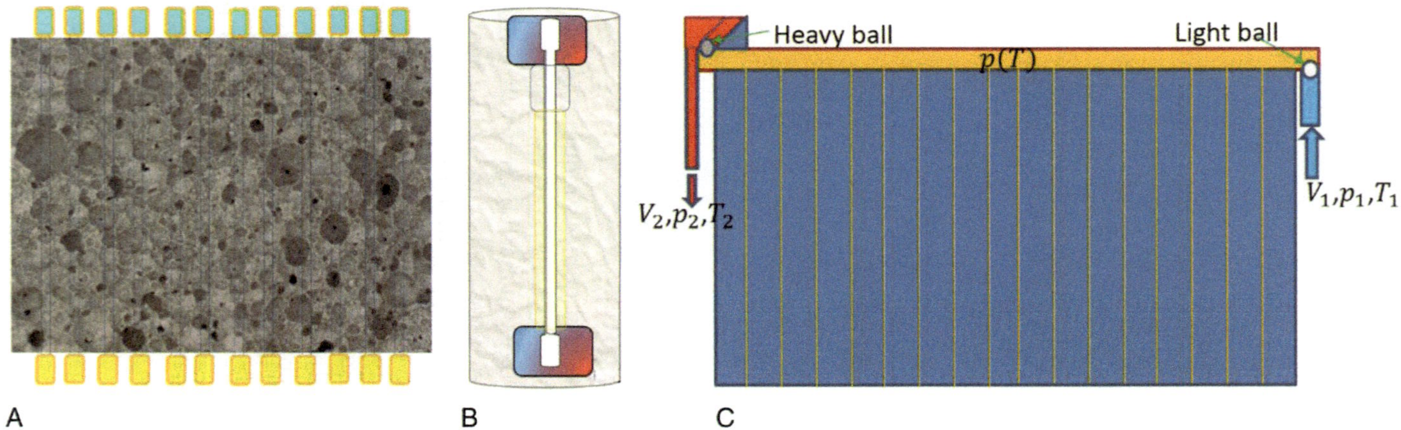

FIGURE 6.10 Innovations of the project: (A) lightweight concrete wall panel containing heat pipes. (B) A geothermal well with a heat pipe exchanger. (C) A passive thermal fluid flow is pumped by the thermal expansion of the fluid.

Innovation 1. The heat pipes in the wall panel include heat exchangers at both ends for both heating and cooling (Fig. 6.10A). The heat exchanger at the ends of the heat pipe plugs into two manifolds with frictional contact. The top manifold is for cooling as a condenser and the bottom for heating as an evaporator. It is like a radiator but no flow through the wall panel.

Innovation 2. A geothermal heat exchange will use a heat pipe inserted in PCM for improved heat exchange with the surrounding soil by extending the heat exchanging period (Fig. 6.10B). The top 20′ is thermally insulated to preventing heat loss for seasonal heat storage. The PCM is packaged in metal containers for high thermal conduction and heat storage with its latent heat.

Innovation 3. A passive thermal fluid flow pumped by thermal expansion of fluid within the elastic manifold can achieve passive cooling with zero energy consumption. The working mechanism is illustrated in Fig. 6.10C. In the beginning, the pressure and temperature in the manifold is approximately equal to p_1 and T_1 . The light ball locked the inlet and the heavy ball locked the outlet, which requires a Δp_2 to lift up the heavy ball and Δp_1 to suck up the light ball ($\Delta p_2 \gg \Delta p_1$). The manifold is elastic. With temperature increase, $p(T)$ increases. When $p(T) - p_2$ reaches Δp_2, the heavy ball will roll away, and the hot liquid will flush out. The heavy ball rolls back, the pressure wave passes from left to right and sucks the light ball, and cold liquid will fill into the manifold. Overall the continuity of the liquid makes $V_2 \gg V_1$, and the hot liquid will be quickly pushed to the geothermal well and become cold liquid.

Heat pipe technology combines the principles of thermal conductivity, phase change, and mass transfer with convection to effectively transfer heat between two solid interfaces with high performance. For an example of the heat pipe in the geothermal well, in summer hot fluid is transferred to the bottom of the well, the refrigerant in the heat pipe in contact with a thermally conductive solid hot surface turns into a vapor. The vapor then rises along the heat pipe to the cold interface and condenses back into a liquid, releasing the latent heat to the rock wall. The liquid then returns to the hot interface through gravity, so that the cycle repeats and the thermal fluid is cooled down by the heat pipe flow back. In winter, the heat exchanger works at the top of the heat pipe to extract heat for wall and roof heating in a similar fashion. The wall panels and roof panels use similar technology but use aluminum to replace copper for the lightweight and lower cost. Because the vapor flow in the heat pipe is much faster than heat conduction along the depth direction, and the heat transfer coefficients for boiling and condensation are very high, heat pipes are highly effective thermal conductors. For example, the effective thermal conductivity, varying with heat pipe length, can approach 100 kW/(mÅK) for long heat pipes, in comparison with approximately 6.4 kW/(mÅK) for copper (Wikipedia, 2021).

TABLE 6.2 The Feature-Advantage-Benefit (FAB) table for the geothermal-BIPVT system.

Features	Advantages	Benefits
Heat pipes used in wall/roof panel	No fluid circulation within the panel	Easy installation, maintenance, replacement
Heat pipes used in wall/roof panel	Uniform temperature of panel	Reduce thermal crack of PV cells for longevity and efficiency
Heat pipes used in wall/roof panel	Heat pipe is strong and relatively stiff	Reduced weight and thickness for workability and overall costs
Heat pipes used in geothermal well	Rapid heat transfer in geothermal well	Less drilling depth is needed for the same capacity
Heat pipes used in geothermal well	Modular design for fluid circulation and heat exchange	Easy to replace the parts and optimize their design
Phase change materials used in geothermal well	High heat storage by latent heat at a specific temperature	Lower temperature variation and extended heat exchange window
Passive fluid pump by pressure change caused by temperature	Automatic operation powered by the heat exchange when a temperature difference exists	Make building operated like a basement
Passive heat exchange without heat pump	Quiet heating and cooling with minimal ducts and equipment	Excellent fitting for both retrofitting and new construction

Compared with the traditional geothermal-BIPVT systems, the novel design exhibits the following features, advantages, and benefits in Table 6.2.

6.4.3 Technical and economic feasibility

The proposed disruptive technology can achieve the equivalent indoor comfort in a new way: using one heat exchanging cycle to achieve both cooling/heating and ground heat exchange. A backup heat pump system is only needed when customized air conditioning is needed, and it can be easily integrated into the system. The use of a heat pump system will be significant reduced because the passive system keeps running in the background. For example, a 10,000-sq. ft building space generally requires a geothermal heat pump of ~20 tons by drilling ~3000 ft (8–10 wells). Using the proposed passive heat exchange system, drilling ~1200 ft can be sufficient by enhanced passive heat exchange with a 6-ton backup geothermal heat pump system. The O&M costs can be further reduced as the heat pump system is less used. As extreme cases for some geographic areas, a stand-alone passive geothermal system can sufficiently meet the cooling demand without any heat pumps, which creates a game-change technology to break the industry criterion based on the COP as no energy input is required for the operation in those cases.

This technology will (1) significantly improve the performance of geothermal energy system by incorporating thermal energy storage into geothermal wells, (2) increase low-temperature geothermal energy harvesting and utilization efficiencies via highly efficient heat transfer of heat pipes and thermal storage by nonlinear low-temperature heat flow and PCMs, and (3) improve building energy efficiency and drive down the costs and risks associated with geothermal energy. The research is expected to provide a cost-effective means to reduce the total energy demand of a system and allows for more efficient heating and cooling of buildings while renewable sources such as solar energy can be incorporated to cover the residual demand. Therefore, the research provides an economically viable pathway for achieving zero-energy buildings in the future.

The temperature of a geothermal well diminishes with time near the heat exchangers, and in the long run around the boreholes in a radius as large as 10 m (Desideri et al., 2011). However, in the current proposed system, with the heat being injected into the earth in the summer, the temperature of the well will stay constant or even slightly increase after the first year of operation.

The economic feasibility of this technology was analyzed based on the cost of goods and services, the available electricity, and the incentives in the locations of the project. In general, a geothermal system is capital intensive and requires high upfront investment due to the nature of the project requiring underground exploration, special equipment for drilling boreholes, the piping and connections used, and the heat pump units and control system. In the service, a heat engine cycle is used to remove heat from the indoor environment with electricity at a certain COP, which requires energy input for both heat engine for air cooling/heating and water pump for heat exchanging with the ground.

In general, geothermal systems have a low operation cost and high lifetime. However, a yearly inspection of the pressure/water level and the flow rate of the fluid is recommended. A geothermal well is typically designed for a 30-year useful life but can last even longer with proper operation and maintenance services

The primary requirement for a geothermal well is the adequate amount of heat captured through the loop pipes. In addition to technical viability, the other factors that will affect the economics of a geothermal system are exploration costs, available resource depth which affects the drilling costs, raw materials, and goods used in addition to the labor costs.

The following presents three scenarios for comparison: (1) regular geothermal well linked with a heat pump to provide forced air; (2) enhanced passive geothermal well with wall panel presented in this proposal; and (3) conventional forced-air system used with a furnace which is the common heat, ventilation, and air conditioning system in the United States.

For this analysis, the cost data were collected from the literature and some of the interviews that were done during the customer discovery process of the NSF I-Corps project from Columbia University. PVT clean energy—a Columbia University-based startup—is carrying out the commercialization process. This project will collect the performance data for a rigorous life cycle cost analysis. In this study, whenever the data was missing, additional assumptions were made to fill the gaps, based on the conservative estimates.

The results of this analysis are summarized in Table 6.3. The components of the geothermal well are listed with the cost of each item. For the drilling cost, \$20 per foot is considered as an average number in the state of NY. The designed heat pipe for the geothermal well is 1 inch in diameter made of copper with a unit price of \$10/ft. For both wells, based on the

TABLE 6.3 Physical parameters for each scenario for installation cost analysis.

	Regular geothermal	Passive geothermal
Number of wells	20	8
Drilling costs	60,000	24,000
Heat pipe in the well	0	12,000
Grout	7609	0
PCM	0	33,694
Insulation	0	16
Casing (steel)	700	700
Heat exchanger	0	1000
1 1/4″ high-density polyethylene tubes	6000	2400
Manifold	0	200
Heat pump	15,000	0
Ductwork	12,000	0
Heat pipes in the wall panel	0	10,000
Federal incentives	−23,220	−19,243
State rebates	−24,000	−24,000
Total (\$)	54,088	40,767

geological data of the location of the project, the first 20 ft is assumed to be an overburden layer consisting of sand and soil, followed by the bedrock afterward. For the regular well, a mixture of water and bentonite clay is used to seal the initial 20 ft walls from collapsing. For both wells, a 6.5″ steel pipe is considered to be used as a casing for 20 ft long. In the enhanced well, polystyrene foam is used for insulation, and the rest 130 ft of the well is filled with PCM. The ground loops are assumed to be high-density polyethylene with a diameter of 1 ¼″ and with the length of the well (150 ft*2). For the regular well, a heat pump and ductwork are considered in the design, while in the enhanced well, two heat exchangers and a manifold are used in the design. Finally, 26% federal tax incentive and $1200 per ton of cooling capacity for NY state rebates are applied to the total cost. Maintenance costs, additional labor, and permits are not included in the calculations.

It is observed that the conventional system needs 20 tons, the regular geothermal cost is $13,321 more expensive than the passive one even though in the latter, some improvements in the design have been made. This cost difference is mainly due to the high drilling cost of the 20 wells, whereas in the enhanced system, the total number of wells is reduced to 8.

Table 6.4 shows the economics of the project. The consumption of the heating and cooling parts of the furnace in the conventional forced-air system is divided into two parts: (1) 200 million BTU of heating by Natural Gas for 7 months; and (2) cooling for 5 months using 1000 kWh. However, due to the passive nature of the passive geothermal system, the energy consumption is assumed to be zero.

For the regular geothermal system, the energy consumption of the 20-ton heat pump is assumed to be 10,000 kWh annually, at a rate of 15 C/kWh. For the regular and passive geothermal systems, the federal

TABLE 6.4 Economic analysis and comparison.

	Regular geothermal	Passive geothermal	Conventional forced air
Gross system size (kWh)	70.33	70.33	70.33
Natural gas consumption (CCF)	0	0	300×10^6
Annual energy use (kWh)	14,000	0	NA
Annual operating expenses	1500	0	3300
System cost without incentives	101,309	84,010	19,000
Federal and state incentives	−47,220	−43,243	−550
System cost with incentives	54,088	40,767	18,450
Payback (yrs)	30.0	12.4	5.6

and state incentives are applied as Table 6.3, while for the conventional forced-air system, $150 federal tax for the high-efficiency furnace (95% efficiency) and $400 rebate from the Central Hudson utility company are applied.

The results for the final cost of the forced air systems show a significantly less amount compared with the geothermal system, as expected. However, while the payback time is 5.8 years for the forced air system, the passive geothermal system will payback for its cost at slightly double this number (12.7 years), while it will take 31.8 years for the regular geothermal system to offset the initial capital. Therefore, based on the results of the techno-economic analysis, and considering that the payback time of the passive geothermal system is less than half the regular system, it seems that the development of an enhanced passive geothermal system would be economically feasible.

Based on the above rough estimate, it is feasible to achieve a 20% reduction of the initial cost and 50% energy saving compared with the baseline system by using a backup heat pump with the passive geothermal system.

6.4.4 Energy consumption and efficiency analysis

A case study to demonstrate the energy efficiency of the proposed passive geothermal system is presented below to compare the passive geothermal system shown in Fig. 6.9 with a single 3-ton unit. Only one hot summer day is analyzed. The parameters are assumed based on our current knowledge. The analysis can be extended to year-round applications and will be updated with future measurements and experience as shown in Table 6.5.

TABLE 6.5 Energy analysis and comparison.

Twenty four hours operation	Regular geothermal	Passive geothermal
Water flow rate	10 gmp	4 gmp
Entering water temperature	65°F	56°F
Leaving water temperature	105 °F	70°F
Heat pump	10.5 KW	0
Water pump	0.5 KW	0
Coefficient of performance	3	Infinite
Cooling power	31.5 KW	8.24 KW
Running rate	25%	100%
Cooling amount	189 KWh (0.65MBTU)	198 KWh (0.68 MBTU)
Electric consumption	66 KWh	0

Temperature conditions: ground temperate 54°F, average ambient temperature 88°F, target room temperature 72°F.

For a regular 3 ton geothermal system with the geothermal well total length/depth at 450′, a 500 W water pump will be used to generate a water flow at 10 gpm with the entering water temperature at 65°F and the LWT 105°F. The system is running at 25% of the period at a COP = 3. The total energy consumption on heat and water pump is (10.5+0.5) × 24 × 25% = 66 KWh to remove 10.5 × 3 × 24 × 25% = 189 KWh = 0.65MBTU of heat from the room.

Using the passive geothermal system, we assume the entering water temperature at 56F and LWT at 70F at a rate of 4 gpm for 24 h. It remove (70–56) °F × 5/9 °C/°F × 4 gal × 3.785 kg/gal /60s * 4.2 * 1000 J/kg/C × 24 h = 198 kWh heat from the room. Actually, the impact of the passive system produces much less impact to ground temperature because much less heat is rejected to the ground, which will improve the system performance in another aspect.

Notice that the LWT of the passive geothermal system means the liquid flow back from the room with the room temperature at 72°F, which has different meanings from the one for regular geothermal.

There are two issues with the passive geothermal system compared with the regular one:

1. The passive geothermal system cannot suddenly change the temperature setting because of the lower cooling power. A backup heat pump can be added to achieve this need at reduced cooling power.
2. The key is to generate a flow rate at a level of 4 gpm or higher by the passive flow pump with our invention. Although the working mechanism is sound and practical in the laboratory test environment, significant research is required to scale it up for field application and long-term operation.
3. The moisture and temperature of the room should be carefully sensed and monitored to avoid condensation caused by solid heating or cooling

6.5 Case study 1: A net–zero energy house at Wisconsin

This house (Fig. 6.11) is located in Wisconsin Rapids, Wisconsin, and was designed net-zero, including the electricity bought and sold to the grid. Fig. 6.12 shows the layout othe whole BITERS system on the roof. It is a Leadership in Energy and Environmental Design (LEED) certified house in Wisconsin (Fig. 6.13). The house is 2500 square feet without the basement, and the basement is 900 square feet. The house was built in 2009 and the owner received the LEED certification in 2011. The house is net-zero energy running on solar and geothermal energy. This system has 500 Sunslates

FIGURE 6.11 The net-zero energy house in Wisconsin.

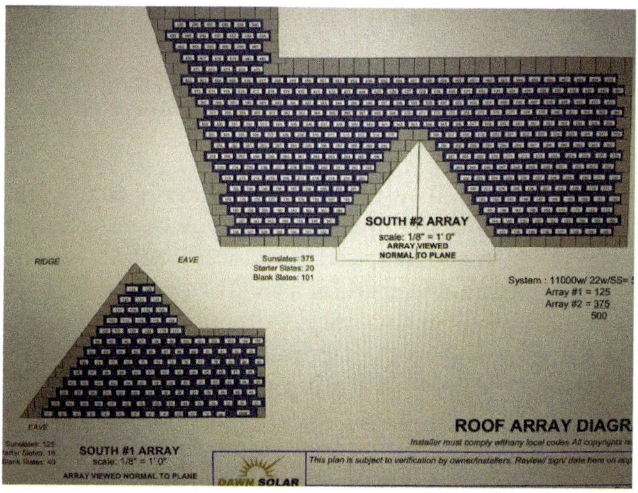

FIGURE 6.12 Layout of the whole BITERS system for the net-zero house.

with each shingle having an output of 19 W which results in the production of 10,000 KWh annually on a fully sunny day.

Figs. 6.14 and 6.15 show the construction and installation of the BITERS system. The thermal management system is made of cross-linked polyethylene (PEX) tubes (Fig. 6.16), which pass through the roof and connect to a water tank in the basement (Figs. 6.17–6.19). The Sunslate panels are installed onto the tube system (Figs. 6.20–6.23). Besides solar and geothermal energy, other practices that the owner has adopted to make the house even more energy-efficient include better isolation of the house walls, sensor control of faucet, LED lights automatic on/off, shades

FIGURE 6.13 LEED certificated of the house. *LEED*, Leadership in Energy and Environmental Design.

FIGURE 6.14 Installation of the bottom frame for the Sunslate panels.

FIGURE 6.15 View of the BITERS system in construction on the roof.

FIGURE 6.16 PEX tubes inside the BITERS system for transferring heated water.

for windows. The owner learned about the solar roof design 10 years ago and decided to install it for his house. He regularly does online research for up-to-date sustainable technologies that can be employed in his house.

Since the house was built in 2009, the owner has not paid any electricity bill because they use the solar electricity combined with grid

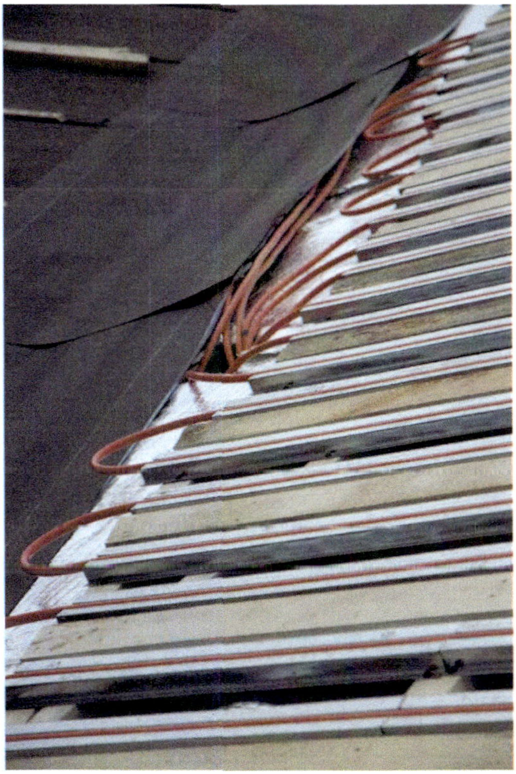

FIGURE 6.17 The tubes are all passed through the roof to a water tank in the basement.

FIGURE 6.18 A layer of insulation is placed under the PEX tubes to avoid heat waste to the roof.

FIGURE 6.19 Outlook of the BITERS system on the roof.

FIGURE 6.20 Finished BITERS system ready to install the Sunslate panels.

electricity but the cost from the grid is balanced out from selling the solar electricity to the grid. Todd sells solar electricity to the grid at 25 cents per KWh and pays grid electricity at 9 cents per kWh, although in 2020 the conversion rate has changed to sell the electricity at the same rate that the customer buys. Over 8 years, he has produced 106,000 kWh of electricity. Where the house is located, the temperature in November is around 40°F and in February goes down as low as −15°F, while in the summertime is between 80 and 90 °F. The owner only turns on the air

FIGURE 6.21 Sunstale panel on the roof.

FIGURE 6.22 Sunslate panels are placed on top of the BITERS system.

conditioner in summers for 10–20 days for humidity and to avoid the pollens from trees. His system has two inverters of Selectria PV I-5300. On a normal sunny day in October, when the temperature outside is 40°F, the temperature from the BITERS can reach 94°F, and the system produces almost 45 kWh. At this time of the year, the sunrise is at about 7 AM and the sunset is at about 6:45 PM. On such days, the system has an average current output of 6 amps, frequency of 60 Hz, and energy output of 23 kWh. The BITERS system is separate from the geothermal system. The

FIGURE 6.23 Finished net-zero house with the BITERS system on the roof.

heat collected from the BITERS system heat the city water from 45–90°F, and then it will be further heated via a hot water heater which runs on a gas or geothermal heat pump the water temperature gets to 125°F which is the ideal temperature of domestic hot water use. The loops on the roof are also connected with the geothermal well. When the temp difference between the solar loops and the geothermal loops is greater than 30 degrees, the solar loop would passively mix in a separate hot water heater to bring up the temp of the ground loop prior to entering the Geothermal furnace.

In this house, there is also a floor heating system where plastic PEX pipes were laid down first followed by lightweight concrete and then floor materials. The owner chose to have a concrete driveway because it would last about 20 years for the weather in his area, and the lifespan of a concrete driveway is longer than asphalt concrete. He does not have a heated driveway because he thinks he and his family members do not mind shoving snow and he wanted to save the cost of an expensive heated driveway.

Below, you can find the data for the past 4 years of usage and solar production of the BITERS system in Tables 6.6–6.9. This system is designed for roughly 10,500 kWh of electricity annually. As you can see 2018 had low output. This was due to problems with shorted-out Sunslates that were replaced in June 2018. The performance monitoring and system maintenance are key factors to assure the system is running appropriately and yields energy as designed.

TABLE 6.6 Consumed electricity and generated solar electricity of the house for 2017.

2017	Consumed (kWh)	Generated (kWh)
Jan.	3102	157
Feb.	1958	393
Mar.	1289	499
Apr.	1077	439
May.	1081	605
Jun.[a]	1769	602
Jul.	2603	992
Aug.	2018	1192
Sep.	2331	1154
Oct.	1274	698
Nov.	1645	715
Dec.	2908	379
Total energy	23,055	7825

[a] *Fifty percent of the system was not working until Jun. when it was fixed.*

TABLE 6.7 Consumed electricity and generated solar electricity of the house for 2018.

2018	Consumed (kWh)	Generated (kWh)
Jan.	3230	723
Feb.	2520	791
Mar.	1870	1278
Apr.	1661	1024
May	1478	1243
Jun.	2215	1152
Jul.	2834	1309
Aug.[a]	2694	1048
Sep.	1818	981
Oct.	1090	726
Nov.	2036	460
Dec.	2505	652
Total energy	25,951	11,387

[a] *The rate of electricity sold to the grid changed from $0.22 to $0.09*

TABLE 6.8 Consumed electricity and generated solar electricity of the house for 2019.

2019	Consumed (kWh)	Generated (kWh)
Jan.	3598	428
Feb.	3119	99
Mar.	1848	843
Apr.	1423	906
May	1245	1078
Jun.	1553	1102
Jul.	2747	1266
Aug.	2336	1163
Sep.	1595	803
Oct.	1357	715
Nov.	1603	459
Dec.	1939	381
Total energy	24,363	9243

TABLE 6.9 Consumed electricity and generated solar electricity of the house for the first half of 2020.

2020	Consumed (kWh)	Generated (kWh)
Jan.	2313	181
Feb.	2258	842
Mar.	1980	941
Apr.	1979	1175
May	2039	1181
Jun.	2888	1260
Jul.	3540	1236
Aug.	2714	1178
Sep.[a]	–	–
Oct.	–	–
Nov.	–	–
Dec.	–	–
Total energy	19,711	7994

[a] *Data not available for months Sep.–Dec.*

6.6 Case study 2: BIPV curtain wall building

The project was built in 2006 at the Kankakee Community College in Illinois (Fig. 6.24). Around that period, geothermal and storage technologies were not quite ready. To obtain more sunlight on buildings is always very attractive especially applying insulated glass with photovoltaic. Figs. 6.25 and 6.26 show the layout and connection of the PV modules with the top view shown in Fig. 6.27. Though in this case, the photovoltaic portion is quite minimal, it could be helpful to tie in on different systems in the building. However, in the insulated glass situation, in the summer and wintertime situation, it shields off heat and cold that save energy consumptions. In Kankakee, the insulated glass uses air to fill as an insulator while the Rock Valley, the glass uses argon gas which is the best gas insulator as compared to any other gases available. The production power of this system is 2 kW which can provide enough electricity for the electronic pumps of the restroom faucets in the building.

In another project, curtain walls were used at Rock Valley Community College in Illinois in the south and west-facing orientation at the right angle (Figs. 6.28 and 6.29). The module layout on the façade is shown below (Figs. 6.30–6.32). The Rock Valley Community completed the Karl J. Jacobs Center of Science and Math in 2012. It was awarded LEED Gold by US Green Building Council. As one can see that the whole system in the

FIGURE 6.24 Kankakee Fine Arts Building: insulated glass/glass curtain wall (courtesy of Kankakee Community College).

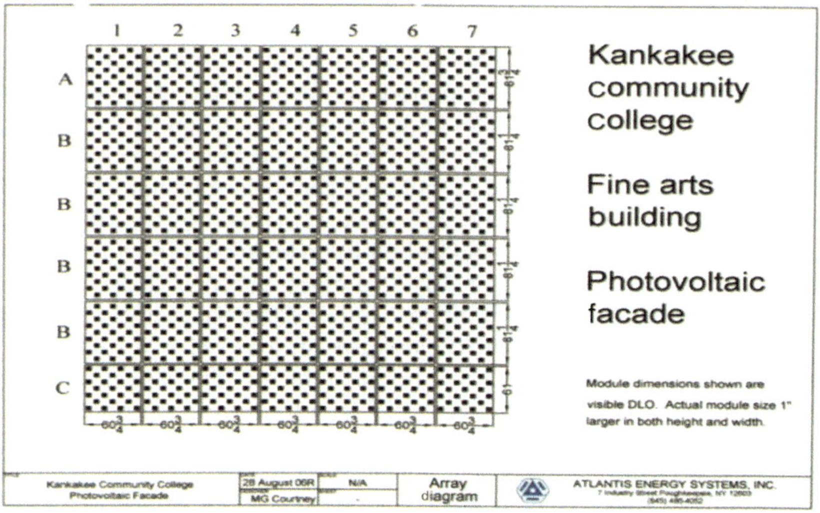

FIGURE 6.25 Layout of the PV modules used in the project.

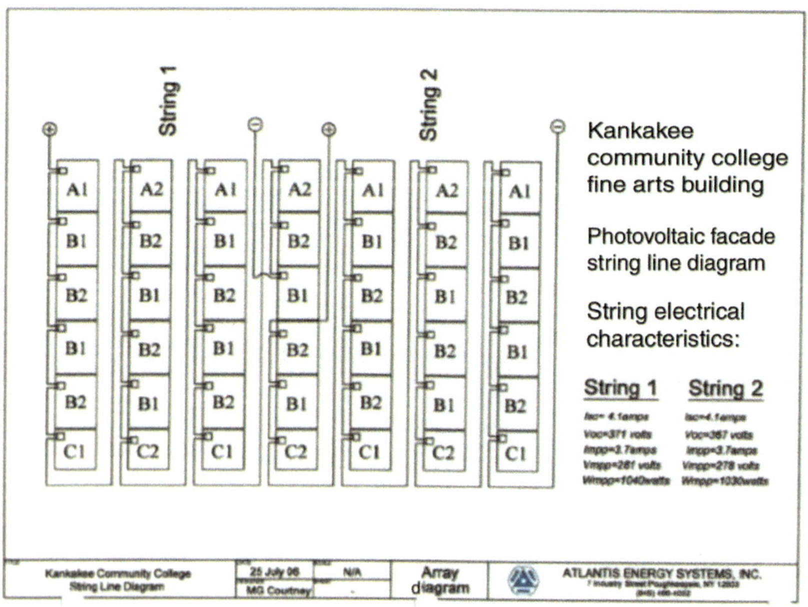

FIGURE 6.26 String line diagram of the PV modules.

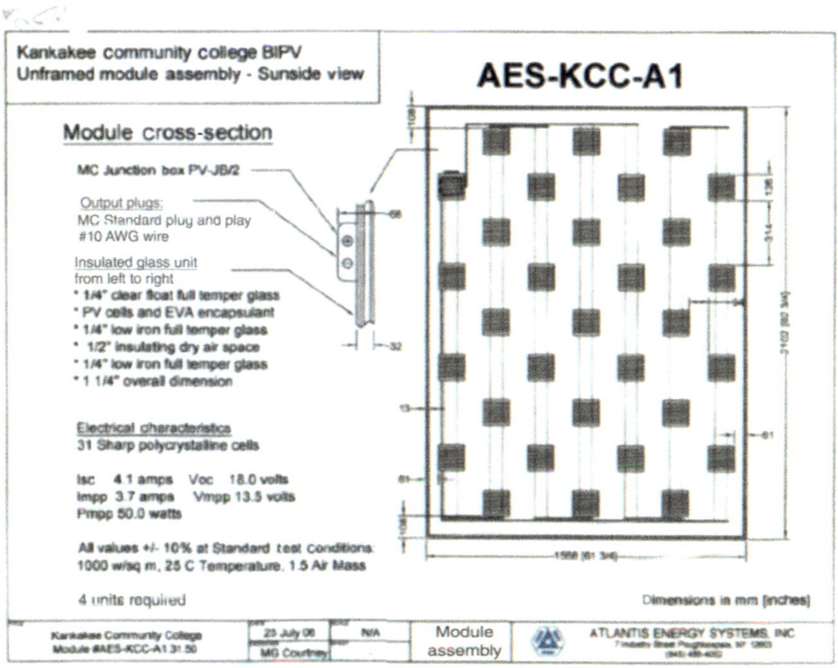

FIGURE 6.27 Top view of the module assembly.

building has a substantial number of energy-saving technology systems. Geothermal systems (Fig. 6.33) also play a major role. In this case, solar electricity not only produces energy but also allows sunlight in, which brightens the interior as well as making it aesthetically pleasing. The bill of materials is provided in Table 6.10. The system has an annual saving of $76,000.00. In light of the recent development of the storage and generation technology system, the energy and cost savings could be higher.

Apart from the BIPV curtain wall, the Rock Valley system consists of the items below:

1. Four hundred and twenty ton closed-loop geothermal heating and cooling.
2. Water-cooled heat pumps.
3. Chilled beam cooling system.
4. Energy recovery on the air handling system.
5. Variable volume heating and cooling pumps.
6. Thermally efficient glazing (low-E coatings, ceramic fritting).
7. White membrane roof W/R insulation.
8. Geothermal domestic water heating.

FIGURE 6.28 Exterior view of the Karl J. Jacobs Center of Science and Math (RockValley).

FIGURE 6.29 Interior view of the building at Rock Valley Community College.

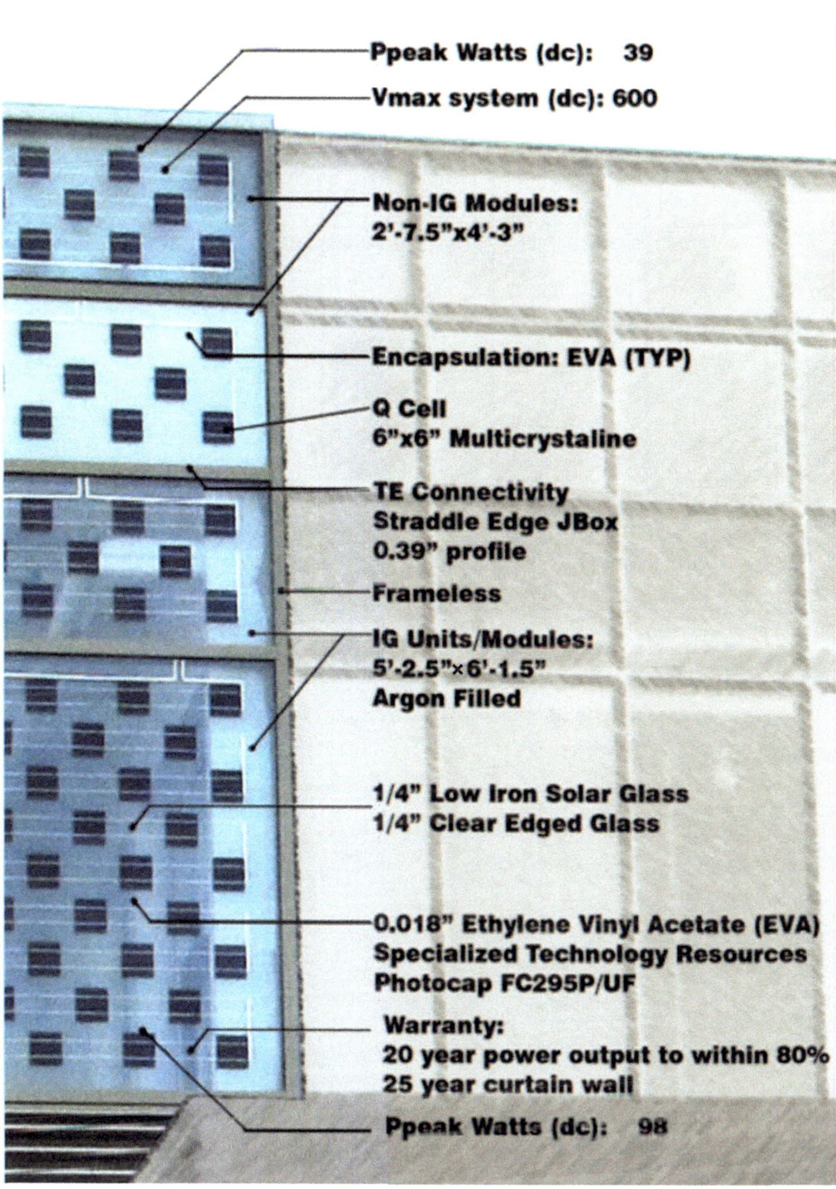

FIGURE 6.30 Detailed specifications of the modules.

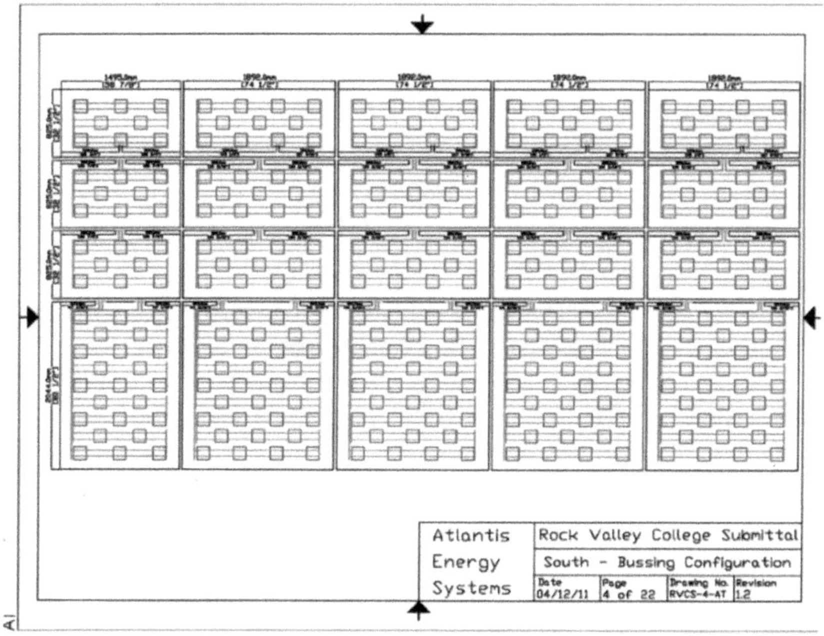

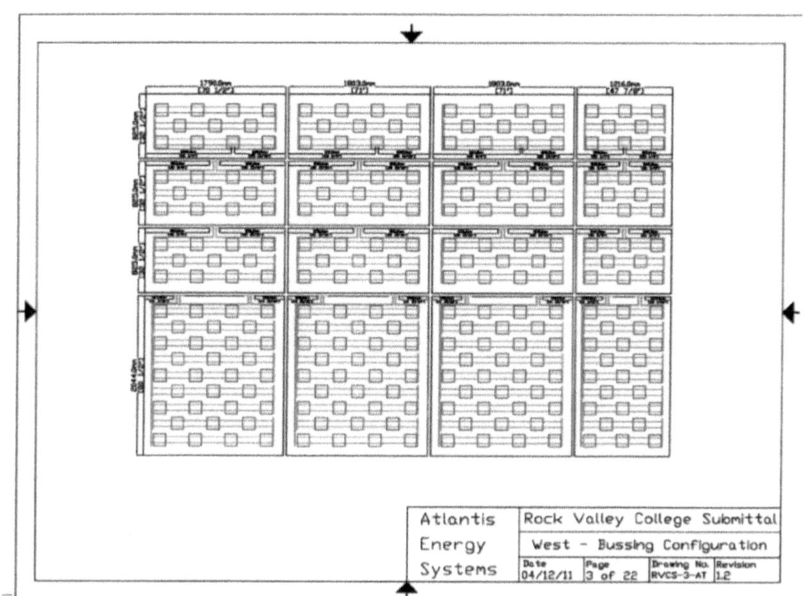

FIGURE 6.31 Layout of the south and west facing modules.

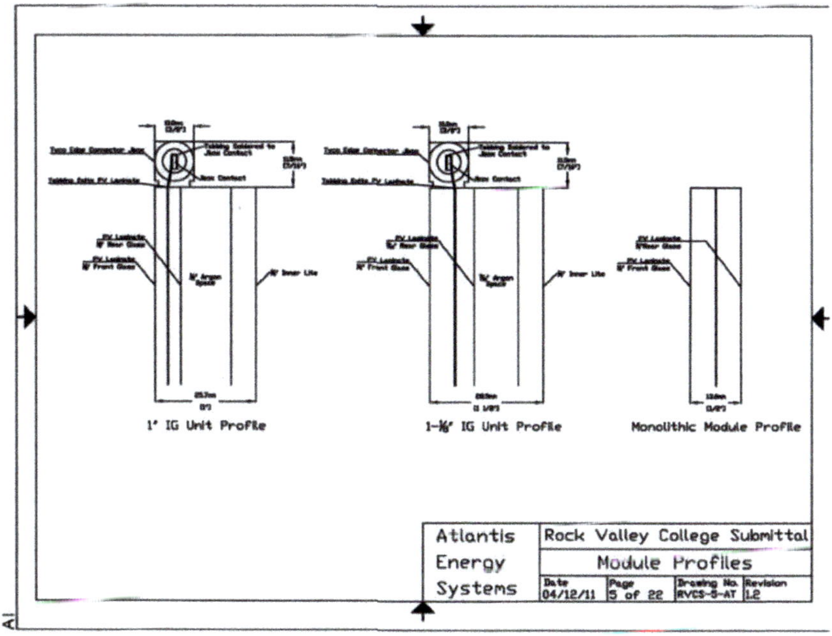

FIGURE 6.32 Rock valley module insulate glass design.

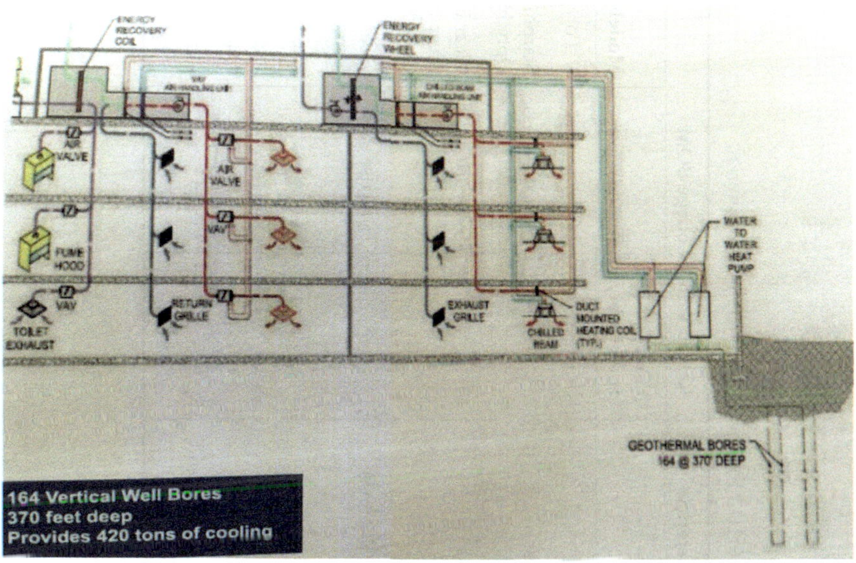

FIGURE 6.33 Rock valley geothermal system.

TABLE 6.10 Rock valley module bill of material.

Bill of materials	ROCK Valley College BOM						Version 1.1 3/2/11
Item	**QTY. required for assembly**	**Vendor**	**Model #**	**Dim. width (mm)**	**Dim. length (mm)**	**Area (m^2)**	**Notes**
Cells	Varies	LOF	6″ Poly 3.89W	156	156	0.024	
Front glass	1	Syracuse glass	Starphire Tempered 1/4″ Optwhite Tempered 1/8″	Varies	Varies	N/A	
Rear glass	1	Syracuse glass	Clear Float Temperd 1/4″ Clear Float Temperd 3/16″ Clear Float Temperd 1/8"	Varies	Varies	N/A	
Inner lite (IG unit)	1	Syracuse glass	Clear Float Temperd 1/4″ (some with white frit)	Varies	Varies	N/A	
Encapsulant	3	STR	Photocap 15295P/UF	Varies	Varies	N/A	
Tabbing/bussing matrial	Varies	Ubrich	7746-9464 7746-9574	2.3 8	N/A	N/A	
Solder	N/A	Kester	#14-6337-0062 SN63PB37 SLD	1.5	N/A	N/A	

(continued on next page)

Insulation strips	N/A	Madico	InsuPatch (Transparent)	N/A	N/A	N/A	
Tape	N/A	Coverre	Dy Mat E	19	30	N/A	
Blue tape	N/A	3M	3M Polyester Tape 8901	36	N/A	N/A	
J box	Varies	Tyco Multicontact	1971569-1 PV-JB/2/N	N/A	Varise	N/A	
Connectors	Varies	Tyco Multicontact	1971131- (Plus Socket) 1971131-2 (Minus Socket) 1971131-1 (END socket) PV-KST4/61-UR (Male) PV-KBT4/61-UR (Female)	N/A	N/A	N/A	
Diodes (for MC J-box)	Varies	Global Semisolutions	90SQ045	N/A	N/A	N/A	
Insulation sleeves	N/A	Dabum	D200-2	N/A	N/A	N/A	
Glue	N/A	Shin-Etsu Dow Corning Tonsan	KE-45-T PV 804 1521	N/A	N/A	N/A	

9. Variable frequency drives.
10. High-efficiency lighting fixtures:
    - Daylight/occupancy sensors.
    - Multiple switching zones.
    - Daylight harvesting sensors.
    - Use of low mercury lamps.
11. Variable frequency-drive for motors.
12. Enhanced commissioning.

References

Badache, M., et al., 2016. A new modeling approach for improved ground temperature profile determination. Renew. Energy 85, 436–444.

Beckers, K., Young, K.R., 2018. Technical requirements for geothermal resource confirmation. GRC Trans. 42.

Boyle, G., Energy for a Sustainable Future: Renewable Energy. Open University Press, 2004.

Brown, G.O., The history of the Darcy-Weisbach equation for pipe flow resistance, in *Environmental and water resources history*. 2003. pp. 34–43.

Carslaw, H.S., Jaeger, J.C., 1992. Conduction of Heat in Solids. Clarendon press, Oxford, England.

Carvalho, A.D., et al., Ground source heat pumps as high efficient solutions for building space conditioning and for integration in smart grids. Energy Convers. Manag., 2015. 103: pp. 991–1007.

Desideri, U., et al., 2011. Feasibility study and numerical simulation of a ground source heat pump plant, applied to a residential building. Appl. Therm. Eng. 31 (16), 3500–3511.

Dickson, M.H., Fanelli, M., 2013. Geothermal Energy: Utilization and Technology. Routledge, T&F Group, London, UK.

Emmi, G., et al., An analysis of solar assisted ground source heat pumps in cold climates. Energy Convers. Manag., 2015. 106: pp. 660–675.

Esen, H., et al., Modelling and experimental performance analysis of solar-assisted ground source heat pump system. J. Exp. Theor. Artif. Intell., 2017. 29(1): p. 1-17.

Fridleifsson, I.B.J.R., 2001. Geothermal energy for the benefit of the people. Renew. Sustain. Energy Rev. 5 (3), 299–312.

Glassley, W.E., 2014. Geothermal Energy: Renewable Energy and the Environment. CRC press, Taylor and Francis Group, London, UK.

Hakkaki-Fard, A., et al., A techno-economic comparison of a direct expansion ground-source and an air-source heat pump system in Canadian cold climates. Energy, 2015. 87: pp. 49–59.

Hepbasli, A.J.E.C., 2010. A review on energetic, exergetic and exergoeconomic aspects of geothermal district heating systems (GDHSs). Energy Convers. Manag. 51 (10), 2041–2061.

Hepbasli, A.J.E.S., Current status of geothermal energy applications in Turkey. Energy Sources, 2003. 25(7): pp. 667–677.

Kräuchi, P., T., Schluck, M., Sulzer, 2015. Modelling of low temperature heating networks with IDA-ICE. In: Proceedings of International Conference CISBAT 2015 Future Buildings and Districts Sustainability from Nano to Urban Scale. LESO-PB, EPFL.

Luo, J., et al., Heating and cooling performance analysis of a ground source heat pump system in Southern Germany. Geothermics, 2015. 53: pp. 57–66.

Metz, P.D., Design, construction, and operation of the solar assisted heat pump ground coupled storage experiments at Brookhaven National Laboratory. In Presented at the 4th Ann. Heat Pump Technol. Conf. 1979.

Nouri, G., Noorollahi, Y., Yousefi, H., 2019. Solar assisted ground source heat pump systems–a review. Appl. Therm. Eng. 163, 114351.
NREL, 2018. https://www.nrel.gov/geothermal/resource-assessment-mapping.html (accessed: 2020).
Ozgener, L., et al., Energy and exergy analysis of geothermal district heating systems: an application. Build. Env., 2005. 40(10): pp. 1309–1322.
Prasanna, A., V. Dorer, and N.J.E. Vetterli, Optimisation of a district energy system with a low temperature network. Energy, 2017. 137: pp. 632–648.
Ramey Jr, H.J., 1962. Wellbore heat transmission. J. Petrol. Technol. 14 (04), 427–435.
Rees, S., 2016. Advances in Ground-Source Heat Pump Systems. Elsevier Science, Netherlands.
RockValley. Rock Valley Brochure provided by Rock Valley Community College through Kelly Andereck, Architect.
Ruiz-Calvo, F., et al., Development and experimental validation of a TRNSYS dynamic tool for design and energy optimization of ground source heat pump systems. Energies, 2017. 10(10): p. 1510.
Sakellariou, E.I., Axaopoulos, P.J., Wright, A.J., 2021. Energy and economic evaluation of a solar assisted ground source heat pump system for a north Mediterranean city. Energy Build. 231, 110640.
Spitler, J.D., Gehlin, S.E.J.R., 2015. Thermal response testing for ground source heat pump systems—an historical review. Renew. Sustain. Energy Rev. 50, 1125–1137.
Verda, V., et al., Second law analysis of horizontal geothermal heat pump systems. Energy Build., 2016. 124: pp. 236–240.
Vieira, A., et al., Characterisation of ground thermal and thermo-mechanical behaviour for shallow geothermal energy applications. Energies, 2017. 10(12): p. 2044.
Watson, A., 2016. Geothermal Engineering. Springer, Switzerland.
Weather.gov, 2021. Average monthly & annual temperatures at Central Park. https://www.weather.gov/media/okx/Climate/CentralPark/monthlyannualtemp.pdf (accessed: 2021).
Wikipedia, Heat Pipe. 2021. https://en.wikipedia.org/wiki/Heat_pipe (accessed: 2020).
Yildirim, N., Toksoy, M., Gokcen, G.J.E., 2010. Piping network design of geothermal district heating systems: case study for a university campus. Energy, 35(8): pp. 3256–3262.
Yin, H., Pao, F., Zadshir, M., Lou, J., Wei, Z., 2021. Building integrated photovoltaic thermal systems. U.S. Patent Application No. 63/248,502.

This energy-independent building integrates solar, wind, and wave energy harvesting, and provides extended space for living or business. The underwater structure provides a unique view of the sea world and the top structure in the air is surfaced with BIPVT and supporting wind turbines. The water surface layer houses wave energy converters. This novel structure is not a simple superposition of the three energy harvesting devices but renders synergistic functions and benefits by avoiding existing problems for the individual methods. Solar energy harvesting requires both a large land surface due to its low energy density and large energy storage or backup system due to its partial daily availability, whereas wind and wave energy supply is desynchronized with higher energy density than solar energy. The building can easily balance the energy use with minimal energy storage, which can be realized by pumping and releasing water between the top pool and ocean. Because of the huge pressure difference at the bottom of the building, the desalination of seawater into freshwater can also be done. The ocean provides natural temperature conditioning to the building. This technology can turn the remote unlivable islands into vacation resorts rapidly.

CHAPTER

7

BIPVT coupling with wind and wave energy

7.1 Introduction

Solar energy has been a promising solution for energy and environmental challenges and global warming threats (Crowley, 2000; Green and Ratner, 2004; Lewis et al., 2005; Hemminger et al., 2008; www.nature.com, 2009). While we are suffering from energy crisis and environmental pollution related to fossil fuel combustion, the sun delivers to the earth 120K TW, which is the largest energy source, renewable or nonrenewable, and is about 8000 times the rate at which human civilization currently produces and uses energy (Crabtree and Lewis, 2007). Table 7.1 shows the renewable energy sources. Although the availability of solar energy is good, it spreads thin and requires a considerable land area to reach a large scale of power generation, which commonly leads to a high installation cost but is susceptible to damage in extreme weather. Moreover, the PV system only works in the daytime, a large energy storage or backup electric generation system is required for nighttime electricity use. So far, most PV electricity is fed into the grid for routine use together with other electrical power supplies and cannot independently supply renewable electric power yet.

Wind energy is the second-largest renewable energy source. It is partially another form of solar energy. Winds are caused by the heating difference of the atmosphere by the sun, the rotation of the Earth, and the Earth's surface and air moisture irregularities. Because the wind is caused by the air pressure difference, it is not synchronized with solar energy. Instead, wind energy harvesting together with solar power may be a complementary combination to cover longer time and significantly save energy storage capacity. Wind power is cost-effective. The land-based utility-scale wind is one of the lowest-priced energy sources available today, costing 1–2 cents per kilowatt-hour after the production tax credit (Energy.gov, 2020). Because the electricity from wind farms is sold at a fixed price over a long period (e.g., 20+ years) and is independent of

Building Integrated Photovoltaic Thermal Systems: Fundamentals, Designs and Applications
DOI: https://doi.org/10.1016/B978-0-12-821064-2.00007-9

TABLE 7.1 Order-of-magnitude estimates of theoretical/extractable/technical world potentials, and current supplies, of renewable energy resources (Sandia, 2006).

	World				
Energy resource	**Theoretical potential (TW)**	**Extractable potential (TW)**	**Technical potential (TW)**	**2001 Supply (TW)**	**United States 2001 Supply (TW)**
Hydropower	12 TW	3.5	1.2	0.23	0.056
Ocean Wave	34 TW	8.5	0.62	~0	~0
Ocean surface currents	8.1 TW	2.0	0.012	~0	~0
Ocean thermal gradient	3.9 TW	0.033	0.0033	~0	~0
Ocean salinity gradient	3.0 TW	0.74	0.074	~0	~0
Ocean tidal	2.4 TW	0.60	0.037	0.000050	
Wind	1,000 TW	250	14	0.0050	0.0023
Geothermal	44 TW	2.8	1.9	0.0050	0.0016
Solar electricity	89,000 TW	58,000	7500	0.00015	0.000025
Solar fuels	89,000 TW	61,000	2500	0.19	0.088
Solar thermal	89,000 TW	19,000	5600	0.00060	0.00018

any fuel, wind energy mitigates the price uncertainty caused by the fuel price change. Wind energy is a clean energy source and does not produce atmospheric emissions that cause acid rain, smog, or greenhouse gases. The wind supply is abundant and inexhaustible over many areas. Over the past 10 years, US wind power capacity has grown 15% per year, and wind is now the largest source of renewable power in the United States. Wind turbines can be built on farms, ranches, deserts, seashore, and off-shore, where most of the best wind sites are found. However, the installation cost of a wind farm is relatively high and electricity transmission lines must be built to bring the electricity from the wind farm to the grid. If a wind farm is close to cities, noise, and visual impacts caused by turbine blades may become a big issue.

About 70% of the earth's surface is covered by ocean, which may provide enough area for solar energy harvesting avoiding the land competition under the population explosion since the last century. Instead, considering the sea-level change, it is foreseen that humans need to explore more living space and food supplies from the ocean as well. At the same time, ocean waves carry abundant energy, which is estimated to be 34 TW globally in Table 7.1. Wave energy is stored in the form of water particle motions under the waves. Compared to solar and wind energy, wave energy has a much higher energy density. According to the recent report (R.E.H. Sims, 2007), 2% of the world's coastline exceeds a wave power density of 30 (kW/m). As the huge potential of wave energy, the first patented wave energy converter (WEC) was invented as early as 1799 (Clément et al., 2002). Since then, various types of WECs have been invented with different working principles and locations. The energy harvesting efficiency could commonly reach 25%–50%, which is higher than solar or wind energy conversion. However, wave-power generation is not a widely employed commercial technology compared to other established renewable energy sources such as wind and solar power due to the potential impacts to the biophysical environment and marine animals as well as the relatively high installation, operation, and maintenance costs.

Considering the land and energy availability, one can see combined solar, wind, and wave energy harvesting in the ocean including onshore, nearshore, and offshore exhibits great potentials for large energy capacity and less land and environmental restriction. The chapter cover illustration shows an excellent example of the application of this concept to ocean resorts, ocean fishing farms, or future apartments. The energy-independent building integrates solar, wind, and wave energy harvesting together, and provides extended space for living or business. The underwater structure provides a unique view of the sea world, and top structure in the air surfacing with BIPVT and supporting wind turbines. The water surface layer equips WEC.

The three types of energy sources can be relatively independent: solar devices are mainly spread in the horizontal plane, but wind in the vertical plane, and wave below the air surface. They are available at different timing. The energy harvesting by three sources provides an excellent combination for redundancy and resilience of energy supply. For a peaceful sunny day, solar energy may power the building; on a stormy night, wind and wave energy may take the role. This novel structure is not a simple superposition of the three energy harvesting devices but renders synergistic functions and benefits by avoiding existing problems for the individual methods. Solar energy harvesting requires both a large land surface due to its low energy density and large energy storage or backup system due to its partial daily availability, whereas wind and wave energy supply is desynchronized with higher energy density than solar energy, the building can easily balance the energy use with minimal energy storage, which can be realized by pumping and releasing water between the top pool and ocean. Because of the huge pressure difference at the bottom of the building, it enables the desalination of seawater into freshwater. The ocean provides natural temperature conditioning to the building. This technology can turn the remote unlivable island into a vacation resort rapidly. It can also create zero energy, zero water living, and farming space for food-energy-water supplies of our next generation. Because the three energy harvesting systems share the same foundation of the building with multiple additional functions, the average initial investment will be significantly reduced. When multiple buildings are constructed in an area to form an energy farm, their interactions may exist, such as sun shading, wave, and wind interaction, etc., which will require further investigation and modeling.

Because the natural energy inputs in extreme weather, such as a hurricane, heatwave, or storm, have been intercepted by those systems and well addressed in the design, the safety and durability of the structure will be much reinforced. In line with the thoughts of Shaun Donovan, former Secretary of Department of Housing and Urban Development, "When Hurricane Sandy devastated communities in the region, we were reminded of the importance that climate change will have in all development and planning for our communities to become more resilient and sustainable" (Team, R.b.D. 2013; Gilfillan, 2013). This novel design will improve the resilience of the infrastructure and mitigate potential risks to public health and safety. Such risks are often associated with climate change and extreme weather, taking the form of rising sea levels, lowered lake levels, more intense storms, but also increasing periods of drought and the rising frequency of forest fires. The reliability and safety of infrastructure systems (including their interconnected ecosystems and social systems) to absorb such disturbances and still retain their basic function and structural capacity will be quantitatively evaluated. The principle of resilience will be established for risk mitigation. Sustainability, the companion goal to resilience, relates to longer-term consequences that arise from aging infrastructure and environmental change. The novel structure design will be very useful to achieve the goals.

7.2 Fundamentals of wind energy harvesting

Wind energy is conventionally harvested by a Horizontal-Axis Wind Turbine (HAWT) whose rotor is mounted on top of a tower and is usually equipped with several blades (Manwell et al., 2010; De Lellis et al., 2018). Fig. 7.1 shows the schematic illustration of the HAWT including the rotor with blades and the supporting hub; the drive train with gearbox, brake, and generator; the nacelle and frame with the yaw orientation system; the tower with the foundation and electric system. When the airstream passes the blade plane through either upwind or downwind configuration, the aerodynamic force will rotate the rotor, which is similar to a traditional windmill. The mechanical energy will pass to the electric generator by the gearbox and is converted to electricity. Because the most wind energy

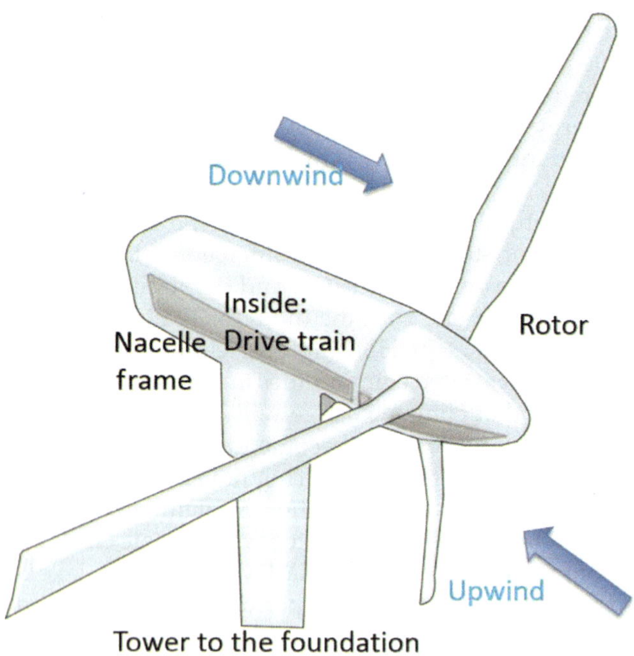

FIGURE 7.1 Schematic illustration of a horizontal axis wind turbine (*HAWT*).

is captured by the outer side of the blade, the inner part, nacelle, frame, and tower are mainly supporting parts to enable the energy harvesting process, airborne wind energy has been investigated in the past decades by removing these supporting parts, which can save the initial costs significantly. However, efficiency may be much lower than HAWT because of less support. This section will introduce the mechanics' theory and key factors leading to the efficiency of HAWT.

The efficiency of the wind energy harvesting process depends on how much mechanical energy can be captured by turbine blades, and how much electricity can be converted from the mechanical energy by the drive train with the generator. Because the efficiency from mechanical energy to electricity can be very high (>95%) with modern electromagnetic transducers, the overall efficiency mainly depends on the first step—how much mechanical energy the turbine blades can capture.

The wind is an airflow driven by four kinds of forces: the pressure force caused by air pressure difference, the Coriolis force caused by the rotation of the Earth, the frictional force with the Earth surface, and the centrifugal force by the curvature of isobar lines. The pressure force is the major driving force to move air particles. It can be defined by unit mass as:

$$F_p = -\frac{1}{\rho}\frac{\partial p}{\partial n} \tag{7.1}$$

where ρ is the air density; n the directional norm along the force direction, which is typically taken at the perpendicular to the constant pressure line for the maximal value; and p the air pressure. The four forces control the motion of air particles.

When air is moving at a certain speed v, it carries kinetic energy. Fig. 7.2 shows an airflow through a rotor disk.

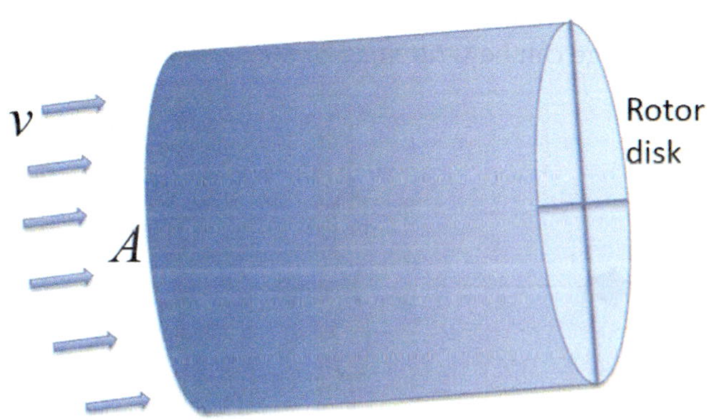

FIGURE 7.2 Airflow at velocity v through a rotor disk of area A.

The kinetic energy per unit time, or power P, of the airflow, can be written as

$$P = \frac{1}{2}\frac{dm}{dt}v^2 = \frac{1}{2}\rho A v^3 \tag{7.2}$$

where the continuity of airflow $\frac{dm}{dt} = \rho A v$ is used. The wind power density can be obtained

$$\frac{P}{A} = \frac{1}{2}\rho v^3 \tag{7.3}$$

Notice that this is the overall energy that is available for the rotor to harvest. However, because the airflow is not constrained by the disk and the air particles keep moving after passing through the disk, the wind power will not be harvested thoroughly. Typically, a modern HAWT can take less than half of available wind energy although theoretically, the Betz limit reaches 59%, which can be derived from a one-dimensional (1D) momentum theory as below (Betz, 1926).

Consider an airstream that passes through a wind turbine in Fig. 7.3. The pressure and velocity will decrease as the kinetic energy is taken away, which also leads to the cross-sectional area increase due to the continuity of the stream. Let the upstream velocity and cross-sectional area are v_1 and A_1, the downstream v_2 and A_2, respectively.

Assume the flow here is a homogenous steady-state flow without frictional drag, and the turbine blades are ideal without any energy loss. The velocity and cross-sectional area at the rotor disk are v and A. For simplicity, consider the air stream as an incompressible flow with a constant density ρ. The continuity equation is written as

$$\frac{dm}{dt} = \rho A_1 v_1 = \rho A v = \rho A_2 v_2 \tag{7.4}$$

In a unit time, the mass passing the rotor will be $\Delta m = \rho A v$. Its linear momentum change can be written as $\Delta m.(v_1 - v_2) = \rho A v(v_1 - v_2)$. Based

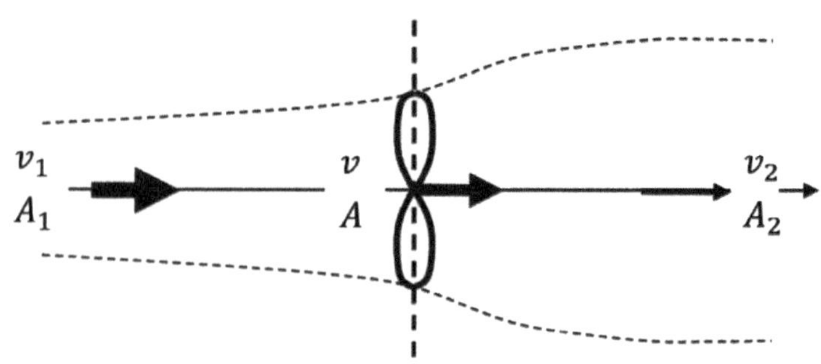

FIGURE 7.3 A simplified 1-D momentum model for airflow passing a wind turbine.

on the conservation of the linear momentum, the above change is caused by the force F from the rotor on the unit time as

$$F.1 = F = \rho Av(v_1 - v_2) \tag{7.5}$$

Therefore, the extractable power from the wind on the wind turbine is

$$Fv = \rho Av^2(v_1 - v_2) \tag{7.6}$$

which should be consistent with the wind power loss from the upstream to the downstream as

$$\Delta P = \frac{d}{dt}\left(\frac{1}{2}mv_1^2 - \frac{1}{2}mv_2^2\right) = \frac{1}{2}\frac{dm}{dt}\left(v_1^2 - v_2^2\right) = \frac{1}{2}\rho Av\left(v_1^2 - v_2^2\right) \tag{7.7}$$

The power expressions in the above two equations are equal. So the following is obtained

$$v = \frac{1}{2}(v_1 + v_2) \tag{7.8}$$

Therefore, the wind turbine saves as a brake to slow down the wind speed from uniform upstream velocity v_1 to downstream velocity v_2 and collects the wind energy in this process. The original available wind power through the rotor disk is $P_0 = \frac{1}{2}\rho Av_1^3$. Due to the disturbance of the air stream, not all air particles referring to the undisturbed one can pass through the rotor disk as $A_1 < A$. In addition, the energy of air particles passed the rotor disk is not completed taken away $v_2 > 0$. Therefore, a significant part of wind energy cannot be harvested for this unguided air stream. The efficiency can be written as

$$C_p = \frac{\Delta P}{P_0} = \frac{\frac{1}{2}\rho Av\left(v_1^2 - v_2^2\right)}{\frac{1}{2}\rho Av_1^3} = \frac{v\left(v_1^2 - v_2^2\right)}{v_1^3} \tag{7.9}$$

Now introduce a downstream velocity factor as

$$\beta = \frac{v_2}{v_1} \tag{7.10}$$

Eq. (7.9) can be written as

$$C_p = \frac{1}{2}(1 + \beta)(1 - \beta^2) \tag{7.11}$$

where C_p depends on the downstream velocity factor $\beta \in (0, 1)$. Notice that for β approaches zero, which means the downstream air becomes static even if it is not physical, C_p approaches 50%, because 50% of air particles referred to as the undisturbed configuration do not pass the rotor disk at all. Therefore, it may not be the highest efficiency. Taking $\frac{dC_p}{d\beta} = 0$, one can obtain

$$\frac{1}{2}(1+\beta)(1-3\beta)=0 \tag{7.12}$$

which shows that when $\beta = 1/3$ or $v_2 = \frac{1}{3}v_1$, the wind turbine reaches the highest efficiency at $16/27 = 59.3\%$. This is the Betz's limit of ideal HAWT for unguided airflow. Compared with the case of $\beta \to 0$, more air particles will pass through the rotor disk and contribute to higher efficiency. However, the Betz's limit is not feasible to reach because of the energy loss for friction, air vortex-induced vibration, the gap between air blades in rotation, among other factors. The actual efficiency will depend on the size, tip speed ratio (TSR) of the rotor, pitch angle, the number of paddles, etc. Modern wind turbines may reach ~40%, which typically requires the maximization of energy conversion from air flow to rotor and the minimization of the frictional loss by drag force and mechanical motion. A large scale of wind turbines typically exhibits a higher efficiency.

Notice that the Betz's limit is calculated based on a free air flow. When the airflow is guided through a channel, the loss of air particles can be avoided, and the Betz's limit will not be applicable anymore. When energy-independent buildings are constructed together, the building or solar panels can be used to guide the airflow, so that higher efficiency can be reached through some creative design and construction.

7.3 Fundamentals of wave energy harvesting

Wave energy harvesting can be conducted in a similar process to wind energy harvesting: the wave energy is collected by a power take-off (PTO) system in form of mechanical energy and then is converted to electricity by a generator. However, the wind energy is in the form of a flow with a certain direction (similarly to river and ocean currents), whereas the wave energy can be in form of both vertical and horizontal motion of water particles. Because wave energy exhibits much higher density and diverse motion modes, various WECs have been designed and developed in the past decade. Among many existing WECs, submerged pressure differential wave energy converter (SPDWEC) is merged under the water and driven by the oscillating water pressure force. Compared to floating WECs, SPDWECs have higher survivability because the water force near the seabed is much smoother than the wave impact force on the ocean surface. As a result, the costs of the structure and the mooring system can be reduced significantly. It will provide benefits for the integration with buildings.

The first example of SPDWEC is the M3 WEC (Morrow, 2013) (Fig. 7.4). Because the power generator is driven by an air turbine, which is installed

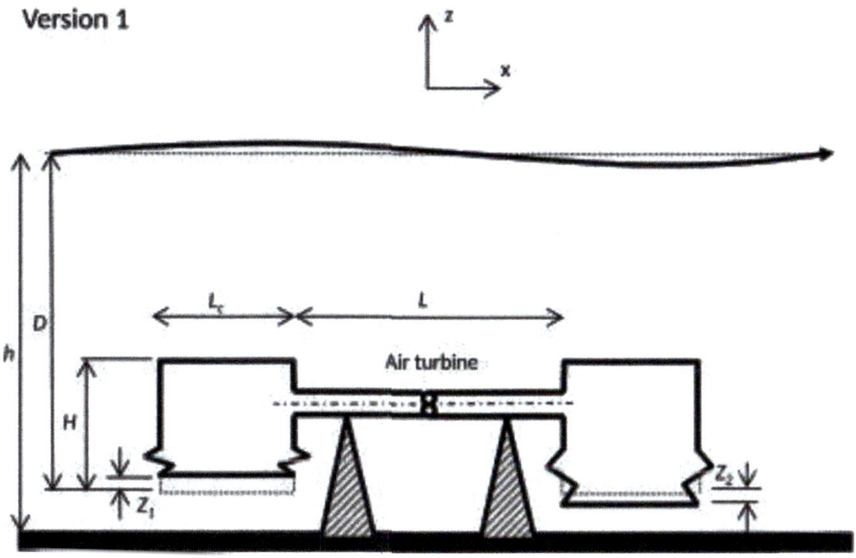

FIGURE 7.4 Schematic of M3 WEC (adapted from Babarit et al., 2017). *WEC*, wave energy converter.

inside a pipe connecting the two deformable air chambers, the PTO system has significantly less power absorbing capacity and efficiency compared to PTO systems driven by pistons.

Two examples of piston-driven SPDWECs are the WaveCarpet (Lehmann et al., 2013) (Fig. 7.5) and the WaveSwing (Valério et al., 2007) (Fig. 7.6). The working principles of WaveCarpet and WaveSwing are

FIGURE 7.5 Schematic of WaveCarpet (adapted from Lehmann et al., 2013).

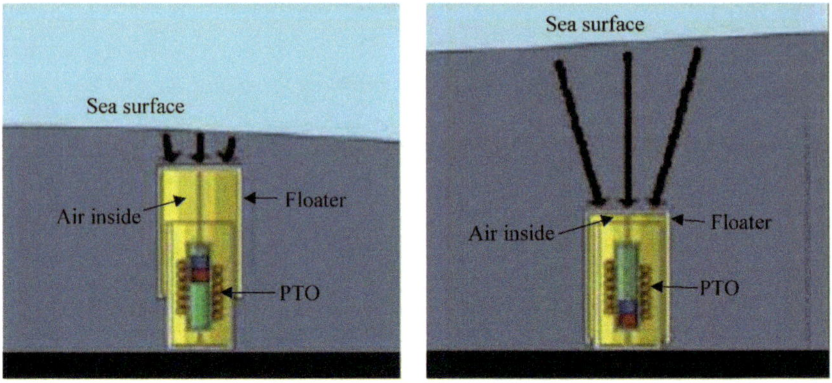

A under wave trough B under wave crest

FIGURE 7.6 Schematic of WaveSwing (adapted from Valério et al., 2007).

different. For the WaveCarpet, the membrane of the device is activated by the wave-induced water pressure difference between the top and bottom surfaces of the membrane. For the WaveSwing, the floater of the device moves up and down as the air inside the device is compressed differently under oscillating wave-induced water pressure.

However, these two SPDWEC designs have limitations. While the WaveCarpet can have high wave-energy absorbing efficiency, it is susceptible to corrosion and biofouling on the PTO system when exposed to seawater. The WaveCarpet also has relatively large dimensions (>180 m long, 3~4 wavelengths), and may cause high manufacturing and installation costs for each device. While the WaveSwing has relatively small dimensions (<60 m, smaller than one wavelength) and low manufacturing and installation costs for each device, it has relatively low wave energy-absorbing efficiency. This is caused by the relatively small displacement amplitude of the floater caused by the compressibility of the air inside.

Many WECs use the motion of a mass interacting with the waves to take off the wave power. The mass can be a buoy, piston, rotating board, or water body itself, and the motion can be oscillatory linear motion or rotation or their combination. Different mechanisms often share the same fundamental principle: use the water particles of the wave to force the oscillation of a mass, and pass the kinetic energy of the mass to a generator, which is schematically illustrated in Fig. 7.7 using a buoy as an example.

The mass m moves with waves along coordinate x; the supporting structure with a stiffness K provides an elastic force Kx; the system, exhibiting viscous effect with damping coefficient c_m, also provides a damping

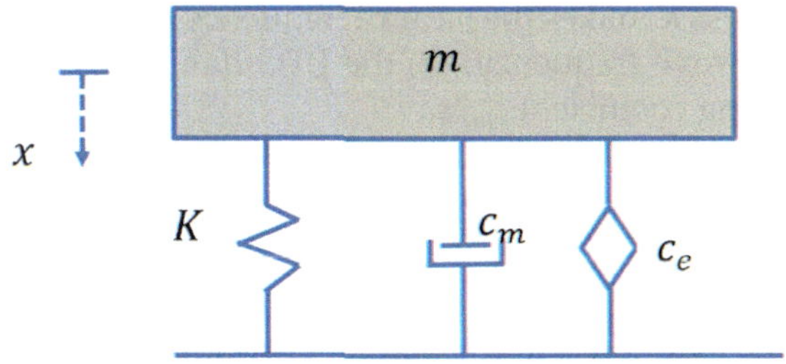

FIGURE 7.7 Schematic of wave energy conversion through the oscillation of a mass.

force $c_m\dot{x}$; the PTO system will slow down the motion of the mass with a damping force $c_e\dot{x}$; the mass will be excited by hydrodynamic oscillatory force $F(\omega)$ from the wave at a frequency ω. The governing equation for a mass motion can be expressed as

$$m\ddot{x} + (c_m + c_e)\dot{x} + Kx = F(\omega) \tag{7.13}$$

where $\dot{x}$ and $\ddot{x}$ are the velocity and acceleration of the mass, respectively. The instant power capture of the PTO system is:

$$P_{\mathrm{PTO}}(t) = F_e x(t) = C_e\dot{x}^2(\mathrm{t}) \tag{7.14}$$

where F_e is the force passed to the generator as $F_e = c_e\dot{x}$. Under regular waves of frequency ω, the excitation force $F(\omega)$ can be written as a sinusoidal function

$$F(\omega) = F_0\cos(\omega t) \tag{7.15}$$

Given an initial condition, the equation of motion, Eq. (7.13), can be solved for the motion of the mass (Cargo et al., 2012), which includes the transient response and the steady-state response. The transient response will decay to zero exponentially so that the energy output depends on the steady-state part

$$x(t) = \frac{F_0}{\sqrt{\left(k - m\omega^2\right)^2 + \left(c_m + c_e\right)^2\omega^2}}\cos(\omega t - \theta_0) \tag{7.16}$$

where θ_0 is the phase angle of the mass motion lag behind the excitation. The energy capture over one period can also be calculated by the integral of the instant power capture (Cargo et al., 2012), and then the average power output over the period can be derived as

$$\overline{P_{\mathrm{PTO}}} = \frac{1}{2}\frac{c_e F_0^2}{\left(\frac{K}{\omega} - m\omega\right)^2 + \left(c_m + c_e\right)^2} \tag{7.17}$$

By adjusting the spring stiffness K and PTO damping coefficient c_e, the average power can be maximized under the following two conditions: (1)

the spring stiffness K makes the natural frequency of the PTO system equal to the incident wave frequency; (2) the PTO damping c_e is chosen at the system's damping coefficient c_m, say

$$\overline{P_{\max}} = \frac{1}{8}\frac{F_0^2}{c_m} \tag{7.18}$$

The maximum power calculated in the above equation can be an overestimated value compared to the real power capture of the WEC because Eq. (7.13) is based on linear wave theory assuming the motion response of the wave energy device does not affect the incident waves. However, in reality, the WEC does affect the waves. As the device extracts energy from the waves, the wave height becomes smaller therefore the wave excitation force also becomes smaller.

7.4 Design of a PV–wind–wave coupled system

Many marine structures have been built with the foundation at the seabed, such as tunnels, bridge piers, oil drilling platforms, whose heights can be in the range from 0 to 10,000 ft. They can be much higher than the highest building of the Burj Khalifa tower in Dubai at 2717 ft. As water might carry some weight of the structure, the height limit of an underwater high-rise building will be much higher than the counterpart on the land. Although some underwater buildings have been built around the world, such as museums, restaurants, and hotels, they commonly require high operation and maintenance (O&M) costs and rely on high energy inputs to the building. The scale is small, and they are not ready for routine residential living or business office yet.

For example, Ithaa underwater restaurant has been built 5 m below the sea surface, which is a 275-ton structure with the acrylic envelope, stabilized with sand ballast, and fixed into position near Conrad's Sunset Grill Restaurant in the Maldives. Guests dine underwater with panoramic 180-degree views of the ocean and marine life with colorful tropical fish and even sharks. Ithaa is highly popular as a top-ranked restaurant but relatively small as it can host only 14 people at a time.

The future PV-wind-wave coupled system can integrate energy harvesting, freshwater production, and food supplies into the building system, it can carry multiple functions with significant incomes in its life cycle to offset the high O&M costs. It is very promising to make it available for routine residential and business use.

In the following, we focus on the design of a high-rise building on the cover page of this chapter, which can be nearshore or offshore at different scales.

7.4.1 Overview of the structural design and construction

The structure of the building includes three sections of the underwater, interface, and air parts, respectively, which are vertically connected by the transport system with elevators in the center of the building, which shall be first constructed following the procedure of fixed platforms. It hosts people transport, plumbing, ventilation, heat exchange, electric, and information communication systems, and provides structural support for other building parts.

Horizontally, the underwater wall, frame, and building envelope shall be preassembled on the platform and sink into the water layer by layer. Notice that because the weight of conventional building materials in each layer is much smaller than that of the volume of seawater, although the building will be stabilized by the self-weight of the whole building including the air part, for waterproof and maintenance, it is critical to have each layer self-balanced with a waterproof system. For reference, the Empire State Building weighs 365,000 tons and its volume is 37 million cubic feet, which is equivalent to >1 million tons of seawater or ~3 times of its self-weight. Therefore, the underwater part will require a thick floor and ceiling as well as the curtain wall and even extra weight for balancing the buoyant force. When the structure in the air is constructed, it will provide extra load onto the underwater structure and fully stabilize the whole building.

The underwater part can use a pyramid shape with additional anchors to the seabed for stabilization under the lateral load from water motion and wind. Because the wave energy is rich in the ocean surface, the interface layer will be used for wave energy harvesting. Depending on the wave characteristics, this layer may have a larger height, say ~30ft, to host the WEC as well as the building access, utility, and maintenance components. This is the novel part that requires creative design and research. We will show our recent design of Wavehouse in Section 7.5 as a case study.

The BIPVT system will be installed on the structure surface in the air. At the top layer, a water reservoir will be built to store freshwater collected at the bottom of the building, which also stabilizes the building by increasing the self-weight and damping effect. It can be released to the water tank at the bottom as well to use the water gravity to generate electricity with a decent water head (say over 500 ft) when it is needed. Therefore, it serves as both a routine water supply and energy storage and backup system in addition to structural stabilization.

Wind turbines are installed on the top of the building. Because it will produce lateral loads to the building, a bracing system will be built to create an additional load path from the building top to the foundation. Notice that the uprising force is mainly from the bottom layer of the building as the lateral wall panels are subjected to horizontal pressures which are balanced in each layer. The bottom can use a thick steel plate tightly anchored into the seabed.

It is anticipated that the energy and water generated in the building will be more than the building needs, which creates feasibility for local ocean farming or provides energy and water supply to other communities. For example, Innovasea Systems (Innovasea, 2021) have created offshore fishing farm systems but the energy supply relies on diesel. The present technology can provide support for their business with the onsite building and energy supply. Chapter 9 will introduce more of BIPVT technology in farms for future food production.

7.4.2 BIPVT design and maintenance

As cold seawater has a higher density, the bottom of the building surface generally exhibits a lower temperature, while the structure in the air is more susceptive to the ambient temperature. The temperature variation along the vertical direction of the building surface will lead to the temperature difference of rooms on different floors, which requires flexible temperature conditioning at different floors. However, if an appropriate heat exchange system can be developed to average the temperature difference as well regulate the temperature by existing heat source and sink, it may significantly save the heat, air ventilation, and cooling (HAVC) cost compared with the conventional HAVC systems.

An air-based BIPVT system can be integrated into the air circulation system in the following way: when the Sun heats the BIPVT system, the cold air from the bottom layer is pumped to the BIPVT panels and forms an airflow under the solar panel to collect the solar heat and cool down the solar panel for higher PV efficiency. The heated air to a certain temperature can be delivered to different floors for temperature conditioning. The water reservoirs on the top and bottom layers can also serve as a heat sink as needed.

Because the BIPVT system is constantly exposed to ocean wind and water, additional consideration of corrosion resistance and periodic inspection and cleaning maintenance is required. There are three main ways salt mist can affect solar panel performance:

- Salt corrosion
- Galvanic corrosion
- Sea grime

A standard for salt mist resistance for solar panels has been set by the International Electrotechnical Commission (IEC). Panels have to meet a standard called IEC 61701 to be suitable for solar panel installation near the sea. The IEC 61701 standard has six levels of corrosion resistance.

Each level must survive a certain number of days of testing to simulate decades of exposure to salt with an allowance of a certain amount of power decline. Considering the average panel warranty is 25 years under this application, this highest level at Level 6 is recommended—rating panels have to survive 112 days of testing that simulates decades of exposure to a salty environment with less than 2% decline in power output. Corrosion will also affect the other parts, such as electrical connection, junction box, panel frame, and its supporting systems.

When two different metals are in contact with each other with different electrode potentials, one metal (that is more reactive) acts as an anode and the other (that is less reactive) as a cathode. The electro-potential difference between the reactions at the two electrodes is the driving force for an accelerated attack on the anode metal, which dissolves into the electrolyte. In dry conditions, this process is slow and is often not a problem, but add in salt mist and the more vulnerable of the two metals can quickly corrode. A classic example is when the solar panel aluminum frame is in contact with steel mounting hardware. Corrosion may quickly erode the aluminum frame. This problem can be avoided by carefully choosing the materials for each part.

Sea grime can also stick to solar panels and reduce solar panel efficiency and lifetime. More frequent panel maintenance or cleaning is required to recover the solar panel's efficiency and reduce its degradation process.

7.4.3 Freshwater generation

Freshwater for a building is necessary for drinking and cleaning among other uses. For a building in the ocean, it is hard and costly to bring in freshwater but the availability of seawater is unlimited. The reverse osmosis (RO) process is the leading process for desalination in terms of installed capacity and yearly growth. It uses semipermeable membranes and applied pressure (on the membrane feed side) to induce water permeation through the membrane while rejecting salts so that freshwater can be separated from seawater. Typically, high pressure of 50–80 bars is required to enable the RO process. The creation of such pressure is the main energy consumption for water production.

The water pressure at the bottom layer of the building can be high enough to enable the automatic, continuous RO process so that the freshwater can be collected and stored in the bottom water reservoir in the building and be pumped to the top water reservoir.

RO uses a thin-film composite membrane, which comprises an ultrathin, aromatic polyamide thin-film. This polyamide film gives the membrane its transport properties, whereas the remainder of the thin-film composite membrane provides mechanical support. The polyamide film is a dense, void-free polymer with a high surface area, allowing for its high water permeability. It can be installed onto a perforated ring around the building for continuous freshwater production.

When the height of the building is not enough to provide such a high pressure, we can use a WEC to feed pressured seawater from the surface to the bottom. Because of the existing static water head, a small extra pressure will be enough to enable the RO process without any other energy input.

7.4.4 Power generation and management

A microgrid will be set up to manage energy generation, storage, and consumption. Because the sea breeze and wave are available most time with changing intensity, the wind and wave energy may meet the regular use in the building. The shortage can be matched with solar energy and hydropower stored in the water reservoir on the top of the building. The extra energy can be used to pump water and feed into the macrogrid or other microgrids as possible.

Because of the variation of the energy input from wind and wave over time, appropriate synchronization or regulation is necessary to boost the voltage or normalize the frequency deviation.

The majority of the solar energy is generated when the sun is shining brightly, but virtually none is captured at night or in substantial cloud cover. The direct current of PV can be converted to alternating current by an inverter with a small conversion loss or be directly used by a direct current water pump for energy storage.

When a big load is added into a microgrid, it may cause brownout or blackout by the drop of voltage. Therefore, timely communication with the microgrid from energy generators (wave, wind, solar, and backup hydropower) and energy consumers (HVAC, lighting, transportation, communication, electronics, etc.) will be critical. A sensing and control system is integrated with the microgrid. The resilience of the power management system also depends on the energy redundancy in long term and appropriate storage for energy equilibrium in short term, which should be well analyzed in the design phase.

The energy equilibrium analysis of the energy harvesting and the system power consumption can be conducted over a period with statistical data of energy harvesting and use, which can be used to determine the size of water reservoirs. The amount of energy to be stored, E_{sto}, is decomposed into the harvested power P_{har} and the consumed power P_{con} as below

$$E_{sto} = \int_{T_0}^{T_1} (P_{har} - P_{con})dt \tag{7.19}$$

where P_{har} and P_{con} change with time. Giving the functions of $P_{har}(t)$ and $P_{con}(t)$ over a period, say 365 days with a repeating pattern, one can find that when $P_{har} > P_{con}$, the energy will be stored; otherwise, the storage will be discharged. In long term, E_{sto} is always positive for energy redundancy with $P_{har} > P_{con}$. However, one can find a series of times, at which P_{har} and P_{con} switch the order, which means that storage starts to discharge energy. Starting at any such a moment, say T_0, the above integral in Eq. (7.19) will be negative and decrease with the increase of $T_1 (> T_0)$ at the beginning, but will eventually increase and become positive. The lowest value of E_{sto} determines the energy demand from the storage at this time. Repeat this process for all discharging moments, one can find the highest energy discharge demand, which can be considered as the required energy storage capacity, E_{sto}^{cap}. The size of the topwater reservoir can be calculated by:

$$V = \frac{E_{sto}^{cap}}{\eta \rho g H} \tag{7.20}$$

where ρ is the density of the seawater; g the acceleration of gravity; H the height difference of the two reservoirs or the height of the building; and η the efficiency of the hydropower conversion efficiency which is typical in a range of 0.75–0.9. A certain safety factor can be applied to adjust the size of the top reservoir. The size of the bottom reservoir shall be double the top one. Simulation shall be conducted to evaluate the dynamic response of the building under seismic excitation with water damping, to finalize the reservoir size.

To reach higher confidence in energy resilience, batteries can also be added for extra redundancy.

7.5 Remarks for prospective development and applications

With the population increase and sea level rise, people are moving from shore for disaster mitigation, which will create more pressure on sustainable development of the world. However, the oceans are the indispensable resources of energy, water, and food, which are life supplies for human's future. It may create space for human activities. It is an engineering challenge to turn the dangerous area into a safe, nutritious, and usable space to feed and power our next generations. By design and construction of strong structures on shore or in ocean, which can resist hurricanes and storms, we can control, harvest, and use the destructive energy and tame the ocean for a livable place.

As people are thinking about immigration to the outer space or other planets, the ocean is not as far as the moon or Mars, but we need to creatively resolve the critical problems under the extreme weathers. There are huge challenges and opportunities for PV-wind-wave energy harvesting on, near or off-shore:

1. High density and strength materials, which are stable and durable under the ocean environments, are required for the building blocks. The oil drilling technologies can provide good examples to start with. Particularly, as more oil platforms are built and discarded, we can directly use the structure as the support to the future PV-wind-wave energy buildings and increase the incomes during and at the end of the service lifetime of oil platforms.
2. The fluid-building interaction in the ocean environment with horizontal and vertical water flow/motion should be accurately characterized under extreme weathers. It could be similar to wind design of high-rising buildings. However, it exhibits different physics and mechanics on the density, speed, and energy of the flows. With the current understanding in deep ocean exploration with vessels or drilling structures, the technical feasibility can be well established. However, to scale up the application, the economic feasibility shall be quantitatively studied and optimized for different working parameters and conditions.
3. The energy harvesting systems should be designed for high energy density with strong sunlight, wind and wave, which will require much less area to reach the same scale of energy harvesting on the land. Particularly, the energy flows from the environment to the energy converters are more dynamic with a large range. If the energy is not well used or controlled, it may damage or destroy the energy converters or building structures. It requires engineering design, characterization, and optimization to tune energy converters for maximized energy harvesting efficiency and lifetime under the constraint of the cost.
4. Energy storage, distribution, and utilization need to be systemically designed and managed. With the advancement of blue economy in deep ocean, there are growing needs for ocean-derived food, materials, energy, and water supplies as well as beneficial carbon and waste recycling. The PV-wind-wave energy building will facilitate different methods and fuel up those processes as it is expensive to bring power to ocean in a safe manner. For example, ocean farming needs much energy to distribute food to fish and requires space for fish cleaning and storage. The PV-wind-wave energy building can host all activities and improve the safety and productivity of the ocean farm.
5. Because the energy density reaches the highest in the extreme weathers such as hurricanes but they do not occur frequently, if the system and structure are designed according to the extreme weathers, the cost will be high and a large energy harvesting capacity will not be used. Different operational modes should be designed to balance the costs

and benefits. For example, at the normal energy level, all activities can be allowed as designed; when the energy level is higher than the alert level, human should be evacuated and the system should be automatically running, the smart sensing and control system are necessary to monitor and manage the operation, and; when the energy level is extremely high, the energy converters may be destroyed, so that they should be stopped not to catch any energy for the safety.

Indeed, it will be a high-risk, high return application which requires intensive research and development in future. We have formed a small business company E-Wave Technologies to start this research and development initiative.

7.6 Case study: Wavehouse

The case study introduces a recently invented SPDWEC, namely Wavehouse (Yin et al., 2021), for high power absorption efficiency and high performance in corrosion and biofouling resistance for durability and longevity of the structure (Folley et al., 2007) which will be critical for solar–wind–wave building applications.

To assure the lifetime and safety requirement of a building, a WEC is required to exhibit high corrosion and biofouling resistance and necessary wave energy-absorbing efficiency with a reasonable size. A novel SPDWEC (Folley et al., 2007) was designed in Fig. 7.8, named "Wavehouse," which was also inspired by the merits of the existing technologies. It could be mounted to the building's interface layer or nearshore seabed. The top structure of Wavehouse comprises one air tank with a number of air bags, each connected to a PTO cylinder. The size of the PTO cylinders and spacing between them can be designed and optimized with local wave conditions and structure dynamic properties for minimized hydrodynamic interactions and maximized energy harvesting potentials. In this paper, we choose the PTO cylinder with a radius of 4 m and a spacing of 15 m apart in the wave propagation direction and 15 m apart perpendicular to the wave direction.

Inside each cylinder, there is a piston separating the air inside an air bag and the air tank. When incident waves pass through the WEC, the dynamic water pressure surrounding the air bag changes the air pressure inside, and as a result, activates the piston movement. When the air bag is under a wave crest, the piston moves down because the above air pressure p_1 is larger than the underneath air pressure p_0. When the air bag is under a wave trough, the piston moves up because the above air pressure p_1 is smaller than the underneath air pressure p_0.

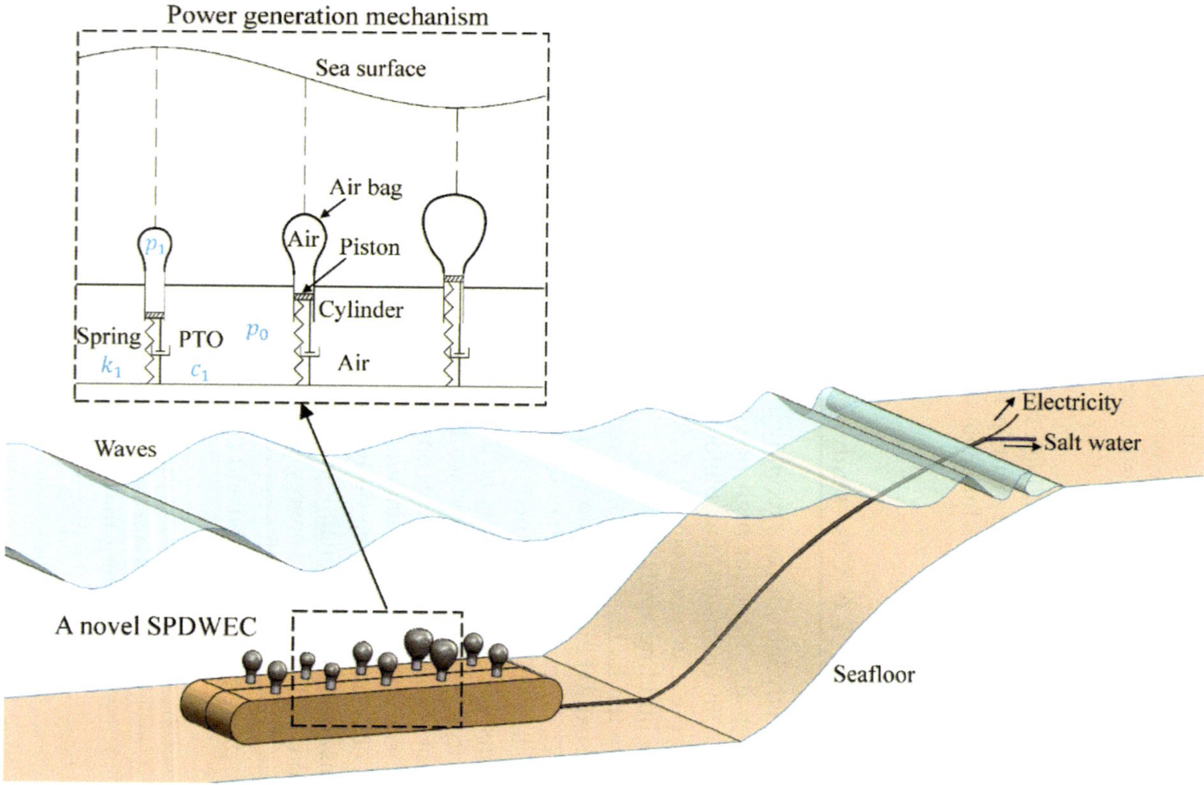

FIGURE 7.8 A novel submerged pressure differential wave energy converter of Wavehouse. By adjusting the spring stiffness $\mathbf{k_1}$ and the power take-off damping $\mathbf{c_1}$, the pistons can move in resonance with dominant ocean waves, so that the electrical power capture can be maximized. In addition, the WEC is protected by the air bag without corrosion and biofouling. *WEC*, wave energy converter.

Overall, the design of Wavehouse exhibits the following benefits:

- High energy harvesting efficiency with fluid-structure interaction optimization.
- Low O&M cost due to the air bag protection.
- Long lifetime of the system due to the minimized corrosion and biofouling effects.
- Wave separation and protection for the underneath structure with relatively uniform average air pressure in the air bag.

It can be implemented by an energy farm facility for isolated islands, self-powered ocean farm applications, and roof structures for underwater buildings or structures. As more underwater restaurants, hotels, and apartments are envisioned, Wavehouse provides an excellent solution for the power supply and structure protection of the energy-independent building.

A numerical model is developed to predict the energy harvesting of individual cylinders, assuming each cylinder is not affected by others. In the numerical model, the interaction between the WEC cylinder and waves is simplified with the linear wave theory. In Fig. 7.9, the piston inside the cylinder is assumed to be activated by the linear wave forces of excitation force, diffraction force, and radiation force. Then the radiation force is converted to the forces of radiation damping and added mass.

Therefore, the governing equation for a piston motion can be expressed as

$$(m_1 + m_a(\omega))x + (c_1 + c_r(\omega))\dot{d} + k_1 x = f_i(\omega) + f_d(\omega) \tag{7.21}$$

where x is the displacement of the piston; m_1 is the mass of the piston; $m_a(\omega)$ is the hydrodynamic added mass at wave frequency ω; c_1 is the PTO damping coefficient; $c_r(\omega)$ is the radiation damping coefficient; k_1 is the spring stiffness; $f_i(\omega)$ is the incident force (Froude–Krylov) force; and $f_d(\omega)$ is the diffraction force. Note that the hydrodynamic parameters of $f_i(\omega)$, $f_d(\omega)$, $c_r(\omega)$, and $m_a(\omega)$ are calculated using a boundary element method program.

To estimate the power capture of Wavehouse, a single-cylinder is firstly considered by disregarding the hydrodynamic interactions between cylinders. The instant power capture of the PTO system is

$$P_{\mathrm{PTO}} = c_1 \dot{x}^2 \tag{7.22}$$

For each cylinder, as mentioned earlier, the governing equation of mo tion, Eq. (7.21) is based on linear wave theory, which means the wave excitation force f_e and the diffraction force f_d are sinusoidal. Under regular waves of frequency ω, the average power out can also be calculated

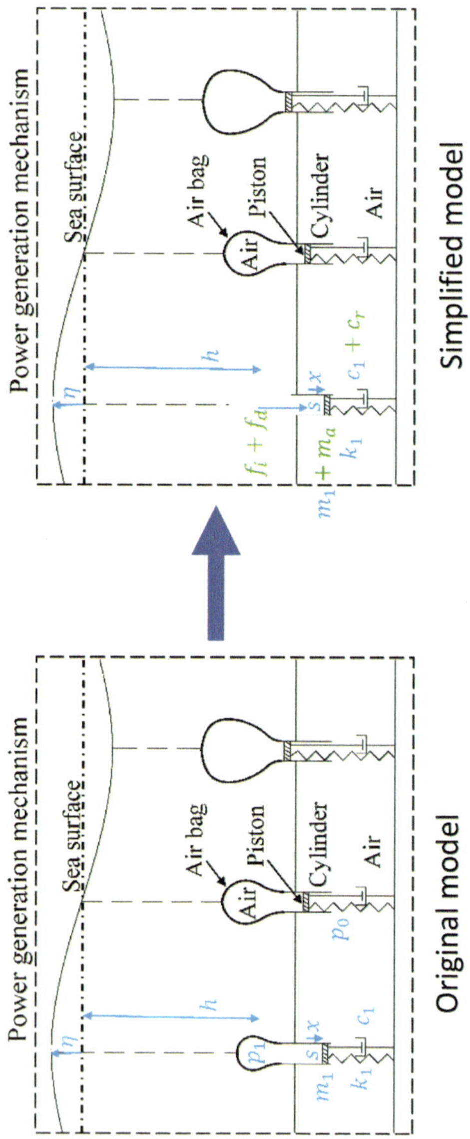

FIGURE 7.9 Simplification of a numerical model for each piston.

analytically (Cargo et al., 2012):

$$\overline{P_{PTO}} = \frac{1}{2} \cdot \frac{c_1\left(\overline{f_i(\omega) + f_d(\omega)}\right)^2}{(c_1 + c_r(\omega))^2 + \left(\omega(m_1 + m_a(\omega)) - \frac{k_1}{\omega}\right)^2} \tag{7.23}$$

where $\overline{f_i(\omega) + f_d(\omega)}$ is the amplitude of $f_i(\omega) + f_d(\omega)$.

By adjusting the spring stiffness k_1 and PTO damping c_1, the average power can be maximized. When the maximum power is obtained (1) the spring stiffness makes the natural frequency of the PTO system equal to the incident wave frequency; (2) the PTO damping c_1 is equal to the radiation damping coefficient c_r:

$$\bar{P}_{\max} = \frac{1}{8} \cdot \frac{\left(\overline{f_i(\omega) + f_d(\omega)}\right)^2}{c_r(\omega)} \tag{7.24}$$

where $\bar{P}_{\max}$ is the maximum average power under the condition of $k_1 = k_{\text{opt}} = (m_1 + m_a(\omega))^* \omega^2$ and $c_1 = c_{opt} = c_r$.

For the design of a submerged WEC, the first parameter to consider is the water depth at which the device is deployed. The effect of water depth on the power capture performance was studied for a surging flap-type WEC by analytical algorithms (Folley et al., 2007). This section develops similar analytical algorithms for the pressure differential WEC for the relationship between the power capture and the water depth.

For the SPDWEC, the incident wave-induced force on a single piston is

$$f_i = (p_1 - p_0)s \tag{7.25}$$

where p_1 is the air pressure in the air bag, p_0 is the air pressure in the air tank, and s is the area of the piston.

Based on the linear wave theory, the WEC body is assumed to be still when the incident wave force f_i is calculated. Therefore, p_1 inside the air bag is equal to the dynamic water pressure around the air bag, which is written as

$$p_1 = \rho g h + \rho g \eta \frac{1}{\cosh kh} \tag{7.26}$$

where h is the water depth; k is the wave number; and η is the free surface elevation, namely $\eta = \frac{H}{2}\cos(\sigma t)$, in which H is the wave height and σ is the wave angular frequency.

The air pressure p_0 under the piston is a constant value $p_0 = \rho g h$ because the volume of the air in the wave-house is much larger than the volume of the air in the air bag. Even though the displacement of an individual piston is large, the total volume change of the air below the multiple pistons can still be small when other pistons move in opposite directions over the whole wave period. Therefore, the piston motion of Wavehouse does not change the average air pressure under the piston. As a result, the piston motion amplitude of Wavehouse could be significantly larger than that of conventional SPDWECs, where the piston motion reduces the air pressure difference on the piston. Therefore, the incident wave force in Eq. (7.25) is reduced to

$$f_i = \rho g \eta K_p s \tag{7.27}$$

where the pressure response factor $K_p = \frac{1}{\cosh kh}$.

As the dimension of a cylinder is much smaller than the wavelength, the incident wave force f_i is assumed to be much larger than the diffraction force f_d. The maximum power capture in Eq. (7.24) is approximated as:

$$\bar{P}_{\max} \approx \frac{1}{8} \cdot \frac{\left(\overline{f_i(\omega)}\right)^2}{c_r(\omega)} \tag{7.28}$$

The substitution of Eq. (7.25) into Eq. (7.28) yields

$$\bar{P}_{\max} \approx \frac{1}{8} \cdot \frac{(0.5\rho g H K_p s)^2}{c_r(\omega)} \tag{7.29}$$

where the maximum power capture is shown to be approximated proportional to the pressure response factor K_p squared, given constant values of H, s, and c_r.

At wave frequency ω, K_p is a function of water depth. In deep water, K_p is close to zero, which means the water pressure on the seafloor is not affected by the water waves on the sea surface, as a result, there is no wave-induced pressure difference to drive the piston. In shallow water, K_p is close to one, which means the water pressure on the seafloor is fully affected by the water waves on the sea surface. For example, for a wave period at 10 s, the pressure response factor K_p decreases quickly from 0.9 to 0.25 as the water depth increases from 5 to 50 m. Therefore, the WEC should be deployed at the smallest water depth to maximize the incident wave force and power capture. However, wave breaking occurs at a water depth smaller than $1.3H$ causing significant wave energy loss. Therefore our WEC is designed to be deployed at a water depth of 5-10 m.

Below the power capture performance of Wavehouse mounted at a water depth of 10 m is studied. Each cylinder of the WEC has a diameter of 8 m. First, the hydrodynamic parameters of $f_i(\omega)$, $f_d(\omega)$, $c_r(\omega)$, and $m_a(\omega)$ of the air bag are calculated using a boundary element method program AQWA. As shown in Fig. 7.10, the air bag is simplified to be the top plain surface of the cylinder. The air bag is not modeled as a curved surface

FIGURE 7.10 Mesh for a single cylinder of the Wavehouse in the BEM model. *BEM*, boundary element method.

because the corresponding hydrodynamic parameters change with the shape of the curved surface. As the focus of the case study is to investigate the effects of spring stiffness and PTO damping on the power capture, we use the plain surface as a simple model.

The ideal and actual power captures of the WEC are investigated under regular waves of 1 m wave height with different wave periods.

Fig. 7.11 plots the change of the maximum power capture $\overline{P_{PTO}}$ with respect to the parameters of PTO stiffness k_1 and damping c_1 based on Eq. (7.24). Under the regular waves of 1 m wave height and 10 s wave period, the power capture of the cylinder can be maximized by adjusting the two parameters. When $k_1 = k_{opt}$, i.e., the natural period of the PTO system is the same as the wave period of 10 s, and $c_1 = c_{opt}$, i.e., the PTO damping equals the radiation damping, the power capture reaches its maximum value of 90 kW.

The maximum power captures are then calculated for regular waves with typical periods ranging from 5 to 14 s based on Eq. (7.29). As shown in Fig. 7.12, the maximum power capture increases from 20 to 190 kW as the wave period increases. As listed in Table 7.2, the corresponding optimal stiffness k_{opt} decreases from 180 to 40 kN/m as the wave period increases; the corresponding optimal damping coefficient c_{opt} ranges from 70 to 45 kN*s/m (the maximum value occurs at the wave period of 7 s).

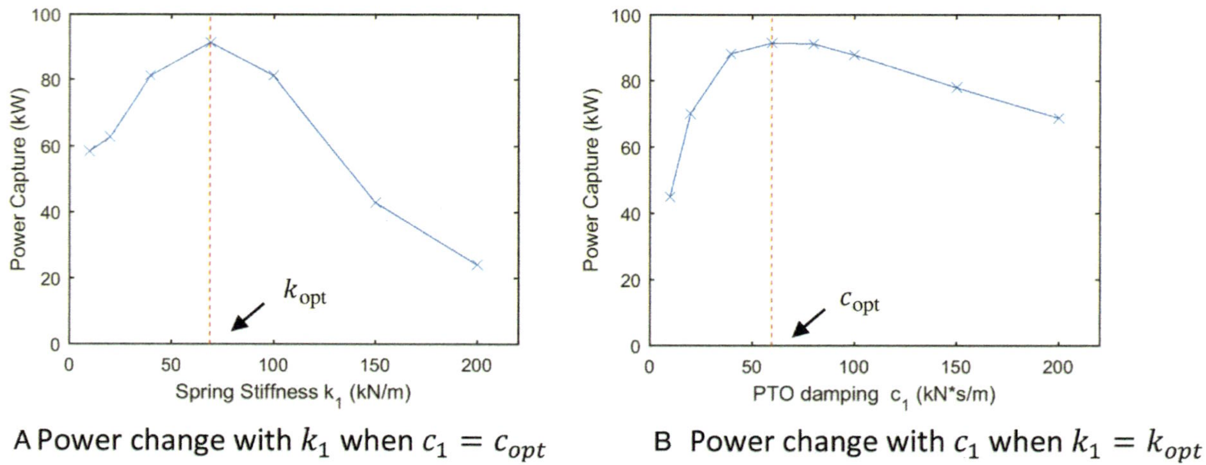

FIGURE 7.11 Change of power capture with respect to spring stiffness k_1 and PTO damping c_1 (wave height $H = 1$ m, and wave period $T = 10$ s). (a) Power change with k_1 when $c_1 = c_{opt}$; (b) power change with c_1 when $k_1 = k_{opt}$. *PTO*, power take-off.

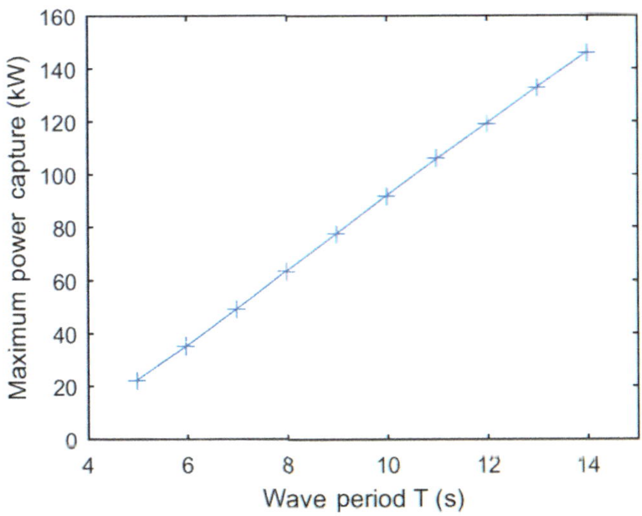

FIGURE 7.12 Maximum (ideal) power capture at wave height $H = 1$ m.

TABLE 7.2 List of optimal spring stiffness and power take-off damping.

j	1	2	3	4	5	6	7	8	9	10
Wave Period $T(j)$ (s)	5	6	7	8	9	10	11	12	13	14
Spring stiffness $k_{opt}(j)$ (kN/m)	186.1	139.1	113.5	94.9	80.3	69.0	60.0	52.3	46.2	41.2
Power take-off damping $c_{opt}(j)$ (kN*s/m)	53.1	67.3	70	67.8	63.9	59.8	55.7	52	48.6	45.6

The prototype development and tests are planned to validate the modeling results and explore different material selections for optimal performance.

References

Babarit, A., et al., 2017. Investigation on the energy absorption performance of a fixed-bottom pressure-differential wave energy converter. Appl. Ocean Res. 65, 90–101.

Betz, A., Wind-energie und ihre ausnutzung durch windmühlen. Vandenhoeck & Ruprecht, The University of Michigan, 1926, Vol 2.

Cargo, C.J., et al., 2012. Determination of optimal parameters for a hydraulic power take-off unit of a wave energy converter in regular waves. Proc. Inst. Mech. Eng. Part A J. Power Energy 226 (1), 98–111.

Clément, A., et al., 2002. Wave energy in Europe: current status and perspectives. Renew. Sustain. Energy Rev. 6 (5), 405–431.

Crabtree, G.W., Lewis, N.S., 2007. Solar energy conversion. Phys. Today 60 (3), 37–42.

Crowley, T.J., 2000. Causes of climate change over the past 1000 years. Science 289 (5477), 270–277.

De Lellis, M., et al., 2018. The Betz limit applied to airborne wind energy. Renew. Energy 127, 32–40.
Energy.gov, https://www.energy.gov/eere/wind/advantages-and-challenges-wind-energy. (Accessed: 2020).
Folley, M., Whittaker, T., Henry, A., 2007. The effect of water depth on the performance of a small surging wave energy converter. Ocean Eng. 34 (8-9), 1265–1274.
Gilfillan, B.C., 2013. Donovan Launches Regional Design Competition to Promote Resilience for the Sandy-Affected Region. U.S.D.o.H.a.U. Development, Washington, DC.
Green, M., Ratner, M., 2004. Books-third generation photovoltaics: advanced solar energy conversion. Phys. Today 57 (12), 63–72.
Hemminger, J., Crabtree, G., Kastner, M., New science for a secure and sustainable energy future. US Department of Energy, Washington, DC (2008).
Innovasea, 2021. https://www.innovasea.com/. (Accessed: 2021).
Lehmann, M., et al., 2013. An artificial seabed carpet for multidirectional and broadband wave energy extraction: theory and experiment. In: Proceedings of 10th European Wave and Tidal Energy Conference.
Lewis, N.S., et al., Basic research needs for solar energy utilization. Report of the Basic Energy Sciences Workshop on Solar Energy Utilization. 2005, DOESC (USDOE Office of Science (SC)).
Manwell, J.F., McGowan, J.G., Rogers, A.L., 2010. Wind Energy Explained: Theory, Design and Application. John Wiley & Sons.
Morrow, M., Delos-Reyes, M., 2013. Submerged wave energy converter. World 586 patent WO 2013/019214A1.
Sandia, https://www.sandia.gov/~jytsao/Solar%20FAQs.pdf. (Accessed: 2021).
Sims, R.E.H., Schock, R.N., Adegbululgbe, A., Fenhann, J., Konstantinaviciute, I., Moomaw, W., Nimir, H.B. et al., Climate change 2007-mitigation of climate change. Energy Supply. Clim. Chang (2007): 253–315.
Team, R.b.D., What is Rebuild by Design? http://www.rebuildbydesign.org/what-is-rebuild-by-design/. 2013. (Accessed: 2021).
Valério, D., Beirão, P., da Costa, J.S., 2007. Optimisation of wave energy extraction with the Archimedes Wave Swing. Ocean Eng. 34 (17-18), 2330–2344.
www.nature.com, 2009. Editorials: A pledge for immediate action. Nat. Mater. 8 (2), 81, www.nature.com/naturematerials.
Yin, H., Pao, F., Zadshir, M., Lou, J., Wei, Z., 2021. Building integrated photovoltaic thermal systems. U.S. Patent Application No. 63/248,502.

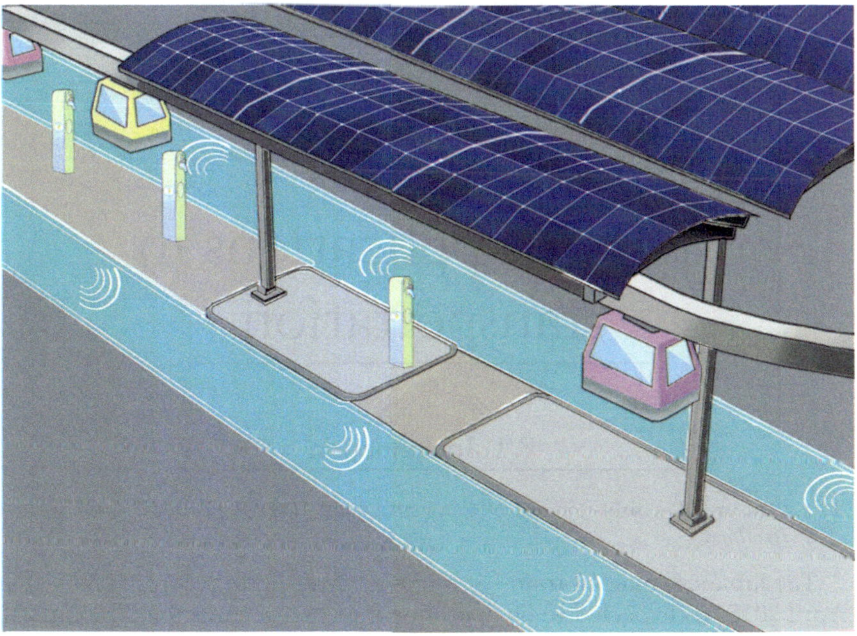

A BIPVT roof covers the roadway through a structural framework, which forms a tunnel under the sunshine, i.e., a sun tunnel. The structure supports lightweight vehicles moving within the tunnel and also hosts power transmission system for electric vehicles and traffic sensing/control systems for smart transit. Through multimodal integration with existing transportation and power systems, this new infrastructure will use the land of civil infrastructure to collect solar energy to power transportation with zero emission and form a physical Internet to transport people and goods with computer control. It significantly increases the resilience of the infrastructure. For example, when an accident occurs, first aid can be delivered by vehicles in the air and the surface traffic can be quickly resolved. Overall, this zero-emission sun tunnel system will fulfill the following synergistic multiple functions: computer-controlled vehicles for smart transit, solar energy harvesting for zero-emission power, energy storage, and transmission and distribution. As for energy security, coverage of roadway surface for roadway lifetime extension, and water collection by roof system avoids the second pollution.

CHAPTER

8

BIPVT applications in transportation

8.1 Introduction

"Dream today, reality tomorrow" ***Claire Marchand, Managing Editor, IEC, Geneva, Switzerland.***

The Japanese maglev train—it broke its own world record (603 kph) in April 2015—is expected to be operational in 2027, connecting Shinagawa, South of Tokyo, and Nagoya in Central Japan. Trains such as the maglev will increase mobility and permit daily long-distance commutes. In the United States, Tesla CEO Elon Musk is behind the Hyperloop project (Fig. 8.1), a futuristic transportation system that would whisk passengers through vacuum tubes at supersonic speed, the companies, and consortia are looking into the development of similar ventures.

Long hours and traffic congestions are the banes of today's commuters all over the world. To offer alternatives to road transport, municipalities are turning to means of transport that have proved to be reliable and efficient over nearly a century. Cable cars (Fig. 8.2), traditionally associated with mountains and skiing, are gradually becoming regular features in cities around the world. Being cost-effective, environmentally friendly, safe, and requiring little infrastructure, cable cars can cross rivers or scale hills, and thus complement other urban/suburban means of transportation.

Mankind has been dwelling on earth relying on a two-dimensional (2D) surface transportation network for centuries. However, at the local scale, our urban and suburban transportation has remained to be a one-dimensional (1D) roadway. To some extent, the subway helps a bit but it is getting very costly to implement. Consequently, people must look to the sky for the solution. The main reason that airplanes have become so popular and heavily used is that they are in a three-dimensional (3D) mode that multiple flights can take off in parallel and reach different destinations at the same time, which breaks the rule of surface transportation that vehicles are squeezed into the connected roadways in sequence or

Building Integrated Photovoltaic Thermal Systems: Fundamentals, Designs and Applications
DOI: https://doi.org/10.1016/B978-0-12-821064-2.00010-9

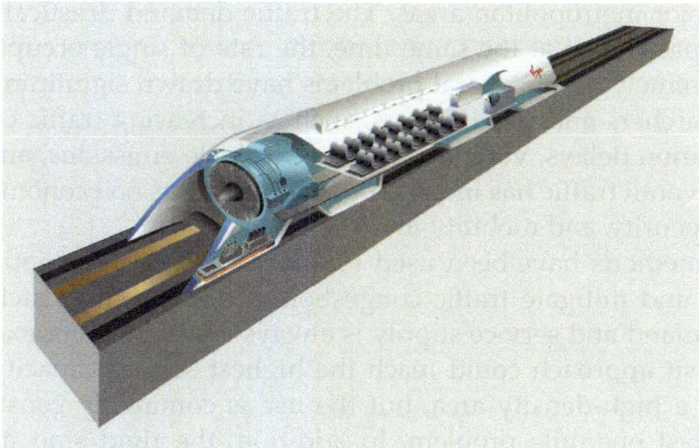

FIGURE 8.1 Hyperloop on high-speed travel.

FIGURE 8.2 Cable Car in London a way to avoid traffic congestion (Claire Marchand, 2015).

series. For example, the John F. Kennedy International Airport can directly connect with any airports on earth but the New York Penn Station needs to reach other cities by several highways and railways through a dozen of tunnels and bridges. One of them screws up, the whole system may slow down or stop. Although weathers and air turbulences are concerns for air transportation and transition between 1D and 3D can be tricky sometimes, diversification and optimization of transportation modes are critical for future life.

With a marked increase in population and the number of private cars, traffic has become a severe problem for most people living in the US drivers who waste time and money in traffic jams. The problem is especially dire in major metropolitan areas. The traffic demand drastically exceeds design capacity, but at the same time, the rate of single-occupancy vehicles is extremely high. Several problems have drawn significant attention from researchers and practitioners, such as increasing traffic congestion, transportation delays, volatile gas prices, vehicle emissions, and security threats. Overall traffic has imposed tight constraints on economic growth, national security, and mobility in metropolitan areas.

Many methods have been used to improve the transportation system efficiency and mitigate traffic congestion. However, the match between transit demand and service supply is always a problem. For example, the mass transit approach could reach the highest service capacity to move people in a high-density area, but the use is commonly constrained by the first/last one-mile problem. In addition, the rigid stop mechanism forces all people to wait and significantly reduces the system efficiency. With the lower ridership, mass transit systems exhibit very low efficiency. Inspired by the Internet technology of data packaging and communication, if small groups of people can be assigned to packages or small vehicles, and let them move through the roadway network automatically, we can form a physical Internet for smart transit and traffic mitigation.

This concept can be traced back to the personal rapid transit (PRT) in 1953. Fichter (1964), a city transportation planner, began research on PRT and alternative transportation methods and proposed an automated public transit system for areas of medium to low population density. Although the concept of automatic control essential to PRT had been firmly established by the early 1950s, the evolution of these systems has been continuous but fluctuating for reasons that had little or nothing to do with the technical feasibility of the idea or its potential value of the urban society. A bill was introduced but never made it through the House of Representatives. After John F. Kennedy, the 35th President of the United States came to the office, the bill was reintroduced, passed, and signed into law on June 30, 1961. In 1962, President Kennedy guided Congress to investigate transit situations. He stated, "To conserve and enhance values in existing urban areas is essential. But just as important are steps to promote economic and livability in areas of future development. Our national welfare requires the provision of good urban transportation, with the properly balanced use of private vehicles and modern transport to help shape and serve urban growth" (Wikipedia, 2021d).

The development of automated urban transportation systems, among which PRT is considered to be the goal, has been a highly interactive process among a wide variety of professionals, politicians, and dedicated citizens. In examining the writings, it is clear that these people saw the need for a viable complement to the automobiles, and they understood that such a complement could not be just of more conventional transit types. They were willing and able to invest freely of their own time and treasure to realize a dream.

Let us initially take a look at the history of PRT. The interest in the PRT came in the late 1960s as the HUD (Department of Housing and Urban Development) of the Federal Government did numerous studies. At that time, the Department of Transportation did not exist yet and the Urban Mass Transportation Act established the Urban Mass Transportation Administration (UMTA) as a unit of the Department of Housing and Urban Development. The new UMTA followed the direction of Congress and initiated a series of studies. In all the studies, the most influential of all the HUD reports were two items: (1) a study by Stanford Research Institute whose task was to develop on paper various new concepts from moving sidewalks to PRT to dual-mode and to estimate their economic benefits for the United States; (2) a study by the General Research Corporation (GRC) of Santa Barbara. GRC's major task was to model alternative transport systems in actual cities to determine how they would perform compared to conventional systems. A team of 17 specialists in various fields chose Boston as a typical large transit-oriented city, Houston as a typical large auto-oriented city, Hartford, as a small transit-oriented city, and Tucson as a small auto-oriented city. The results of these computer modeling studies strongly favored new transit systems. They showed that with projected population growth and growth of the use of automobiles if only conventional transit systems were developed, the problems of the cities would continue to worsen. Only by deploying personal transit systems would it be possible to reverse the direction of worsening congestion in our cities.

The GRC study has been the most influential of the HUD studies for two reasons: The first is that the results are summarized in a very readable article in Scientific Americans. This article has become a classic source and has been the starting point for much more thinking about the problems of new transport technology. The second reason is that the GRC work convinced its Chairman, Ben Alexander, of the importance of trying to create a national commitment to develop these new transport technologies. He talked to politicians and testified before congressional committees, thus, bringing PRT and Dual Mode more strongly into political thinking in Washington.

Unfortunate consequences on the development of PRT occurred due to the change of administration when the HUD studies were released. The Research and Development (R&D) played a minor role in UMTA's agenda. The main task was to prevent the collapse of the existing transit system in the country, and in doing so by providing capital grants for the purchase of buses and rapid rail systems. Throughout the years from the 1950s to the 1970s, there were numerous countries like Sweden, Germany, UK, Japan, Canada, and the United States that had investigated PRT systems and development of them one way or another.

8.2 Early personal rapid transit systems

8.2.1 UK: Cabtrack

The British Cabtrack System, a true PRT system, was initiated by the activities of L.R. Blake, who at that time worked in the Brush Electric Company had come to the United States and examined the Alden staRRcar, Urbmobile, as well as automated transit systems. Sometime around 1967, he wrote an article where he described the synthesis of his findings into a system that, as he believed, would be suitable for towns and cities in the UK; he called his system "Autotaxi." Consequently, he started a company and later sold it to Brush Electric. Their executives later managed to convince the Minister of Transportation to continue the idea which resulted in a joint arrangement being made with the National Research and Development Board to fund 50% of the work of developing Catrack to the test track stage. The total budget was 250,000.00 pounds. Later on, the Royal Aircraft Establishment at Farnborough Hants set up an urban transport group and they renamed it "Cabtrack" (Anderson, 1996b). The initial phase was a 9-month study with a comprehensive report issued in December 1968. Following it, the Royal Aircraft Establishment signed an 18-month contract and further contracts that culminated in the testing of a one-fifth scale model. In March 1974, the last report was issued. Their work was the first comprehensive system study of PRT by a large government organization and was considered not only a technical development but extensive demand and layout analysis. It included a large variety of control schemes as well as became confident of operating at a minimum headway of 0.6 s. They awarded a contract to Robert Matthew, Johnson-Marshall & Partners, a large British architectural firm, for the study of the integration of Cabtrack into a section of London. The results of that study were reported in the May 1971 issue of Architects' Journal. It was the earliest serious study of the visual impact of the overhead-guideway automated transit systems. In 1972, unfortunately, after a new election in the UK and the appointment of a new Minister of Environment, the Cabtrack program was discontinued. The Cabtrack program in the UK was the earliest serious development program in the world on high capacity PRT and the final reports are still of great value both in methodology and results. It would be better if all the documents on the Cabtrack program have been summarized in a book which could be valuable information.

8.2.2 Japan: Computer-controlled vehicle system

The Computer-controlled Vehicle System was developed by a Japanese industrial consortium during the 1970s. The original design was initiated with a 2000 lb four-passenger PRT carriage system in 1968. To learn more about the system, they come up with a scale model of a 1000 vehicle network to simulate a study. In 1972, outside in the suburb of Tokyo, they built a 4.8 km guideway with 60 vehicles. After the completion of two phases in 1978, the consortium started looking for deployment opportunities, developing a serious proposal for an installation in Baltimore in the late 1970s. After a series of studies, they decided not to continue the program due to the reasons listed below (Anderson, 1996a):

1) The cost and the visual impact on the guideway were a concern because it is 3 m wide and 1.8 m in depth.
2) Concerns on harsh weather conditions.
3) Concerns on the rough rides.
4) Concerns on the loading and unloading of passengers and carriages capacities.
5) Weight and the cost were also problems of concern.

Later, the Japanese came to the United States as part of their study regarding the progress of the PRT. Their concern was the barriers that they have to overcome and the market demand which had kept them from continuing the project.

8.2.3 Germany: Cabinentaxi

It was in 1970, the German Ministry of Science and Technology got interested and aware that the two organizations, Messerschmitt-Bölkow-Blohm and Demag, had independently been working on the PRT concepts, relatively similar to one and another. They were both inspired by the HUD reports in the United States. Consequently, the Ministry asked them to cooperate as well as sharing resources with funding and support from the Bundesministerium für Forschung und Technologie (BMFT, the German Ministry of Research and Development).

They developed a program for analyzing various designs on suspension, switching, motor, cabins, and tracks which allowed them to come up with different size cabins and how they could support each other under a beam as well as above the beam. Also, the vehicle runs on rubber tires and they were driven by linear induction motors on both sides of the cabin. In this case, it allows the operation of one cabin to be relatively close to the other which could be in the order of seconds. According to their study on the control, they select asynchronous instead of synchronous or quasi-synchronous because the asynchronous control would be more flexible to handle especially adjusting the change of speed and stopping the vehicles.

In May 1973, they tested the system full scale and in October 1974, they were successful in the demonstration to the Minister of Science and Technology together with the German press. They tested the reliability together with the human factor interface and maintenance, which brought interest for deployment from some cities. The team also looked into the potential of deploying the Cabinentaxi to Freiburg and Hagen which they believed could be successful as well as to other German cities (Babarit et al., 2017).

Raytheon Missile System Division in 1975 examined numerous PRT development and they decided they would try to license the deployment of Cabinentaxi in the United States. However, during the reconsideration of Raytheon's Missile System Division plans on their core business, they decided to cancel the plan. Meanwhile, DEMAG and Messerschmitt-Bölkow-Blohm continue to promote the PRT in the US market like Minnesota. A comprehensive study on the automated guideway was conducted for the Central Business District of Indianapolis. In addition, around the same time, the Germans were in the process of building a demonstration of the 12 passenger cabin version in Hamburg. However, due to the economic situation in the early 1980s, the German government had to cut the project's budget which eventually resulted in canceling the project completely. It was unfortunate that the project was terminated because it could demonstrate that PRT works for improving the traffic and transportation conditions in many cities.

Some unique features and advantages of the Cabinentaxi system are listed below:

1) Vehicles are small with seats available at stations and ready for use upon demand.
2) Goes from origin to destination operation only without stops.
3) Average speed at 30–40 kph.
4) Being on the elevated guideways, it is separated from other modes of transportation.
5) Totally automated operation system.
6) Applying linear induction propulsion to better control vehicle separation when operating at headways under 10 s.
7) Low noise levels and no air pollution.

8.2.4 France: Aramis

In France, they started with a four-seated cabin PRT system using a rubber tires type wheel. The cabin ran on a unique type of variable reluctance motor. Gerard Bardet began working on it in 1967 with a budget of 10,000.00 Francs. In May 1970, Engins Matra of a French aerospace company bought the patent and started developing the PRT system. In the late 1970s, Matra received a contract on Aramis from Datar, a French Agency. In April 1973, they started full-scale testing at the Orly International Airport in Paris, and in the Summer of 1974, they managed to test the first phase of the system on the basic concept. In addition, in the same year, Matra received a contract from the Paris Metro authority to make preparations to have a public demonstration on the PRT system in the suburb of Paris (Fig. 8.3). In the initial phase, it proved that system was safe and reliable (Wikipedia, 2021c).

FIGURE 8.3 Aramis test site in Paris, France (Wikipedia, 2021c).

The Aramis system is unique as compared to other PRT because the vehicles were trained electronically in platoons where each vehicle was controlled to be separated by 30 cm through ultrasonic and optical sensing. Additionally, any vehicle could be switched out of a platoon into a station by the in-vehicle switch. The vehicle could leave the stations between platoons to catch up to the last platoon vehicle. The main concept in this arrangement is to show that it would be able to rapidly increase the station capacities to load and unload passengers at the same time.

The platoon concept was designed to meet the circumferential system around the city of Paris. However, the system is not suited in a network condition like the PRT. It would be hard to control the brakes, especially in bad weather conditions. Later on, they decided to increase the vehicle capacity to a 10 passenger cabin which was a mistake that resulted in security concerns as well as serious concerns in station operation. Consequently, the Aramis program was eventually discontinued. Otherwise, other PRT systems could be implemented in other major cities in France, like Lyon, Marseilles, and Bordeaux.

8.2.5 Sweden: Gothenburg

In Sweden, the Gothenburg Transport Authority took the lead on the PRT system and was inspired by the British Cabtrack system. They thought of subway systems but in Gothenburg, most of the ground is of solid rock. Consequently, the PRT solution looked attractive to them. In March of 1973, the Gothenburg authorities had studied PRT system international standings and concluded that none of them are ready for deployment in Sweden. Under that situation, they decided to expand the tram system while anticipating how other countries made advancements on the PRT. They also made substantial inquiries around the world and obtained significant knowledge. In recent years, Sweden is again showing some interest in the PRT.

In 1994, a study was launched to examine the design and possibilities of adapting a PRT technology with elevated guideways to the central area at an old Swedish town like Gävle, Sweden (population 90,000). The report provided beautiful graphics for a city-wide PRT network of Gävle. A map of this PRT network was provided for those interested in an overview. This study includes examples of the design of different types of support columns and guideways as well as plan view drawings, sections, and perspective drawings for three station areas, town models (3D), and color photos of the existing city. The study covers technical dimensions of columns and load-bearing structures, architectural design of columns, guideway elements, and stations as well as the environmental design of those streets that would be affected. The results show that an elevated PRT system could lead to a significant visual encroachment upon the city's environment. The major visual intrusion is caused by the stations, track girders, and carriages that move along the tracks. The PRT system's encroachment in the form of noise disturbance and as a barrier are minor. However, the study did not result in an actual construction of this project.

Vectus is the latest PRT system that has been under development through collaboration with South Korea (Anderson, 1996b). A full-scale test track demonstrating all the elements of a commercial application has been in operation in Sweden since 2007 with Vectus. It is fully certified and approved by the Swedish Rail Agency and has been open to the visitor. Because the electric cars are lightweight, the elevated rail structure needed to safely support them is less expensive than adding, say, a new subway line or elevated roadway for cars. In 2013 Vectus broke ground on a new track at South Korea's Suncheon Bay, a coastal wetland that needs a transportation system with minimal environmental effects.

8.2.6 United States: Morgantown personal rapid transit

The effort of PRT in the United States started in the late 1960s by the Aerospace Corporation, which is a not-for-profit corporation established by the US Air Force. During the period from 1968 to 1971, the Aerospace Corporation developed the entire system concept to an advanced state and proved the feasibility of large PRT networks with many thousands of vehicles operating at headways as low as one-sixth of a second at 60 mph.

They performed economic and patronage analyses of PRT for Los Angeles and Tucson, Arizona, and lectured widely on the advantages of PRT. The White House OST (Office of Science and Technology), Dr. Lawrence A. Goldmuntz, who was the Director of Civilian Technology, has shown great interest in developing the PRT together with the proposal from the Aerospace department which resulted as a lead technology announced by President Nixon on January 21, 1972. UMTA was asked to provide $20 million to fund the development of the high-capacity PRT system. The Department of Transportation in the Fall of 1972 approved the program and work together with the National Aeronautics and Space Administration. However, due to the election and change of personnel after the election, the project was delayed, and by 1974, it was eventually canceled because many organizations lobbied against the development of the project.

At the state level, Morgantown's PRT system (Fig. 8.4) has been the most successful one. The project was initially looked into by Prof. Samy Elias, Head of Industrial Engineering at the University of West Virginia around the late 1960s. He was already aware of the PRT and learned that there were some test tracks around the country. He believed that PRT could be the appropriate solution to transport students between the campuses in the University and could be less costly as well as punctual on arrival. Luckily, due to the support of the UMTA, including congressional

FIGURE 8.4 The Morgantown PRT guideway and station (Wikipedia, 2021b). *PRT*, personal rapid transit.

delegation together with the University and the City of Morgantown plus political pressure from the state which give them the opportunity to get the project going. At that time, the Secretary of Transportation was John Volpe. He followed up on the proposal and the Jet Propulsion Lab of National Aeronautics and Space Administration in Pasadena, California, became the system manager, and signed the contract in 1970 with an industry team including:

(A) Boeing at Seattle was assigned to handle the manufacturing of the vehicles.
(B) Bendix Company at Ann Arbor, Michigan, was assigned to handle the control system.
(C) F. R. Harris Engineering Company at Stamford, Connecticut, was assigned to handle the guideway, stations, and surrounding areas.

All of these organizations had never had this experience in handling or building these systems or operations which resulted in substantial cost overrun which disappointing to Congress and foreign countries. Despite that happened, the Morgantown PRT has operated for more than 40 years and has been accident-free until the recent collision in 2016. On top of that, the operation is still in great condition with relatively low maintenance cost. Therefore, we should consider a long-term sustainability system. The PRT in Morgan is also called automated people mover. The system has been in operation since 1975. It presents a golden opportunity to evaluate the effectiveness of the performance to meet the demand of transporting the student and the local population around the area. In the past 40 or so years, the main transport in the country has been our automobiles even in a great portion of the urban areas and highways. The PRT could be an alternative to handle our transportation demands because especially our buildings and dwellings have been 2D for centuries but our transportation systems have been largely 1D. Although the subway has helped to some extent, the PRT system could be another solution.

As to the Morgantown PRT system, there were numerous important innovations that we should pay attention to.

1. Started the guideway transit switching through the vehicles instead of on the track switching like the train.
2. Initiated the demand mode fixed guideway transit service.
3. Started the control where the central control handles the automated vehicle communication control applying inductive wire loops hidden in the guideways.
4. Started the use of "moving slot" control system.
5. Started the automated re-distribution on passenger unloading and meeting the demand of the vehicles in other locations.

Though it overrun the budget substantially, considering how much it managed to accomplish as well as all the innovations been developed to the commercial standards, the whole project is really a worthwhile endeavor. In general, the commercialization of innovations is very risky and costly and often ends up a failure as it happened in the early PRT systems in other countries. However, the Morgantown PRT is an accomplishment and it should be used as a model on the PRT system in the coming years. On the Morgantown PRT system, it carries around ~18,000 passengers per day and around 3 million annually. The system operates in the same configuration as when it got started back in 1975. There were some changes, especially in the areas of the electronic component. Some of the original parts were no longer available that resulted in having to make substitute, though the system design is still based on the same principle. Throughout the years, there have been increment advancements in the system in order to be more up-to-date on some technologies.

In recent years, aside from the Morgantown, West Virginia, PRT system, couple more countries have come up with a somewhat similar system, like Heathrow Airport using the Ultra PRT and Masdar City in Abu Dhabi, UAE, and Park Shuttle in the Netherland, and Skycube in South Korea. Table 8.1 provides some brief information of current PRT systems.

In the following section, the details of the Morgantown PRT system will be introduced as an example for general PRT systems.

8.3 Integration and operation of the Morgantown personal rapid transit

The Morgantown PRT system covers a length of 8.7 miles of guideways that link with five stations together with two maintenance locations. Morgantown is built along the river valley. The transportation infrastructure was built according to the terrain along the Monongahela river. The PRT guideway goes parallel with the river in the older part of the town around the Walnut and Beechurst stations. On the northern portion of the guideway, it curls back to the northeast through a small valley. Land space at one time was limited and the university and the town increased the space through development. The fleet consists of 71 vehicles with rubber tires. Each one could carry 21 passengers with seats for only eight passengers. All the vehicles managed to log in around 1.5 million miles annually. The majority of the guideways are elevated. The highest point of the guideway rises to 350 feet, located between Walnut Street station and the Medical Center. To operate the vehicle from origin to destination, each of the three middle stations has by-pass lanes and turnaround channels. This is a design concept in transportation that would allow the vehicle to bypass the middle three stations was the latest advancement at that time. Meanwhile, being in a linear operation, it does not require a by-pass lane. Considering how today's buses and trains operate, they have to stop at every single assigned station.

TABLE 8.1 Current personal rapid transit system worldwide.

System	Manufacturer	Locations	Guideway	Station/vehicles	Notes
Morgantown personal rapid transit	Boeing	Morgantown, West Virginia, United States (1975)	13.2 km (8.2 mi)	5 /73	Up to 20 passengers per vehicle
ParkShuttle	2getthere	Rivium, The Netherlands (November 2005)	1.8 km (1.1 mi)	5	The second-generation group rapid transit vehicles accommodate up to 24 passengers (12 seated)
CyberCab	2getthere	Masdar City, Abu Dhabi, UAE (November 2010)	1.5 km (0.9 mi)	2 passenger, 3 freight/ 10 passenger, 3 freight	As of 2013 the freight vehicles had still not been put into service because they had not worked out how to get freight to and from the stations
ULTra personal rapid transit	ULTra	Heathrow Airport, UK (June 2011)	3.8 km (2.4 mi)	3/21	The Heathrow personal rapid transit system became operational in 2011, connecting Terminal 5 with a long-term car park
Skycube	Vectus	Suncheon, South Korea (September 2013)	4.64 km (2.9 mi)	2/40	

Source: Wikipedia, 2021a.

8.3.1 System operation

There are three operation modes in the system. (1) demand mode, (2) schedule mode, and (3) circulation mode. At the peak time, the demand and schedule mode operates at that period and the circulation modes operate at the off-peak period. Demand mode tries to capture the on-demand situation of the PRT. In optimizing the volume of the vehicle at the peak period or limit the operation cost at the off-peak period, these two functions are being handled by the other two modes. In the case of the demand mode, it reacts dynamically to the passengers' service demand and is controlled by the algorithm that checks between the passengers' wait time and vehicle capacities (Raney and Young, 2005).

8.3.2 Personal rapid transit operation and communications

The system is controlled by the vehicle control and communications subsystem (VCCS) which is part of an automatic control system installed in all the vehicles. The system handles the vehicle movements and operation by communicating with the stations' control signals through the station control and communication subsystem (SCCS) so that both sides know the position of the vehicle at the guideway. The SCCS identifies and transmits information on the location of the vehicles. An inductive communication link through the guideway control and communications links interexchange information between the VCCS and SCCS through important signals that are transmitted by tones while non-essential signals are through digital messages. The VCCS System includes the following components:

1) Antennas: it is a two antennas system that provides a two-way communications loop system installed embedded in the guideway. One dual antenna assembly is used for transmitting while the other is being used for receiving from the vehicle through low-frequency electromagnetic signals. All the antennas are physically mounted on each vehicle.
2) The communication system: it communicates from the receiving antenna through low-frequency signals. The information is then transmitted to the data handling system and onward to the antenna onto the guideway.
3) The data handling system: it receives the logic signals from the Commu nication system. The received logical information is decoded into logical instructions which allow the system to initiate appropriate logical commands to the vehicle.
4) The control system: it responds to the information from the data handling system to handle the vehicle operation.
 (a) Brakes.

(b) Steering.
(c) Doors.
(d) Propulsion.

5) The support system: it synchronizes the logic signals on the system, electrical energy conditions, electronic circuit isolation, and the interactions between the transmitter and the receiver.

8.3.3 The station vehicle management system

All the stations have a computer system that handles the vehicle's movements through a protocol from the control center regarding the routing of an incoming vehicle in load and unloading passengers, which is subject to

1. The station vehicle inventory and the channel availability.
2. Berth open space.

Directing the vehicle to the appropriate channel is implemented through the routing logic decision system in the station. The speed of the vehicle and the space availability is crucial as it enters or leaves the station. The correct channel switching and the time delay control system operation and the steering response of the vehicle to command verifications are also vital implementations of the system.

Once the switching section is cleared, the vehicle slowly decelerates to the point of a complete stop. The PRT automatic system could have an accurate and precise stop to the accuracy of ±6 inches. To accomplish it, the system onboard the vehicle interface with guideway stopping loop and the commands of the station's computer that instructed of the stopping loop at the channel location.

After the vehicle comes to a complete stop, the door is opened which allows the passengers to get off. Afterward, the door is closed again automatically and according to the instructions, it moves up to the forward position. Subject to the preprogrammed command, the vehicle takes appropriate action whether to wait for passengers or go to another station or go to an assigned channel. In case, there are additional passengers demand at the current station, instructions to allocate vehicles coming into the station will be sent according to customers' requests on the destination. However, in a situation when there is no empty vehicle in the station, it will be sent to another one. Meanwhile, after the passengers are all on board the vehicle, the door is automatically closed and it is ready to go. In case, there are any problems with closing the door, the sensors will instruct to delay until all the conditions have been cleared. Before moving, the vehicle will instruct the station that will instruct the central control system as to where the vehicle's destination is. Once the time to destination is determined, the vehicle keeps moving accordingly. The instructions provide the vehicle at the platform channel to speed up at the ramp as it moving onto the guideway by a detector.

8.3.4 Vehicle guideway operations

The computer at the station monitors the vehicle as it starts moving onto the guideway and controls the vehicle through the information listed below.

1. Vehicle ID.
2. Vehicle location at all time.
3. Vehicle destination and switch condition.
4. Vehicle speed.
5. Door situation.
6. Vehicle braking condition.
7. Operation anomaly.

All detailed information is being transferred between stations by the guideway detector through the central control regarding the vehicle identification, time, destination, status, and the guideway assignment position. The central control system also provides information to the guideway and the destination station to prepare to receive the vehicle. Any faults or abnormalities will be monitored to prevent unforeseen problems. The software in the computer will be able to detect as the vehicle approaches the station so that appropriate space will be allocated at the berth. In case, there is no space available, the vehicle will be asked to wait at the ramp until space is being provided in the berth.

8.3.5 Passenger and personnel safety

In the design and development of the Morgantown PRT system, substantial efforts have been put to make it safe for all. Observed the potential hazards to passengers and system personnel were identified and eliminated during the design and operation development process including the usage of components that could affect safety. In addition, the system is also looked into the potential of human errors and how to prevent them which resulted in able to achieve more than 7 million miles of injury-free record of Phase 1B operation. The system also allows handicapped access to the PRT. All the stations have elevators. The vehicles have a spot to confine to wheelchairs (Babarit et al., 2017).

After the success of the PRT operation which started in 1975 at the University of West Virginia, throughout the 1960s and 1970s, as mentioned, there have been numerous countries that had tried developing similar systems, due to design problems, economics, and political situations, all those projects were not able to continue. As to this PRT project, it not only demonstrated the automatic vehicle on a guideway is possible, it also has proofed that is safe because, in 45 years, it has not got a single fatality as well as managing to deliver passengers to their destination on time. Besides, all students at the university, get the rides for free.

The operation cost is around $3.00 million annually. Passenger fare is $50.00 and the operating employees would be around $63.00 per semester. The collected fares cover around 50%–60% of the operating cost. Considering the size and performance capabilities, the PRT operation is quite reasonable (Babarit et al., 2017).

Morgantown's PRT system has been a showpiece of the city and boosted the economy in the 2000s with the lowest municipal unemployment rate in the United States (Corio, 2009). However, with the outstanding service record with zero accidents for >40 years, the infrastructure system has been aged and more inspection and maintenance were required for accident prevention. On November 30, 2016, the system had its first reported accident when two PRT cars were involved in a crash between the Beechurst and Walnut stations. On February 10, 2020, two PRT passengers, and WVU students, were taken to Ruby Memorial Hospital for injuries due to a boulder that dislodged from the nearby hillside striking a PRT car. The long-lasting glory and recent accidents sent a clear message: PRT is a sustainable solution to increasing traffic congestion and transportation demands yet with critical challenges and exciting opportunities.

8.4 Challenges and opportunities

In the 1960s and 1970s, the United States was the eras after the completion of the Apollo Moon Landing Program and the need to find alternative government-funded projects. In 1972, President Nixon said: "If we can send three men to the moon 200,000 miles away, we should be able to move 200,000 people to work three miles away." For a variety of institutional reasons, it turned out to be much more difficult. According to the Congressional Study report done back in 1975, the UMTA failed to understand the capabilities of the private sector and the local transportation authorities which resulted in discontinuing the Automated Guideway Technology deployment. However, on a positive note, this study suggested two areas that the Federal Government could move to eliminate barriers for Automated Guideway Technology and PRT innovations: (1) contractual practices, (2) capital grants (TRANSIT A.G., 1975).

Recently, the PRT concept was revisited by a large senior-design task force at the Massachusetts Institute of Technology. A report called Project Metran was published that embodied most of the basic ideas of PRT and influenced the development of PRT. It was around that time the United States was boycotted by the Arabian countries on the oil supply, which resulted in severe shortages of gasoline nationwide, and an over 100% increase in gas prices that substantially triggered inflation. Although the PRT system was already in the works in West Virginia, in light of the situation, the action on the development of the system was enhanced. One of the main purposes was to be less dependent on foreign oil. Even though the Morgantown PRT was highly over budget which discouraged many organizations and foreign interested entities, it has been a success considering how the system has performed throughout all these years. It has been a great investment, especially since it is the first-ever built in the United States, thus being a fairly impressive achievement.

From the lessons from the early PRT systems, although the early concept attracted significant interests from governments and peoples of many nations, the efforts were often discontinued due to the high investment in a wide range of $30–80M per mile and long payback period. As the climate change, technology advancement, and energy and material price change, the economic metrics of early PRT systems have been significantly changed in comparison with other alternatives under today's market. Much less material could be required with lightweight vehicles; the modernized infrastructure can perform multiple functions and yield more returns for a short payback period, and the technology can rejuvenate the aging infrastructure for high efficiency and longevity. It is about the right time to reevaluate the technology.

We all know that oil and natural gas are finite and depleting rapidly based on the rate that is being consumed. Though there have been legislations on reducing the use of fossil fuel in recent years by numerous countries and states, it hardly makes a dent in consumption. In recent years, there have been new inventions on automobiles such as using electric motors instead of using internal combustion engines which helps somewhat but not to fully resolve the problem. As the world's population grows, so do the cities and the traffic congestions, and dealing with these problems are very serious issues.

Making a shift of paradigm always has many obstacles to overcome, e.g., legislations, political, economics, public sentiments, and technical issues. All of them will take substantial time and resources to resolve. Any one of these issues that could not come through could hinder the success of the implementation of a new type of transportation system like the PRT. Many countries have done numerous studies and have agreed that PRT will solve the congestion problems, however, none of them were willing to take the chance. It is about time to change: more organizations need to step up to the plate.

Currently, climate change issues are very serious concerns that will affect all our lives in the years ahead. In addition, droughts, extreme weather situations and devastations, food shortages, an increase in pandemics like SARS, Ebola, and COVID-19 plus numerous other concerns could affect our lives. Climate change is an immense problem to tackle and we should all work together to overcome it. One of the solutions is to massively deploy solar technologies and use them in transportation by applying them to the operation of the PRT which could substantially reduce the consumption of fossil fuels and has a great potential of solving traffic congestion problems.

PRT is relatively light in weight; being at a few hundred pounds as compared to an automobile that is more than a ton. The number of passengers in each cabin would be between 4 and 6 which would result in its total weight being about one thousand pounds. The solar energy collected can be sufficient to satisfy the required energy to operate, and in case there is additional energy available, it is possible to be fed into the grid. Basically, the solar system could be a grid as well. Once it has been installed, with proper maintenance, it could provide energy for the next 20 years, instead of having to be concerned about any energy shortages or price fluctuations. On top of everything is freedom from relying on energy sources from hostile countries which could potentially be a major problem. Energy is one of the major operating costs and is powered by solar, there is one less item to worry about that could reduce transportation costs and make it more affordable to the public.

In recent years, there have been increasing interests in autonomous vehicles or self-driving cars, which share some features of PRT in electrification, smallness, and autonomy. However, parking and traffic would still be their problems. Table 8.2 shows the comparisons of PRT and autonomous vehicles.

TABLE 8.2 Comparison of personal rapid transit with autonomous vehicles.

Personal rapid transit	Autonomous vehicle
No Parking required	Requires parking
Only travel cost	Upfront vehicle cost
Avoids traffic congestion	Cannot avoid traffic congestion
Guarantees arrival time	Cannot guarantee arrival time
Individual maintenance not required	Requires individual maintenance
Individual insurance not required	Requires individual insurance
Less likely on a breakdown	Breaks down could happen more
Requires no individual fuel cost	Requires individual fuel cost
Requires going to the station	Does not require going to the station
Massive decrease in accidents	Potential of getting into an accident
Technically is simpler to build	Technically could be more complex

There could be some concerns about job opportunities because the PRT system is automated. However, to some extent, certain types of jobs could be eliminated but instead, there could add a substantial number of jobs, mainly high-paying professional positions like engineers especially electronic and civil engineers, system analysts, system designers, and system managers. Each location has different requirements. When PRT grows to a certain size, it can perform like the internet. It is a physical network instead of a cyber internet. It was the wildest dream that the cyber internet had opened so many opportunities. The physical internet could be even greater. The areas that could lose some jobs would be Uber, Lyft, and Taxis because they would be low on-demand. However, they might be able to do better because the roads will be less congested after PRT is being installed.

On PRT, it is an automatic pod car that goes directly to one's destination nonstop. A lot of time traveling in an urban city, one not only has to make stops but also has to change buses or trains to reach the final station or stop. Waiting between connecting vehicles could also take some time as well as stops on each bus. Adding the time and energy costs could be very substantial. PRT could avoid all these issues. A rush-hour commute to and from the city could be another concern on public transportation and automobiles. Considering vehicle stopping means that the system already uses less energy which would result in a less damaging effect on the environment.

PRT could play a vital role to improve the traffic situation. It could reduce the number of cars going to and from the city or commercial centers and it also saves parking spots. However, planning on the location of the PRT stations in a busy downtown area could be a challenge though the stations do not require too much space. A good layout design could make a big difference in the performance of the traffic flow.

8.5 BIPVT for sun powered personal rapid transit

Transportation accounts for 71% of the petroleum products consumed in the United States today, and this number has been increasing in the past decade (Fig. 8.5). Unfortunately, the oil stock in the world is inherently finite and will be exhausted in ~40 years if we continue the current practice, which leads to a serious energy crisis in the near future. Moreover, 29% of greenhouse gas emissions are from vehicles, which contribute to global warming and climate change. As a result, parts of the transportation system may be subjected to an accelerated aging and weathering environment.

To promote the sustainability of the transportation system, transformative technologies are needed for reducing gas consumption and protecting

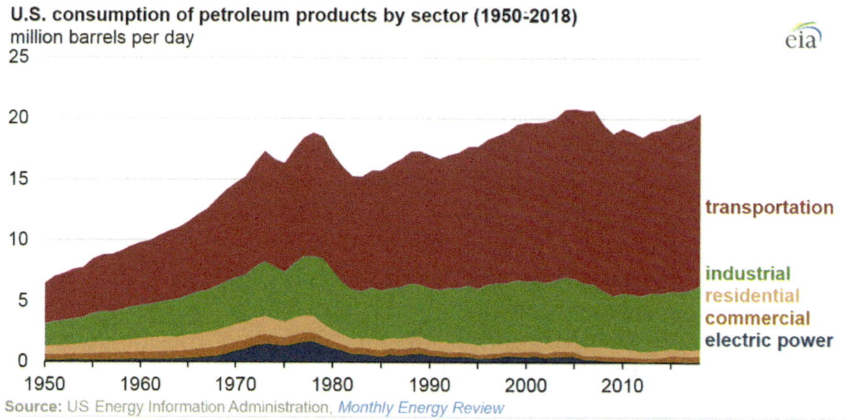

FIGURE 8.5 US oil consumption from 1950 to 2018 by the five sectors (EIA, 2019).

the infrastructure. A novel transportation system—sun tunnel (Yin et al., 2021) was proposed to transform the existing railway and highway infrastructure for higher material, energy, environmental, and economic sustainability (Fig. 8.6). In this sun-powered PRT (SPPRT) system, the sun tunnel employs state-of-art green technologies with low material and

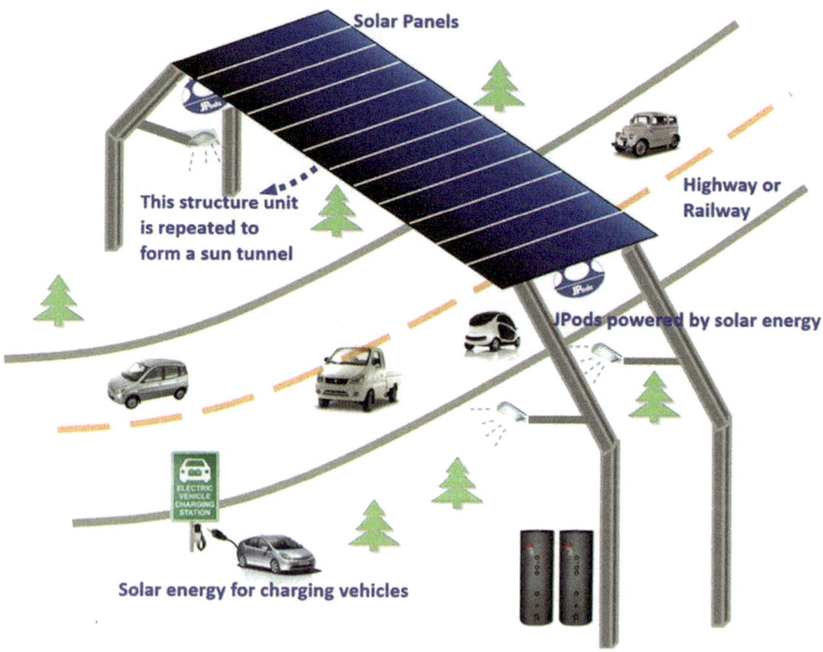

FIGURE 8.6 Schematic of JPods moving in a sun tunnel with a BIPVT coverage.

construction costs. It performs multifunction for solar energy harvesting by BIPV, additional PRT service by JPods, which is to be elaborated in Section 8.5.3, road protection by roof coverage, electricity supply for electric cars, water collection by the roof, and smart traffic monitoring and control through vehicle-infrastructure connection by sensing and information technologies. It helps with the energy-saving and gas emission reduction from the transportation system and provides a clean energy platform for hybrid-electric and electric vehicle applications. In addition, this sun tunnel platform enables lifetime extension, modular construction, operation cost reduction, and safety and security enhancement of the future road system.

8.5.1 Motivation and design philosophy

Due to rapid urbanization and population expansion nationally and globally, the addition of new construction in a built environment is not trivial while the existing transportation infrastructure is constantly subjected to increasing demand. Any major disruption, caused by either natural or man-made actions, could have a strong impact on a large part of our nation. Therefore, protection of existing transportation infrastructure, including roads, highways, bridges, etc., and enhancement of their performance and service-life become of extreme importance from the viewpoint of national security and economy. The durability of transportation infrastructure is most devastatingly affected by solar radiation, temperature, moisture, and other environmental conditions that lead to aging and degradation observed in corrosion, cracking, and spalling of surface materials in these infrastructure elements. Research aimed at securing the performance of these systems, and indeed, improving their performance is an essential part of sustaining a healthy, functional urban area. Synergistic research is needed to rapidly develop and implement new technologies to renew infrastructure to improve its durability and extend its lifetime, particularly for urban and sub-urban transportation infrastructure.

The main approach is to develop durable, multifunctional, smart skins, or envelopes to protect infrastructure with self-healing, temperature regulation, self-powered sensing, and control methods, as well as meeting other needs in energy, water, and the environment. The technologies and methods to be developed could also be extended to transportation infrastructure in rural areas because the energy harvesting and water supply can fit into the lifeline infrastructure system although the PRT has less demanding.

Though the PRT system consumes relatively less energy as compared to trains and automobiles, the pods are light enough to be solar powered that is a big plus to be able to use alternative renewable energy. Solar photovoltaic (PV) energy could not only converge the energy to electricity, but it could also split the water from hydrogen and oxygen. The hydrogen could also be used on fuel cells which could generate and store electricity as well. As these technologies are getting more developed and popular, prices on the products could be reduced for better economical metrics.

As shown in Fig. 8.7, the PRT lane also indirectly protects the bicycle lane. Bicycles, especially at rush hours, could be very venerable to all other vehicles in the streets. Meanwhile, even with one less lane, it would make motorists driving a lot easier because of the reduction in traffic congestions especially when PRT could manage to offload it. Also, it could be a scenic ride along the Greenway.

FIGURE 8.7 Rendering image of the solar-powered PRT on the Rose Kennedy Greenway, Boston, MA. *PRT*, personal rapid transit.

To establish the SPPRT system, the design, construction, and operation of this new infrastructure system should address the following aspects with unique advantages:

Energy harvesting, storage, and utilization: using road surface for energy harvesting will improve the yielding of the land, which depends on its energy efficiency and storage. The appropriate design will significantly improve the resilience and reliability of energy infrastructure. The onsite utilization also avoids the electric transmission reduction.

Transportation service and traffic mitigation: the lightweight vehicle moving along the tracks on the roof provides a new dimension of transit mode with a very small headway distance by computer control. It may send rescue teams timely for emergency response and reopening surface transportation. The sensing and monitoring system can report traffic situations and prediction for intelligent transportation systems.

Structural integrity and weather adaptation: the individual panels will have sufficient strength to carry loads due to gravity, wind, and other loads. The strong support for lightweight vehicles and the BIPVT system will protect the roadway and vehicles from weathering effects and weather influence.

Water collection and management: the infrastructure provides feasibility to collect rainwater from the roof avoiding the second pollution with interact with the road surface. It can be coupled with a BIPVT system for thermal management and energy storage.

Sound absorption: the envisioned roofing panel will be designed for optimum sound absorption and reduction over the audible sound spectrum for noise control in surrounding urban areas and to travelers and vehicles inside the sun tunnel.

Chemical and physical compatibility and longevity of pavements: the various materials and components will be designed for chemical and physical compatibility to assure a durable end product. With moisture and solar irradiation separated and temperature-controlled, the road materials and structures will have a much longer lifetime and thus lower life cycle cost.

Architectural aesthetics: with the glazed surface, the BIPVT will significantly alleviate the visual encroachment or intrusion that the earlier PRT systems were attacked. It shows a modernized appearance to the area with many design flexibility for architects to use as an expressive tool with the surrounding environmental styles.

Prompt maintenance and construction: the maintenance and construction will no longer be affected by weather or temperature. Once a problem is found, it can be fixed immediately. The modular design of the BIPVT system will simplify the infrastructure inspection and management procedure.

Integrated lifeline network: with this infrastructure, the electrical, water, information, and transportation network can be integrated into this infrastructure forming a critical lifeline and thus reduce the risks of powerline breakdown in storms.

Intelligent highway infrastructure: Using the structure of a sun tunnel, cameras and sensors can be installed and protected, and infrastructure-vehicle interaction can be implemented through the information communication network.

The success of the SPPRT system will turn the roadway from the huge gas consumer into an energy plant, move people at much higher transportation capacity and efficiency, serve as a shelter for humans under global warming and climate change, and renew our aging transportation infrastructure with extended lifetime.

8.5.2 The sun tunnel of solar canopy

It is a very vital part of the PRT system. So far, the energy for all the PRT systems comes from the grid and they are mostly generated by fossil fuels. The solar canopy locates above the tracks. Being in this position, not only it stays dry and is protected from the rain or storms, but also prevents the snow from getting on the track. Normally, the snow accumulated on the solar canopy would eventually slide off because it is slightly curved and pitched.

New cost-effective surface technologies will be used to protect materials and structures for improved durability and longevity of transportation infrastructure systems. Because most aging, degradation, and failure start at the surface, renewing the surface layer or adding a protective envelope to the infrastructure can not only enhance the existing function but also provide possibilities for new functions, such as energy harvesting, water collection, and self-sensing, etc. Therefore the overall cost of the structure will be leveraged by multiple systems if they were installed separately in a conventional way. The BIPVT system perfectly fits the need for the sun tunnel roof. It intercepts sunlight and rain that accelerates road degradation for valuable use.

First and foremost, the envelope serves as a space enclosure to protect the interior infrastructure from environmental distresses. Secondly, a separate but related requirement is that it can safely carry all external loads applied to it, such as wind, snow, and self-weight, and transfer these to the structure. In urban areas, it is also necessary to separate the interior noise from outside noise, especially in the vicinity of major traffic arteries and airports. The foamed concrete panel can be a good candidate material.

Fig. 8.8 shows the solar canopy prototype for parking lot applications. Using it on the roadway in Fig. 8.9, one can form a sun tunnel under the sunshine along the Boston Seaport.

Depending on the needs, a water collection system can be installed with a gutter and drainpipe connecting to a water reservoir. The rainwater can directly be released to the river without the second pollution with road surface or be used for irrigation, cleaning, or other purposes. A water-based circulation system can be integrated into the BIPVT for roof cooling.

FIGURE 8.8 Solar canopy prototype.

FIGURE 8.9 Layout of the Sun Tunnel along the Boston Seaport.

8.5.3 Personal rapid transit system operation

In Morgantown, the vehicles communicate with the central station control room. To form a physical Internet in the future, it is important to implement a decentralized system with distributed control and communication for extensibility and flexibility. JPods (Inc. J., 2021) has spent many years developing the new concept of PRT and designing the vehicles. It uses distributed collaborative computer networks to route transit in a manner similar to the data trafficking of the Internet. In Fig. 8.10, a JPod vehicle is hung onto the rail network through a connector like "J". The wheel carries the connector and drives the ultralight vehicle with riders or cargo to move through the guideway.

FIGURE 8.10 Personal rapid transit (Jpods) demonstration at Boston City Hall.

The lightweight of JPods' small vehicles allows smaller guideways and support structures than mass transit systems like light rail. The smaller structures translate into lower construction costs, smaller easements, and less visually obtrusive infrastructure. JPods vehicles communicate with each other, a data server, and other devices, so that they move through the guideway network as an independent vehicle without a driver, and can be dispatched by the server to meet the transit demand. They have their own sensors so they can execute their own behaviors even if they cannot communicate with each other or the reporting center. The procedure would be as following:

1. Initially, a vehicle requests a trip from a Negotiator based on the input from a passenger, cargo, or network need (the network will shuffle empty vehicles to the anticipated needs).
2. The Negotiator reads the "time-load-map" from a load balancer to pick a route (series of lines and switches) between the origin and destination based on load, distance, travel time, customer requests, etc.
3. As soon as the Negotiator has chosen a trace route, it passes the trace route to the vehicle, every network device on the route (switches, reporters, etc.), and a load balancer.
4. All the affected devices add future events to their internal database. If a switch, reporter, etc., calculates the added load will exceed the comfort level (load design and the other considerations), it will complain to the Negotiator and a load balancer. The Negotiator and load balancer will loop through alternatives until an acceptable trace route is agreed to be within parameters for all devices to be affected.
5. As the vehicle navigates the trace route, the associated devices will chat with each other to coordinate their behaviors. They will use their internal sensor in those collaborations. The better the sensor readings between merging vehicles and switches agree, the lower the headway between the vehicles.

The JPods system can grow with the current research in autonomous vehicles with much lower complexity as the guideway significantly simplifies the direction and speed options in driving control.

8.5.4 The sensing and control system

As a JPod drives on the network independently, does anyone know where one JPod is at any certain time? Yes, a sensing and control system will know where the pod is at any time. In the Morgantown PRT, a central reporting center (Load Balancers) controls all the activities of all the devices (vehicles, switches, reporters, etc.) on the network. However, for a scalable physical internet, because the complexity and uncertainty grow with the scale exponentially, it will be impossible to use one computer to control everything. Similarly to the Internet, the SPPRT will have distributed servers to manage the JPods and match vehicles with riders.

The coverage area of SPPRT can be divided into many zones with a zoning server, which allows new zones to be added as a scalable system. For each zone, a sub-server will manage all JPods and guideway sections in its domain. Basically, a JPod is associated with a sub-server, and when a JPods crosses the borderline of two zones, the sub-server will be switched to the new one. A certain overlap of neighboring zones is allowed for a smooth transition. Therefore, each JPod can be traced and located in real-time (Inc. J., 2021).

Today the sensing and information communication technologies are improved rapidly. Mechanical, optical, or acoustic sensors can be used to detect and identify JPods by the guideway, JPods is also equipped with GPS. The redundant information will assure the stability and robustness of the system. With G4 and G5 data communication, the routing and dispatching commands can be transferred to JPods immediately. The distance between JPods on the same guideway will be sensed by a laser distance sensor, which triggers the brakes of vehicles automatically when an emergency occurs although it is unlikely to happen. Notice that the two recent accidents in Morgantown PRT can be avoided if this mechanism can be implemented in their system.

8.5.5 Energy storage and operation system

Energy storage technologies have grown substantially since the exponential growth of the PVs industry. Due to the intermittent nature of solar energy and the constant use of energy for transportation, it is vital to have sufficient energy storage to balance energy harvesting and utilization. As shown in Fig. 8.11, solar energy is collected from the roof of the PRT system,

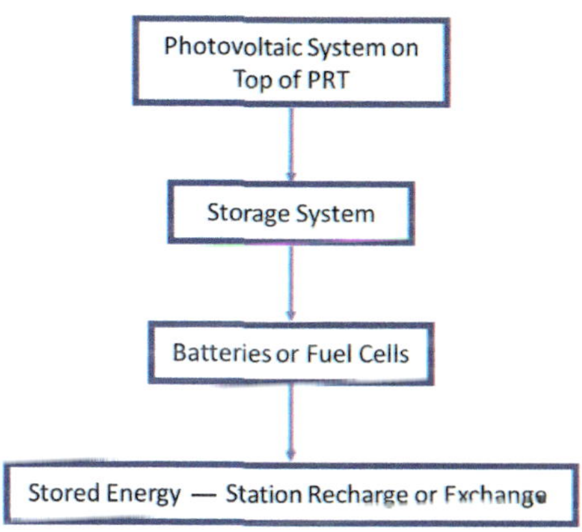

FIGURE 8.11 The energy flow in an SPPRT system. SPPRT, sun-powered personal rapid transit.

flew to a storage system, released to batteries or fuel cells based on the energy demand, and eventually used by the PRT vehicles. Depending on the design energy storage and operation system, the batteries or fuel cells can be merged either into the storage system or the PRT vehicles. With high efficient wireless recharge technologies, PRT vehicles can be charged at the station during the stop period.

Initially, lead-acid batteries were common in the solar on PV applications especially when installations were off-grid or when systems required backup. In recent years, lithium-ion batteries are getting more popular with the electric auto industry booming in spite of their higher cost because they are more compact and have a higher energy density. However, batteries generally have a certain number of charging cycles in their lifetime, which has been a concern. In addition, the disposal of end-life batteries is a serious problem due to the environmental consequences.

However, fuel cells could be an alternative. It has been a setback due to the cost. As popularity increases, the price will fall. Also, hydrogen availability will be growing more rapidly as more countries are investing in technology and generation capabilities. Fuel cells use hydrogen to react with oxygen and generate water and energy. It is different from an internal combustion engine because it relies on the continuous redox reaction. The number of charging cycles of fuel cells could reach a million as compared to batteries is a few thousand.

Both batteries and fuel cells can be used in JPods to power a motor and drive a vehicle.

In going through the operation of the energy storage, the current plan on the PRT would be using batteries or fuel cells. As the energy needs for a lightweight JPod is low, one can exchange or charge the batteries at the stations. All the stations allow for charging the batteries.

For the large-scale application of SPPRT, a dedicated energy storage system or access to the grid is still favorable as the energy harvesting from the PRT roof will be too much for batteries or fuel cells. At daytime, it may generate much more energy than the system needs but at night it relies on the electric supply from the storage system. Using the grid, the SPPRT performs like the combination of a utility PV generator and electricity customer, and let the grid deal with the energy balance issue, which could be a huge impact on the existing grid.

For dedicated energy storage, it can go through chemical, mechanical, or thermal storage. Batteries or fuel cells are in the chemical category. The energy can be stored in hydrogen. The system will use the collected electricity from the overhead PV system and split the water into hydrogen and oxygen. The hydrogen will be the source for the fuel cell to generate electricity and charge the batteries when needed. Chemical energy storage is costly and can reach a high efficiency $>70\%$.

For mechanical energy storage, when the roadway passes the hilly region, using water reservoirs at different altitudes can store and release the gravitational energy through the water pump and turbine. Modern hydroelectric technology can reach a high efficiency >70% as well depending on the scale. An example will be provided in Chapter 9 for farm applications. Recently some groups are considering use the gravitational energy of solids for the same purpose. It is possible because solids could exhibit higher density than water.

For heat energy storage, it depends on the temperature that the thermal fluid can reach in energy harvesting. As described in Chapter 2, the Carnot efficiency will be the limit. Different from the chemical or mechanical energy storage that goes through electricity-storage-electricity, the thermal energy storage can go directly from solar heat to storage and then electricity. Therefore, the overall efficiency may follow different mechanisms. For a BIPVT system with 25% PV efficiency, the actual efficiency from the sun to JPods will be ~18% after the mechanical or chemical storage. For heat storage, if a high temperature of the heat storage can be reached by a cascade heat pump, it is possible to reach a comparable efficiency as well theoretically. It is hard to manage the high volume of heat storage.

The combination of the three approaches may provide redundancy and reach higher capacity.

8.5.6 The layout of the station operation

The station is most likely to be in a city in a relatively dense area. Because it would be difficult to install a sloping roof in such a location, the building-integrated thermal electric roofing systems will be applied. On the PV side, in a facade or shade, the solar energy can still be collected regardless of the time of the day or if the sun is blocked by some neighboring building. The thermal system will circulate the liquid to collect the thermal energy to warm the building and the water tank for hot water.

Looking at the systems in the stations, they could not only store energy but also produce it. As one can see, the solar-powered PRT system not only could support the operation of JPods, but also provide energy on other applications.

In summary of SPPRT, a solar canopy covers the roadway or railway through a structural framework, which supports lightweight vehicles—JPods—moving within the tunnel and a power transmission system for both the smart transit vehicles as well as traffic sensing/control systems. Through multimodal integration with existing transportation and power systems, this new infrastructure will use the surface of civil infrastructure to collect solar energy to power transportation with zero emissions and transport people and goods by computer control. This technology will immediately create new research opportunities in advanced materials, novel solar collectors, smart sensors and control systems, and information communication and integration algorithms. Overall, this zero-emission SPPRT system will fulfill the following synergistic functions: computer-controlled vehicles (JPods) for smart transit, solar energy harvesting for zero-emission power, energy storage as redundancy for energy security, coverage of roadway surface for roadway lifetime extension, and water collection by the roof to reduce runoff-driven pollution. The large scale application of this system will form not only a self-powered physical Internet to grouped individuals or cargo for nonstop transportation but also a new infrastructure network to accommodate the emerging needs in transportation electricity supply and transmission, roadway-vehicle communication and traffic monitoring, climate adaptation, and management, as well as lifeline protection and lifetime extension.

8.6 Perspectives of BIPVT in transportation

The significant benefits of the BIPVT application to the SPPRT system can spark other applications in transportation using some components to achieve a certain goal. For example, BIPVT panels are used for highway noise barriers, smart signs and advertisement boards, energy independent parking and recreational facility in rest areas or remote areas, etc. Today the Columbia team is developing a self-heated pavement technology using solar and geothermal energy.

Inspired by human skin, keeping the surface temperature of pavements within a certain temperature range allows avoiding pavement distresses caused by extreme climates while engendering synergistic energy benefits. The self-heated pavement system contains three components: BIPVT harvesting both electricity and heat, a road surface heat exchanger control the surface temperature, and a geothermal well to store and release heat. Using seasonal heat storage and self-powered smart sensing, we can narrow down the temperature range of pavement significantly to reduce rutting in summer and avoid snow/ice removal in winter, so as to improve the safety and longevity of pavements, particularly, for some dangerous road sections, such as highway ramps, slope areas, and turning points.

Some applications were tested but may not be successful. For example, directly using PV modules on road surfaces may be challenging to maintain required roughness for road safety and clean surface for energy harvesting efficiency. Using solar panels to power a vehicle can never reach the energy equilibrium for the solar energy inputs and car's power needs because of the huge mismatch of the energy density.

Now it is a critical time to revisit the PRT concept with solar technology. The marriage of solar power and PRT will catalyze a revolutionary change in our future life. "PRT is a concept that has been evolving for over 50 years. With today's innovative technologies, PRT is emerging for the future because it has the potential for contributing significantly to the solution of the fundamental problem of modern society including congestion, global warming, and dependence on a decreasing supply of cheap oil" notes Frost & Sullivan Analyst Poonam Tamana. Dec. 2008 (Anlauft, 2008). As one can notice, the whole PRT system is totally green. Starting from the collecting and then to the storage and generation which is a closed-loop system. The building-integrated thermal electric roofing systems use clean energy sources and materials. In addition, the massive deployment of PRT will result in the massive installation of BIPV solar which is the way to deployment of solar technologies because apart from being aesthetically more pleasing, the solar panel is part of the building and will be well protected from storms and any severe weather conditions. Meanwhile, it also opens a new set of building materials and products that could be called "active building materials and products." There is still room for advancements that will be developed as more installations progress. Currently, it is in the initial phases and as more green technology systems are available, they will be implemented as they become applicable.

Considering the current existing transportations modes such as subway, buses, trains, or even automobiles, they all could be a problem as well as a concern on virus situation especially COVID-19 which is a very deadly and contagious one and has affected millions of people worldwide. According to a recent study by Prof. Jeffrey at Massachusetts Institute of Technology (Harris, 2020), the New York City subway system is a major disseminator of the virus and by Prof. Lara Gosece, University of London: analyzing the link between public transport use and airborne transmission mobility and contagion in London underground (Goscé and Johansson, 2018). Wearing masks would help but social distancing could be a problem. PRT is essentially a self-driving car with tracks that could prevent all these concerns. Our planet's situation on climate change and the pollution is getting worse. If nothing is done about it, the place will be inhabitable.

8.7 Case Study: Personal Rapid Transit (PRT)

8.7.1 Overview

The PRT system technology has been around for decades, although currently, there are only some limited size operations in the world such as Heathrow airport outside of London and at Morgantown, West Virginia, which handle mainly the traffic between the campuses of the university.

The Morgantown system to some extent demonstrated the performance of the PRT which proved that it is safe, injury-free since 1975, on-time arrival, automatically operated, and low maintenance. The initial cost was high because it was never built before in the country nor were there any professional engineers having any experience with it. However, considering after 45 years of operation with such great performance, it has paid well for itself. In addition, due to the population growth which has resulted in traffic congestions in all major cities in the world, many countries have done quite a few studies on the PRT and its advantages to solve this issue. In the remaining, we will look at two examples of these studies.

The main concerns on PRT have been the initial and maintenance cost. Prior to going any further, let us look at the current automobile and bus cost. Currently, most buses and cars run on fossil fuel. However, in a few decades from now, most of them would be converted to electric and autonomous vehicles. In light of what we have learned about electrical vehicles, we expect them to be less costly which is a great advantage. However, adding more cars would not help solve the congestion problems. Nowadays, the younger generation is avoiding owning cars mainly due to the financial and liability costs. Consequently, more and more people will be depending on taxis, and ride-sharing services such as Uber and Lyft as well as other means of transportation such as trains, buses, and trams. In an unfortunate situation such as what happened in the year 2020, in which the whole world underwent a serious pandemic of COVID-19, it caused a hard hit on public transportations because the virus was highly contiguous. With growing concerns about Climate Change and the resulting consequences, many scientists believe that a similar virus would be more prevalent in the coming years.

In terms of current public transportation alternatives, there are monorail and light rails, however, their initial costs are substantially higher than the PRT. Here is the comparison of three of these systems:

(a) Monorail $ 50 – 70.00 million per mile—subject on location.
(b) Light rail $150 – 300 million per mile—subject to location.
(c) PRT rail $ 10 – 30 million per mile—subject to location.

Looking at the above information on cost comparison among other rail systems, one can say that the PRT is more cost-effective, especially if it is powered by solar energy, the initial investment will pay for itself for the followed 20 years after installation. As to the cost of the light rail, the base price is around $150–300 million. However, price could increase substantially in places like Boston and New York. The recent budget on the light rail as part of the green line extension to the suburb of Boston was considering 1 billion a mile, for which the state has expressed no problem affording it. Somehow, its cost in European cities is about 40% lower than in the United States.

8.7.2 Intra Dwarka Trips PRT Project

Worldwide, considering the traffic congestion and pollution problems, there has been quite a bit of interest throughout the past two decades on PRT. Numerous major cities had done studies on its feasibility. One of these studies has been done in New Delhi which is a densely populated metropolitan area with a population of over 28 million. The section of the city that the study has picked is Dwarka, situated in the southwest part of the National Capital Territory of Delhi. Dwarka is divided into 29 sectors. The study concentrated in 15 sections of Dwarka and they called it the "Intra Dwarka Trips," as shown in Fig. 8.12.

The PRT network route that was proposed in this case study was 18 km in length (Fig. 8.13). To get started, they carefully examined the population on their interests in shifting from their current means of transportation to PRT. The options could be cars, two-wheelers, cycle rickshaws, bicycles, and walk. Out of the total of 607,496 trips per day, there was 36.6% interest in shifting to use PRT. Therefore, they predicted that the demand for the

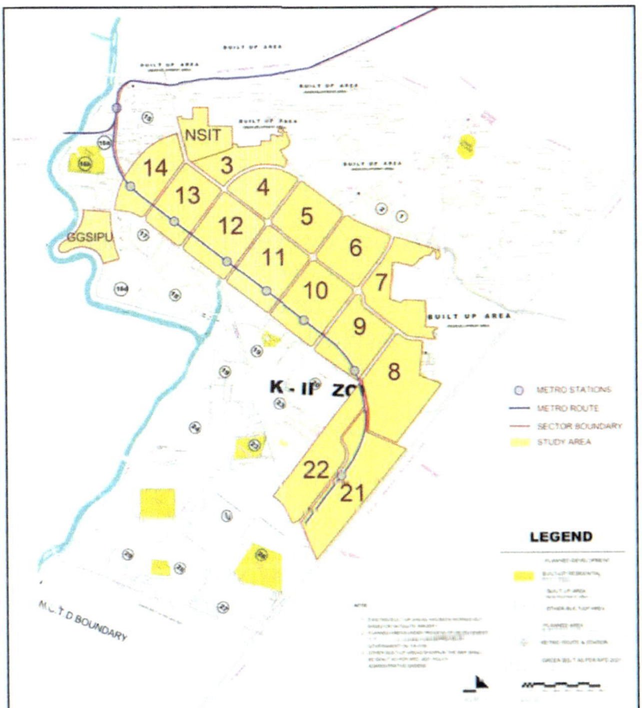

FIGURE 8.12 Intra Dwarka Trips map of the study (Sarkar and Jain, 2017).

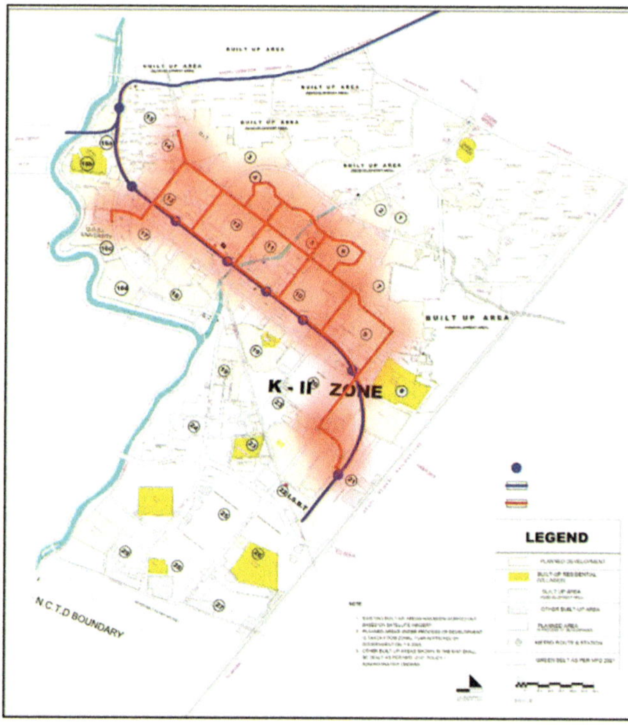

FIGURE 8.13 The PRT proposed route at Dwarka (Sarkar and Jain, 2017). *PRT*, personal rapid transit.

PRT would be 222,465 trips/day. However, the study neglected the transit bus because it plays an insignificant role.

Based on their peak hour factor, the demand for the PRT will be 21,165 per hour. The operational characteristics are as follows:

1. Distance headway 80 m.
2. PRT speed 30 miles/h.
3. No. of passengers 4 per Pod
4. Average trip length 1.6 km.
5. Route length 18 Km.

The study estimated that the capacity of the PRT system at Dwarka would be in the order of 27,000 passenger trips per hour which exceeded the peak hour demand capacity. The number of pods that they anticipated is 450 in order to meet the demand.

They came up with these parameters: demand, operational, and financial in Fig. 8.14. In it, they were based on net present value (NPV) and internal rate of return (IRR).

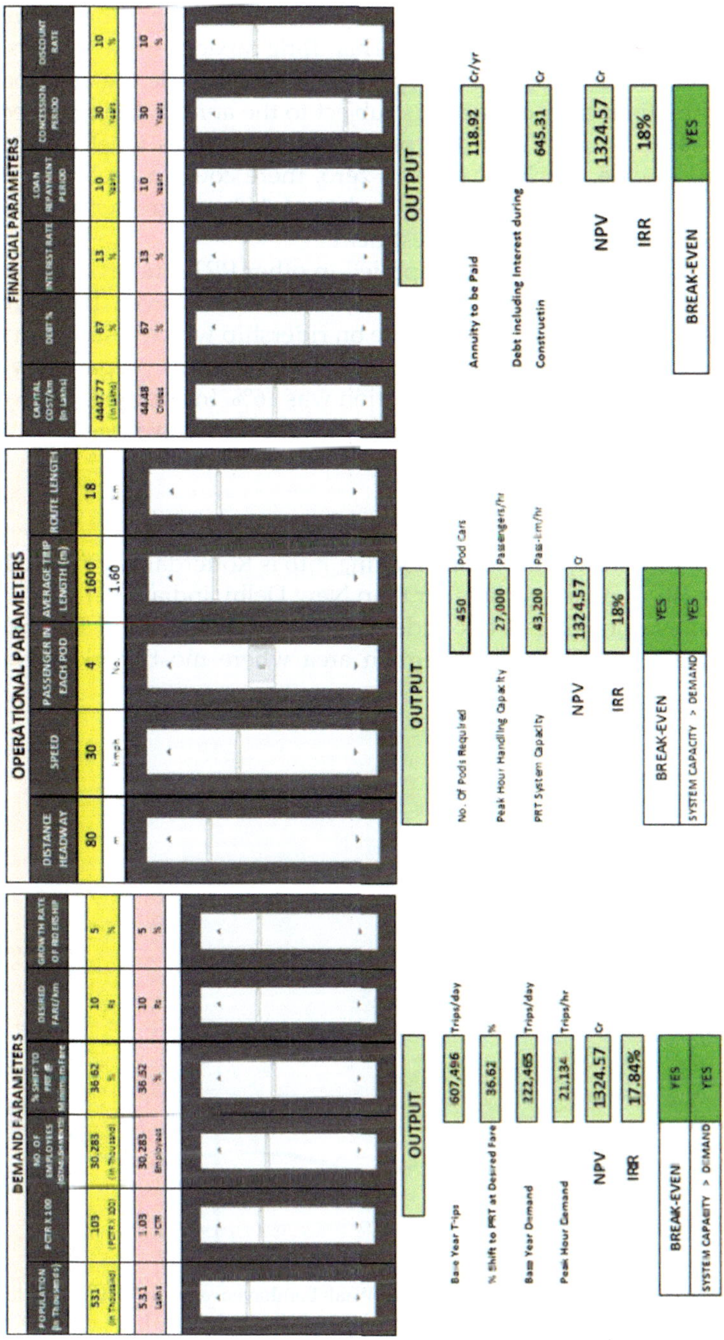

FIGURE 8.14 The layout of the demand, operational, and financial parameters (Sarkar and Jain, 2017).

The demand parameter is the base year trips, base year estimated demand, and peak hour demand. The operational parameter is based on the number of required pods including the handling capacity in passengers per hour. It also verifies whether the operation could handle the anticipated demand. The financial parameter is subject to the annuity being collected each year, IRR, and NPV on the project which verifies whether would breakeven. In case, the NPV goes to zero, there could a problem. They believe that a project with an IRR value of 16% or greater would be considered worthwhile on the investment.

Similar parameters have been applied in other PRT studies in the past. Amritsar and Trivandrum had done PRT studies. However, Trivandrum used different fare and the growth rate on ridership which resulted in IRR at 2% and 12%. As to the Amritsar situation which was prepared by the Fairwood Group and their IRR projection was 16%. In addition, Fairwood is based on the Ultra PRT system which a relatively costly system.

8.7.3 Rotterdam PRT Project

Another PRT study that we are looking into is Rotterdam in the Netherlands. The city is relatively smaller than New Delhi, India.

The case was looking at two older harbors Waalhaven and Eemhaven (Fig. 8.15) that are close to the urban area where most workers travel

FIGURE 8.15 Initial design of PRT in the Waal–Eemhaven area in connection to the existing public transport and road system (*blue* dot and *blue* dash-line) (Li et al., 2010). *PRT*, personal rapid transit.

by car or carpool to their work locations. The initiative of the Port of Rotterdam was to investigate the possibilities to enhance the public transportation for improved accessibility of parts of the port area. Because the transportation demand changes signicantly with time and space in this area, as during the peak hours or busy events, passenger car traffic causes congestion and long delay even though regularly it is not busy at all, the public transportation plays a minor role and companies in the port area serve their employees working in shifts. They compete for the limited road transportation capacity at low efficiency, which can be improved by PRT.

A feasibility study was launched to improve the accessibility of the area by PRT. As shown in Fig. 8.15, the area is divided by some nodes with connections to neighbor nodes, which forms a network. The existing public transport stations serve the entrance points of the network. With the aid of the existing transportation model containing the information on network infrastructure and multi-modal demands covering the entire Netherlands, PRT is added into the multimodal model for traffic simulation and modeling. A stated preference survey has been conducted to assess how travelers switch their transportation mode to PRT. Fig. 8.16 presents the average waiting-time and maximum waiting-time of passengers for different fleet sizes. Both maximum and average waiting-time decrease rapidly as the fleet size increases from 100 to 200. The further increase of fleet sizes will

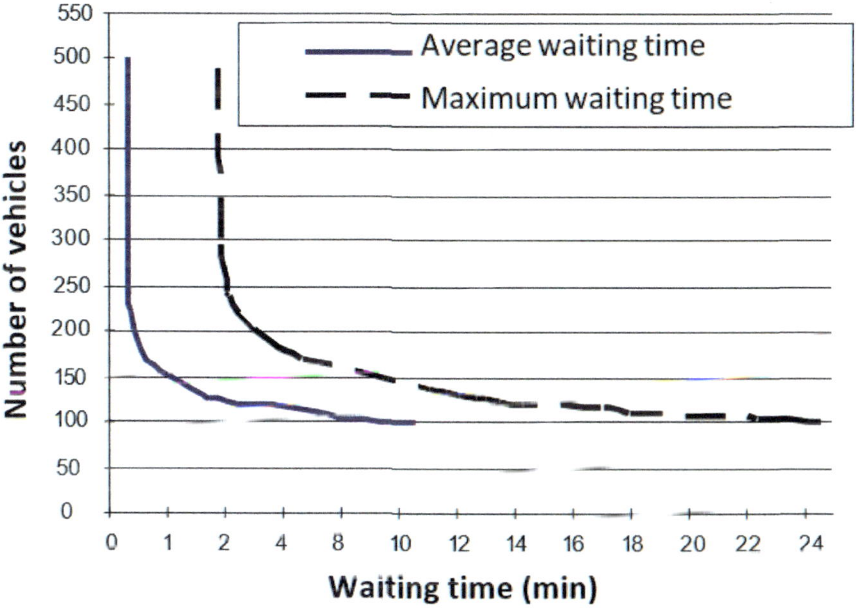

FIGURE 8.16 Waiting time versus fleet size (Li et al., 2010).

not produce much effects with an average waiting-time of 0.7 min and maximum of 3.8 min.

Going through the two case studies discussed here, there are a couple of concerns that have to be considered because they are only feasibility studies and the actual construction projects have not been done yet. Without having physical knowledge of the system, it would be hard for someone to imagine and understand all the advantages or appreciate the convenience. Though one can describe the system through pictures or articles, it would not be comparable to the actual system. In addition, there are many unknowns that are being assumed which resulted in not an accurate evaluation of the system. To get started it would be better to investigate the PRT as a feeder system to the main public transportation system.

Let us go back to what the Morgantown system had done. The Mayor of Morgantown has pointed out all these points listed below.

- It has proven to be a reliable system of automated transit that is relatively inexpensive to operate.
- It has offered on-time service rates for better than the bus system it replaced.
- It has eliminated much of the gridlock of traffic which existed in the hub of downtown Morgantown.
- It has proven to be safe, with no serious injuries reported since the operation began in 1975.
- Approximately 16,000 riders take advantage of the system daily.

8.8 Remarks for future development and applications

In the past couple of decades, there had been numerous experimental projects throughout the whole world. Nations and cities had been taken extensive actions on case studies on the viability of PRT. Although all the studies showed that it would most likely resolve the traffic congestion problems as well as cost savings on transportations, there have hardly been any actions on implementing the PRT systems except in some demonstration examples like the ones shown at Heathrow Airport by Ultra. It is an awfully expensive system as one of the case studies pointed out that each vehicle cost 140,000.00 Euro. To make it practical, we should be looking at vehicle costs in the range of $10K–$20K. In addition, the case study was done in New Delhi, India mentioned that the vehicle should be 80 m apart which should not be a reality to automatically controlled systems. Most designers should be looking at less than 1 s apart or 30 m apart. One meter apart should be the minimum. Meanwhile, instead of just making all these assumptions which might be a bit hypothetical, we should consider looking into deploying PRT in places like airports and the local hotels or local train stations where passengers are commonly stuck in the traffic going to the terminals through the hotel bus services or airport buses. PRT could potentially handle the situation very well because it does not go through the regular traffic and only runs on-demand which results in saving energy cost and time.

It is prudent to give it a try. Any time there is a change of paradigm, it is always bound to have oppositions. However, considering the current situations on climate change, serious environment concerns, and continuous traffic congestions problems in our urban areas, we should give a chance to test the PRT on an automated two-dimensional system.

References

Anderson, J., 1996. Some lessons from the history of personal rapid transit (PRT). In: Conference on PRT and Other Emerging Transit Systems.

Anderson, J.E., Some lessons from the history of personal rapid transit (PRT), 1996. Version 2. https://faculty.washington.edu/jbs/itrans/history.htm. (Accessed. 2020).

Anlauft, A., Frost & Sullivan, The Global Emergence of Personal Rapid Transit, London, December 1, 2008. /PRNewswire/– https://www.investegate.co.uk/frost—38–sullivan/prn/the-global-emergence-of-personal-rapid-transit—/20081201133100NN589/.

Babarit, A., et al., 2017. Investigation on the energy absorption performance of a fixed-bottom pressure-differential wave energy converter. Appl. Ocean Res. 65, 90–101.

Claire Marchand, I., 2015. Dream today, realty tomorrow. e-tech on transportation, https://storage-iecetech-prd-iec-ch.s3.eu-west-1.amazonaws.com/etech_issues/etech_03-2015.pdf. (Accessed: 2020).

Corio, E., 2009. Morgantown, W.Va., Bucks Unemployment Trend NPR. https://web.archive.org/web/20090501234821/, http://www.npr.org/templates/story/story.php?storyId=100890701.

EIA, In the United States, most petroleum is consumed in transportation. https://www.eia.gov/todayinenergy/detail.php?id=40752, 2019.

Fichter, D., 1964. Individualized automatic transit and the city. Providence [R.I.].

Goscé, L., Johansson, A., 2018. Analysing the link between public transport use and airborne transmission: mobility and contagion in the London underground. Environ. Health 17 (1), 1–11.

Harris, J.E., 2020. The subways seeded the massive coronavirus epidemic in New York City. https://www.nber.org/papers/w27021. (Accessed: 2020).

Inc., J., 2021. Solar powered PRT. https://www.jpods.com/. (Accessed: 2021).

Li, J., Chen, Y.S., Li, H., Andreasson, I., van Zuylen, H. (2010, September). Optimizing the fleet size of a Personal Rapid Transit system: A case study in port of Rotterdam. In 13th International IEEE Conference on Intelligent Transportation Systems, IEEE. pp. 301–305.

Raney, S. and S.E. 2005. Young, Morgantown people mover: updated description.

Sarkar, P.K., Jain, U., 2017. Benchmarking of personal rapid transit system (dynamic model). Trans. Res. Proc. 25, 4207–4218.

TRANSIT A.G., 1975. The automated guideway transit assessment on PRT by the Office of Technology Assessment at the US Congress, https://www.princeton.edu/~ota/disk3/1975/7503/7503.PDF. (Accessed: 2020).

Wikipedia, 2021a. Personal rapid transit. https://en.wikipedia.org/wiki/Personal_rapid_transit. (Accessed: 2021).

Wikipedia, 2021b. Morgantown personal rapid transit. https://en.wikipedia.org/wiki/Morgantown_Personal_Rapid_Transit. (Accessed: 2021).
Wikipedia, 2021c. Aramis (personal rapid transit). https://en.wikipedia.org/wiki/Aramis_(personal_rapid_transit). (Accessed: 2021).
Wikipedia, 2021d. Transit System funded by The Urban Mass Transportation Administration (UMTA), HUD reports. https://en.wikipedia.org/wiki/HUD_reportsMass. (Accessed: 2021).
Yin, H., Pao, F., Zadshir, M., Lou, J., Wei, Z., 2021. Building integrated photovoltaic thermal systems. U.S. Patent Application No. 63/248,502.

This mountain farm converts the marginal land into an arable farm for food supplies. At the bottom, there is a water reservoir (natural or man-made) accumulating rainwater and underground water. Another large water reservoir is built at the mountaintop for irrigation and energy storage. An innovative greenhouse, which has building integrated photovoltaic thermal systems, covers the greenhouse and is extendible to the reservoir surface. The photovoltaic electricity is used to pump water from the low reservoir to the high reservoir, which will then be used for irrigation as well as hydropower generation by releasing water from the high to the low reservoir. Solar heat from building integrated photovoltaic thermal systems will be used for greenhouse heating through a seasonal underground heat storage system. This farm enables versatile water and energy management toward the optimal food supplies through the greenhouse operation, a fish farm at a water reservoir, and arable farmland extension.

CHAPTER

9

BIPVT applications in farms

9.1 Introduction

According to the report from the United Nations Department of Economic and Social Affairs (R.b.D. Team, 2013), the current world population of 7.3 billion is expected to reach 9.7 billion in 2050. Such rapid population growth forecasts a food production increase of 70% in order to meet the food demand. More importantly, agriculture's environmental footprint must shrink drastically to safeguard the ecosystems that we humans rely on, as we have learned that the production and distribution of food have been a major contributor to greenhouse gas emissions and energy consumption (Gilfillan, 2013; Morrow and Delos-Reyes, 2013). Overwhelming scientific evidence shows the human release of greenhouse gases is changing the Earth's climate (Royal Society and Royal Academy of Engineering, 2018). Like food, energy, and water (FEW) have become more increasingly interlinked with some demanding requirements and constraints, there is a growing need for new, holistic, whole-system approaches to address the FEW nexus challenges.

One key challenge is that the arable land per person is shrinking with the increase of the population, which motivates us to consider those undulating land that has not been well cultivated before. The farmlands, particularly those in the rural areas of the United States, can be net exporters of food, but they often need to import energy and water resources to do so. Globally, food production accounts for about 70% of water abstractions and 90% of water consumption, respectively (Allan et al., 2015). Primary energy production and power generation account for roughly 10% of total worldwide water abstractions. About 30% of global energy use is for food production and its supply chain, and 8% for water withdrawal and transportation (Liu et al., 2017). In the United States, agriculture accounts for approximately 37% of total freshwater withdrawals and 81% of water consumption, considering >97% of the Earth's water is saltwater, and among the 3% freshwater 77% is locked in polar ice caps and glaciers (Rost et al., 2008). Crop production across the high plains aquifer (HPA)

Building Integrated Photovoltaic Thermal Systems: Fundamentals, Designs and Applications
DOI: https://doi.org/10.1016/B978-0-12-821064-2.00004-3

region in the central United States has an annual market value greater than $20 billion—approximately 10% of the entire US crop value (NASS-USDA 2012). Irrigation is essential to much of this crop production, as irrigated agriculture across the high plains accounts for 30% of all irrigated farmland in the United States, and 97% of high plains irrigation water is extracted from the HPA (Haacker et al., 2016). Due to extensive and inefficient irrigation that lasts for a long time in a region, groundwater levels across large sections of the HPA have been declining for decades, particularly in the southern section where the aquifer is thin and irrigation demand is high (Scanlon et al., 2005). Future decades are forecast to bring more widespread groundwater declines, effectively depleting broad regions of the HPA if current practices continue. Major reductions in water availability would result in enormous consequences for food and energy production.

Agriculture in the United States has long been a huge consumer of energy, directly in the form of gasoline and electricity, for the operation and maintenance of farms and growth and production of crops; and indirectly in energy-intensive inputs such as fertilizer and pesticides. Large amounts of fossil fuels are often needed to supply energy to sustain the farming industry in the United States, which results in annual emissions of 600 million tons of greenhouse gas (Allred et al., 2015). The US agricultural sector consumed 1714 trillion British thermal units (Btu) of energy in 2014 (Allouche et al., 2014). Although farms can be a major energy consumer, they can produce renewable energy. For example, corn is used as an energy input for ethanol production. Other on-farm renewable resources including photovoltaics (PVs), wind turbines, and methane digesters have been used in recent years as well (Hitaj and Suttles, 2016). As both a consumer and producer of energy, the agriculture sector plays an important role in achieving the goals set by the clean power plan for carbon emission reductions toward America's clean energy future by maximizing crop yields and minimizing the consumption of fossil energy. Technologies that improve energy efficiency and boost renewable energy development on farms are in high demand now and will be in the future.

Even though energy consumption varies across principle commodities, crop operations consume much more energy than livestock operations, and energy expenditures for crops account for a higher percentage of farm operating costs (Ling and Poon, 2013). Therefore, in recent years greenhouses are being increasingly adopted to extend the growing season and increase food production. Greenhouses also allow farmers to grow a greater variety of food, which saves a significant amount of energy on long-distance food transportation. All greenhouses collect solar energy to maintain a suitable indoor environment (e.g., temperature, humidity, light) for the healthy growth of the plants within. However, a wide variety of greenhouse designs, materials, construction, and production methods are available, which determines the cost, operation, and performance of the greenhouse. A large number of greenhouses use solar power in a passive matter—letting sunlight passing through transparent plastics or glasses (e.g., maximizing solar gain and minimizing heat loss) yet without energy redistribution in the greenhouse. Such passive solar greenhouses do not work well in colder climates or areas with long periods of cloudy weather (Tong et al., 2013). Semiactive solar greenhouses built in recent years often use water or air (as a thermal mass) to store solar energy at high temperatures and release or redistribute the energy at low temperatures (Ling et al., 2014). However, such heat exchange systems take a lot of space and are not energy efficient. The structure of the passive or semiactive greenhouses often consists of single or double layers of ultraviolet-protected plastic, fiberglass, or glass over either high-carbon steel pipes or aluminum or wood frames. These greenhouses are easy to build; however, they mostly do not integrate any PV panels.

Considering that solar energy is abundant yet with large temporal and geographical variations in most areas of the world, optimized use of solar energy in farms can be very beneficial for both food production and energy sustainability. As compared to the traditional plastic or glass greenhouses, greenhouses integrating PV systems most likely have a high upfront construction cost, but they are more durable, energy-efficient, and can reduce fossil fuel energy usage significantly in the long run. Therefore, the concept of net-zero energy greenhouse has motivated industrial experts in PV systems and agronomists to work together on the design of more sustainable greenhouses that integrate solar PV systems and actively use solar energy. In recent years, a few pioneer companies such as Onyx Solar (Agrawal and Tiwari, 2010), Artigianfer (Skoplaki and Palyvos, 2009), and Soliculture (Solar Panel Temperature – Facts and Tips, 2015) have started exploring various advanced technologies to manufacture PV-integrated greenhouses. These companies share the same goal of providing the agricultural industry with a clean source of renewable energy and sustainable technologies that are symbiotic with optimized growing practices.

Solar energy is usually collected and utilized in the form of electricity and heat; therefore, scientists have never stopped conducting research on how to improve solar panel efficiency and how to store thermal energy more effectively. In the aspect of developing more energy-efficient solar panels, Lumb *et al.* created a multijunction solar cell on a GaSb substrate that can effectively convert the long-wavelength photons and therefore gave an implied cell efficiency of 44.5%, much higher than 25% from the commonly used solar panels (Lumb et al., 2017). Mandal *et al.* developed a room-temperature "dip-and-dry" technique that exhibits excellent, wide-angle solar absorptance and promising mechanical and thermal stabilities at temperatures up to 200°C. This technique is claimed to be simple, inexpensive, and environmental-friendly (Mandal et al., 2017). These latest developments in solar panels and absorbers would greatly improve solar cell efficiency and lower the cost of solar energy; however, it still takes time to commercialize these technologies. On the other hand, if these new technologies can be ready in a year or so, the advanced BIPVT panels developed by this team can also quickly adapt them for higher-efficiency solar panels.

In terms of how to store thermal energy and redistribute it to when and where heat is needed, ground to air heat transfer (GAHT) system was developed by CERES Greenhouse Solutions (Ibrahim et al., 2011). The GAHT system uses a fan to draw the excess heat underground where it is stored in the soil; when the greenhouse needs heat at night or over the winter, the fans draw cool air through the warm soil back into the greenhouse. However, the main components of the GAHT system including fans, airflow, and underground pipes are not as efficient as our PCM-geothermal heat well combined with a fluid circulation system.

Greenhouse heating is often carried out by either passive or active methods. Passive heating uses water, concrete, or stones as thermal mass to store thermal energy at high temperatures for later use at lower temperatures (Santamouris et al., 1994). Examples of active heating for greenhouses include ground geothermal water and earth-air heat exchanger (Bargach et al., 2000; Jain and Tiwari, 2003). These heating methods help extend the growing season to some extent, but none of them are effective enough to ensure temperature and humidity uniformity in the greenhouse, especially those in colder climates, due to the limited thermal storage capacity of the thermal mass or localized storage/dissipation units. In fact, many greenhouses still need to use fossil fuels as supplementary heating sources to protect plants from extreme cold weather. Secondly, most PV modules in production are based on crystalline silicon wafer technologies. The electricity conversion efficiency of silicon solar modules for commercial applications is commonly about 12%–25% (Müller et al., 2015). However, the majority of the incoming solar energy is absorbed as heat (He et al., 2006).

The BIPVT system, on the other hand, appears as an exciting new technology because it harvests both electrical and thermal energy in a synergetic fashion (Agrawal and Tiwari, 2010). The most common BIPVT systems are realized through a heat transfer fluid in an open-loop (usually air) (Nagano et al., 2003; Yang and Athienitis, 2015; Yang and Athienitis, 2014) or closed-loop (usually liquid) configuration (Ghani et al., 2012; Corbin and Zhai, 2010, Anderson et al., 2009). Generally, the closed-loop configuration is more efficient than the open-loop due to the high thermal conductivity of liquid compared to air (Kumar et al., 2015). Most closed-loop BIPVT systems employ water tubes for cooling and thermal energy collection. Normally, the water tubes are embedded in insulation material and covered by an absorber that is in contact with the PV elements above (Ibrahim et al., 2011). However, such designs commonly exhibit poor heat conduction due to the small contact area between the absorber and water pipes, and thus unable to efficiently cool the PV cells to increase the PV efficiency.

The above drawbacks limit the energy efficiency of a greenhouse and delay its break-even point. Advanced BIPVT systems coupled with a high-capacity energy storage/dissipation system can be used to design year-around, net-zero energy greenhouses for the production of healthy, economic food. Fig. 9.1 illustrates a sustainable greenhouse system using solar and geothermal energy, which exhibits the dual synergetic benefits: (1) improved energy harvesting efficiency toward net-zero energy agriculture and; (2) smart indoor environment control of light, temperature, and moisture for enhanced farm productivity. This novel greenhouse design consists of five components: (1) advanced building-integrated photovoltaic thermal (BIPVT) panels for electricity and heat harvesting; (2) a geothermal system for heat storage and dissipation using phase change materials (PCMs) in the geothermal heat well; (3) a battery array for electricity backup and storage; (4) a fluid circulation system with glycol for environment control and thermal management with a geothermal system; and (5) a sensing and

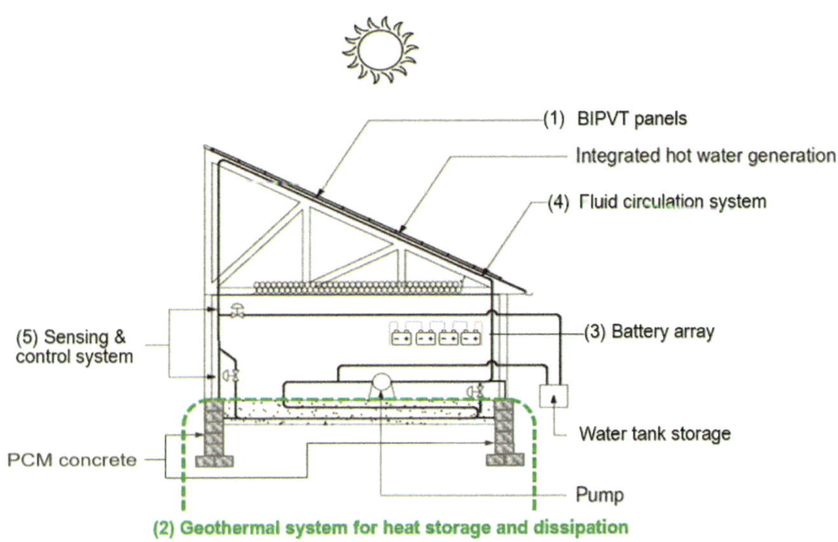

FIGURE 9.1 Schematic illustration of the net-zero energy greenhouse using BIPVT panels and geothermal energy. *BIPVT*, building-integrated photovoltaic thermal.

control system for smart indoor environment monitoring and management with a heat pump.

9.2 Solar powered greenhouse

9.2.1 Background

For a new technology to be effective and economically feasible, a holistic approach is essential to assure that the interactions and potential synergies among the various physical, social, and economic aspects are properly understood, modeled, and where possible exploited while eliminating or minimizing any potentially detrimental consequences or interactions. Specifically, innovative FEW systems have to start with recognizing and specifying various requirements and constraints on farmland, understanding the fundamental relationships among the design variables, and then exploring the possible solutions. Considering multiple constraints, our design philosophy is to maximize the FEW yields while minimizing resource inputs that are external to the farm. Therefore, our ultimate objective is an energy-water independent farm; ideally, the farm can supply FEW to the community without external inputs.

Over the past decades, solar energy harvesting has evolved as a promising solution for energy and environmental challenges as well as climate change threats (Chen and Yin, 2016; Chen et al., 2016; Choi and Fthenakis, 2010). Taking a holistic view of the farm operation and incorporating solar energy technologies within the greenhouse and water infrastructure surface will overhaul today's farm operations and ultimately lead to a strategy that achieves a zero-energy farm. Solar energy has mainly been harvested in form of PV and heat. The water reservoirs and greenhouses on farmland become ideal places for solar energy harvesting. The BIPVT system has become a very promising technology, which not only harvests the heat but also increases the PV efficiency by maintaining the PV's temperature in an ideal range (Anderson et al., 2009; Debbarma et al., 2017; Gautam and Andresen, 2017). However, there are still two problems with the BIPVT systems:

1) The PV system can only harvest solar energy in the daytime, and energy storage often becomes an issue for the continuous operation of farming activities. Currently, solar energy typically feeds back to the grid, which relies on external power infrastructure and is therefore not aligned with our goal of a net-zero energy farm. It is also possible to use batteries or fuel cells to minimize the interruption of farming work; however, such practices are expensive and commonly limited to small-scale applications. Instead, we propose to use the excessive PV electricity to pump

water from a lower reservoir (LR) to a higher reservoir, and the water will be used for irrigation by simple gravity. This is an ideal solution for energy use and farm operation.

2) The PV efficiency exhibits a bottleneck at 20%˜30% due to the heat generated by the silicon solar cells despite some research-based solar cells reaching a higher efficiency. The heat generated from the PV cells (accounts for 60%–70% of the solar energy absorbed by the PV cells) is often not utilized and dissipated to the air. We propose to harvest the wasted heat energy and store it in the earth, which is then used to promote plant growth and save energy in greenhouse heating.

Fig. 9.2 shows a BIPVT solar panel that the Author Yin has developed in his NSF Career Project CMMI 0954717, which integrates a transparent waterproofing protective layer, PV cells, functionally graded material (FGM) layer with tubes cast inside, and a lightweight concrete substrate into a composite system. The FGM layer consists of composite materials that gradually transform from a metal-dominated region to a polymer-dominated phase. The tubes are cast in the top part of the FGM layer, where the high aluminum concentration creates high thermal conductivity so that heat can be immediately transferred to the working fluid in the tubes in all directions. The double serpentine configuration of the tubes makes the temperature distribution in each solar panel as uniform as possible, thus maximize the PV cell efficiency. The proposed solar panel will be integrated into the covers of the reservoirs, which will be connected with a fluid circulation system and a flow control unit. In combination with the geothermal system, the BIPVT system will harvest both the electricity

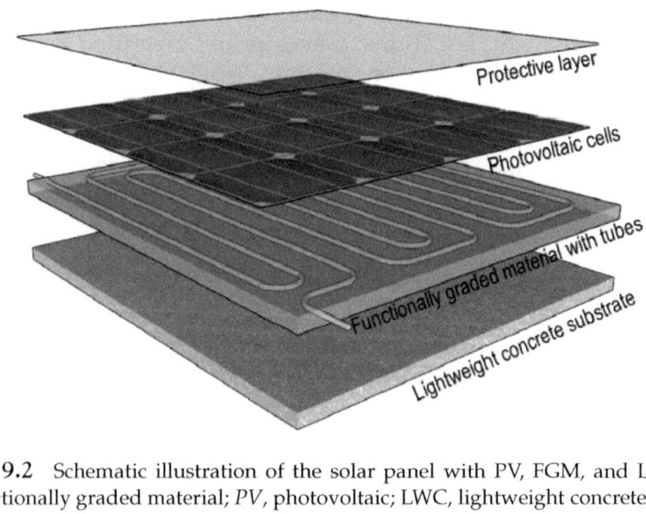

FIGURE 9.2 Schematic illustration of the solar panel with PV, FGM, and LWC layers. *FGM*, functionally graded material; *PV*, photovoltaic; LWC, lightweight concrete.

and the heat at high efficiencies in both summer and winter, with better utilization of renewable solar and geothermal energy. Notice that for the growth area of the greenhouse, sufficient light is required and a new BIPVT system will be developed to balance the needs of energy harvesting and plant growth.

Compared with energy independence, the water independence of a farm will be more challenging. While we can improve the energy independence of the farm system, a farm will never be "water independent" from its surroundings. Instead, we aim to improve: (1) efficiency of on-farm water use and (2) on-farm water storage strategies to buffer against droughts. The greenhouse itself will allow us to optimize the water use, while we accelerate and maximize crop production. Water loss from the system will be minimized through precision irrigation techniques, which will reduce soil evaporation loss and leakage of water out of the crops' root zone. Our reservoir design will be optimized to deal with site-specific climate characteristics, considering both historical and projected precipitation variability.

Farming and agriculture consume a great amount of water and energy for crop growth and production. Because water/precipitation is often not evenly distrusted spatially and temporally, irrigation is often needed at various growth periods of the crops to satisfy their moisture need (Allan et al., 2015). An irrigation system often consists of conduits, pipelines, sprinklers, and water is delivered to the point of interest either by a pump powered by electricity or using the natural terrain (i.e., the potential energy). Such traditional farming practices have a low energy utilization efficiency because the crops/vegetation constantly interact with the open space environment for heat, moisture, and light exchange. To increase food production, greenhouses are being increasingly adopted because they collect solar energy to maintain a suitable indoor environment (e.g., temperature, humidity, light) for better growth of the plants within. However, a large number of greenhouses use solar power in a passive matter – letting sunlight passing through transparent plastics or glasses (e.g., maximizing solar gain and minimizing heat loss) yet without energy redistribution in the greenhouse (Hassanien et al., 2016; Xue, 2017).

Overall, given the rapidly increasing world population, climate change, and limited arable land resources, we are under severe pressure for innovative solutions that improve the sustainability, efficiency, and resiliency of the FEW systems. A synergistic solution that utilizes renewable, ubiquitous solar energy to repower the microwater cycle as part of FEW systems on a farm will be of high value to our next generation. A BIPVT coupled with a geothermal system can be integrated with an active solar greenhouse. In addition, low-cost, high-efficiency PV panels can also be used to cover the reservoir surface (instead of taking the space off the limited land) for electricity generation and to reduce evaporation. The PV systems collect electricity and heat simultaneously: the electricity can pump water from an LR to a high reservoir (HR) for irrigation, power the greenhouse for general operation (e.g., lighting, sensing, and control), or be stored in a battery array for emergent needs under inclement weather conditions. Heat can be transferred and stored in the underground geothermal storage/dissipation unit for daily, seasonal, year-round greenhouse heating. Together with smart sensing and control systems for water and energy management throughout the farmland, our system would enable more sustainable farming and maximize crop yields.

9.2.2 Solar greenhouse design

Although greenhouse can be designed with different sizes and styles, Fig. 9.3 shows a schematic design of a small-size energy-efficient greenhouse that utilizes BIPVT coupled with geothermal energy for the dual synergetic benefits: (1) improved energy harvesting efficiency toward zero-net-energy agriculture, (2) smart indoor environment control of light, temperature, and moisture for enhanced farm productivity and extended growing season. It can be scaled up depending on the farm's configuration such as the arable land area, water availability, topographic form, as well as the demands of food and energy production.

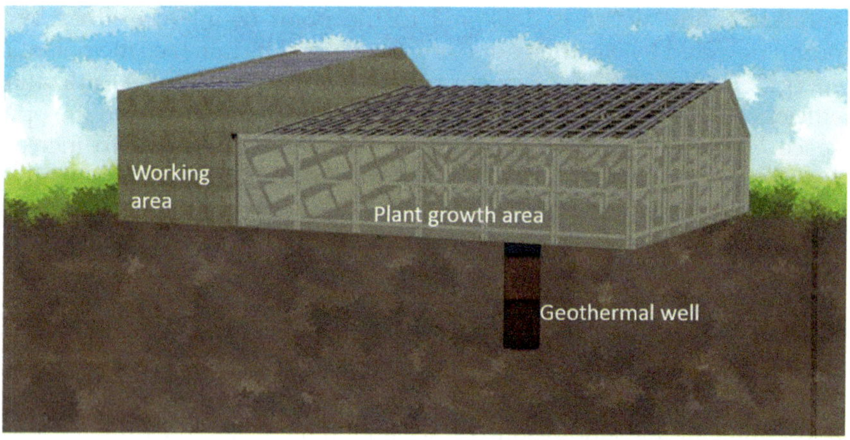

FIGURE 9.3 A greenhouse with a working area (10 m × 5 m) and a growth area (10 m × 15 m) covered by a BIPVT and equipped with a geothermal well for seasonal heat utilization. *BIPVT*, building-integrated photovoltaic thermal.

This greenhouse (10 m × 20 m) includes a growth area of 10 m × 15 m which is covered by a special BIPVT system, which will be elaborated in the next section, and a work area of 10 m × 5 m which is covered by a conventional BIPVT system. The BIPVT systems are connected to a geothermal well to store energy for seasonal heat utilization. The BIPVT harvests both electricity and heat for farm energy supply. As shown on the cover page of this chapter, the PV electricity will be used to pump water from the LW to HW for irrigation and energy storage. In the evening, the water can be released for hydropower generation, which can power the greenhouse for the 24/7 growth of the plant for accelerated food production.

Because the plant growth requires light, it will compete with the energy harvesting for the roof area. However, the optimal light intensity and spectrum for plant growth may change with different species of plants, using artificial lights may provide flexibility to tailor the light inputs for the maximum growth potential. As the energy loss in the process of sunlight-electricity-storage-light is high, it is more economical to combine nature sunlight and supplementary light for the optimal benefit-cost ratio in a greenhouse operation. Semitransparent BIPVT panels can be used.

9.2.3 Building-integrated photovoltaic thermal panels

The "semi-transparent" roof consists of columns of BIPVT panels and transparent glass panels arranged alternatively (Fig. 9.4), with LED lights

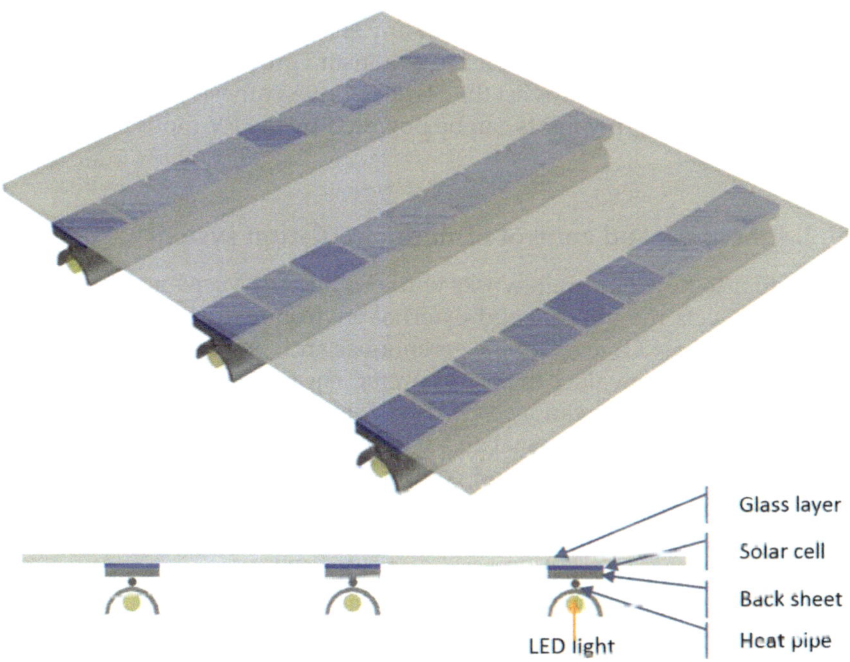

FIGURE 9.4 A building integrated photovoltaic thermal panel for greenhouse application.

attached to the ceiling for tailored needs of the plants inside the greenhouse. Because 25%–30% of roof area is taken by PV cells, lighting deficit may happen to some plant growth. The LED lights compensate the light inputs with the specific spectrum as the plant needs.

The BIPVT panel integrates a protective layer, PV cells, and transparent back sheet and will be developed at Sustainable Product Development Lab (SPDL). The BIPVT panel is designed to be multifunctional: the solar cells generate electricity, and the heat pipes under the backsheet harvest heat, which is dissipated into the ground by a geothermal system. The foundation of the greenhouse is constructed by PCM concrete, which can store excess heat during the day and release it at night when the temperature drops.

As compared to a regular solar panel, our BIPVT solar module offers higher PV efficiency by regulating its temperature by thermal energy harvesting and serves as both the roof structure and the electricity & heat harvester for the greenhouse. To allow and maximize natural sunlight to pass through the roof for plant growth, BIPVT panels and transparent glass panels will be assembled in alternative columns perpendicular to the east-west direction. The area ratio of the BIPVT versus transparent glass will be optimized based on: (1) primarily the electricity and thermal energy needs of the greenhouse at a geographical location, and (2) secondarily the light the plants require. To compensate for any shortage of light the plants may need at night or on cloudy days, we will install grow lights that emit blue- or red-spectrum lights to assist the plants' photosynthesis process (Ibrahim et al., 2011.). The grow lights can be powered by the PV modules.

9.2.4 Sensing and control of fluid circulation system

Similar to how our brain works with our body, the sensing, and control system detects the internal and external environment (e.g., temperature, humidity, light, CO_2 level) of the greenhouse and then adjusts these values according to preset values the plants prefer. The temperature and humidity on leaves of the crop can also be used for growth health monitoring as they are reported to be associated with crop diseases (Park et al., 2011). Components in the sensing and control system will include temperature sensor, humidity sensor, leaf temperature sensor, leaf humidity sensor, wireless sensor node, relay nodes for automatic control, and data server. A wireless sensor network (Mekki et al., 2015) will be developed to monitor and control not only parameters of the environment (e.g., temperature, humidity, light, CO_2 level) but also the fluid circulation and electronic system (water pressure, flow rate, thermal energy input, and output, electricity input and out, battery life, etc.). Manual operation of the sensing and control system will be available for maintenance purposes.

The fluid circulation system and the electronic circuits connect the various components of the greenhouse, and they should be connected seamlessly for greenhouse operation. The fluid circulation system runs through the BIPVT panels, the PCM-concrete wall structure, and the PCM-concrete geothermal heat exchanger. The electrical components include the BIPVT panels, the inverters, light switches for grow lights, and other illuminative lights, battery array, pumps, etc. Connections among these different fluid circulation components need to be done carefully to ensure continuous heat transfer and avoid leakage. Should there be any leakage during the greenhouse operation, the sensing and control system will send an alarm to the operator for repair. The same principle applies to electronic circuits.

Once the key components (i.e., the BIPVT panel, wall structure, and geothermal heat exchanger) are all connected, the operation of the greenhouse (i.e., generation and consumption for electricity, thermal energy, and water resource consumption) will be evaluated systematically and then optimized for each crop type growing in the greenhouse. Once the growth curve with the environmental control and fertilization is established and validated, it can scale up for mass accelerated production of food.

9.2.5 Geothermal heat well system

Geothermal energy is derived from the thermal energy generated and stored in the earth. The energy is accessible since groundwater transfers the heat from rocks to the surface either through boreholes or natural cracks and faults. Geothermal energy is a renewable resource because there is a constant heat flow to the earth's surface and the atmosphere from the immense heat stored within the earth. Geothermal energy is a commercially proven renewable form of energy that can provide constant power and heat. Direct use of geothermal fluids with temperatures from a few degrees C to 150 degrees C can be found in space and district heating, greenhouse heating, snow melting, bathing, and swimming. For heating, direct use applications depend on technical advances of heat exchangers in other sectors as the geothermal fluids are often not suitable to be distributed to district heating networks.

In recent years, district heating systems incorporated with shallow geothermal energy have attracted increasing research interests (Ozgener et al., 2005; Hepbasli, 2010; Yildirim et al., 2010). Ground source heat pumps (GSHP) use shallow geothermal energy which is available almost everywhere. They convert the low-temperature geothermal energy to thermal energy at a higher temperature which can be used for space or water heating (Spitler and Gehlin, 2015; Luo et al., 2015; Emmi et al., 2015). Usually, a refrigerant is used as the working fluid in a closed cycle, and an antifreeze solution is circulated inside a closed coil and exchanges heat with the heat source/sink through the ground heat exchanger. Despite the successes in the past and continuous growth, research and development in GSHP are mainly focusing on increasing the efficiency of GSHP systems and reducing costs. The main development areas include ease of maintenance and repair, improved control systems, more efficient working fluids, and increased efficiency of auxiliaries, such as pumps and fans. Ground collectors need to be improved by optimization of design and grouting material. Currently, plastic tubes are commonly used for ground collectors which offer low cost and corrosion resistance but show low thermal conductivity. To overcome the low-efficiency issue of GSHP use of geothermal energy, bidirectional low-temperature networks have been proposed in recent years to lower heat fluid temperature in the networks to ambient temperature level (Kräuchi et al., 2015; Ruiz-Calvo et al., 2017). In the case of heating demand, the circulation pump of the building withdraws water from the warm line, uses it in a heat pump to reach temperatures suitable for space heating, and then discharges the cooled water to the cold line. In the case of cooling demand, the system works in the other direction. Depending on the heating and cooling demands of the connected buildings, the fluid flow in the network can change direction.

Despite the previous efforts, there is a significant need for further research to optimize the operations of low-temperature geothermal-based bidirectional energy storage (Prasanna et al., 2017). For example, the total difference between heating and cooling flows in a bidirectional thermal energy system needs to be balanced by external sources (e.g., central heat pumps and chillers, solar thermal plants, or seasonal storage facilities). The requirement for external energy input is also reflected in the flow of fluids in the pipe systems, which is often driven by considerable amounts of electricity especially when there are significant diurnal and seasonal heating and cooling demands. In addition, in many cases, the heat transfer between the pipe system and the surrounding soil and the thermal fluid is limited by the low thermal conductivity and temperature difference.

The temperature of a farmland changes with the season, climate, and geographic conditions. Generally, the air temperature is much more dynamic than the ground temperature, which may reach a nearly constant value beyond a certain depth. Because both air and ground temperatures have significant contributions to plant growth, understanding the heat transfer processes (e.g., air to the ground, and BIPVT system to the ground) is critical to the design and operation of the BIPVT greenhouse. Using geothermal energy for seasonal energy storage and utilization may provide an economical and sustainable way for temperature control of greenhouse.

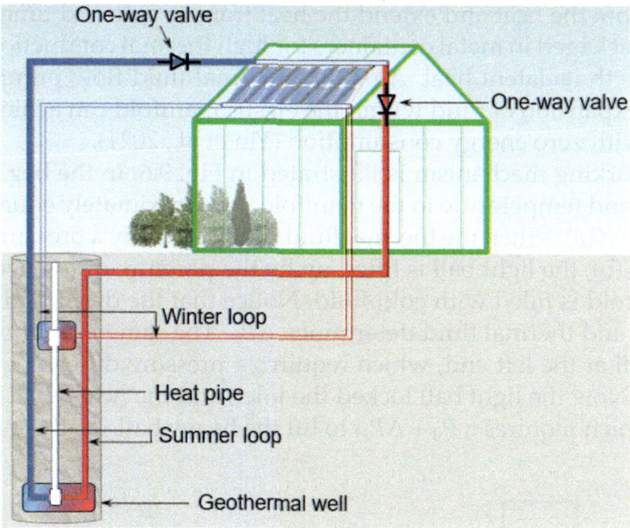

FIGURE 9.5 Schematic illustration of the heat pipe-based geothermal system.

Fig. 9.5 schematically illustrates the connection of a geothermal heat well system. For a 100′ shallow geothermal well, the top 20′ will be insulated for heat leakage, and the bottom 80′ heat pipe installed below the insulation pipe. For wintertime, the heat transfer follows the red and blue unfilled loop; for summertime, the heat transfer follows the red and blue filled loop where the one-way valves allow automatic control of heat exchange.

Inspired by human skin and circulation systems, the proposed technology will introduce a thermal fluid circulation system into the greenhouse and building skin and connect it with a geothermal well for thermal management of the greenhouse environment. In summer, our technology can take heat off from the greenhouse and building surface, and store the excessive thermal energy to the ground reducing greenhouse temperature. In winter, it can extract and direct the heat from the underground geothermal up to the greenhouse because hot air rises naturally. Although additional heat pumps and refrigeration cycle systems can tune the system performance for active control, a novel bidirectional passive heat exchange system can automatically condition greenhouse temperature by exchanging thermal energy with the ground using the volume change with temperature.

Because the thermal conductivity of the soil or stone underground is very low, the heat transfer between the heat pipe and ground is slow. To improve the heat transfer performance, PCMs will fill in the geothermal well to store the heat and extend the heat transfer rate and time span. The PCM is packaged in metal containers for high thermal conduction and heat storage with its latent heat. A passive thermal fluid flow pumped by the thermal expansion of fluid within the elastic manifold can achieve passive cooling with zero energy consumption (Yin et al., 2021).

The working mechanism is illustrated in Fig. 9.6. In the beginning, the pressure and temperature in the manifold is approximately equal to P_0 and T_0 in Fig. 9.6(a). When the thermal fluid is turned on by a pressure $P_0 + \Delta P_L$ in Fig. 9.6(b), the light ball is lifted up by the pressure difference ΔP_L, and the manifold is filled with cold fluid. Notice that the density difference of light ball and thermal fluid determines ΔP_L. The same is applicable to the heavy ball at the left end, which requires a pressure difference of ΔP_H to lift it up. Now the light ball locked the inlet, and the heavy ball locked the outlet, which requires a $P_0 + \Delta P_H$ to lift the heavy ball. As the temperature

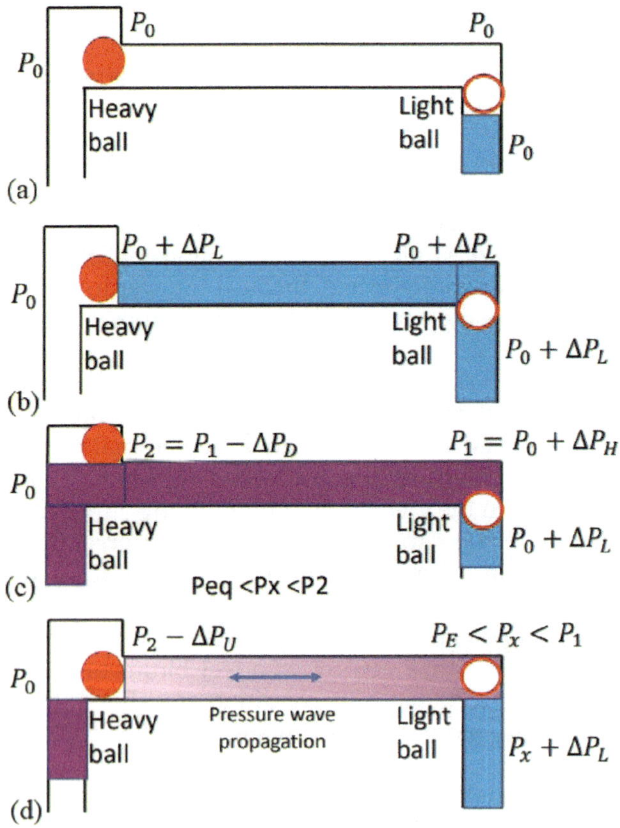

FIGURE 9.6 A passive thermal fluid flow pumped by thermal expansion or phase change of fluid.

increases in the manifold caused by heat energy harvesting, the pressure in the manifold increases to $P_1 = P_0 + \Delta P_H$ in Fig. 9.6(c), so that the heavy ball is lifted up and hot fluid is squeezed out and pressure at the left starts to drop down to $P_2 = P_1 - \Delta P_D$, and the ball falls down and locks the outlet again in Fig. 9.6(d). Because the manifold is elastic, the pressure wave will pass over from the left to right to release the pressure in the manifold. When the low pressure front reaches the right end at P_E, the pressure difference will suck up the light ball again, and cold fluid will fill into the manifold again. Therefore, the pressure change caused by the thermal expansion of the fluid will power the fluid circulation and the hot liquid will be quickly pushed to the geothermal well and become cold liquid for continuous operation.

The geothermal and BIPVT systems work together in the proposed greenhouse complementarily: in summer, the heat from the roof will be taken by a thermal fluid down to the ground which cools down the roof temperature for higher PV efficiency; in winter or at night, the heat will be pumped back to the ground surface and roof for temperature conditioning in the greenhouse, which can also be used for snow removal or defrosting on the roof to recover the PV efficiency.

9.2.6 Energy storage integrated in farm's microwater cycle

Similar to the sun driving the grand water cycle in nature, the BIPVT integrated farm system reproduces the microwater cycle in the mountainous farmland. Surface and underground water are collected constantly at the LR. BIPVT covers the reservoir surface, which reduces the water evaporation, and provides electricity to pump water from LR to HR. The water at HR can be used for irrigation automatically by gravity and the excess water may return to LR again. This microcirculation system also provides an opportunity for high efficient, economic energy storage and generation by water pump and regeneration. The efficiency can be comparable to that of leading battery technologies but can reach a much higher scale. It is important to understand the boundaries and correlation of energy and water transport among the FEW components.

For a medium or large water pump (>10KW), its power efficiency can reach >90%, and similarly, a water turbine can transfer potential energy to electricity at 80%~90%. Therefore, overall solar energy storage and release efficiency by the microwater cycle can reach >70%.

For example, given the water head between LR and HR, for example, $H = 100$ m, a BIPVT system with $P = 10$ KW may power the water flow at

$$Q = \frac{\eta P}{\rho g H} \approx \frac{0.9 \times 10^4}{9.8 \times 10^3 \times 100} \approx 9 \times 10^{-3}\ \mathrm{m^3/s} \tag{9.1}$$

where the water pump efficiency of 90% is used. Assume the equivalent sun power availability is 6 hours at the full capacity, the total water volume reaches

$$V = Q.T = 0.009 \times 3600 \times 6 = 194.4\ \mathrm{m}^3 \tag{9.2}$$

which is equivalent to the size of a regular water tank at a diameter of 10 m and height of 2.5 m. The stored potential energy of water is:

$$E = V\rho gH = 194.4 \times 10^3 \times 9.8 \times 100 = 1905\mathrm{MJ} = 53\mathrm{Kwh} \tag{9.3}$$

Assuming the water turbine's efficiency at 80%, 42KWh of electricity can be used in the evening time for a 10KW solar system, which in general covers a roof area of ˜600 sq ft. The calculation can be scaled up. Notice that if the water reservoir surface is used for BIPVT installation, the power capacity can be much higher with the higher efficiency of the water pump and turbine.

If the BIPVT covers ˜1.5 acres of the water reservoirs, the farm may provide ˜1MW power supply for farm operation and the neighboring community. The technology can transform the marginal land into energy and food farm.

9.3 Nexus of food-energy-water system

The working principles of the BIPVT system in a farm are as follows: (1) the advanced BIPVT panels will be installed on the south-facing roof of a solar greenhouse to take ˜25% of plant growth area and 100% of working area as well as partially covering water reservoirs, for PV and thermal energy harvesting; (2) the electricity can be stored in water reservoir or a battery array for plant growth lighting at night; (3) the hot water will run through the fluid circulation system down to the underground geothermal unit (where the thermal energy will be stored in summer and released in winter) and then be circulated back to the BIPVT panels for roof temperature control; (4) sensing and control systems will be installed to monitor energy usage and distribution for optimal operation and production of the greenhouse by controlling the crop-environment dynamics. Notice that for cold climate zones, the fluid circulation system can use glycol as a thermal liquid, which has been widely used for anti-freezing purposes. Thermal energy harvested in the form of hot water from the BIPVT panels will be captured and transferred to the PCM-based concrete blocks placed underneath the greenhouse, and then absorbed and stored by the semi-infinite earth body (Ibrahim et al., 2011).

Fig. 9.7 shows the schematic of water and energy flow and circulation for food production through the FEW systems. The temperature of a farmland

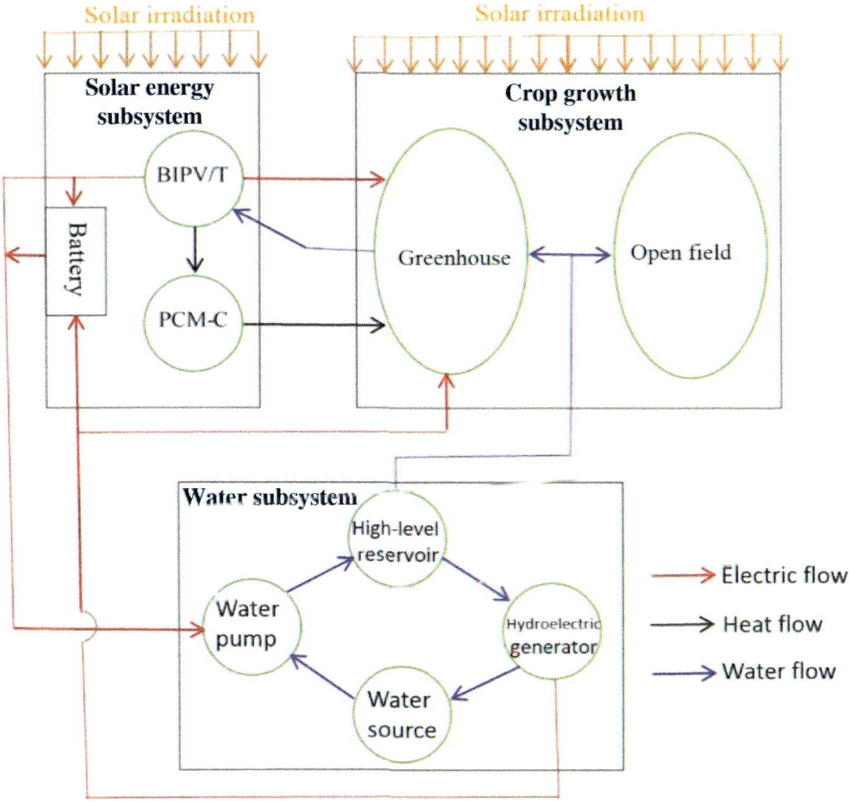

FIGURE 9.7 Schematic illustration of the relationship of the BIPVT integrated FEW systems. *BIPVT*, building-integrated photovoltaic thermal; *FEW*, food, energy, and water.

changes with the season, climate, and geographic conditions. Generally, the air temperature is more dynamic than the ground temperature, which may reach a nearly constant value beyond a certain depth. Because both air and ground temperatures have significant contributions to plant growth, understanding the heat transfer processes (air—the ground, and BIPVT system—the ground) is critical to the design and operation of the proposed greenhouse. Based on the calculation and measurement of the energy flow in the FEW systems, the physical mechanism of the microcirculation among the components of the FEW system will be established in order to optimize the energy and irrigation efficiency and food productivity in the system. In summary, the advanced BIPVT technology combined with the powerful PCM concrete geothermal system will work together coherently to achieve optimized electricity generation and heat storage as well as food production. The BIPVT-geothermal system targets clean-energy resources for a year-round operational, energy-efficient greenhouse. Hydroponics can be used in greenhouse for large-scale applications. Depending on the scale, more water reservoirs and greenhouse can be added for system resilience.

On the other hand, crop productivity is in general the most important indicator of the effectiveness and efficiency of the FEW systems. As such, the target of system optimization is to determine the maximum crop production under energy constraints or to determine the minimum energy consumption for a given crop production level. The influences of the BIPVT material and its resulting shading on the spectral distribution of solar radiation entering the greenhouse, and the radiometric and thermal properties of the BIPVT panel are important environmental factors to affect the greenhouse's internal climatic condition. These factors are also related to the energy demand for supplementary lighting, heating, and ventilation of the greenhouse that can be extracted from solar harvesting. Optimization shall be conducted for environmental control, energy consumption, and food yielding.

Parametric studies can be carried out for sensitivity analysis to determine the key parameters that affect the performance of the FEW systems. The main constraints that need to be considered include the energy input and crop output, the capacity of the water reservoir, the capacity of the geothermal system, the power of the BIPVT panel, and the power of the energy storage system. Plants adapt to the available light by responding to absolute and relative amounts of incident light over specific wavebands. Many developmental processes (which contribute to the productivity of a plant) are mediated by a set of blue, red, and far-red photoreceptors. In addition, light quality also affects plant morphology, which determines density and aperture control of stomata and eventually regulates photosynthesis and water uptake of the plant through the rate of transpiration. As such, the quality and quantity of incident light in the greenhouse will be controlled by tuning the relative amounts and intensity of blue, red, and far-red elements for improved food production.

9.4 Future 3D farming

To feed a growing number of people with reducing arable land, traditional farm based on the land surface may not be efficient. As population density is significantly increased with the mature of the high-rise building, farming in the urban area can similarly grow in the vertical direction. Three-dimensional (3D) farming was firstly implemented in the ocean to produce different food (fish, seaweed, shell, etc.) at different depths of water so that the whole volume of the water body can be used to maximize the yielding of the ocean area.

Similarly, on the land, vertical farming has been proposed to grow crops in vertically stacked layers, which often requires environmental control with supplementary lighting, temperature, moisture, and CO_2 levels. It can be combined with water-based farming techniques such as hydroponics, aquaponics, and aeroponics. Vertical farming was first proposed in 1999 by Dr. Dickson Despommier from Columbia University. Despommier and his students came up with a design of a skyscraper farm that could feed 50,000 people but had never been built yet. The huge start-up and operational costs are the main challenges to implement the design. BIPVT can be a solution to this problem.

The company of Schuco-USA introduced a novel idea of a Versatile Envelope Garden by growing leafy greens and herbs in the curtain wall of a high-rise building (Nathaniel Barlam and Silverman, 2020). The plants growing in buildings and spaces provide access to fresh, local food and create a natural, healthier living/working environment. It could be interesting for a high-rise building to devote a couple of floors for local food supply with vertical farming while improving indoor air quality and absorbing CO_2 from the surrounding city, entertaining people with a connection with plant farming, and changing the architecture. To scale up the idea, extra energy inputs are required. The installation of BIPVT will be an immediate solution to avoid the extra load to an electric power supply.

9.5 Case study 1: Thermal and lighting management by BIPV at University of British Columbia

The project was conceived by Prof. John Robinson of the University of British Columbia (UBC) with the architect of Peter Busby of the Perkins and Will, and was motivated by the UBC Center for Interactive Research on Sustainability (CIRS)'s goal for the most innovative and high performance building in North America. The solar system shares the unique aspects of BIPVT in farm applications, such as solar energy harvesting, thermal and lighting management, water circulation, and energy storage and management. It sets an excellent model for sustainable engineering and management in energy and water.

The design of this project began in 2008 and the site is located at the Vancouver campus of the University of British Columbia. The BIPV system with the skylight and solar shade module which is about 25 kW covers around 10 percent of the total electricity needs. The rest of the energy comes from the British Columbia Hydro grid. The total size of the building is about 5,700 meters. As to the solar shade, apart from generating electricity, it is being used to control the glare and the heat gain (Fig. 9.8). On the domestic hot water, an array evacuated tube was installed on the rooftop to capture the solar thermal.

FIGURE 9.8 An outlook of the UBC CIRS sunshades.

The prime objective of the facility is to serve as an interdisciplinary hub together with sustainable building research which demonstrates a Living Laboratory on campus. Should anyone visit Vancouver, visiting the center is highly recommended. One could learn a lot about various areas of sustainability. The project team applied IDP (Integrated Design Process) to assist with the design and they worked closely to make it happened. In addition, the team used the BIM (Building information modeling) software to lay out the drawing plans of the construction documents that resulted in developing a very well thought design.

The BIPV system at UBC CIRS (Center of Interactive Research on Sustainability) includes two types of large modules as follows:

1) Skylight Module (Fig. 9.9)
Size: 1490mm × 1814mm
Quantity: 21
Strings: 8 × 10 cells string = 80 cells
Specified power = 315 Watts

2) Sunshade Module (Fig. 9.8)
Size: 1513mm × 720mm
Quantity: 130
Strings: 9 × 4 cells string = 36 cells
Specified power = 140 Watts

FIGURE 9.9 A view of BIPV panels on the rooftop of the UBC CIRS center.

Sunshade (Small) module
Size: 1235mm × 720mm
Quantity: 6
String: 7 × 4 cells string = 28 cells
Specified power = 110 Watts

The skylight and sunshade module specifications and layout drawing are illustrated in Figs. 9.10 and 9.11, respectively. The skylight system is connected and mounted in accordance with Fig. 9.12 and the whole system layout is shown in Fig. 9.13.

The building has its own water collection system that makes it totally water self-sufficient through rainwater collectors at all the rooftops. The water goes through a filtration system, sanitized on-site prior to being distributed to different parts of the building. In addition, the indoor air

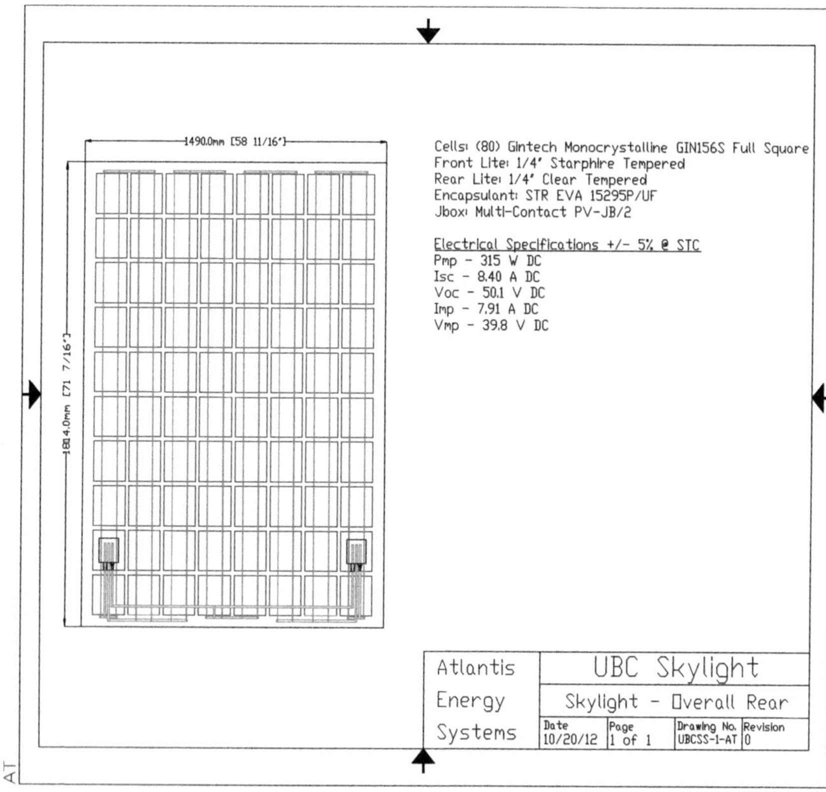

FIGURE 9.10 Skylight module specifications and layout drawing.

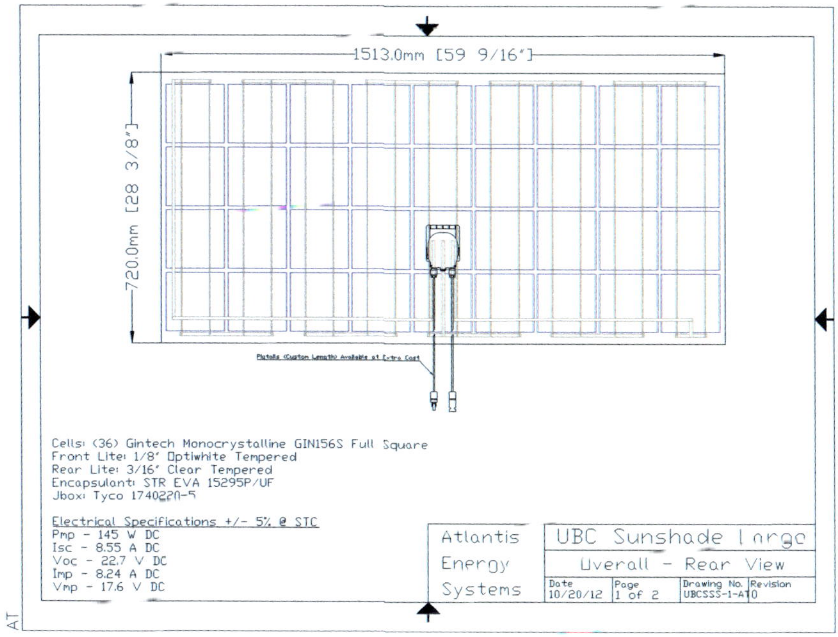

FIGURE 9.11 Sunshade module specifications and layout drawing.

ventilation system is being monitored to control and maintain air quality. As to lighting areas, the building is designed to optimize the use of daylighting in order to reduce the use of artificial indoor lighting as much as possible.

The project after it was opened in November of 2011, in addition to being awarded the first LEED Platinum certification of the Living Laboratory, received a long list of prestigious awards over the years including from some of the world-renowned organizations, including but not limited to the following:

- The project was awarded the first LEED Platinum certification of the Living Laboratory, 2012
- Architectural Innovation Award: Architectural Institute of British Columbia (AIBC)
- Award for Engineering Excellence: Association of Consulting Engineering Companies
- Excellence in Structural Engineering Award: National Council of Structural Engineers

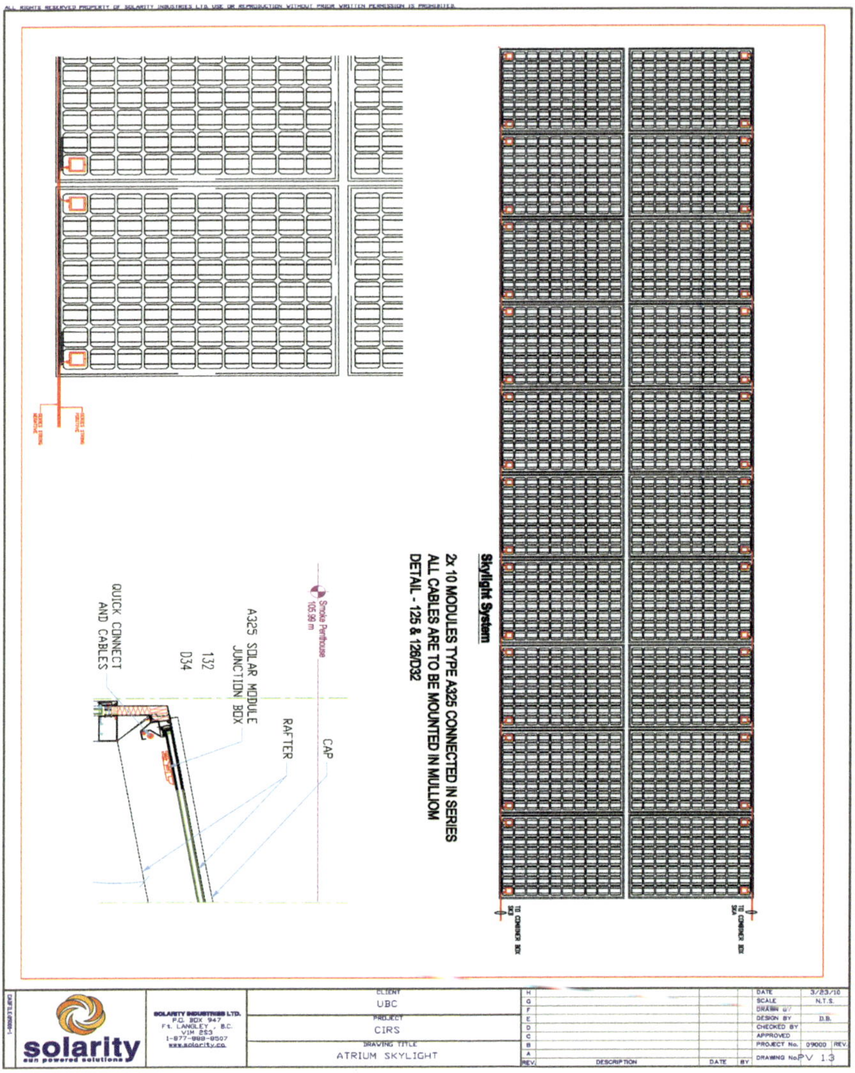

FIGURE 9.12 Skylight system connection and mounting drawing.

- Perkins and Will Design Biennale – Design Merit Award, 2013
- Sustainable Building of the Year: World Architecture New (WAN)
- Sustainable Development Award: Golder Associates, 2014
- Canadian Green Building Award: SAB Magazine, 2015
- ISCN 2015 Sustainable Campus Excellence Award, Excellence in Building: International Sustainable Campus Network

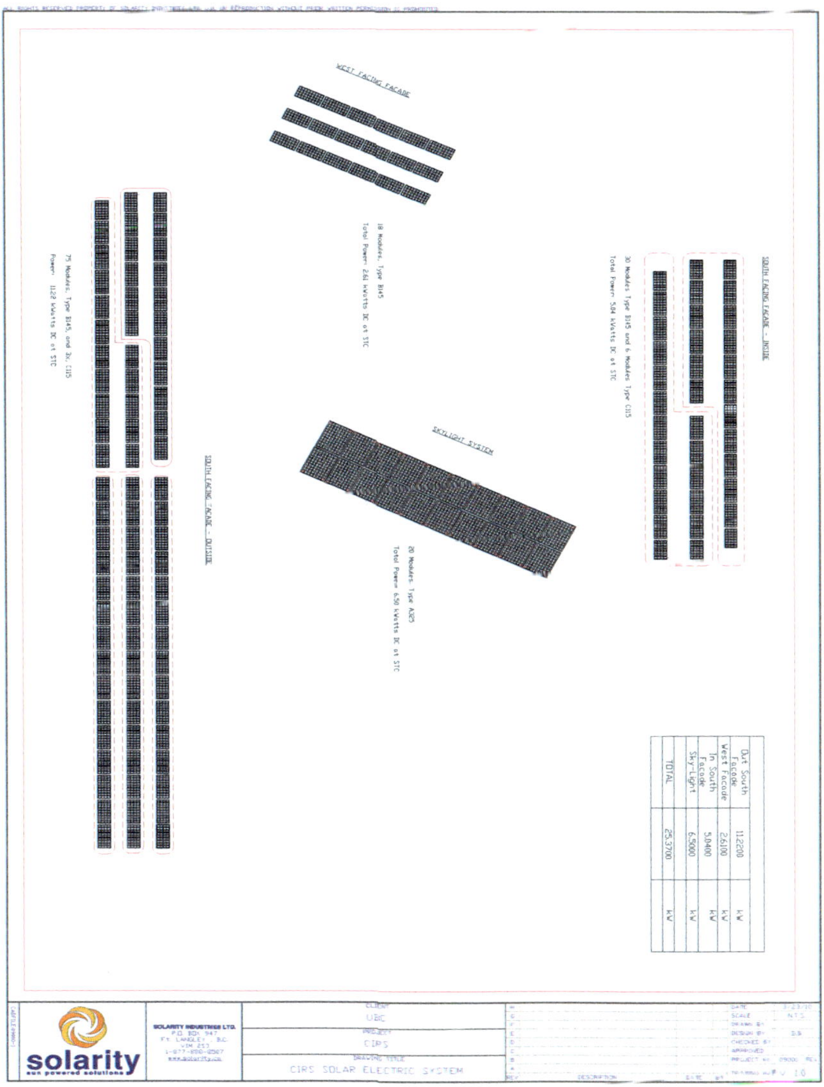

FIGURE 9.13 BIPV system layout drawing.

- Green Building Award: Royal Architectural Institute of Canada (RAIC)

Reference: UBC CIRS building manual 1) Energy System section, 2) Water Re-claim section, and 3) Day Lighting System section.

9.6 Case study 2: A greenhouse project at Como Park

Architect: Linda Morrissey of HOK Architecture Group.

In 1845, the construction of Como Park was started which is claimed to be the treasure of St. Paul, Minnesota. It was not until around 1914–15 that they started to build the conservatory where the Dome stands to this day (Fig. 9.14). Around early 2000, the McNeely provided a sizeable donation which resulted in building the new wing of the conservatory where the Atlantis modules were installed (Figs. 9.15 and 9.16). Inside, there is a display of new orchids and bonsai. These modules were installed in different parts of the greenhouse. The prime objective of that the solar modules were designed with partial shade was to have Bonsai all year

FIGURE 9.14 The new extension opened in 2005 in Como Park.

FIGURE 9.15 Como Park Conservatory with the new extension at St. Paul, MN.

FIGURE 9.16 Como Park Conservatory on the other side of the building.

round, Tulips and Canna Lily in the spring, Chrysanthemum in Autumn and Poindexter in the Holiday Seasons.

9.6.1 Introduction of the greenhouse modules

Module A and module B are both glass/glass modules in Figs. 9.17 and 9.18, respectively. Module C exhibits a different pattern of solar cells. The front and back views of the modules are shown in Figs. 9.19–9.23, and the pigtail connection shown in Fig. 9.24. They looked simple but were problematic in handling and the manufacturing processes, especially during lamination due to the flow of the encapsulant in such a narrow channel. Several methods were taken, and the AES team was eventually able to overcome the problems which resulted in producing perfect products.

FIGURE 9.17 Como Park module A.

All the modules were insulated glass (Fig. 9.25) with multiple layers because being in such a cold climate such as Minnesota, some of the tropical plants and bonsais will have problems. The recommended temperature range for bonsai is from 61 to 75°F in the daytime and 57 to 62°F at nighttime. Figs. 9.26 and 9.27 show the installed solar modules on th roof. Table 9.1 provides the specifications and bill of materials for the three types of modules.

Looking at the three different types of panels, the main unique part of all of them is that the design allowed a substantial amount of sunlight to come throws the panels, which is vital for the plants in a greenhouse. Considering the total amount of energy that all these panels generate is

FIGURE 9.18 Como Park module B.

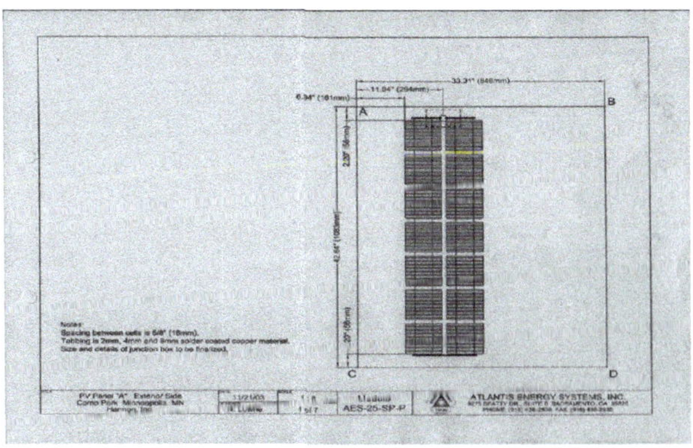

FIGURE 9.19 Drawing of module A (front view).

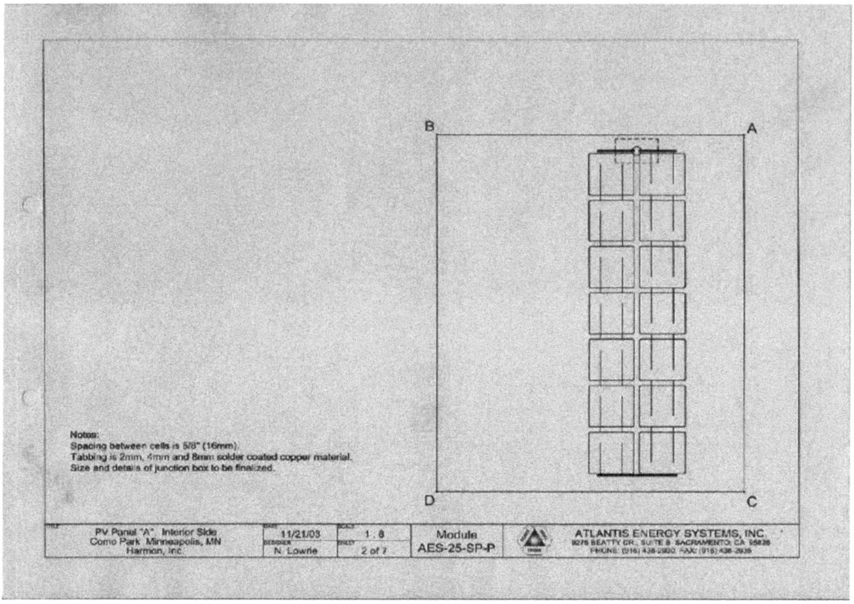

FIGURE 9.20 Drawing of Module A (back view).

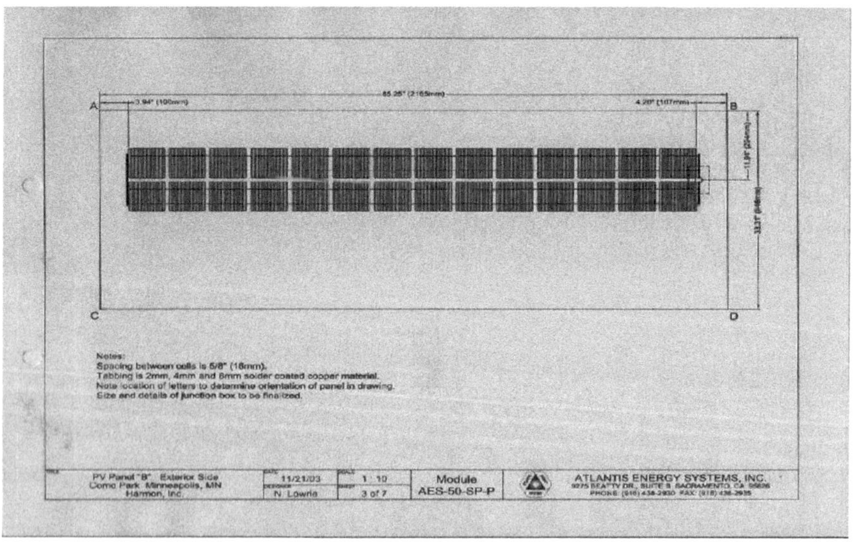

FIGURE 9.21 Drawing of module B (front view).

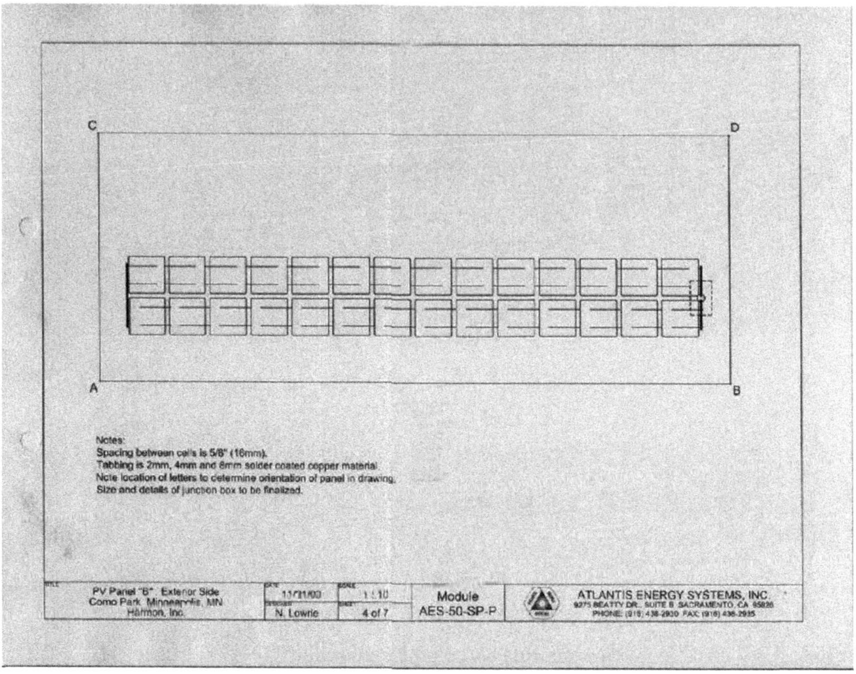

FIGURE 9.22 Drawing of module B (back view).

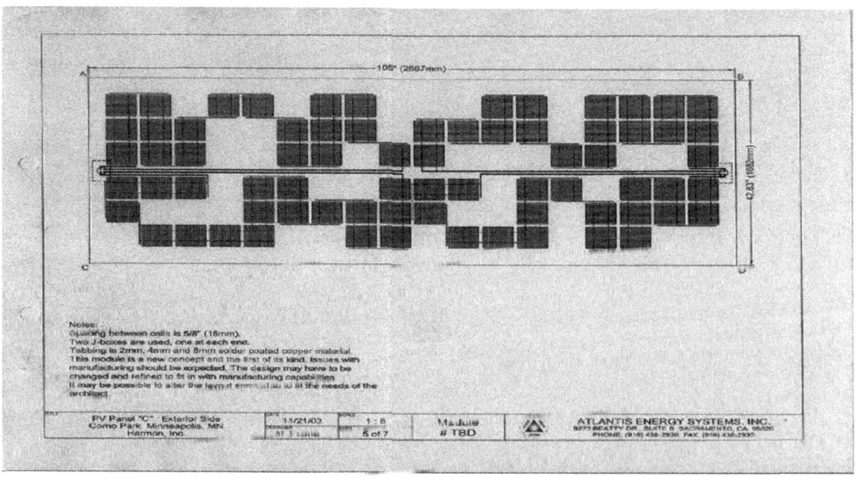

FIGURE 9.23 Drawing of Module C (from view).

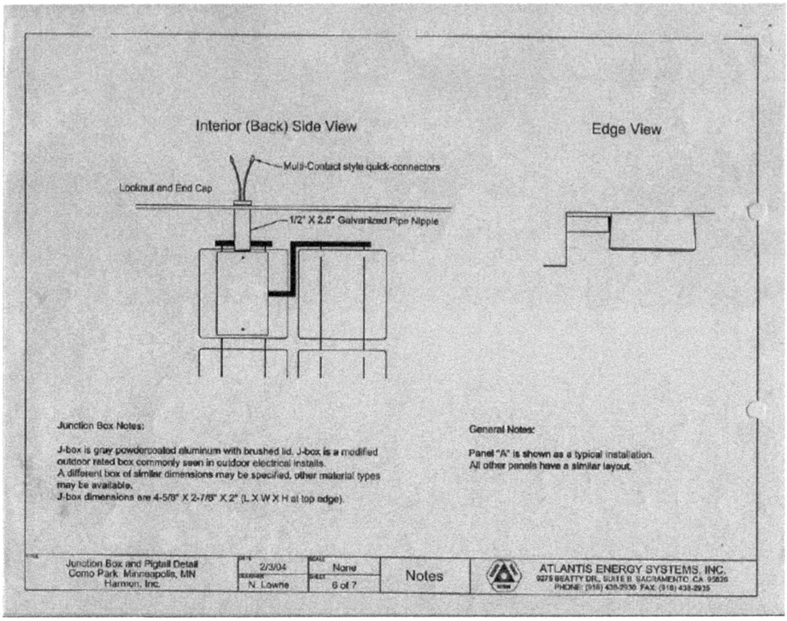

FIGURE 9.24 Como Park module and pigtail connection.

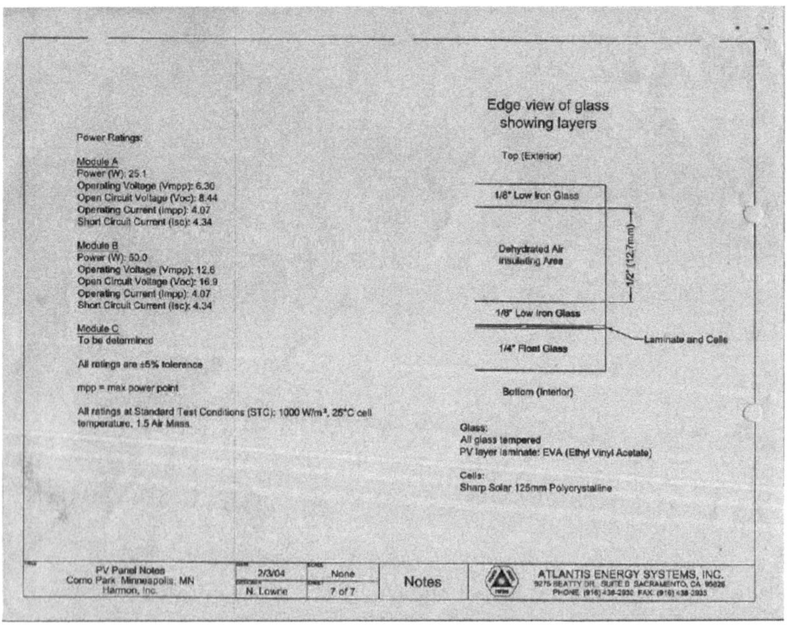

FIGURE 9.25 Insulated glass module.

FIGURE 9.26 Como Park module B installed.

FIGURE 9.27 Como Park module C installed.

TABLE 9.1 Atlantis Energy Systems (AES) glazing—specification sheet and bill of materials.

Project: Como Park, Minneapolis, MN			
Estimated Delivery: Spring 2004			
Construction details	Module "A"	Module "B"	Module "C"
Module quantities	49	163	8
Spares	0	0	0
Total quantities	49	163	8
Dimensions			
Width (in.)	33.31″	33.31″	42.64″
(mm)	846	846	1083
Length (in.)	42.64″	85.25″	105″
(mm)	1083	2165	2667
Photovoltaic Superstrate			
Thickness (in.)	1/8″	1/8″	1/8″
(mm)	3.2	3.2	3.2
Glass type	Solatex II	Solatex II	Solatex II
Strength	Tempered	Tempered	Tempered
Laminate			
Type	EVA	EVA	EVA
Tedlar insert	None	None	None
Tedlar color	N/A	N/A	N/A
Transparency	clear	clear	clear
% Translusence	76.1%	76.1%	
Photovoltaic cells			
Manufacturer	Sharp	Sharp	Sharp
Grade	Poly-A	Poly-A	Poly-A
Type	poly-crystalline	poly-crystalline	poly-crystalline
Size (in.)	5″	5″	5″
(mm)	125.5 × 125.5	125.5 × 125.5	125.5 × 125.5
Cell matrix	2 × 7	2 × 14	Random
# Cells	14	28	72
Vert. spacing (in.)	0.625″	0.625″	
(mm)	16	16	

(continued on next page)

TABLE 9.1 Atlantis Energy Systems (AES) glazing—specification sheet and bill of materials—cont'd

Project: Como Park, Minneapolis, MN			
Horz. spacing (in.)	0.625″	0.625″	
(mm)	16	16	
Cell tabbing (in.)	0.08″	0.08″	
(mm)	2	2	
Total length (in.)	No extra	No extra	
(mm)	required	required	
String tabbing (in.)	0.31″	0.31″	
(mm)	8	8	
Total length (in.)	17.4″	17 4″	
(mm)	442	442	
Hole tabbing (in.)	Same as	Same as	
(mm)	string	string	
Total length (in.)	Tabbing	Tabbing	
Quantity	2	2	
	Module "A"	Module "B"	Module "C"
Edge isolation			
Top (in.)	2.20″	4.20″	
(mm)	56	107	
Bottom (in.)	2.20″	3.94″	
(mm)	56	100	
Left (in.)	11.94″	11.94″	
(mm)	294	294	
Right (in.)	16.5″	16.5″	
(mm)	419	419	
Photovoltaic substrate			
Thickness (in.)	¼″	¼″	¼″
(mm)	6.4	6.4	6.4
Glass type	Clear float	Clear float	Clear float
Strength	Tempered	Tempered	Tempered
Hole diameter (in.)	1.5″	1.5″	1.5″
(mm)	38	38	38

(continued on next page)

only around 12 kW which is fairly low. It could probably only be enough energy to cover the light requirements which would be subject to the design of the building and the specifications. However, it is a great showcase for applying building-integrated PV technology on an iconic landmark building that solar could be aesthetically pleasing. As to Como Park, we have learned that it is the second most well-visited place in Minnesota.

Greenhouses will be quite important to us in the coming years as we have a look into taking care of food supplies due to Climate Change issues. The Como Park project was a great start and we should consider taking more opportunities and actions on improving the design of greenhouses.

9.7 Remarks for prospective development and applications

Because both energy demands and availability in farms are large, it is critical to manage the energy for the maximized yield. As the energy needs are often mismatched with the available energy in time or space, energy harvesting, storage, distribution and efficiency are the key factors to achieve sustainable farming technologies towards net-zero energy farming. In near future, there are several technologies that one could think of applying, such as collecting heat energy from geothermal wells, the seasonal energy storage and utilization systems, and hybrid energy generation systems, and solar shading systems. Particularly, nowadays, there have also been sensor systems that automatically control the greenhouse and regulate the required operations. Integration of the BIPVT with these technologies may create dynamic and novel applications on our food producing industry, including but not limited to the following:

1. BIPV-geothermal system in greenhouses enables seasonal energy storage and extended plant growth seasons.
2. BIPV-water system integrates energy storage, irrigation, and water reservation, which can immediately turn more marginal lands to arable lands.
3. BIPV and BIPVT systems applied in isolated areas may create new opportunities for energy independent agriculture.
4. BIPV technologies can be integrated into 3D farming for high yielding with limited land use with controlled environment agriculture.
5. BIPV and BIPVT systems on off/near/on shore energy harvesting may facilitate the fish farming.
6. BIPV and BIPVT system on animal farms may create well controlled environments for animal growth as well as energy supplies.

TABLE 9.1 Atlantis Energy Systems (AES) glazing—specification sheet and bill of materials—cont'd

Project: Como Park, Minneapolis, MN			
Insulated glass			
Insulation gap (in.)	½″	½″	½″
(mm)	38	38	38
Insulation medium	Dehydrated air	Dehydrated air	Dehydrated air
Insulating superstrate			
Thickness (in.)	1/8″	1/8″	1/8″
(mm)	3.2	3.2	3.2
Glass type	Solatex II	Solatex II	Solatex II
Strength	Tempered	Tempered	Tempered
Tabbing connections			
Wire gauge	16 AWG	16 AWG	16 AWG
Heat Shrink	Polyolefin	Polyolefin	Polyolefin
Potting material	Silicone	Silicone	Silicone
Junction box			
Material	Aluminum	Aluminum	
Dimensions			
Length(in.)	4″	4″	
(mm)	102	102	
Width (in.)	2″	2″	
(mm)	51	51	
Height (in.)	2″	2″	
(mm)	51	51	
Cover material	Aluminum	Aluminum	
Cover color	Unpainted	Unpainted	
# Diodes	1	2	
Diode mounting	PCB	PCB	
# Input terminals	2	3	
AWG input terminals	16 AWG	16 AWG	
# Output terminals	2	2	
AWG Output terminals	10 AWG	10 AWG	
Potting/adhesive	silicone	silicone	

(continued on next page)

TABLE 9.1 Atlantis Energy Systems (AES) glazing—specification sheet and bill of materials—cont'd

Project: Como Park, Minneapolis, MN			
	Module "A"	Module "B"	Module "C"
Ext. wire run (pig tail)			
AWG (+) wire	10 AWG	10 AWG	
Color (+) wire	Black	Black	
AWG (−) wire	10 AWG	10 AWG	
Color (−) wire	White	White	
Wire length (in.)	60″	60″	
(mm)	1524	1524	
Conduit type	Flexible metal	Flexible metal	
Conduit material	Liquid-tight pvc	Liquid-tight pvc	
Conduit diameter (ID)	½″	½″	
Conduit diameter (OD)			
Conduit length (in.)	48″	48″	
(mm)	1219	1219	
Connector type	Liquid-tight	Liquid-tight	
Connector material	Stainless steel	Stainless steel	
Connector angle	0″	0″	
Electrical parameters: cell Mfg. ratings			
*V*oc (V)	0.652	0.652	0.652
*I*sc (A)	4.34	4.34	4.34
*V*mp (V)	0.488	0.488	0.488
*I*mp (A)	4.07	4.07	4.07
*P*max (W)	2.09	2.09	2.09
Tolerance (+/−)	3%	3%	3%
Module maximums			
*V*oc (V)	9.13	18.26	93.9
*I*sc (A)	4.34	4.34	4.34
*V*mp (V)	6.83	13.66	70.25
*I*mp (A)	4.07	4.07	4.07
*P*max (W)	29.26	58.52	295.35
Tolerance (+/−)	3%	3%	

(continued on next page)

TABLE 9.1 Atlantis Energy Systems (AES) glazing—specification sheet and bill of materials—cont'd

Project: Como Park, Minneapolis, MN			
AES module ratings			
*V*oc (V)	8.44	16.89	86.6
*I*sc (A)	4.34	4.34	4.34
*V*mp (V)	6.3	12.6	64.1
*I*mp (A)	4.07	4.07	4.07
*P*max (W)	27.1	54.1	278.2
Tolerance (+/−)	3%	3%	
% Max	92.5%	92.5%	92.5%
STC rating (W)	27.1	54.1	278.2
PTC rating (W)	21.7	43.3	221.3
AES model #	AES-27SP	AES-54SP	

Of course, the cost is always a serious constraint especially in the production of food and products which are competitive. Life cycle cost analysis should be conducted to demonstrate the economic viability for large scale applications.

References

Agrawal, B., Tiwari, G.N., 2010. Building Integrated Photovoltaic Thermal Systems: For Sustainable Developments. Royal Society of Chemistry.

Allan, T., Keulertz, M., Woertz, E., 2015. The water-food-energy nexus: an introduction to nexus concepts and some conceptual and operational problems. Int. J. Water Resour. Dev. 31 (4), 800.

Allouche, J., Middleton, C., Gyawali, D., 2014. Nexus Nirvana or Nexus Nullity? A dynamic approach to security and sustainability in the water-energy-food nexus, STEPS Working Paper 63, Brighton: STEPS Centre.

Allred, B.W., et al., 2015. Ecosystem services lost to oil and gas in North America. Science 348 (6233), 401–402.

Anderson, T.N., et al., 2009. Performance of a building integrated photovoltaic/thermal (BIPVT) solar collector. Sol. Energy 83 (4), 445–455.

Bargach, M.N., et al., 2000. Survey of thermal performances of a solar system used for the heating of agricultural greenhouses in Morocco. Renew. Energy 20 (4), 415–433.

Chen, F., Yin, H., 2016. Fabrication and laboratory-based performance testing of a building-integrated photovoltaic-thermal roofing panel. Appl. Energy 177, 271–284.

Chen, F.L., He, X., Yin, H.M., 2016. Manufacture and multi-physical characterization of aluminum/high-density polyethylene functionally graded materials for green energy building envelope applications. Energy Build. 116, 307–317.

Choi, J.-K., Fthenakis, V., 2010. Design and optimization of photovoltaics recycling infrastructure. Environ. Sci. Technol. 44 (22), 8678–8683.
Corbin, C.D., Zhai, Z.J., 2010. Experimental and numerical investigation on thermal and electrical performance of a building integrated photovoltaic–thermal collector system. Energy Build. 42 (1), 76–82.
Debbarma, M., Sudhakar, K., Baredar, P., 2017. Thermal modeling, exergy analysis, performance of BIPV and BIPVT: a review. Renew. Sustain. Energy Rev. 73, 1276–1288.
Emmi, G., et al., 2015. An analysis of solar assisted ground source heat pumps in cold climates. Energy Convers. Manag., 106, pp. 660–675.
Gautam, K.R., Andresen, G.B., 2017. Performance comparison of building-integrated combined photovoltaic thermal solar collectors (BiPVT) with other building-integrated solar technologies. Sol. Energy 155, 93–102.
Ghani, F., Duke, M., Carson, J.K., 2012. Effect of flow distribution on the photovoltaic performance of a building integrated photovoltaic/thermal (BIPV/T) collector. Sol. Energy 86 (5), 1518–1530.
Gilfillan, B.C., 2013. Donovan launches regional design competition to promote resilience for the sandy-affected region. U.S.D.o.H.a.U. Development, Washington, DC.
Haacker, E.M.K., Kendall, A.D., Hyndman, D.W., 2016. Water level declines in the high plains aquifer: predevelopment to resource senescence. Groundwater 54 (2), 231–242.
Hassanien, R.H.E., Li, M., Lin, W.D., 2016. Advanced applications of solar energy in agricultural greenhouses. Renew. Sustain. Energy Rev. 54, 989–1001.
He, W., et al., 2006. Hybrid photovoltaic and thermal solar-collector designed for natural circulation of water. Appl. Energy 83 (3), 199–210.
Hepbasli, A., 2010. A review on energetic, exergetic and exergoeconomic aspects of geothermal district heating systems (GDHSs). Energy Convers. Manage. 51 (10), 2041–2061.
Hitaj, C., Suttles, S., 2016. Trends in U.S. agriculture's consumption and production of energy: renewable power, shale energy, and cellulosic biomass. Eco. Inf. Bull., USDA Eco. Res. Serv. (No.159) 47 p. iv +.
Ibrahim, A., et al., 2011. Recent advances in flat plate photovoltaic/thermal (PV/T) solar collectors. Renew. Sustain. Energy Rev. 15 (1), 352–365.
Jain, D., Tiwari, G.N., 2003. Modeling and optimal design of ground air collector for heating in controlled environment greenhouse. Energy Convers. Manage. 44 (8), 1357–1372.
Kräuchi, P., Schluck, T., Sulzer, M., 2015. Modelling of low temperature heating networks with IDA-ICE. In: Proceedings of International Conference CISBAT 2015 Future Buildings and Districts Sustainability from Nano to Urban Scale. LESO-PB, EPFL.
Kumar, A., Baredar, P., Qureshi, U., 2015. Historical and recent development of photovoltaic thermal (PVT) technologies. Renew. Sustain. Energy Rev. 42, 1428–1436.
Ling, H., et al., 2014. Active heat storage characteristics of active–passive triple wall with phase change material. Sol. Energy 110 (Supplement C), 276–285.
Ling, T.-C., Poon, C.-S., 2013. Use of phase change materials for thermal energy storage in concrete: an overview. Constr. Build. Mater. 46, 55–62.
Liu, J., et al., 2017. Challenges in operationalizing the water-energy-food nexus. Hydrol. Sci. J. 62 (11), 1714–1720.
Lumb, M.P., Mack, S., Schmieder, K.J., González, M., Bennett, M.F., Scheiman, D., ... Walters, R.J. (2017). GaSb-based solar cells for full solar spectrum energy harvesting. Adv. Energy Mat., 7 (20), 1700345.
Luo, J., et al., 2015. Heating and cooling performance analysis of a ground source heat pump system in Southern Germany. Geothermics, 53, 57–66.
Mandal, J., et al., 2017. Scalable,"Dip-and-Dry" fabrication of a wide-angle plasmonic selective absorber for high-efficiency solar–thermal energy conversion. Adv. Mater. 29 (41), 1702156.

Mekki, M., et al., 2015. Greenhouse monitoring and control system based on wireless Sensor Network. In: 2015 International Conference on Computing, Control, Networking, Electronics and Embedded Systems Engineering. IEEE.
Morrow, M., Delos-Reyes, M., 2013. Submerged wave energy converter. World 586 patent WO 2013/019214A1.
Müller, J., et al., 2015. Resistive power loss analysis of PV modules made from halved 15.6× 15.6 cm 2 Silicon PERC solar cells with efficiencies up to 20.0%. IEEE J. Photovolt. 5 (1), 189–194.
Nagano, K., et al., 2003. Development of thermal-photovoltaic hybrid exterior wallboards incorporating PV cells in and their winter performances. Sol. Energy Mater. Sol. Cells 77 (3), 265–282.
NASS-USDA, 2012. www.agcensus.usda.gov/Publications/2012/Full_Report/Volume_1,_Chapter_1_US/usv1.pdf. Census of Agriculture, US Department of Agriculture, National Agricultural Statistics Service, Washington, DC. Accessed on: 2021.
Nathaniel Barlam, Y.L., Silverman, B., 2020. The Versatile Envelope Garden. A communal threshold for urban environments. https://www.facadetectonics.org/papers/the-versatile-envelope-garden#!. Facade Tectonics 2020 World Congress. Accessed on: 2021.
Ozgener, L., Hepbasli, A., Dincer, I., 2005. Energy and exergy analysis of geothermal district heating systems: an application. Build. Environ., 40 (10), 1309–1322.
Park, D.-H., et al., 2011. A study on greenhouse automatic control system based on wireless sensor network. Wireless Personal Commun. 56 (1), 117–130.
Prasanna, A., Dorer, V., Vetterli, N.J.E., 2017. Optimisation of a district energy system with a low temperature network 137, 632–648.
Rost, S., et al., 2008. Agricultural green and blue water consumption and its influence on the global water system. Water Resour. Res. 44 (9), 1–17.
Royal Society and Royal Academy of Engineering, Greenhouse gas removal 2018. Issued: September 2018 DES5563_1 https://royalsociety.org/-/media/policy/projects/greenhouse-gas-removal/royal-society-greenhouse-gas-removal-report-2018.pdf.
Ruiz-Calvo, F., Montagud, C., Cazorla-Marín, A., Corberán, J.M., 2017. Development and experimental validation of a TRNSYS dynamic tool for design and energy optimization of ground source heat pump systems. Energies, 10 (10), 1510.
Santamouris, M., Argiriou, A., Vallindras, M., 1994. Design and operation of a low energy consumption passive solar agricultural greenhouse. Sol. Energy 52 (5), 371–378.
Scanlon, B.R., et al., 2005. Impact of land use and land cover change on groundwater recharge and quality in the southwestern US. Global Change Biol. 11 (10), 1577–1593.
Skoplaki, E., Palyvos, J.A., 2009. On the temperature dependence of photovoltaic module electrical performance: a review of efficiency/power correlations. Sol. Energy 83 (5), 614–624.
Solar Panel Temperature – Facts and Tips. 2015, http://www.solar-facts-and-advice.com/solar-panel-temperature.html. Accessed on: 2021.
Solutions, C.G., 2021. Ground to air heat transfer (GAHT^{TM}) systems. http://www.ceresgs.com/climate-control/gaht/#1470867745027-d0beb645-bae2. Accessed on: 2021.
Spitler, J.D., Gehlin, S.E.A., 2015. Thermal response testing for ground source heat pump systems—An historical review. Renew. Sustain. Energy Rev. 50, 1125–1137.
R.b.D. Team What is Rebuild by Design? 2013, http://www.rebuildbydesign.org/what-is-rebuild-by-design/. Accessed on: 2021.
Tong, G., et al., 2013. Passive solar energy utilization: A review of cross-section building parameter selection for Chinese solar greenhouses. Renew. Sustain. Energy Rev. 26 (Supplement C), 540–548.
Xue, J., 2017. Photovoltaic agriculture-New opportunity for photovoltaic applications in China. Renew. Sustain. Energy Rev. 73, 1–9.

Yang, T., Athienitis, A.K., 2014. A study of design options for a building integrated photovoltaic/thermal (BIPV/T) system with glazed air collector and multiple inlets. Sol. Energy 104, 82–92.

Yang, T., Athienitis, A.K., 2015. Experimental investigation of a two-inlet air-based building integrated photovoltaic/thermal (BIPV/T) system. Appl. Energy 159, 70–79.

Yildirim, N., Toksoy, M., Gokcen, G.J.E., 2010. Piping network design of geothermal district heating systems: Case study for a university campus. Energy, 35 (8), 3256–3262.

Yin, H., Pao, F., Zadshir, M., Lou, J., Wei, Z., 2021. Building integrated photovoltaic thermal systems. U.S. Patent Application No. 63/248,502.

Climate change leads to frequent extreme weather. This solar sailing ship will ride the storm with the building integrated photovoltaic and thermal (BIPVT) power and together with wind power. The sails made of solar panels are mounted on masts to harness the power of wind and propel the vessel. BIPVT panels are extended in the horizontal plane for additional solar energy harvesting. This picture demonstrates the design philosophy of novel BIPVT applications to create new functions with solar panels and extend the surface area for improved energy harvesting capacity.

CHAPTER

10

Perspectives of the current, emerging, and future BIPVT technologies

10.1 Introduction

In February of 2021, as this book was being prepared, a winter storm and record-low temperature in Texas caused the state's electric grid to suffer a major crisis and massive statewide blackout (Fig. 10.1). It is estimated that more than 4.5 million homes and businesses were left without power for several days (Wikipedia, 2021). This resulted in food and water shortage, and almost 100 people losing their lives due to severe cold

FIGURE 10.1 Several inches of snow in Texas caused power outage in the largest energy-producing state in the United States (Courtesy of Wfaa) (McLellan et al., 2018).

Building Integrated Photovoltaic Thermal Systems: Fundamentals, Designs and Applications
DOI: https://doi.org/10.1016/B978-0-12-821064-2.00008-0

FIGURE 10.2 People standing in line for grocery shopping during the power outage in Austin (Courtesy of The Guardian) (NPR, 2021).

weather. Fig. 10.2 shows 2 the long line of people standing outside for grocery shopping during the power outage in Austin.

Looking at this incident, we realize critical and necessary changes are required to be taken to our infrastructure to prevent such happenings to occur again in the future. It draws quite a bit of attention as to what needs to be done on our energy infrastructure. Our grid is outdated and needs to be updated. Rather than continue to do it in the same old way, now is the time to look into an alternative way. Dr. Steven Chu, our Secretary of Energy back in 2009 strongly recommended considering insulating the building or home which would reduce energy consumption. Energy transmission is costly both in energy loss and installation. Back then, it cost $3 million a mile on the transmission line. Prices could be higher today. In addition, long-distance energy delivery is very vulnerable especially in the case of snowstorms and hurricanes which has been demonstrated in the Texas situation. It is time to consider an alternative way that could be less costly and energy delivery could be more secure.

Localizing the energy delivery could be a solution. Micro-grid and standalone buildings and houses on renewable are highly recommended. Nowadays, considering the advancements of storage and generation systems as well as the solar electric and thermal systems, it would be economical and long-term security on energy insurance. However, the quality and reliability of the products are vital to assure success on the system installation. Some serious study needs to be done.

In a micro-grid situation, built-in monitoring will offer instant knowledge on how the system is operating which one could know how the system is performing. Applying BIPV or BIPVT in the solar would be highly recommended. Connecting with appropriate storage and generations system would make the whole system work. All these systems would be subject to the location, the setup of the buildings, and the energy requirements. It could be an apartment complex or a hotel or an office or an industrial building.

We need to examine each of these solutions immediately because due to climate change, we could anticipate more severe and frequent storms. Together with more wildfires and more temperature fluctuations that would be increasingly hard for the outdated central grid to handle the challenges. As Texas has demonstrated these outages put us in a very vulnerable position.

Tremendous scientific proofs have shown the connection of the extreme weathers to the increase of the greenhouse gas emissions. As the U.S. is rejoining the Paris Climate Agreement (NPR, 2021), more efforts to reduce greenhouse gas emission will be anticipated, such as renewable energy harvesting, gas power reduction/replacement in engines and burning, energy efficiency and resilience in infrastructure, carbon taxes, and subsidies to building-integrated photovoltaic/building-integrated photovoltaics and thermal (BIPV/BIPVT) technologies. Solar energy will be a key part of this effort, and significant evolution and applications of BIPVT technologies are anticipated. This chapter will share the authors' perspectives of the BIPV/BIPVT arena.

10.2 Current BIPV/BIPVT market and future opportunities

The solar energy industry has been booming in the past decade in the United States and the whole world. The Solar Energy Industries Association (SEIA) (SEIA 2020) reported as of Fall 2019, nearly 250,000 Americans work in solar—more than double the number in 2012—at more than 10,000 companies in every US state. In 2019, the solar industry generated $18.7 billion of investment in the American economy. The installed solar capacity has been under an average annual growth rate of 49% in the past decade and reaches >89 gigawatts (GW) in the United States. The research and technology advancement enables the installation cost to drop 70% in the last decade with higher efficiency and more diverse technology availability, which will boost the rapid growth in the future. For example, in 2019, 40% of all new electric capacity added to the grid came from solar, which is the largest share in history in the United States.

Building-integrated photovoltaic (PV) is the major part of the solar energy industry with three major segments from the residential, industrial, and commercial. Based on the recently released market analysis (Research, 2017), the global building-integrated PV market size was valued at $13.1 billion and is expected to reach $56.2 billion by 2027 under the growth at a compound annual growth rate of 23.1%. With the high pressure of climate change, favorable government policies and guidelines regarding enhanced renewable installations are projected to strengthen the BIPV market growth over the forecast period.

However, high initial investment along with the intermittent nature of renewable solar energy is the main barrier for the adoption of BIPV. As the hardware costs have been reduced to such a level that is very competitive to other energy resources, the soft costs, including labor, quality control/acceptance tests, operation and maintenance (O&M), end-life decommissioning, and other overhead costs, have increased their shares in the total system costs. Moreover, energy storage will be more demanding for local use and decentralized power management. Compared with other BIPV technologies, BIPVT exhibits more severe disadvantages in initial investment and energy storage aspects because of the additional complexity and risks in thermal management requirements, such as leakage, space limit, condensation, etc., which make a higher barrier for BIPVT adoption. Although the benefits of BIPVT are also outstanding in the long term and it may have a shorter investment payback period, more research, and development are necessary to diversify the technology and reduce the risks and costs. To prepare for the future growth of BIPVT, the following research and development opportunities are anticipated:

1. Modular design, manufacture, and construction of BIPVT may shift a significant part of soft costs to hardware costs at the manufacturing phase. The system can be installed in a plug-and-play fashion with self-sensing and control. By integrating the small parts into a big panel before construction, the construction process will be rapid and simple. However, the size, weight, and strength of the panel will be critical for the safety and workability in manufacture, transportation, and construction. Currently, glass-Ethylene-vinyl acetate (EVA)-silicon PV is dominating PV packaging, and the high demand will continue to grow. However, lightweight materials will be favorable to increase the area-to-weight ratio. For example, the durable transparent polymer could be a future alternative for surface coating; high efficient perovskite solar cells (PSCs) has shown great potential for low cost and high flexibility for various application; whereas lightweight concrete could be the back-panel for multifunction in building energy efficiency, acoustic performance, as well as required structural strength. The combination

of them can enable large wall or roof panels for rapid construction of zero or negative energy buildings.

2. As solar energy has become one of the fastest-growing industries in the past decade, some PV modules are approaching the end of their life with reduced low efficiency or material defects or failures. On the other hand, as new high efficient PV cells are emerging, replacement of some existing solar panels becomes a more economic option. Therefore, more and more solar panels will be replaced in the future at an accelerated pace. Recycling and reusing some parts of solar panels in the future BIPVT can promote BIPVT's adoption for lower cost and high sustainability. Most solar panels fail at the end of their life because of encapsulant degradation and delamination. If we can delaminate the PV module and separate the components, we can easily replace the encapsulant and recover the function with a minimal cost and environmental impact. The key to achieving this revolutionary benefit in future PV manufacturing lies in the encapsulating process. For example, the thermoplastic encapsulant may lose adhesive strength at a certain temperature beyond the regular service temperature range, which enables the delamination and replacement of PV cells. It creates feasibility to upgrade the BIPVT system whenever a high efficient PV cell technology is available. This is an excellent example for circular eco-manufacturing (Sica et al., 2018) for high sustainability and life cycle performance.
3. BIPVT paired with a microgrid with energy storage systems such as a battery, high-temperature heat, or hydrogen, etc., may enable off-grid PV systems and save large investment in infrastructure, which will provide leverage to its high installation cost. When a BIPVT system can harvest and store more energy than the building's needs, it creates a large market in vacation houses or buildings in remote or rural areas, which has a restriction of power supply. Even in the metropolitan area, it may increase our power system resilience and safety. When extreme weathers or natural disasters occur, the microgrid system can provide the necessary needs for the community.
4. A higher level of building integration may cover other functions of building system components so as to equivalently reduce the BIPVT costs. New designs, manufacture or construction methods, and building materials are attractive to reduce the life cycle cost, increase the efficiency and lifetime, enhance the building strength, etc. For example, the BIPVT support systems can be integrated with the structural support and plumbing systems. As the BIPVT support system costs >50% of the total cost, the improvement of the support system will make a big difference. For example, active building design cannot only make the building moving with the sun for high BIPVT efficiency but also provides feasibility for earthquake mitigation. Integrating the sensing

and control system with the building management system (BMS) and microgrid can enable smart building and smart city technologies for self-powered, wireless sensing and control systems.

5. Introducing new physics into BIPVT systems may create revolutionary changes to their designs and applications. For example, most BIPVT systems use visible and infrared light to generate electricity and circulate fluid or air for thermal management. The new solar cell also uses ultraviolet (UV) radiation and allows the pass of visible light and infrared for indoor lighting and temperature control. Smart meta-surface materials can tailor the PV harvesting wavelength with its temperature, so as to provide optimal PV-light-heat for the indoor environment.

In the rest of this chapter, some emerging technologies are introduced.

10.3 Emerging thermal management technologies for BIPVT systems

10.3.1 Background

Compared with general BIPV technologies, BIPVT addresses the temperature loss of PV efficiency while harvesting thermal energy. Traditionally air and water are the main media for heat collection, which can be used for space heating, water heating, etc. BIPVT with water (BIPVT-W) is favorable for the higher specific heat and heat transfer coefficient of water, but it brings leakage risks and freezing issues to the building in addition to the material and labor costs of plumbing and heat storage. Moreover, in many scenarios because the temperature of the heat is not high enough for electricity generation and too much water is required to store the energy, it requires extra effort (even extra energy) to dissipate the low-quality heat to the environment to avoid system stagnation or overly heating in hot summer. Therefore, heat collection from the PV panel is not always useful. Moreover, because the heat collector distribution under the PV panel is not uniform and thermal media temperature also changes in the heating process, it will create thermal stress in PV cells, which are often very brittle and prone to micro-cracking. One way using this low-quality heat of heated water is to boost the temperature to a higher one, say from 40–70°C to 150–250°C or even higher by a heat pump and store the heat in a phase change material tank for electric generation at night, which will serve multiple functions of energy storage, heat supply, and thermal management. However, this novel application can be implemented with the existing technologies in Chapter 7 with BIPVT-geothermal systems and will not be elaborated in this section.

BIPVT with air (BIPVT-A) enables the integration of air channels into the building envelope for ventilation and space heating. Because the thermal mass or density of air is low, the system efficiency will be low, and it may not work well in extreme weather. However, because it can operate passively or automatically by moving air with the density change in heating which can be demonstrated in cooling towers, and it may be designed with one-directional heat transfer to improve building energy efficiency, such as Trombe wall (Hu et al., 2017, Yu et al., 2020), it is anticipated that many novel designs and construction methods with BIPVT-A will emerge in the next generation of energy-efficient building, but the existing technologies may enable those applications.

In this section, BIPVT-W and BIPVT-A will not be revisited but some different mechanisms of thermal management will be introduced, which does not require any flow going through the panels.

10.3.2 BIPVT integrated with phase change materials

With currently increasing energy demand, the goal to meet the Paris Agreement with "holding the increase in the global average temperature to well below 2°C above pre-industrial levels" requires innovative materials and methods in energy management to be implemented. For example, in the building section, heating, ventilation, and air condition (HVAC) account for almost 35% of the total energy consumption. A novel practice is the use of phase change materials (PCMs) to decrease energy consumption and improve the thermal management of the BIPVT module. PCMs have been receiving a great deal of attention for various thermal energy storage applications because they help with reducing temperature variation by utilizing PCMs' latent heat (i.e., PCMs absorb heat when there is an energy surplus and dissipates heat when there is an energy deficit) and thus conserve energy and increase energy efficiency. As the name implies, PCM has to deal with the phase change of the materials, mostly from solid to liquids and vice versa where the latent heat of the materials in changing from one state to another is used to store energy or release (Fig. 10.3). For example, water is a common PCM in water–ice. When ice melts, it absorbs energy and when water becomes ice, it releases energy. But since the phase change temperature for water is 0°C, it is not suitable for most applications.

An application of PCM is to use inside the BIPVT to store the heat energy whenever solar energy is available. Since the surface temperature of the solar panels increases as they are exposed to sunlight, the heat can be harvested and transferred to the PCM to melt the material. Later, this heat can be released during nighttime to deliver indoor thermal comfort, and the PCM will solidify again by losing heat. This way, not only the waste heat from the sun exposure can be collected and used to lower

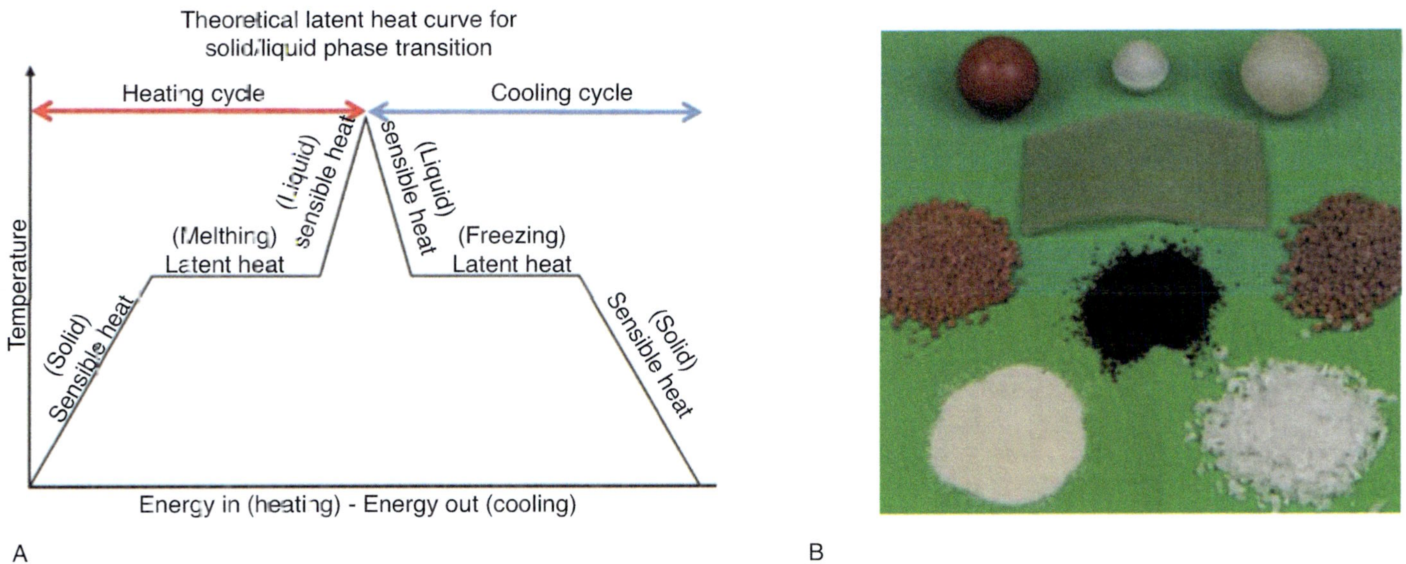

FIGURE 10.3 (A) Phase change materials mechanism based on latent heat exchange (Sutterlin, 2011), (B) different shapes and sizes of PCMs (PCM, 2015). *PCMs*, phase change materials.

the heating and cooling load of the building, but also the efficiency of the solar panel can be kept near its nominal highest value by absorbing the surface heat. Another application for the harvested heat is to warm a fluid (usually water and glycol) inside tubes for partially heating domestic hot water use. For example, Pugsley et al. developed an integrated collector storage solar water heater with BIPVT that employs planar liquid–vapor thermal diodes to regulate absorber's temperature and heat loss as shown in Fig. 10.4 (Pugsley et al., 2020). The planar liquid–vapor thermal diodes consist of two parallel plates separated by a cavity that includes a fluid, where during operation, the evaporator plate becomes hot and in contact with the liquid to form vapor, and from there, the vapor travels to the condenser plate which is at lower temperature and thus, letting the fluid to release energy and become liquid again. This way, the system not only reduces the heat loss in an integrated collector storage solar water heater water storage tank but also improves the thermal and electrical efficiency of the solar panels.

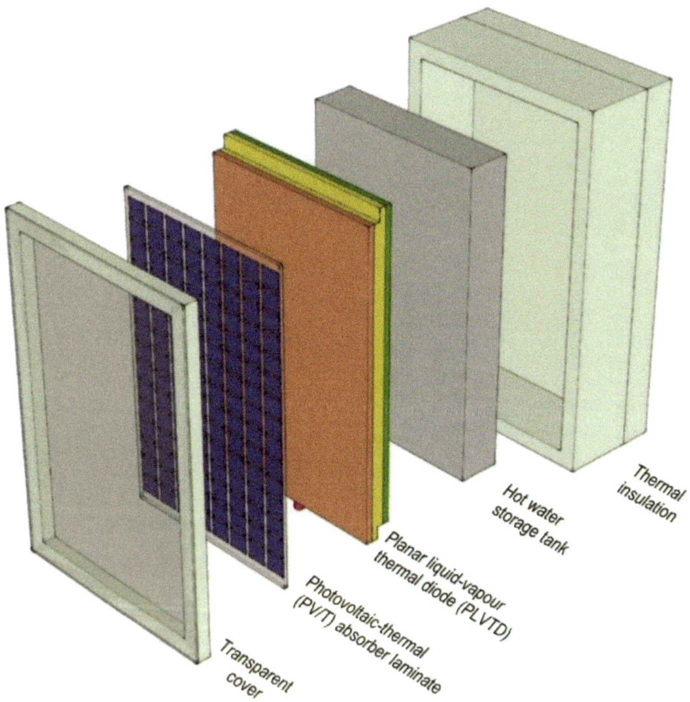

FIGURE 10.4 Schematic illustration and components of the BIPV-PLVTD-ICSSWH system (Pugsley et al., 2020). *BIPV*, building-integrated photovoltaic; *ICSSWH*, integrated collector storage solar water heater; *PLVTD*, planar liquid–vapor thermal diodes.

PCMs are often classified as inorganic, organic, and eutectic. Even though a lot of progress has been made in PCMs-related research and applications, the low thermal conductivity of the PCMs result in poor performance for BIPVT applications. Thus, improving the thermal performance of PCM is important to achieve high-efficiency systems. One way is to disperse thermally conductive materials such as carbon nanotubes, graphene, graphite metal, and metal oxide to achieve nano-enhanced phase change materials (Kumar et al., 2020).

In addition, the majority of the PCMs are considerably more expensive than conventional materials/methods and the most commonly used organic PCMs are petroleum-derived. Therefore, in recent years researchers are looking into the development of bio-derived PCMs (e.g., plant-based, waste cooking oil, waste animal fat residue). Fabiani et al. studied the thermal properties of expired palm oil collected from a local bakery store and further evaluated the potential of using this palm oil as an alternative to commercial paraffin in thermal energy storage systems (Fabiani et al., 2019). Results showed that the palm oil had two separate melting peaks (one at 4°C and the other at 31°C) with melting enthalpies of 33 J/g and 14 J/g, respectively. However, the global latent heat of the expired palm oil is only 1/3 of the commercial paraffin, making it less ideal for any real thermal energy storage applications. Gallart-Sirvent et al. developed a synthesis process to produce eutectic PCM mixtures (i.e., palmitic acid and stearic acid) from non-edible animal fat (Gallart-Sirvent et al., 2017). The eutectic PCM mixtures had melting and solidification temperatures around 55°C and 52°C, latent heat of 172 to 181 J/g, and good thermal cycling stability after 1000 thermal cycles. These PCMs seem to be a good candidate for thermal energy storage. For this project, we will refer to and further develop the methods these researchers developed.

A recent research proposed by Columbia University researchers has been focused on developing a chemical synthesis process that effectively converts the waste animal fat to PCMs of high latent heat and demonstrates the application of these PCMs in a lab-scale thermal energy storage setup (Yin et al., 2021). The research outcomes from this project will benefit other projects (net-zero energy buildings and greenhouses) as well as many other thermal energy-related applications. Since a large portion of the animal body parts such as fats are considered nonedible and end up in waste, thus the success of this project will result in low-cost abundant PCM.

Animal fat is mainly composed of unsaturated and saturated fatty acids. In terms of physical properties, saturated fatty acids have a higher latent heat and melting temperature and are thus more suitable for building applications. Therefore, in this research, our target to achieve the maximum saturated fatty acids from the animal fat and conversion to PCM.

The experimental method has five main stages.

1. Start with bacon oil, characterize the fat to find its molecular composition.
2. Conduct hydrolysis by use of a solution to break the molecules of fat to achieve free fatty acids.
3. Separate the saturated fatty acids (FA) and unsaturated FA by crystallization.
4. Use hydrogenation to transform the unsaturated fatty acids (such as oleic acid) to saturated fatty acids (such as palmitic and stearic acids).
5. Use esterification to convert the saturated FA to esters that are odorless, noncorrosive, and have a high latent heat capacity for building applications.

The first step is to characterize the animal fat by the use of gas or liquid chromatography to achieve a precise composition of the different unstated fatty acids such as oleic acid and saturated ones such as palmitic and stearic acid. Following that the animal fat which is mainly composed of triglyceride molecules is broken apart by hydrolysis to achieve free fatty acids and from there, we go to Step 3, where, by use of crystallization method and a solution, the saturated and unsaturated fatty acids are separated. At this point, the saturated fatty acid that we were aiming for is achieved, however, by taking the hydrogenation process at Step 4, we are able to convert the unsaturated fatty acids from the previous stage to saturated fatty acids as well. This will increase the overall output efficiency and reduce the waste material. And the last stage is to use the esterification process to achieve fatty acid esters. The esters have some unique properties such as providing a wide phase change temperature, high latent capacity, and noncorrosiveness, and odorless to name a few. The final goal is to build a lab-scale BIPVT and use the produced PCMs inside it and evaluate the performance of the system.

The current technology is still at the technology readiness level 3, meaning that there are risks involved in the research feasibility such as the failure to accomplish the expected latent heat or high production cost; however, the success of this research could result in achieving a low-cost PCM from recycled animal fat which results in energy efficiency and savings in civil engineering infrastructures. The proposed project will be different from the existing research in (1) built upon Gallart-Sirvent's work, we will further develop the chemical synthesis process that can separate the saturated FA (usable PCMs) from the unsaturated FA and then explore methods of converting the unsaturated FA to saturated FA so we can maximize the use of the raw FAs; (2) characterize the thermal and physical properties of the produced PCMs; (3) apply the produced PCMs into a lab-scale thermal energy setup to evaluate its performance, which will benefit future large-scale applications.

To be consistent with the 3-stage experimental plan, the milestones for the project are:

1. Develop the chemical processes to convert animal fat to PCMs with maximum yields.
2. Implement the developed chemical procedure to convert raw animal fat (bacon grease) to PCMs and evaluate the PCMs' properties.
3. Apply the produced PCMs to solar PV panel for temperature regulation and assess the performance of the modified solar panel.

10.3.3 Bifacial BIPVT panel

Most BIPV systems monofacial solar cells to harvest the direct solar irradiation, whereas bifacial BIPV modules use solar cells at both sides to produce solar power where one side faces the sun with higher efficiency than regular solar cells and another side absorbs albedo light reflected from the environment. When bifacial modules are installed on a highly reflective surface, such as a white thermoplastic polyolefin (TPO) roof or on the ground with light-colored stones, some bifacial module manufacturers claim up to a 30% increase in production just from the extra power generated from the rear cell. It is often installed on an open roof with good light reflection so that it is not related to the thermal management of the PV module at all. Moreover, for many building applications, the roof is not open and the backside will not have sufficient light, traditional bifacial PV modules are not applicable.

A novel bifacial BIPVT concept is proposed to minimize the intake of infrared to keep the PV cell operating at a lower temperature while reaching a significantly higher energy harvesting efficiency (Yin et al., 2021). Fig. 10.5 shows a novel mechanism to use the bifacial cells for thermal management toward BIPVT applications without any air or water circulation system. The solar cells are embedded between the front glass and backsheet through an encapsulant made of TPO. An anti-reflective (AR) meta-surface coating is applied on the top surface whereas a reflective coating (say aluminum) is applied on the backsheet. When sunlight strikes on an AR coating of the solar panel, >98% irradiation will pass the coating in comparison to ~70% without AR. The solar energy passes front glass with minor loss to heat panel, 15%–25% of energy will be taken by silicon cells on the front side toward to the sun, which mainly concentrates in the visible light range but transparent to infrared that is ~50% of the solar irradiation. The backside will be another type of solar cell, say PSCs, which cover infrared very well. Therefore, another 10%–15% of solar energy can be converted to PV electricity. Because the infrared spread in a wide range, some infrared light will pass the bifacial solar cell and strike on the backsheet, which is built on a structural substrate or is another glass layer. The aluminum coating

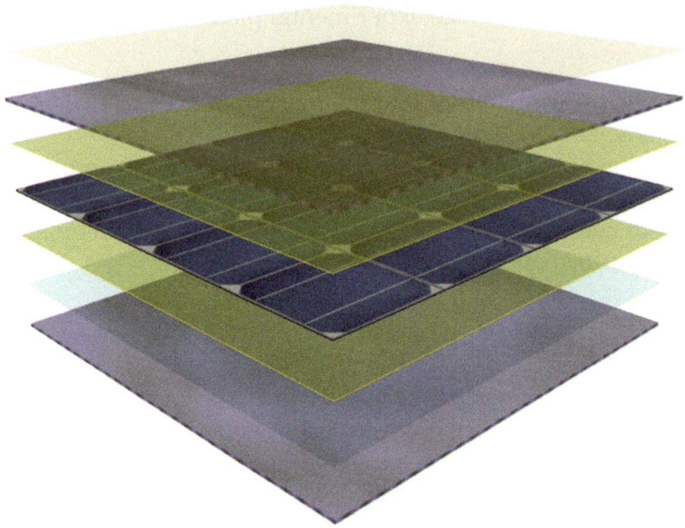

FIGURE 10.5 Schematic illustration of a bifacial PV module with smart coatings. *PV*, photovoltaic.

will reflect ~95% of infrared back to the bifacial cell for another time of PV conversion. It is estimated at 3%–6%. This design takes the advantage of the superior efficiency of silicon cells but recovers the loss by PSCs, which can reach the total efficiency at 28%–46%. Notice that unused solar energy indeed still heats the solar cells and releases more infrared radiation, which will still be reflected to the cell and then to space. In this way, a large portion of solar energy, which could heat solar panels or building roofs, is used by PSCs or reflected to the space. Therefore, the temperature of the roof panel can be lower than the traditional BIPV without thermal management.

A recent prototype of a dual-glass bifacial PV module has been fabricated at Columbia University. Two identical high-mechanical and thermal resistant glasses, and two smart coatings for solar spectrum management are layered up in an approximately symmetric pattern with respect to the bifacial solar cells. This symmetric configuration leads to zero stress in the bifacial cells under bending caused by the external load forces (wind, snow) during its service life. This improved mechanical stability avoids the need for a metallic frame around the module, thus reducing the cost and furthermore, the risk of potential induced degradation. Two smart coatings, one AR coating layer with a tunable meta-surface on the front glass and another reflective coating on the back glass are utilized to manage the solar spectrum that the bifacial PV cells receive in order to maximize their energy conversion.

Although the design of bifacial PV cells has addressed optimizing energy transmission by using an AR coating on the front surface and reflective coating at the back surface, the overall energy efficiency of bifacial BIPV modules may significantly increase with new module packaging methods. One way this may be accomplished is by using the rear side reflectance of solar modules to harvest infrared light that is not absorbed by today's thin crystalline silicon solar cells. By applying a reflective coating on the back glass of the dual-glass module, light that would normally pass through the module can be reflected into the module to be absorbed by the bifacial cell's rear surface.

Utilizing a special AR coating on the front of the module may also enhance bifacial module efficiency. As most bifacial solar cells experience considerable thermal loss at a higher temperature, absorbed infrared energy may not necessarily lead to higher efficiency. There exists an optimal range of infrared energy, which does not make cells overly heated but still supplies sufficient energy inputs. This optimal range can be achieved by utilizing a tunable meta-surface, based on phase transition materials (e.g., $SmNiO_3$) (Li et al., 2016; Yu and Capasso, 2014) on the front glass of the dual-glass module, which will control the transmission of infrared radiation into the module. Since infrared radiation with wavelengths higher than 1200 nm produces negligible electricity under the bandgap for Si cells, we can screen it out with the meta-surface.

The laboratory tests have shown that an additional AR coating on the front glass produced a 20% improvement in current density for bifacial cells compared to no AR coating. Because some bifacial cells only have 40% internal reflection for infrared, there exists the potential to significantly enhance the efficiency by backside reflection. In addition, reflecting solar irradiation reduces the temperature of the solar panel and recovers efficiency loss at a higher temperature.

To ensure the mechanical stability of the PV modules and provide sufficient protection to the cells and metallization, a dual-glass module packaging method is favorable.

Using a reflective coating on the back glass has the potential to increase both the efficiency and reliability of the PV module in a variety of installation conditions. The probability for future success is high given that no upgrades in module manufacturing equipment are necessary and the manufacturing procedure is similar to that of a standard PV module. The tunable short pass filtering properties of the meta-surface coating will be realized independently before integration with other layers of the solar cell. The meta-surface coating will be based on phase-transition material $SmNiO_3$, which provides a large change of optical refractive indices in the visible and near-infrared (NIR) when electrochemically doped with different amounts of electrons.

Spectroscopic measurements of reflectance will reveal the ability of the meta-surface to tune the short pass cutoff wavelength in the NIR.

Concurrently the same measurements will ensure that the reflection at wavelength <700 nm is comparable to commercial AR coatings used widely in today's panels. Spectroscopic ellipsometry will be used to ensure the minimum absorption in the meta-surface layer. Once the meta-surface is characterized, two solar modules using passivated back contact cells with a standard commercial AR coated top glass and the other with the new meta-surface layer will be fabricated and subjected to comparative internal quantum efficiency measurements.

However, internal cell reflection will vary with different cell architectures so a reflective coating on the bottom glass can serve many different kinds of solar cells. Today there are no commercial coatings for solar available that are capable of reflecting light through the rear. Monocrystalline cell panels have openings between cells while certain custom BIPV panels have even larger spacing to allow natural light. The reflective coating at the bottom can therefore allow infrared light to escape through these openings. Hence the reflective coating can be useful for both traditional bifacial monocrystalline panels and BIPV panels in reducing the heating of the mounted panels.

10.3.4 Transparent or translucent BIPVT panel

To avoid overly heating the BIPVT panels, one way is to let a portion of energy pass through them, which may exhibit a lower efficiency (Lunt, 2012) but can be very attractive for benefits in architectural aesthetics, indoor environmental control in temperature and light, and harmony with nature by removal of the visual barrier to the outdoor, etc. Transparent PV (TPV) modules can be used in BIPVT to balance both average visible transmission (AVT) and power conversion efficiency (PCE) and is particularly superior to be used in the glazed facades and windows for additional energy supply.

In the early stage, TPV cells were mainly designed by spatial segmentation to partially cover the surface area with traditional opaque PV cells and achieve semitransparency by tailoring the coverage ratio of the PV cells. The same mechanism has been used for dual glass BIPV projects, such as the Metlife Stadium Solar Ring, which allows light to pass through the space between the solar cells. When the segmentation area is small, the light transmission uniformity can be achieved, and human eyes cannot see the cells. This type of nonwavelength-selective technologies produce electricity from broad absorption of the solar spectrum (including visible photons) and the energy efficiency will be sacrificed with the increase of AVT. Basically, transparency and energy harvesting are competing with the surface area for their individual goals.

In 2011, Lunt and Bulović proposed a novel idea to selectively harvest solar energy based on the wavelength spectrum of the light, which employs photoactive materials that preferentially harvest UV and NIR light, while selectively transmitting the visible spectrum for light (Lunt and Bulovic, 2011). Therefore the novel PV material covers the whole area as well and theoretically it does not affect the AVT as all visible light can pass it. They demonstrated the method with an organic PV cell with AVT >55% although the PCE is low at ~1.7%. The TPV cells use a transparent tin oxide coating on the inner surface of the glass panes to conduct current out of the cell. The organic PV cell uses a molecular organic donor, chloroaluminum phthalocyanine, and a molecular acceptor, C60, that shows peak absorption in the UV and NIR wavelength span of 650–850 nm. Lunt's (Lunt, 2012) analysis showed that the theoretical Shockley–Queisser limit for the PCE of a wavelength-selective TPV single junction with 100% AVT is 20.6%, compared to 33.1% for an opaque PV. In 2016, another MIT team used graphene as the electrodes for an organic TPV and reached a PCE of 2.8%–3.8% at the AVT of 54%–61%. Although the efficiency of organic TPV is still too low for large-scale energy harvesting, it is sufficient for lower-power wireless sensors or electronic devices, particularly at night time. A similar mechanism has also been used for greenhouse shading and solar energy harvesting. Instead of harvesting UV and NIR light, the selective wavelength spectrum can be tailored for maximizing the growth of the crops while harvesting the useless light (Roslan et al., 2018).

Another mechanism to make TPV is to reduce the thickness of a thin film PV to permit some visible light transmission. Perovskite-based architectures have been attracted significant attention in TPV research. Typical PSCs are made out of a sandwich of a metal oxide material and organic transport materials. Perovskites are abundant organic materials that have good electric properties suitable for solar cells, such as a high absorption coefficient, high carrier mobility, direct bandgap, and high stability. Recently perovskite crystal structure can achieve a PCE of over 25% (Wang et al., 2020, BELLINI, 2020), which makes perovskite an excellent alternative to organic TPV although the AVT could be relatively lower.

In general, because TPV cells are compatible with curtain walls, windows, skylights, among other glass structures, and can replace the thermal coating on those building envelope components, if the cost-efficiency-lifetime can reach such a level that is comparable to silicon in opaque PV cell, another huge market on TPV on the glass will emerge. Husain et al. reviewed nine transparent solar cell technologies and summarized the advantages and disadvantages for seven of them in Table 10.1 (Husain et al., 2018).

Although they show different advantages, which may lead to different applications, perovskite PV cells show great potential to lead this emerging market. Currently, stability and durability are the major barriers to large-scale applications in buildings. The availability will become a bottleneck for rapid growth.

TABLE 10.1 Comparison of seven different transparent photovoltaic technologies (BELLINI, 2020; Husain et al., 2018).

Solar cell types	Advantages	Disadvantages
Screen printing dye sensitized solar cell	1. Low complexity in terms of the manufacturing process and final design. Can be produced at an industrial scale 2. Materials used to build the solar cell are organics like titanium and abundant 3. Cost of manufacturing is almost half the price of a silicon-based solar cell. 4. Efficient usage of architectural space 5. Manufacturing process is environmentally friendly 6. Have a homogeneous appearance 7. Shading and high temperatures do not have any major effect on performance	1. Low reliability, live time quality, and power output dropping rate need to be tested and improved 2. Not practical for large-scale deployments where higher-cost and higher-efficiency cells are essential 3. Liquid electrolyte is not stable at variable temperatures
Near-infrared organic photovoltaic (OPV)	1. Organic semiconductor, environmentally friendly 2. High transmission more than 55%	1. The architecture of the device is more complex compared with dye sensitized solar cell 2. Low efficiency not more than 1.7% 3. Low supporting resources. 4. Not ready for Industrial manufacturing
Polymer solar cell	1. Flexibility 2. Low material cost compared with silicon 3. Environmentally friendly 4. Flexible and lightweight 5. Low bandgap 6. Easy manufacturing for industrial requests using the roll-to-roll technique	1. Low efficiency 2. Low stability 3. Short operational lifespan.
Transparent luminescent solar concentrator	1. Improve energy harvesting efficiency 2. Increase electronic device autonomy 3. Reduce the traditional photovoltaic module cost	1. In case in complexity and manufacturing stamps 2. Low efficiency less than 1%

(continued on next page)

TABLE 10.1 Comparison of seven different transparent photovoltaic technologies (BELLINI, 2020; Husain et al., 2018)—cont'd

Solar cell types	Advantages	Disadvantages
Perovskite	1. Close to the market 2. Environmentally friendly 3. Can be made flexible in shape 4. Does not require high-temperature processes 5. Uniform perovskite layers 6. High efficiency	1. Low transmission 2. Most materials used are scarce elements like tellurium (as rare as gold), indium, and gallium 3. Liquid electrolyte has the temperature stability problems 4. Expensive compounds 5. Highly volatile 6. Low stability. Rapid degradation
Electrophoretic technique	1. The transparency can be increased with the use of nanotubes without affecting the efficiency 2. Manufacturing costs reduce with the reduction in the waste of active material used	1. Complicated process that includes chemicals and needs to be standardized for transparent solar cells 2. It is difficult to control the homogeneity 3. It can be used for laboratory-scale however mass-production is complicated 4. Reports supported this process are few compared to other processes
Quantum dot solar cell	1. Capable of absorbing photovoltaic waves from the infrared region of the spectrum 2. Low complexity in manufacturing and installing 3. Better stability than dye sensitized solar cell	1. Low transmission 2. Quantum dots are inorganic

10.4 Smart building with BIPV components

10.4.1 Background

According to the Building America Research to Market Plan (Werling, 2015), high energy consumptions and low comfort performance are the two main issues in today's building sectors, especially for residential homes. The buildings sector is the largest consumer of electricity, it was reported that the buildings sector accounts for about 76% of electricity use and 40% of all US primary energy use and associated greenhouse gas emissions.

The major areas of energy consumption in buildings are heating, ventilation, air conditioning, lighting, and other major appliances (water heating, refrigerators, dryers, etc.). The share of electricity use in building sectors has grown dramatically in the past five decades from 25% of US Annual Electricity consumption in the 1950s to 40% in the early 1970s to more than 76% by 2012. If there is no significant increase in building efficiency, total US electricity demand would have grown much more rapidly than it did during this period.

To address the challenges of transforming the energy use in buildings and to allow for their better integration into the future energy systems, it is necessary to revisit the way building envelopes are designed and manufactured. In order to make remarkable improvements with regard to conserving energy and natural resources, and improving the energy efficiency of buildings, new technologies of efficient and renewable energy supply systems are in high demand. They will also be essential to meet the world's energy and environmental challenges and to reduce costs to building owners and tenants.

According to the National Institute of Building Science, one of the core functions of buildings is to provide healthy and comfortable environments such as HVAC for human activities, which is to "maintain facility performance across its life cycle." In this sense, a building must supply a healthy and comfortable indoor climate to the occupants. An effective HVAC shall achieve three goals: control indoor temperature and humidity at comfortable levels, provide adequate fresh air from outdoors, and remove indoor air odors and pollutants by a combination of air pressure control, filtration, and exhaust ventilation systems.

A properly designed and balanced HVAC makes a huge difference in home comfort and energy efficiency. Human beings are extremely sensitive to temperature changes and even a few degrees of temperature change can make a home uncomfortable and unwelcoming. Overly heating/cooling is also very common over the country, which wastes energy and brings discomfort to occupants. In addition, it is another strategy for energy saving to minimize heating or cooling for space without occupants. It was reported that, depending on the climate zone, each degree increase in the summer saves 0.6% to 0.5% of total building electricity consumption. Each degree the winter setting is lowered saves 1.2%–8.7% of total building natural gas consumption (Gutierrez and Williams, 2016). With a minor revision of the thermostat settings, it is easy to achieve a national saving of 2.5% in all office buildings and restaurants, summing up to national savings of 69.6 million GJ annually, comparable to all 2015 total solar PV generation in the United States. If other types of commercial buildings such as hotels and stores get similar savings as offices and restaurants. Therefore, a robust temperature controller is highly demanded in order that one can make the indoor temperature stable and comfortable and can also make a home a healthier place.

The significance of indoor climate for health and comfort has been emphasized in recent years. People spend about 90% of their time indoors. Therefore, from the viewpoint of health, the quality of indoor air is even more important than outdoor air. It shows that serious health impacts may result from poor indoor air quality (IAQ). More widespread health impacts from exposure to indoor pollutants, and "sick building syndrome" symptoms due to elevated indoor pollutant levels as well as other indoor environmental conditions (ASHRAE, 2009). These more widespread impacts have the potential to affect large numbers of building occupants and are associated with significant costs due to health-care expenses, sick leave, and lost productivity. The potential reductions in health costs and absenteeism and improvement in work performance from providing IAQ in nonindustrial workplaces in the United States are estimated to be in the high "tens of billions of dollars annually" (Mendell et al., 2002). Moreover, when humidity is not effectively controlled, persistent dampness can lead to material damage, corrosion, structural decay, and microbial growth, including mold. Cognizant health authorities have established an association between damp buildings and the increased potential for adverse health effects. Research and practice have shown that the occupants are too often unsatisfied with the building. Complaints concerning modern buildings in respect of health and comfort are common all over the world. Therefore, an effective and reliable sensing and control are of great necessity to save energy cost, enhance indoor air quality, and protect occupants from adverse health effects as well as protect the building.

However, it is not trivial to add a new sensing and control system to an existing building, which was designed and operated in an old fashion because it needs to interface with electric, plumbing, HVAC, and information communication systems of the building and often requires a power supply, wiring, piping, and space for air ducts. Smooth and cohesive integration of the sensing and control system with the existing BMS is very critical to the adoption and implementation of new technology. Using BIPVT to power the sensing and control system through wireless communication can minimize the intrusion of the new system to existing building retrofitting and also provide improved resilience and workability to new construction. Following we will introduce some future smart building technologies with interfaces with the BIPVT system, which show the following intelligent aspects:

1. Adaptive to the environment.
2. Self-power without additional power supply.
3. Wireless communication without wiring.

10.4.2 Solar tracking and active buildings

Because the sun keeps moving, although solar panels exhibit higher efficiency with the technology advancement, the actual efficiency depends on the light intercepted by the panels. When the angle between the solar panel surface and the sunbeam is 90 degrees, the solar energy density can be fully taken by the panel. Otherwise, if there is any angle shift from this position, the efficiency is decreased by the cosine of the angle. If the panel can keep moving with the sun, Kelly and Gibson reported a ~50% increase of power output compared to a fixed-type solar cell (Kelly and Gibson, 2009). It requires a smart design to achieve it and solar tracking is the solution.

The apparent motion of the sun is caused by the rotation of the Earth about its axis by days and the motion of the axis around the sun by years. Therefore, the angle at which the direct component of light will strike the Earth exhibits daily and yearly changes in east–west (E–W) and south–north (S–N) directions respectively. From a fixed location on Earth, the sun appears to move throughout the sky. This apparent motion of the sun has a major impact on the amount of power received by a solar collector. The angle between the sun and a fixed location on Earth depends on the longitude and latitude of the location as well as the time of year and the time of day.

To track the daily and yearly repetitive motion of the sun, essentially two-directional tracking modes should be considered, although the daily one is more important. There are two main types of solar trackers available on the market: single- and dual-axis. Single-axis solar trackers track the sun east to west, rotating on an S–N axis; whereas dual-axis trackers rotate on both the S–N and E–W axes, making panels track the sun directly.

In the actual operation, a group of solar panels can track the sun in centralized or decentralized modes. Centralized or distributed trackers use a single motor to power a driveline between rows that will move an entire segment of panels in unison. Decentralized systems have one motor per tracking row of panels or per panel. There are also instances of trackers with motors present on every set of racking, making rows more adjustable during installation, and in some cases allowing them to track independently of neighboring modules. Active trackers using motors rely on algorithms based on time and location, or their sensing and control mechanisms to track the sun, which may depend on GPS, software, and hardware. Certain installation and operation costs are required, such as hardware and software, periodic calibration and maintenance, power supply and maintenance, etc.

Although the above solar tracking systems have been implemented at the utility-scale as a standard with devoted professionals for O&M services, it is not easy to implement in BIPVT systems because of the constraints in hardware and O&M services as well as the safety of the structure. Traditional buildings are static with their building envelope serving a physical barrier and an interface between inside and outside for solar radiation, precipitation, wind, and extreme temperatures that affect indoor comfort conditions of occupants. In recent years, active buildings (Falkeis, 2015), adaptive facades (Tabadkani et al., 2021), dynamic buildings (Elkhateeb et al., 2018), and smart buildings (Al Dakheel et al., 2020) have been proposed and constructed by introducing solar tracking mechanisms into the building envelopes. However, it is a long way to go to reach a sizeable market. Indeed, introducing the motion of the building components or the whole building will produce big impacts on traditional building designs and construction and require many experiments and a long process to reach a mature technology. Considering the 50% increase of power output, newer experiences in architectural aspects, as well as innovative ways for environmental adaptivity, these emerging technologies will find future customers if the cost and complexity can be significantly reduced. Moreover, it can provide an opportunity to integrate with the seismic design by separation and mitigation of the earthquake waves.

One way to reduce the cost and complexity of solar tracking in buildings is inspired by the sunflower, which moves with the sun elegantly without any cost. If each panel or the whole building can track the sun by itself like a sunflower, it may reduce or remove the market barrier of this technology.

A new passive tracker was developed for Columbia University, which rotates toward the sun using the thermal expansion powered by the sun as well.

Inspired by sunflowers in nature, we designed a new passive solar tracking system (as shown in Fig. 10.6) that orients solar panels toward

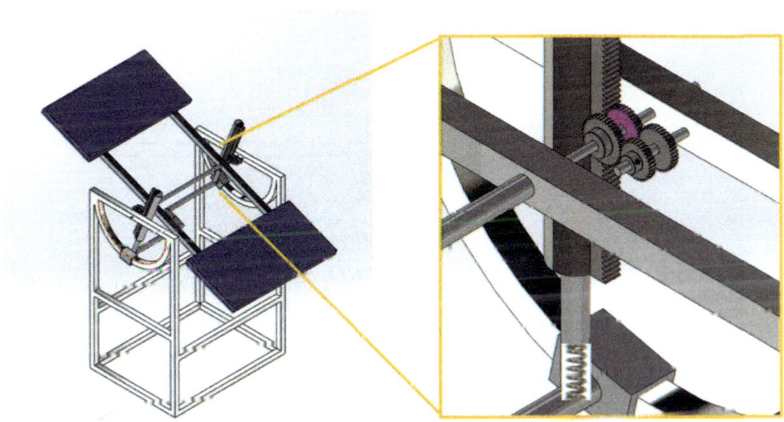

FIGURE 10.6 Schematic illustration of a passive solar tracker developed at Columbia University.

the sun (Yin et al., 2021). This system is driven by two modified hydraulic cylinders which expand differently when the sunlight is not vertical to the solar panel. This solar tracker is designed to be a simple mechanical system, sensitive to the sunlight angle change, and able to hold its position under strong wind. These performances are achieved through inventions of a double-circles supporting system, a spring-supported hydraulic system, and a gear system.

The working mechanism of the passive solar tracker is illustrated in Fig. 10.7 and described as follows. Here θ_1, θ_2, θ_3, and θ_4 are the rotating angles of gear #1, #2, #3, and #4, respectively; and r_1, r_2, r_3, and, r_4 are radius of gear #1, #2, #3, and #4, respectively. When the sunlight is on the right side of thermal rod #1, the absorptive surface on the right side of the rod harvests solar energy and heat the hydraulic fluid inside the rod. As a result, thermal rod #1 extends at a distance of Δl. The rod extension drives gear #1 and makes it rotate at a θ_1 angle. Because gear #1 and #4 are connected, gear #4 rotates at a θ_4 angle. The relationship between θ_1 and θ_4 can be adjust by the gear ratio. As gear #3 is fixed with #4, the rotating angle θ_3 of gear #3 equals to the rotating angle θ_4 of gear #4. Then gear #3 makes gear #2 rotate at an angle of θ_2 because these two gears are connected.

As gear #2 rotates at an angle θ_2, both the solar panel and the thermal rod #1 rotate with gear #2 at the same angle. Under the extension Δl of thermal rod #1, thermal #2, spring #1, and spring #2 will move according

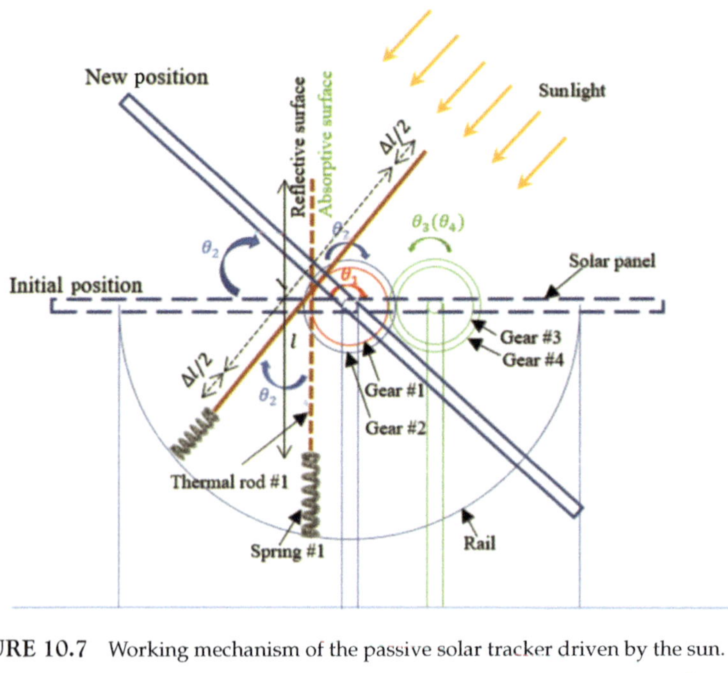

FIGURE 10.7 Working mechanism of the passive solar tracker driven by the sun.

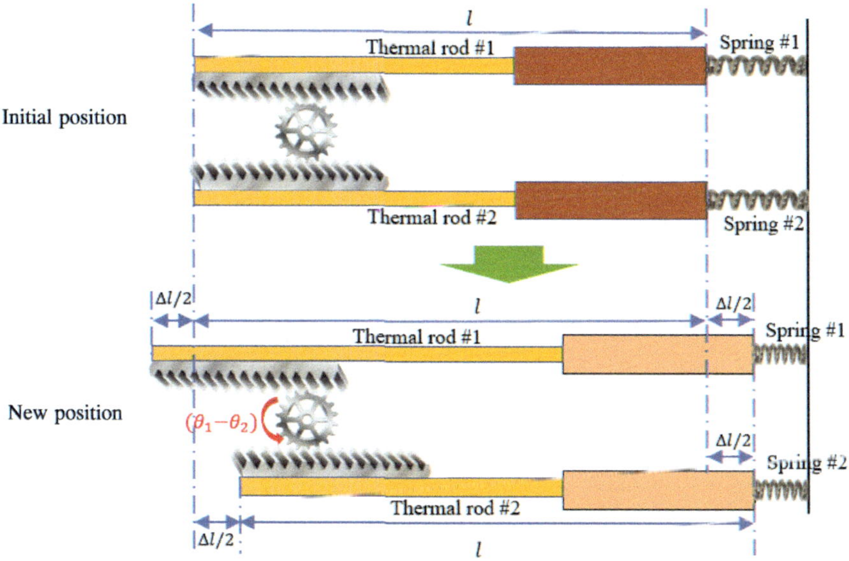

FIGURE 10.8 Displacements of thermal rods and springs.

to the relationship shown in Fig. 10.8. The length of thermal rod #2 does not change. Spring #1 on the end of the rod, which is connected to the rail, is compressed at $\Delta l/2$. Similarly, spring #2 which connects thermal rod #2 and the rail is also compressed at $\Delta l/2$. So, thermal rod #1 moves at a distance of $\Delta l/2$ with gear #1 (Fig. 10.8). Because the rail is in a semi-circle shape. Note that, by adjusting the parameter $\frac{r_2 r_4}{r_1 r_3}$, we can control the sensitivity of the solar panel angle change, i.e. the relationship between θ_2 and thermal rod extension Δl. When $\frac{r_2 r_4}{r_1 r_3}$ is close to 1, θ_2 can be a large angle even with a small Δl.

As the sun moves from East to West (right to left) during the day, the movement of the solar panel has the following stages as illustrated in Fig. 10.9.

At dawn, thermal rod #1 and thermal rod #2 have the same length, and the solar panel is at a horizontal position (Fig. 10.9A).

(a) As the sun comes up in the morning, thermal rod #1 harvests solar energy via its absorptive surface on the right side and expands with increased temperature (Fig. 10.9B). As a result, the solar panel rotates clockwise until it is perpendicular to the sunlight, i.e., the thermal rod #1 is parallel to the sunlight. During the morning, thermal rod #1 is longer than thermal rod #2.

(b) As the sun continues moving to the west (left) during the morning, thermal rod #2 harvests solar energy via its absorptive surface on the

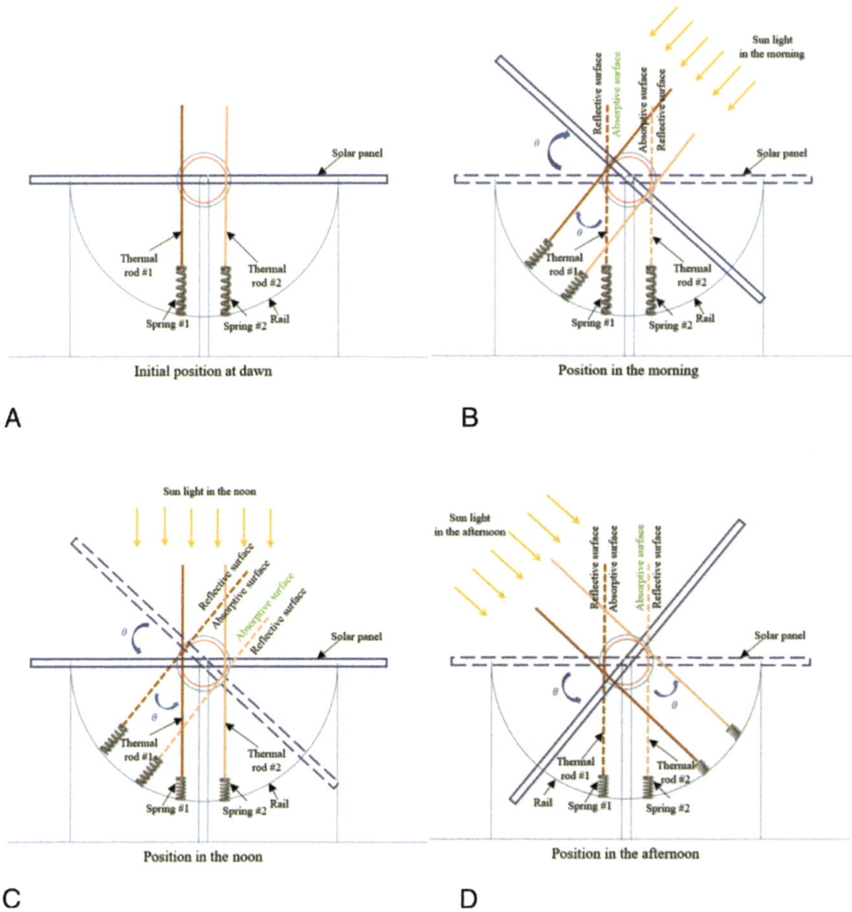

FIGURE 10.9 Movement of the solar panel during the day.

left side and expands with increased temperature (Fig. 10.9C). As a result, the solar panel rotates anti-clockwise. At noon, the solar panel is horizontal; thermal rod #1 and thermal rod #2 have the same length.

(c) As the sun moves further to the west (left) during the afternoon, thermal rod #2 continues harvesting solar energy via its absorptive surface on the left side and expands with increased temperature (Fig. 10.9D). As a result, the solar panel continues rotating anticlockwise. During the afternoon, thermal rod #2 is longer than thermal rod #1.

The passive tracker can be integrated into the building envelope without any wire or motor. In particular when concentrated solar power is used to boost the thermal fluid temperature or for the application of water desalination for zero water houses, this passive solar tracker will be very useful because it is self-adaptive to trace the sun, and will save significant costs for calibration and operation.

10.4.3 Self–powered sensing and control

As the largest consumer of electricity, the buildings sector accounts for about 76% of electricity use but a significant portion of electricity consumption is unnecessary. Reducing unnecessary consumption can increase the building's energy efficiency. As humans are the key for energy consumption in buildings, the humans' presence and preference should be sensed for indoor environmental control, such as moisture, HVAC, and lighting. A large amount of energy could be saved if humans in a building can be accurately detected so that the BMSs will automatically change the indoor environment to meet the indoor comfort of occupants. Such kind of smart building concept has attracted significant attention in the civil engineering industry during the past decade and is estimated to be worth $53.45 billion by 2022 (Research, 2016). Significant attention has been paid to the controlling algorithms of the smart BMS thanks to the boom of artificial intelligence, with a lot of discussions about energy consumption forecast (Wang et al., 2018), smart grid technology (Raza and Khosravi, 2015), and in-room occupancy detection (Chen et al., 2018).

In contrast, the importance of the sensor network was underestimated with fewer research and discussions. Traditional systems rely heavily on wired sensors to monitor the environment, which comes with substantial costs of manpower and time in both installation and maintenance. Therefore, the concept of the wireless sensor network (WSN) becomes more and more popular in the BMSs to ease the installation and maintenance, to enhance flexibility in dangerous areas, and to reduce the cost of sensor deployments. However, most of these sensors are powered by either replaceable or rechargeable batteries, which means they are exposed to limited battery lifetime, environmental hazardous nature, and complex replacement procedures. For certain applications in the building environment, it could be very hard to reach the WSN node for the battery replacement and would introduce an enormous scale of maintenances cost.

The energy harvesting technologies provide a good solution to elongate the battery life and to reduce the maintenance cost of a WSN such that the self-powered wireless sensor network system can be achieved. Energy harvesting technologies use power generating elements such as solar cells, piezoelectric elements, electromagnetic components, and thermoelectric elements to convert light, pressure, vibration, and heat energy into electricity. A comprehensive review of energy harvesting applications in the buildings (Matiko et al., 2013) identified that solar energy and thermoelectric energy are two popular energy sources for the WSN systems in buildings.

Solar energy-based self-powered wireless sensor network was proposed by Tzortzakis et al. for environmental monitoring and smart city design, with the corresponding field test conducted on the campus of the National Technical University of Athens (Tzortzakis et al., 2017). The solar panels provided abundant energy supply up to 3.5 W but also resulted in a relatively bulky size of the prototype. BIPVT can integrate the sensing and WSN into the building envelope and meet the dual goals of both self-powered, wireless smart building control, and architectural aesthetics. By the continuous monitoring of the environment and the smart control of the major building appliances, such as HVAC, lighting, and water heating, etc. Research shows that a potential energy saving of 34.78% could be achieved by the smart buildings comparing to conventional buildings, through the optimal control of major building components like the heating, ventilation, air conditioning, and lighting systems (Berawi et al., 2017).

Fig. 10.10 provides an overview of the proposed self-powered occupancy sensing for BMS. The electricity from the BIPVT panel will be stored in a battery and be used to power the sensors for occupancy detection and microcomputer to transfer sensing data to the data server of the BMS. The sensing platform will be constructed with open access to all other types of sensors for better detection and tracking of occupants and finally a smarter BMS. A microcomputer will be built in each sensing device for controlling and analyzing the data directly from the sensors, after which the data of human detection and counting will be sent to the BMS server for HVAC control. Wireless communication devices will be utilized for the communication of data and remote and automatic control from the microprocessor to sensors and other controlling units. Wi-Fi, Bluetooth, or Bluetooth Smart technique will be used for wireless communications, depending on the balance of transmitting speed and energy consumption.

This BIPVT powered wireless sensing and control system can be easily installed in retrofitting or new construction buildings. Because the hardware is totally independent of each other and information communication is through WSN, the sensing nodes can be flexibly added, relocated, or removed. Machine learning may identify occupants and obtain their preference of indoor environment parameters from historic data so that BMS can operate the system to achieve their desired indoor environment.

10.4.4 Smart building skin

The concept of smart building management has attracted significant attention in the civil engineering industry during the past decade. It provides the solution to achieve optimal interior comfort with minimal energy consumption but would also require smart controlling algorithms and

FIGURE 10.10 Schematic illustration of a self-powdered occupancy sensing system based on image processing and information integration/management for energy efficiency in buildings.

continuously monitored environmental data with Internet of Things devices. Significant attention has been paid to the controlling algorithms thanks to the boom of artificial intelligence, with a lot of discussions about energy consumption forecast (Wang et al., 2018), smart grid technology (Raza and Khosravi, 2015), and in-room occupancy detection (Chen et al., 2018). The importance of the sensor network is underestimated with less research and discussion. Considering the energy consumption of data acquisition, processing, and wireless communication, the bottleneck for sensing the environment is the limited time of operation due to the insufficient battery capacity. For specific applications in the building environment, it could be very difficult to reach the sensing node to replace the battery, which can result in costly maintenance procedures on large scale. Therefore, energy harvesting provides an excellent solution to elongate the battery lifetime and to reduce the cost and time in both installation and maintenance.

On the other hand, because the building envelope is constantly exposed to environmental change, it is in the critical position for environmental sensing and human detection. Therefore BIPVT integration with the sensing and control nodes can make smart building skin. Below we will introduce a recent technology demonstrated at Columbia University with a smart window (Lin et al., 2020), which is extending to smart building skin.

This novel sun-powered smart window blinds (SPSWB) system is designed, prototyped, and tested to improve the energy efficiency of the building envelope by harvesting and storing solar energy, improving the energy dissipation, sensing the surrounding environment, and automatically controlling the window blinds. The SPSWB is designed with flexible PV cells, coated with innovative porous coating for heat emittance, and automatically controlled by the motor and micro-controller based on the sensed data. Compared with the existing alternatives, the present SPSWB system is not only equipped with the energy harvesting component, but also with an innovative cooling system thanks to the porous PVdF-HFP coating, and a smart sensing and control system to actively monitor the environment for optimal performance. This smart window blinds system extends the application of solar energy harvesting to the vast areas of building envelope components, self-powers various kinds of sensors and actuators, and function as an integrated platform to fulfill the building health monitoring and to test controlling algorithms for the next generation of the smart BMS. From the industrial perspective, it also provides a structured design guide and comprehensive test data that can be directly converted to market products. The self-powered functionality provides a clean and self-contained system and leads to tremendous cost savings by eliminating the hassles of electrical wiring and battery maintenance.

The SPSWB system is designed based on an aluminum window system with built-in window blinds, which is schematically shown in Fig. 10.11. The window blinds are covered with flexible PV cells on one side and coated with a porous coating on the other side. The harvested energy from solar cells will self-power the operation and provide additional energy for external features. The heat absorbed by the PV cells will be dissipated through the thermally emissive coating made of porous polyvinylidene fluoride-co-hexafluoropropylene (PVdF-HFP) (Mandal et al., 2018), which can drop the temperature of 5 ~ 8°C lower than the ambient in summer, such that the overall energy efficiency could be increased. Sensors are integrated with continuous environmental monitoring and communicated to a local microcontroller for preliminary data processing. The sensed data will serve as valuable information for the BMS for the optimization of the overall energy efficiency. Although only a temperature sensor is used in the current prototype, the system can be extended to various types of sensors

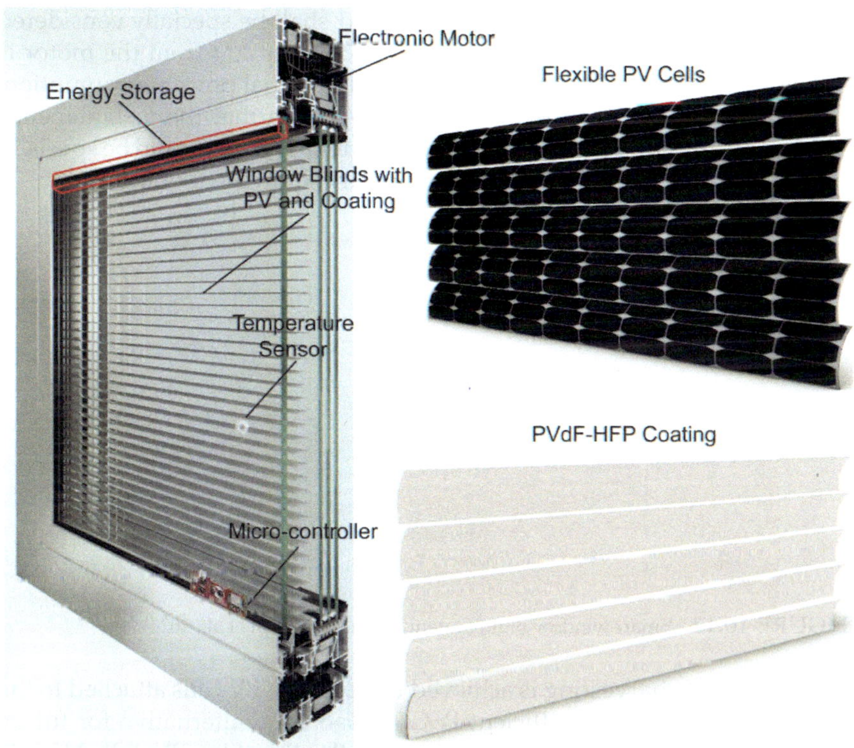

FIGURE 10.11 Schematic drawing of the smart window unit and innovative blind blades (Lin et al., 2020).

for comprehensive environmental monitoring, such as a light intensity sensor to detect direct sunlight, a moisture sensor for the raining leakage detection, a pollution sensor to monitor the quality of air, and an acoustic sensor for the noise level measurement. The corresponding responses of actuators, which is the motor in the scope of this research, are controlled via the microcontroller based on the sensed information, with the goal of achieving the optimal internal comfort with minimum energy expenditure.

The major components of the SPSWB and the relations are schematically shown in Fig. 10.12. The PV cells, the battery, and the power management system form the energy harvesting subsystem, while the sensors, microcontrollers, and the motor compose the energy consumption subsystem. The self-powered design would require the harvested energy larger than the consumed energy within a certain period. The aluminum window system used in this study is the Schuco AWS120 window, where the physical dimensions are 0.9 × 1.5 m. The 24 V DCD22-2-E double shaft motor is integrated into the blind system and can provide a nominal torque of 0.6 Nm. It is worth mentioning that the motor consumes the most power among all the components and shall be specially considered in the engineering. The nominal current consumption from the motor is 320 mA, which corresponds to the 7.7 W of nominal power consumption. The starting current of the motor is 1.45 A, such that the instantaneous power consumption reaches 34.8 W. The battery system should be specially designed to satisfy the instantaneous power requirement of the motor.

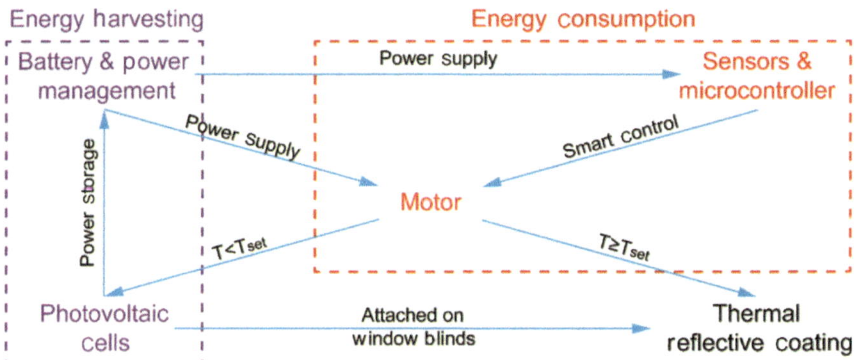

FIGURE 10.12 Smart window blind system architecture (Lin et al., 2020).

Solar energy harvesting is achieved through the PV cells attached to the aluminum blinds. High efficient PSCs are an ideal alternative for future applications. However, in this prototype, the thin-film PV cell, MP3-25 manufactured by PowerFilm, is selected and glued to the window blinds. The MP3-25 is a thin-film (0.22 mm) amorphous solar cell with great flexibility but compromised energy conversion efficiency, which is around

5%. The operating voltage of each PV cell is 3 V and the nominal power output is 90 mW. It is worth mentioning that the surface of the MP3-25 is not UV stabilized and therefore would require special treatment for real applications. Each one of the window blinds is covered by 5 pieces of MP3-25 PV cells and connected in series to boost up the voltage output. The open-circuit voltage is around 14 V under in-room lighting, and around 19 V under the solar simulator. As is shown in Fig. 10.13, the area with the holes could not be fully covered by the PV cells in the current prototype, and a higher voltage output could be expected if tailored cells could be manufactured. 18 pieces of blind slats were fabricated with PV cells and integrated into the window blind system, as shown in Fig. 10.13. The PV-covered blind slats are connected in parallel to boost up the current output for energy harvesting. Around 3 mA of short-circuit current is observed under in-room lighting and 400–850 mA under the solar irradiation of 850–1700 W/m^2. In the scope of this research, 18 pieces of blind slats are covered by PV cells, which corresponds to 30% of the total surface area of the window. More harvested power could be expected with a larger PV-covered area.

FIGURE 10.13 PV integrated smart window blind (Lin et al., 2020). *PV*, photovoltaic.

The other side of the blind slats is covered with the porous PVdF-HFP coating, which has the excellent passive daytime radiative cooling capability. The porous PVdF-HFP coating was recently invented to have a high hemispherical solar reflectance (0.96±0.03), which minimized the solar heat gain, and a high long-wave infrared emittance (0.97±0.02), that maximized the radiative heat loss. The coating was tested in Phoenix and achieved $\sim$6°C lower than the sub-ambient temperature under 890 W/m^2 of solar irradiation, while the corresponding cooling performance of 96 W/m^2 was recorded under 750 W/m^2 of averaged solar irradiation (Mandal et al., 2018). During the PVdF-HFP coating, a primer layer is first applied on the window blind surface to enhance the bonding strength, which takes approximately 10 min to dry. Then, the polymer coating is sprayed onto the blinds via a commercial spray gun Flexio 590 from Wagner Spraytech six times at an interval of 10 min. The polymer coating is initially liquid and turns into a white-colored porous solid film quickly after exposure to the air. Five window blinds are coated with the PVdF-HFP as is shown in Fig. 10.14 and attached with PV cells on the other side to test the cooling performance.

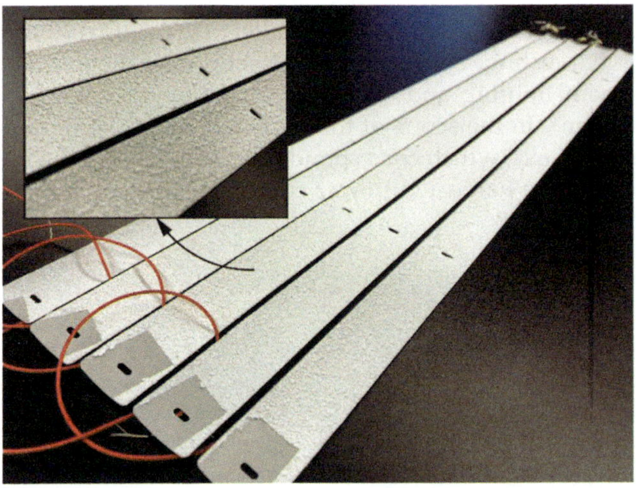

FIGURE 10.14 PVdF-HFP coating on the other side of the window blind (Lin et al., 2020). *PVdF-HFP*, polyvinylidene fluoride-co-hexafluoropropylene.

The majority of the solar energy is generated when the sun is shining brightly, but virtually none is captured at night or in substantial cloud cover. Therefore, an energy storage unit must be introduced to account for the intermittency of solar energy harvesting. As the major energy consumption component, the DCD22-2-E double shaft motor requires a 24 V power input with a start-up power requirement of 34.8 W. In order to meet the start-up current requirement, three 18650 lithium-ion batteries are connected to make a 12 V energy storage unit, as is shown in Fig. 10.15, and are managed by a battery management circuit for overcharge and over-discharge protection. The LTC3780 buck-boost converter is applied to stabilize and regulate the voltage to 12 V for battery charging. It automatically adjusts the internal resistance for the optimal power extraction from PV cells. The voltage is then taken up to 24 V with the help of a voltage booster for the power supply of the DCD22-2-E motor.

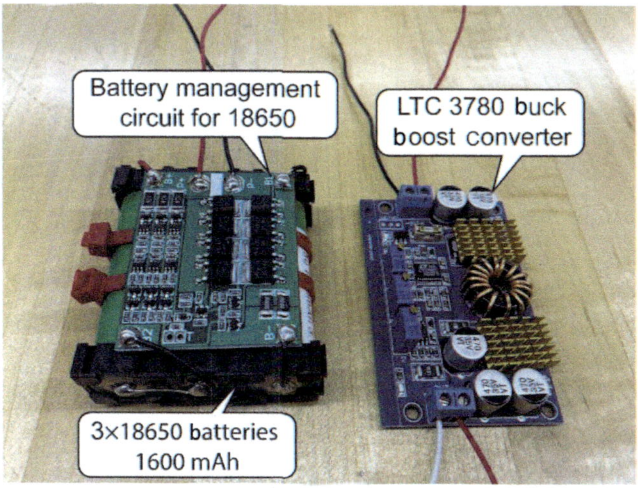

FIGURE 10.15 The 18,650 battery pack and the LTC3780 buck-boost converter (Lin et al., 2020).

The PV cells attached to the window blinds convert solar energy into electricity but also absorb the solar thermal energy and increase the internal temperature. The out-facing solar cells on a hot summer day will significantly increase the temperature and compromise internal comfort. Therefore, a smart controlling system is needed to monitor the surface temperature of the window blinds and to utilize the radiative cooling capability of the porous PVdF-HFP to dissipate the heat when the temperature is above a prescribed threshold. The Arduino Uno R3 development board is used as the microcontroller to collect the temperature measurements from the DS18B20 thermometer and to conduct logical decisions as to when to rotate the window blinds. A dual H-bridge module, L298N, is used to switch the polarity of the applied voltage on the motor, such that the rotating direction of the motor can be digitally controlled by the signal from the microcontroller. The physical components of the Arduino Uno and the H-bridge are shown in Fig. 10.16 and the schematic drawing of the connections is shown in Fig. 10.17. The input signal is connected to the lower side of the H-bridge. Ranging from left to right one can find the 24 V power input from the battery, the ground, and 5 V logic power supply signal from the Arduino Uno. The advantage of having a 5 V signal is that the H-bridge can be turned off completely by the controller, meaning that no current will go through when the motor is not working. The output screw-terminals are located on the left and right of the H-bridge, both of which are connected to the motor with one wire serving as the positive and the other as the ground wire. The output signal is equal to the 24 V power input and the polarity is controlled via the 5 V control signals. The

FIGURE 10.16 The Arduino Uno R3 and the H-bridge (Lin et al., 2020).

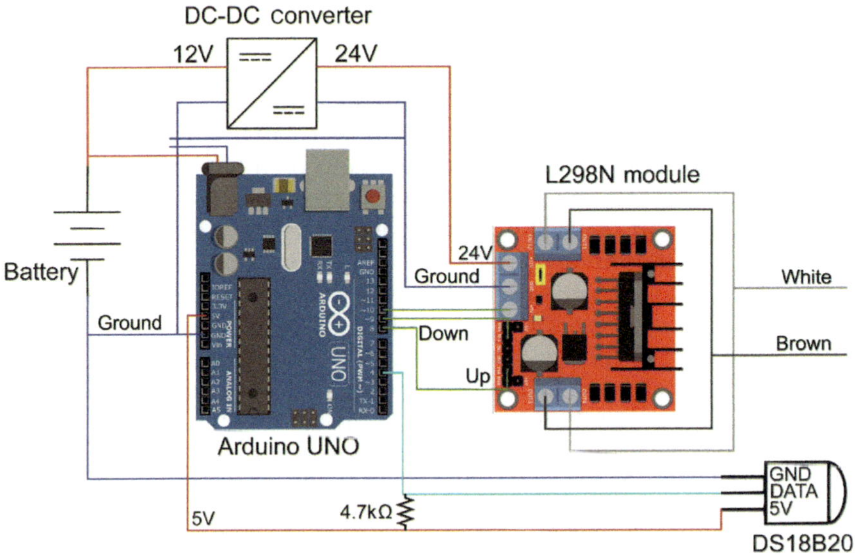

FIGURE 10.17 Schematic drawing of the connections in the controlling system (Lin et al., 2020).

deep sleep mode is activated in the Arduino Uno and every 8 s, the system will wake up from the deep sleep to collect the temperature measurement from the thermometer. It can wake up after the Watchdog-timer finishes or by a manual interrupt signal, as by the example of pushing a button. The microcontroller will then compare the temperature measurement to the predefined temperature thresholds and decide whether to send out the signals for the motor control, after which it will go back to deep sleep and wait for another cycle.

This smart window concept has been demonstrated in the laboratory environment with excellent performance for year-round operation in the New York metropolitan area (Lin et al., 2020) and can be extended to whole building skin. The combined energy harvesting and thermal management with solid physics simplify the installation and operation procedure and enable significant improvement of energy efficiency performance. Particularly, as the federal government and many states of the United States are setting the minimum energy efficiency standard, it is time for this emerging technology of smart building skin to enter the emerging market and grow quickly.

10.5 Circular eco-manufacturing of BIPVT panels

10.5.1 Background

With the exponential growth of the PV industry in the past decade, end-of-life management of solar panels is important for both the existing ones and the future products to ensure that clean energy solutions are a sustainable component of the energy economy for future generations.

Although solar equipment can last for decades, particularly with proper maintenance, different components exhibit different lifetimes. For example, for an EVA dual glass module, the glass may last ~100 years; the solar cells may still be useful after 40 years; but the encapsulant may exhibit yellowness or brawniness in a much shorter period, particularly if there are defects during the lamination process, which reduces the PV cells' light interception rate and fails in overall efficiency. If they can be separated, some components can be reused. For example, PV cells can be reused or refurbished to have a "second life" of generating electricity. The other components of solar systems can also be handled responsibly. Inverters can be recycled as e-Waste and racking equipment can be re-utilized with newer technology or recycled like other metals.

Because a lot of efforts will be involved in the recycling of waste modules, the cost to recycle a solar module is fairly high, which could be out of the expectation of many existing users. For example, a recently formed company—Recycle PV Solar (Recycle PV Solar LLC, 2021)—charges a fee of $25/module for their service, excluding the costs to decommission, package, and ship waste modules. Instead of sending the waste module to the recycling center with extra cost, reusing the valuable components represents the most sustainable way for future solar business. However, the existing manufacturing method makes it very difficult to recycle PV modules by components. The circular eco-manufacturing model can excellently fit the needs, as the circular life cycle employs reuse, repair, refurbishment, remanufacturing, and recycling to create a closed-loop system, minimizing the use of resource inputs and the creation of waste, pollution, and carbon emissions (Geissdoerfer et al., 2017). This regenerative approach will exhibit the great potential to renovate the existing linear process in the BIPVT industry, which has a "take, make, dispose" model of production.

To make a circular life cycle, the straightforward way is to make the end-life product as a feedstock to the beginning of the manufacturing. Meanwhile, it is wise to address the weakest part for frequent replacement. In BIPVT, the encapsulant is the weak part and needs to be replaced most frequently. In the following, we will first review the emerging industry for solar panel recycling, then introduce a new manufacturing process that is much easier for recycling, and finally, present a new technology for modular manufacture and installation of the BIPVT system tailored for circular eco-manufacturing with minimized soft costs.

10.5.2 Solar panel recycling

In 2016, the SEIA of the United States initiated a national voluntary solar module recycling program (SEIA, 2021a). By creating a member-based program, SEIA facilitates a cost-effective and environmentally responsible end-of-life management solution, which enables others in the industry to access recyclers for their disposal needs. This program will include specific benefits for members who utilize the services of SEIA Preferred Recycling Partners. Some SEIA members operate take-back and recycling programs or engage with suitable waste handlers, disposers, or resellers of their products. These recyclers may collect, transport, process, and resell valuable materials from PV modules, inverters, and other equipment. Some companies refurbish and resell PV equipment for secondary market usage such as off-grid or exported outside of the US. End-of-life disposal of solar products in the US is governed by the Federal Resource Conservation and Recovery Act (RCRA), and state policies that govern waste disposal or other disposition.

Currently, most commercial recycling process only recycles the bulky materials in Si modules: glass cullet, Al frame, and Cu wiring. Mechanical methods are used to strip materials from modules as follows: remove junction box from the modules; extract Cu wiring from the junction box; remove Al frame from the modules; and shred the remaining module for glass cullet. The glass cullet contains EVA, Si cells, and backsheet as they are glued together. Some recyclers do not remove the Al frame or junction box but shred the entire module. By grinding the shredded modules, Si can be separated from the glass cullet. The impure Si and glass containing EVA will be used as a feedstock for new Si cell and glass sheet production. The actual cost to collect and process waste modules is much higher than the value of the recycled materials. Therefore, today's recyclers always require government subsidies or fees from module owners to be profitable.

Considering the higher value and longer lifetime of Si cells, alternative recycling methods (Tsanakas et al., 2020) of Si were proposed: (1) refurbish and reuse waste modules; (2) extract and reuse components from waste modules; and (3) extract and reuse materials from waste modules. Because BIPVT's service life is also long and PV cell efficiency increases significantly in recent years, reuse waste modules with lower efficiency may not be economically favorable for the life cycle cost because the remaining lifetime of those refurbished waste modules will be lower too. Reuse the components or materials from waste modules can be a feasible way to increase the sustainability of the BIPVT while maintaining desirable performance.

The efficiency loss of BIPVT is commonly caused by the aging and degradation of the organic parts such as encapsulants, silicone adhesives, and backsheets, the inorganic components of glass and Si PV cells exhibit higher stability and can maintain their good performance for a much longer period than the designated lifetime of the module. If the inorganic part can be separated successfully, the value of waste modules can be significantly increased. Unfortunately, the EVA encapsulant cannot easily be removed because they have been cross-linked during the original application process. New encapsulants and packaging methods, which are suitable for delamination at a certain condition, can resolve this problem and enable the reuse of the inorganic parts and recovery of the efficiency of aging BIPVT panels. When Si cell's efficiency is too low for reuse, it should be recycled by chemical methods to extract the expensive metals such as silver, copper and also to collect environmentally hazardous metal lead first, the remaining silicon can be purified again for PV cell production.

Although the BIPV recycling industry is still in the budding phase, emerging technologies in BIPV and BIPVT will have an increasingly large market as more and more existing solar panels are getting to the end of life phase. More importantly, if future BIPVT can considering the recycling or reusing methods and costs in the manufacturing phase, a circular eco-manufacturing process can be established to eventually stop using more raw materials into this industry, which is similar to the steel manufacture in the United States that most steel is produced from recycled scrap.

10.5.3 Solar panel laminating method for easy delamination

With the technology advancement, Si cell efficiencies are close to the Shockley–Queisser limit, so the pace of efficiency improvement will slow down and cell reuse will become a very realistic option. However, the existing EVA-based packaging method has created a very challenging problem in recycling waste modules. Moreover, the lamination process is prone to create bubbles in the module which will significantly reduce the lifetime of the BIPVT system.

Fig. 10.18 illustrates the existing lamination process of PV module manufacturing. The assembly of glass—encapsulant—PV cell—encapsulant—

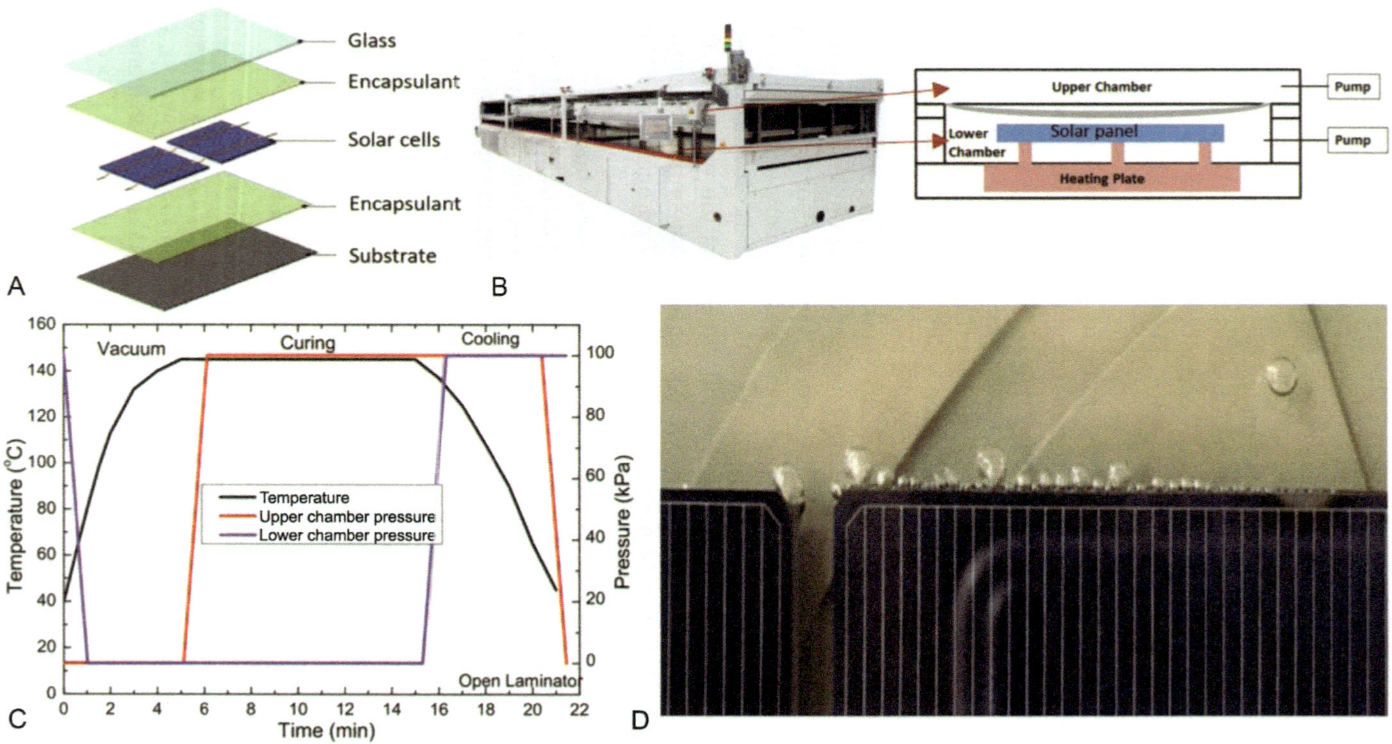

FIGURE 10.18 The lamination process and bubble formation in solar panel manufacturing. (A) The assembly of glass—encapsulant—solar cell—encapsulant—substrate layers before lamination. (B) A laminator for solar panel manufacturing. (C) The lamination recipe with temperature and pressure control. (D) The bubble formation and motion in lamination.

substrate layers in Fig. 10.18A is transported into a laminator in Fig. 10.18B, in which the temperature and pressure are well controlled to allow for crosslinking and curing of the polymeric encapsulant. Fig. 10.18C shows a representative lamination recipe we used recently. The laminator is heated to the curing temperature under a vacuum, following by encapsulant curing under pressure from the upper chamber to ensure strong adhesion between different layers of the PV module. Meanwhile, the lower processing chamber of the laminator is evacuated to remove air bubbles in the module. However, very often bubbles may not be completely removed from a PV module (Fig. 10.18D), which causes hot spots on the module under sun irradiation, PV efficiency loss, and delamination-induced lifetime reduction. Generally, a trial-and-error method is used with temperature and pressure as variables to develop the lamination recipe, which requires making a number of mockup modules and often leads to overdosed encapsulant and longer laminating process to get bubble-free modules. EVA has been a dominant encapsulant in the PV industry. However, UV cut-off and degradation can cause loss of efficiency and reduction of the lifetime. Furthermore, EVA is very 'sticky" creating difficulties in panel recycling at the end of life due to its high cross-link level.

Viscoelastic TPO and polyvinyl butyral have been proposed to replace EVA for future BIPVT packing due to their unique thermomechanical properties. Because the macromolecules of those thermoplastic polymers are crosslinked in the manufacturing, when the temperature increases to ~120°C, it will be melt again and lose the adhesive strength. Therefore, the glass, solar cell, and backsheet or substrate can be delaminated, which would enable the recovery of clean glass and PV cells and therefore increase the recovery value of BIPVT panels (Matiko et al., 2013; Tzortzakis et al., 2017). A new approach is proposed by spray coating of an encapsulant around PV cells as a three-dimensional (3D) encapsulation (Fig. 10.19), which exhibits the following advantage over the existing 2D laminating in Fig. 10.18:

1. Through a precisely controlled dose and uniformity in spray coating of the encapsulant applied all around c-Si cells only, providing better weather protection, and using less encapsulant with optimized coating thickness and targeted coating area, while 2D laminating uses EVA films given a uniform thickness.
2. Less encapsulant leads to less UV degradation of the encapsulant, which slows down the efficiency loss rate and improves the PV module's usable lifetime.
3. Because bubbles can move out at the edge of cells, in contrast to the edge of modules in the 2D laminating, bubble-free modules can be manufactured with high speed and high quality.

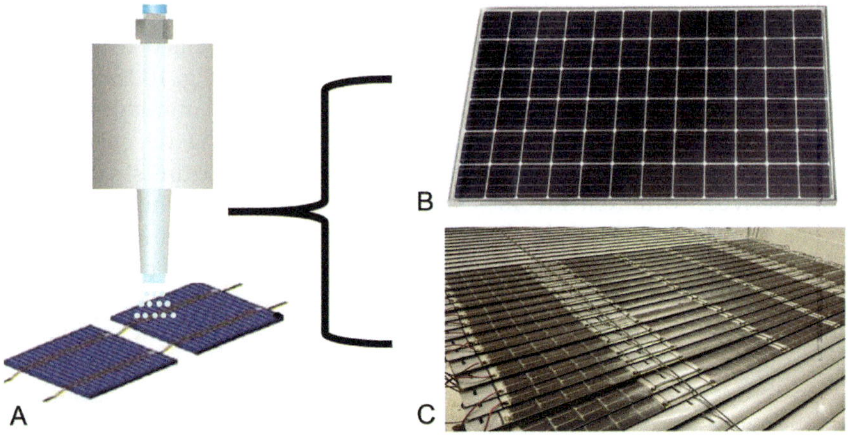

FIGURE 10.19 New spray coating-based encapsulating method toward future manufacturing of glass PV modules and thin-film modules. (A) Spray coating on solar cells with precisely controlled coating thickness and coverage. (B) Laminating cells in glass for regular PV modules. (C) Curing the coating layer for thin-film modules. *PV*, photovoltaic.

4. With a short encapsulating duration, the crosslinking level of the encapsulant can be manipulated for better life-cycle performance and better recyclability by delamination at the end of the lifetime.
5. Using thermoplastics like TPO and polyvinyl butyral instead of thermosetting materials like EVA could reduce efficiency loss and, therefore, increase the longevity of PV modules.

Various spray coating methods, such as ultrasonic spray, air spray, thermal spray, and electrostatic spray, can be used for different encapsulants, leading to different encapsulating processes. The success of the new encapsulating processes shall provide the following advantages in comparisons with EVA 2D laminating: (1) 30% reduction of manufacturing time; (2) 3% efficiency increase by less UV cut-off; (3) 40% projected lifetime increase by less UV degradation caused efficiency loss; and (4) improved recyclability by delamination.

10.5.4 Modular design and construction of the BIPVT building envelope

The adoption and growth of future circular eco-manufacturing of BIPVT not only relies on better life cycle cost performance than the existing linear manufacturing but is also influenced by the initial installation costs and constraints. Particularly, when it enters the market without previous success yet, the initial investment and workability of the project will play a critical role in market acceptance. Because the soft costs including labor, design, permitting, inspection, certification, etc., have a big weight (>65%)

(SEIA, 2019b) in the project cost, reducing and controlling the soft costs in future BIPVT installation is critical to its success.

Modular construction is a process in which a building is constructed off-site, under controlled plant conditions, using the same materials and designing to the same codes and standards as conventionally built facilities, but it can significantly expedite the construction and installation process with much reduced soft costs. The centralized manufacture of modules under a factory-controlled process will generate less waste, creates fewer site disturbances, and allows for tighter construction. The components can be disassembled, relocated, or refurbished for new use, reducing the demand for raw materials and minimizing the number of energy inputs. As most work is completed inside a factory, which mitigates the risk of weather delays and related quality variation, projects can be completed within a shorter, more predictable timeline for better quality control.

Future BIPVT can use modular design and construction with PV integrated into the wall panels already. When a large wall panel is delivered and installed, the BIPVT can be commissioned in a "plug-and-play" fashion. Lightweight concrete can be an ideal material for the BIPVT wall panel. The Columbia team has developed a technology to fabricate a lightweight wall panel made from a unique mixture of foam concrete in Fig. 10.20 (Yin et al., 2021). Infused with PCMs capable of absorbing heat on hot days and releasing stored heat on cold days, the panel can stay cool in the summer and warm in the winter for PV efficiency and indoor temperature conditioning. Lab tests and numerical simulation demonstrate that a 1-inch-thick wall panel could shrink the electricity demand of a 100-m^2 house by ~50% in the summer months of New York City. Larger-scale applications will dramatically improve the energy efficiency of buildings and reduce their carbon footprint.

The panel is made by adding micro-sized and millimeter-sized air bubbles generated by a foaming agent to cement paste. Polymer fibers and metal mesh are used for reinforcement. Next, they mixed and placed the foamy concrete into a mold. Once cured, ~85% of the concrete panel contained small airy pockets. This novel foam concrete exhibits higher performance in thermal and acoustic insulation, as well as good mechanical properties. The capillary pores inside offer an ideal means to infuse PCMs by vacuum for thermal energy storage. The fibers inside the concrete also ensure the composite has an adequate strength (3.3 MPa) to support the workload in service and installation. This novel lightweight panel is applicable in energy efficient wall panels and building integrated PV thermal roof panels.

This disruptive technology is different from the existing process of mixing encapsulated PCMs in cement pastes, which is costly and inefficient. Notice that although most PCMs used in building materials consist of paraffin wax that is underutilized and inexpensive, its encapsulation is expensive. By creating uniformly distributed microstructures in concrete

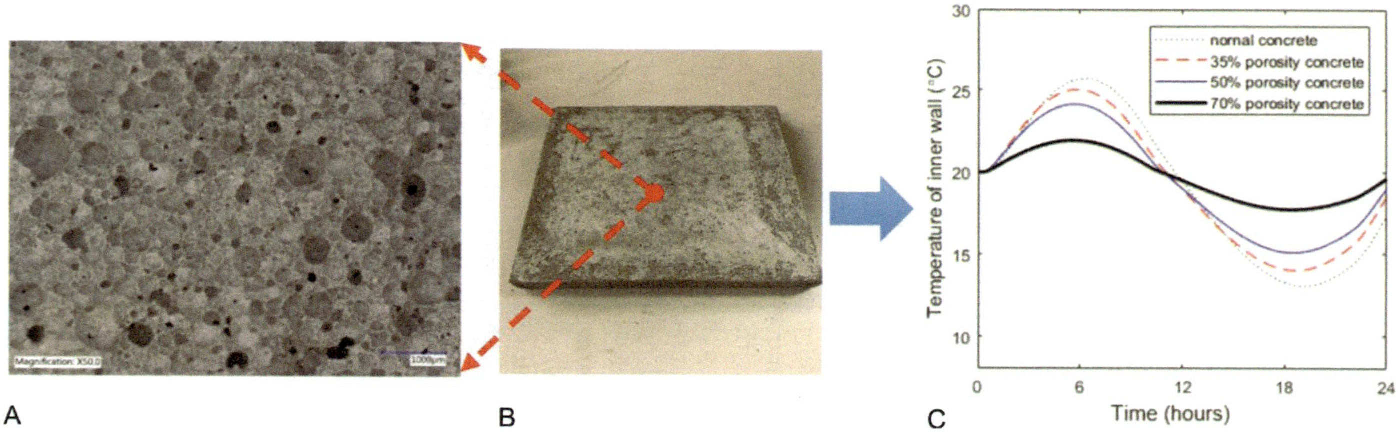

FIGURE 10.20 A phase change material (*PCM*) filled concrete panel for energy-efficient building envelope: (A) a scanning electron microscope (SEM) image, (B) a 12° × 12° × 1° foamed concrete panel, and (C) simulation results of the temperature fluctuation in a PCM filled concrete panel (Aiche, 2020).

for PCM incorporation, we used a multiscale modeling method to guide the design and manufacture of the PCM wall panels. Compared to a commercial foam concrete panel (called UltraliteTM) that costs $43 on the market, this new technology can cut the price in half while adding to its thermal energy storage ability.

The phase-change wall panel can keep room temperature fluctuations to a minimum, which would greatly reduce the energy needed for heating and cooling a room. For instance, Fig. 10.20 illustrates the lightweight concrete panel fabricated at Columbia University. The SEM photo in Fig. 10.20A shows that foamed concrete panel consists of a micro-sized pore structure (medium pore size ~200 μm) suitable for PCM infusion. The panel exhibits a flexural strength of 3.3 MPa suitable for building envelope applications (Fig. 10.20B). By storing heat in the latent heat of phase change, the panel's temperature fluctuation can be significantly reduced. Compared to a regular concrete wall, a foam concrete wall with 70% PCM embedment reduced room temperature swings by 9°C—up or down, which is corresponding to a 50% reduction in energy needed to heat and cool a 100-m^2 house (Fig. 10.20C). These panels can be engineered for specific climates by combining several PCMs with tuned melting temperatures.

10.6 Case study 1: Austin Sunflower

This is an iconic project located at Mueller Development shopping mall along the Interstate I-35 in Austin, Texas. There are totally 15 sunflowers installed in this project, of which eight are made of large modules and seven small modules (Fig. 10.21). The output power for a large sunflower is 1155 W while a small sunflower generates 840 W. The Austin Sunflower produces around 17,000–18,000 watt-hour of energy annually to the Austin Energy grid. During the day, the sunflower provides some shades to the pedestrians and the bikers. At night, it lights up and provides some lighting on the pathway (Fig. 10.22). In addition, it draws attention to motorists. It is through our imaginations on how one could apply PV technologies to these types of creative projects. The sunflower system was designed by Lajos Heder and Mags Harries with partial donation by Applied Material Foundation.

10.6.1 The photovoltaic modules

The PV modules adopt the physical characteristics of Glass—EVA—Solar Circuit—EVA—Glass. Two sizes of sunflowers were designed with 7 modules in each flower with their geometric dimensions shown in Fig. 10.23. The module electrical characteristics are provided in Tables 10.2 and 10.3 for the large and small modules, respectively, at the

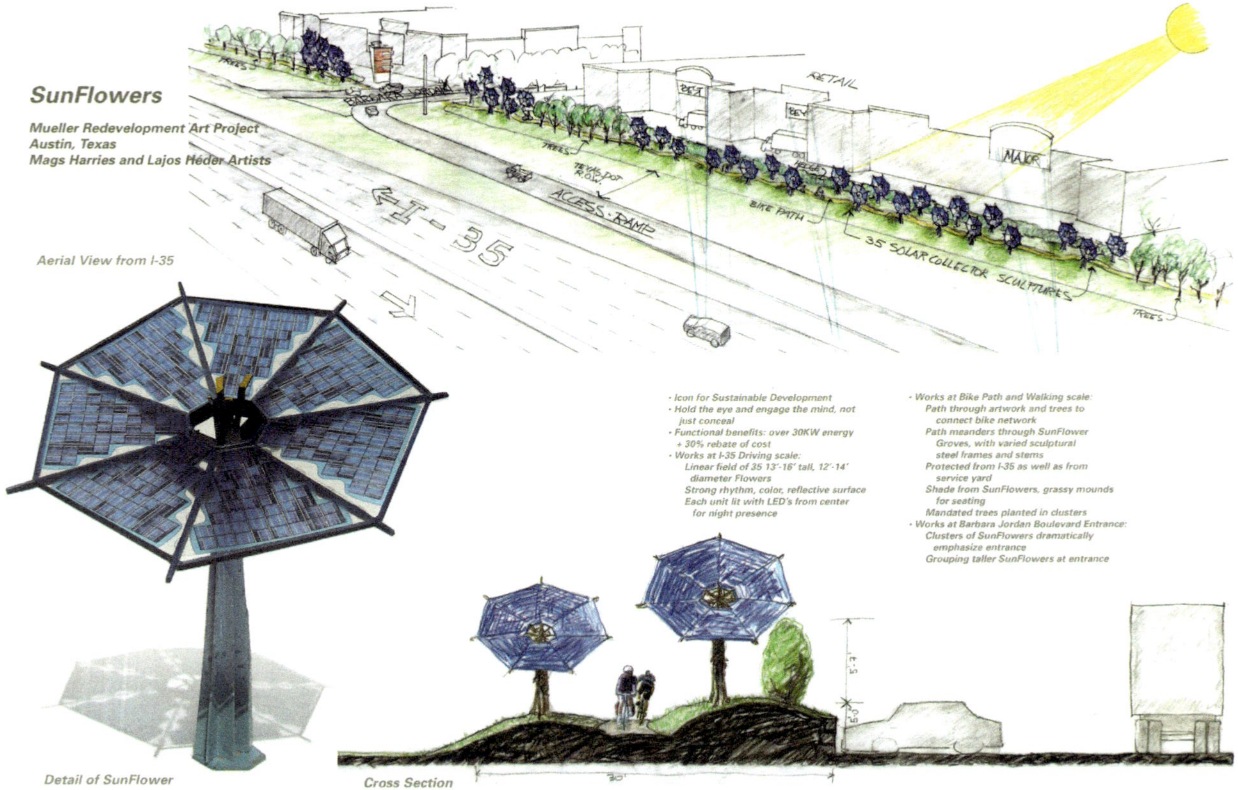

FIGURE 10.21 Schematic of the SunFlower project (Heder).

FIGURE 10.22 Sunflower at night with LED lights (Heder).

TABLE 10.2 The module electrical characteristics of the large module at the standard test condition (STC).

Voc	**26.4 (V)**
Isc	8.0 (A)
Vmpp	21.8 (V)
Impp	7.58 (A)
Pmpp	165 (W)

TABLE 10.3 The module electrical characteristics of the small module at STC.

Voc	**19.2 (V)**
Isc	8.0 (A)
Vmpp	15.9 (V)
Impp	7.54 (A)
Pmpp	120 (W)

standard test condiction, i.e. 1,000 W/m^2 irradiance, 25 degrees C, AM1.5, and ASTM G173-03 standard spectrum. The bill of materials are provided in Tables 10.4–10.6. The end product of two types of modules are shown in Fig. 10.24.

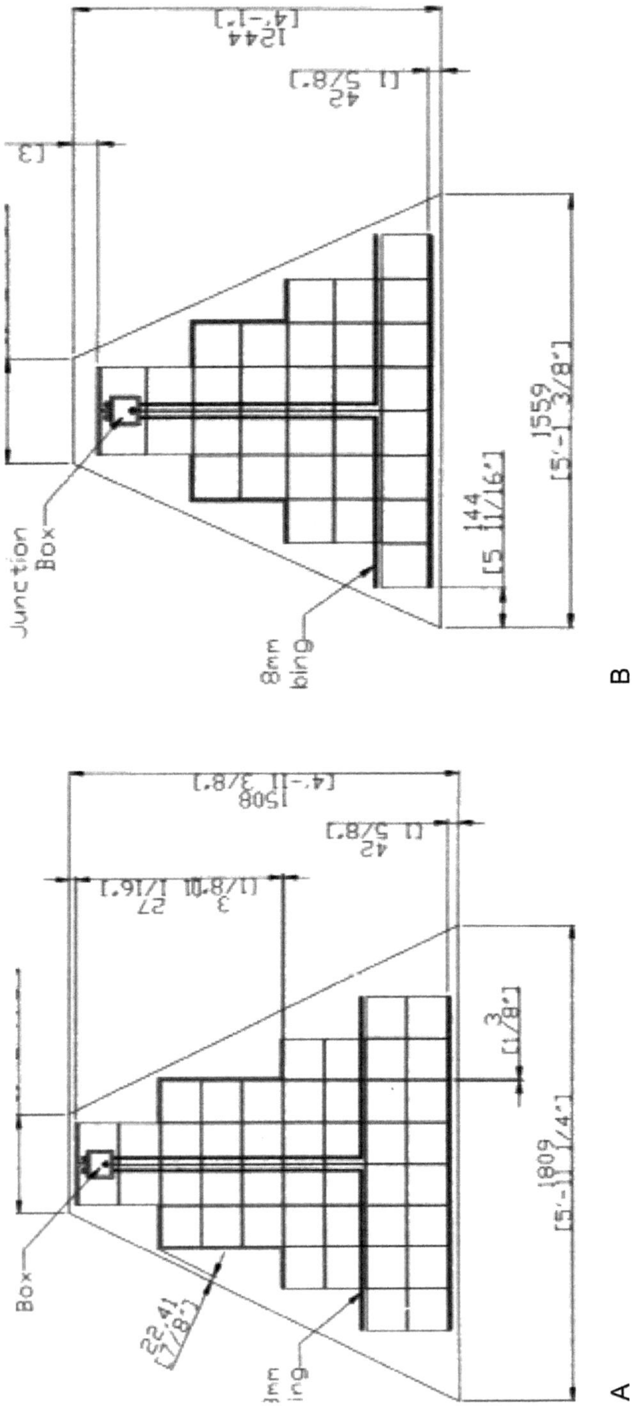

FIGURE 10.23 Left: large module of the SunFlower project, Right: small module.

TABLE 10.4 Sunflower module bill of materials.

Series item#	Identifier	Quantity for assembly	Vendor/contact #	Model or part number	Certification file attached	Width dim (mm)	Length dim (mm)	Area (M^2)	Additional data
100	glass				↓				
101		1	QINGDAO YUESHUICHAUN INTERNATIONAL TRADE QUINDAO CHINA	LARGE TRAPE-ZOID		1509	1096.5	1.655	LOW IRON CLEAR FLOAT; TEMPERED
102		1	QINGDAO YUESHUICHAUN INTERNATIONAL TRADE QUINDAO CHINA	LARGE TRAPE-ZOID		1509	1096.5	1.655	LOW IRON CLEAR FLOAT; PARTIALLY TEMPERED WITH HOLE
103		1	QINGDAO YUESHUICHAUN INTERNATIONAL TRADE QUINDAO CHINA	SMALL TRAPE-ZOID		970	1245	1.208	LOW IRON CLEAR FLOAT; TEMPERED
104		1	QINGDAO YUESHUICHAUN INTERNATIONAL TRADE QUINDAO CHINA	SMALL TRAPE-ZOID		970	1245	1.208	LOW IRON CLEAR FLOAT; PARTIALLY TEMPERED WITH HOLE

(continued on next page)

200	encapsulant				↓				
201	LARGE SUN-FLOWER	SEE CAL-CULATION	SPECIALIZED TECH RESOURCES 10 WATER STREET ENFIELD, CT 06082 1-860-749-8371	1529/UF	YES	1509	1096.5	1.655	AS PER UL1703 CLAUSE 7.1 3 FULL LAYERS PLUS 2 PARTIALS=4.97M^2+ 1.22 M^2=6.19 M^2
202	SMALL SUN-FLOWER	SEE CAL-CULATION	SPECIALIZED TECH RESOURCES 10 WATER STREET ENFIELD, CT 06082 1-860-749-8371	15295/UF		970	1245	1.208	AS PER UL1703 CLAUSE 7.1 3 FULL LAYERS PLUS 2 PARTIALS = 3.624 M^2+1.22 M^2= 4.844 M^2
300	Tabs/buss				↓				
301		SEE CAL-CULATION	ULBRICH PRECISION FLAT WIRE 692 PLANT ROAD WESTMINSTER, SC 29693 1-864-647-9524	P/N7746-9464/-9574	YES	>	>	>	lengths of material ribbon (2.032 mm) large=406mm × 44 cells/total 17.86M/small=406 mm × 32 cells/total 13M buss material (8mm) large=7162 mm or 7.162M small=5689 mm or 5.689M
400	cells				↓				

TABLE 10.5 Sunflower module bill of materials.

Series item#	Identifier	Quantity for assembly	Vendor/contact #	Model or part number	Certification file attached	Width dimension (mm)	Length dimension (mm)	Area (M^2)	Additional data
401		44	Q-CELL GUARDIANSTRABE 16 06766 THALIHEM GERMANY	Q6L	**YES**	156	156	1.06	
402		32	ersol Solar Energy AG Wilhelm-Wolff-Staβe 23 99099 Erfurt Germany	E6		156	156	0.78	
500	backstreet	**N/A**			↓	N/A	N/A	N/A	
600	j-box				↓				
601		1	MULTI-CONTACT USA 5560 SKYLANE BLVD. SANTA ROSA, CA 95403 1-707-575-7575	PV-JB/2-2 BOX	YES	152.4	127	0.019	BLACK IN COLOR

(continued on next page)

700	diodes								
701		2	DIOTEC SEMICONDUCTING AG KREUZMATTENSTR. 4 79423 HEITERSHEM GERMANY 011-49-76-345-2660	P1200 SERIES	YES	N/A	N/A	N/A	
800	connector				↓				
801		3	MULTI-COTACT USA 5560 SKYLANE BLVD. SANTA ROSA, CA 95403 1-707-575-7575	PV-AK-3 TERMI-NAL CLIPS	YES	N/A	N/A	N/A	BLACK IN COLOR
900	cable/ conductor								

TABLE 10.6 Sunflower module of materials.

Series item#	Identifier	Quantity for assembly	Vendor/contact #	Model or part number	Certification file attached	Width dimension (mm)	Length dimension (mm)	Area (M^2)	Additional data
901			US WIRE & CABLE 6301 E, Stassney Ln Expo 6, Suite 100 Austin, Texas 78744 1-800-880-9473	#12 awg	YES	N/A	21″ long times two per unit	N/A	large and small units have same length leads for cables therefore 42″ length of cable per units
1000	Cable ends				↓				
1001		1 each	MULTI-CONTACT USA 5560 SKYLANE BLVD. SANTA ROSA, CA 95403 1-707-575-7575	PV-A-KST3(M)/ PV-A-KBT3(F)	YES	N/A	N/A	N/A	BLACK IN COLOR
1100	frmae	N/A			↓	N/A	N/A	N/A	

(continued on next page)

1200	edge sealant	*none*				N/A	N/A	N/A	
1300	hadware				↓	N/A	N/A	N/A	
1301		blue backsheet	ORACAL	cut to size sheets for back glass coverage		1509	1096.5	1.655	customer required for large SF
1302		blue backsheet	ORACAL	cut to size sheets for back glass coverage		970	1245	1.208	customer required for large SF
1400	label								
1401		1	TECHTAG & LABEL 1901 North Penn Road, Hatfield, PA 19440 1-800,983-8324	TBD 4X4 SIZE BLANK	N/A	102	102	0.0104	
1500	Manuals								
1501			ASSEMBLED BY MANUFACTURER		N/A	N/A	N/A	N/A	

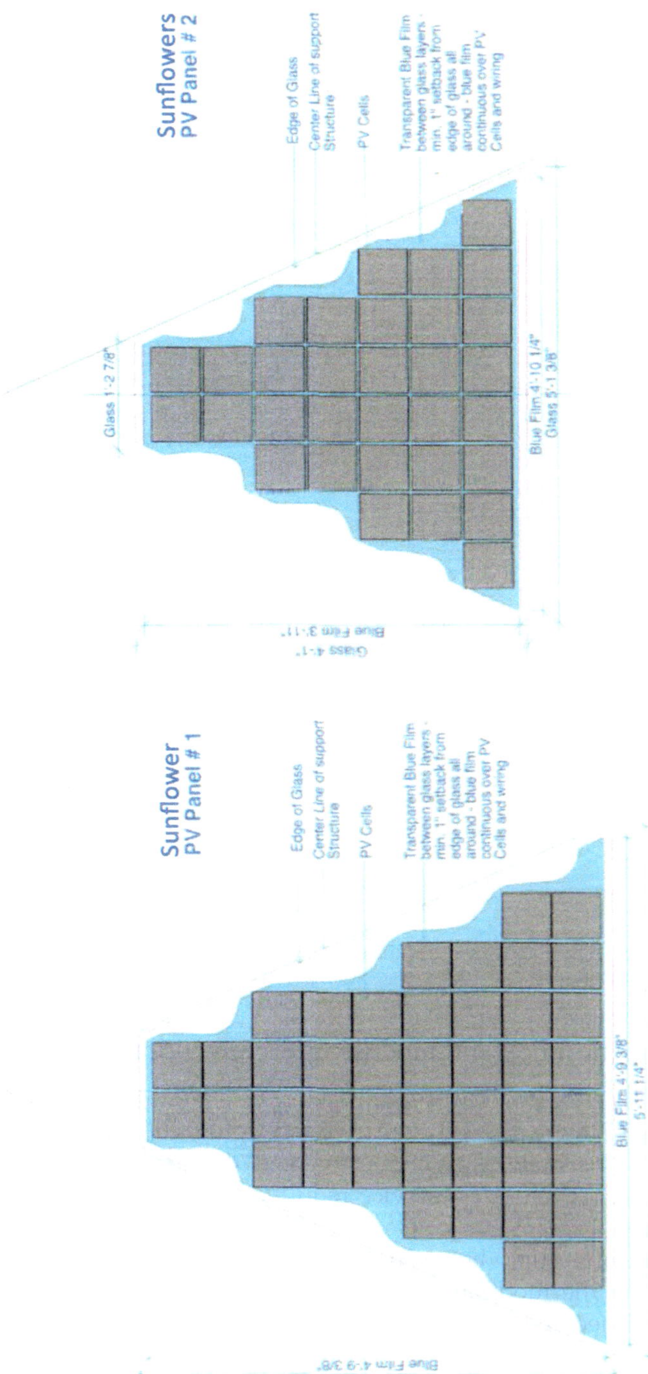

FIGURE 10.24 Sunflower modules.

The total thickness of each module is approximately 13 mm. Each layer of glass is 1/4 inch (6 mm) thick, which is four times the thickness of a standard module. The front glass is fully tempered whereas the back glass is partially tempered and edged. The junction box and connection cables are on the back of the module. The frame and 3D layout of the sunflower are provided in Fig. 10.25 and the actual view in the field is shown in Fig. 10.26.

FIGURE 10.25 The detailed view of the sunflower.

10.6.2 Junction box and module interconnection

Each module has an MC-type junction box mounted on the back of the module. Exiting the box are two MC plug connectors. The positive cable is the male plug connector, and the negative is the female plug connector. There is no service required on the junction box during installation.

The interconnection between modules must be done after physically securing the modular array on the metal structure by connecting the plug connectors to the cables that were pre-installed in the wire raceway.

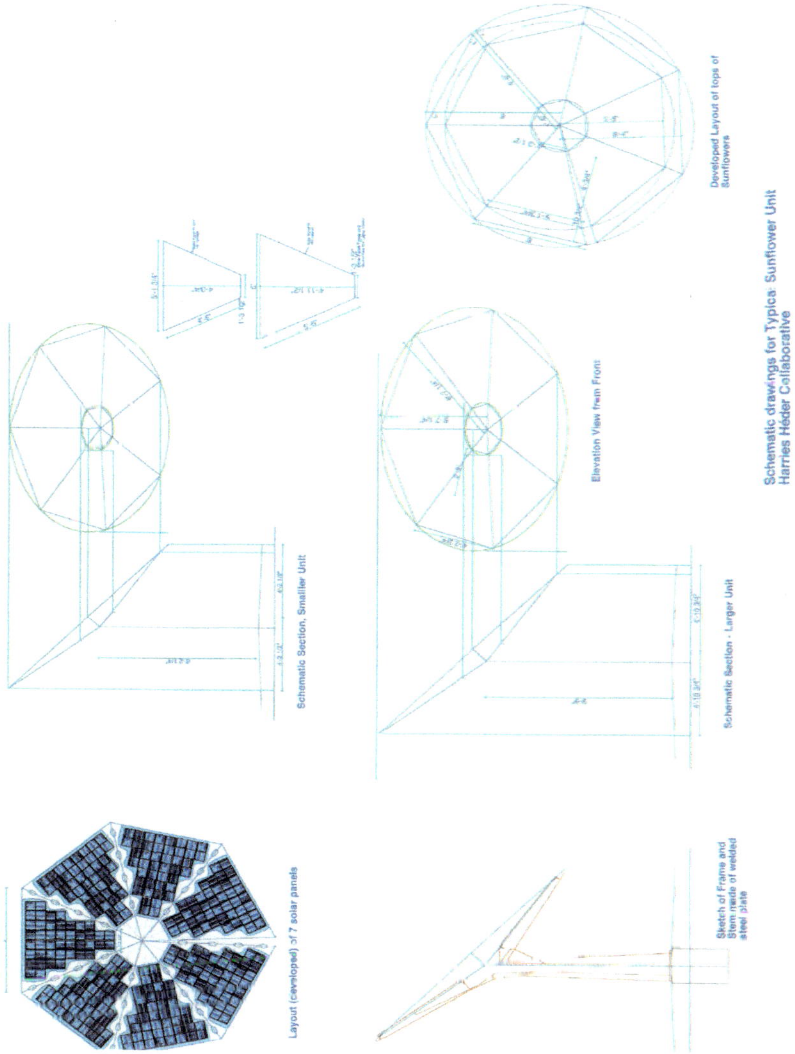

FIGURE 10.26 Sunflower design with the frame and modules.

10.7 Case study 2: Modular manufacture and construction of energy effecient houses by use of sandwich panels

Here, a few examples of houses built with Novasip from Kenpo Sandwich AB panels will be introduced. Kenpo Sandwich is a Swedish company that manufactures and sells building systems for low-energy houses, passive houses, off-grid houses, and plus energy houses. The goal of the lightweight panels is to achieve faster construction times, be cost-effective, and save the environment by use of this technology. These panels are very fast to assemble, economical, due to the high U-value and minimum energy needed for heating and cooling the space, and self-supported for construction from low-weight insulated composite panels. In Fig. 10.27, a 3 floor house was constructed in 16 hours with a crane for assembly of the sandwich panels.

FIGURE 10.27 House build in Stockholm in only 16 h, and with just a crane for assembly.

Shown in Fig. 10.28, this house was built nearby Malmö in Sweden. It uses 7000 kWh per year for heating and warm water. The rainwater is collected in a big tank in the ground. The water is used for the toilets, washing the car, water for the garden, etc. The owner is planning to install solar panels on the roof, which afterward, makes this a plus energy house. Figs. 10.29–10.34 show different houses or buildings in Europe with low energy at different geographic locations for different purposes.

This low energy house technology exhibits unique potentials to change the state of the practice of building construction:

- Dense cities by building on the roof of other buildings. Since this system is extremely lightweight, it can be built on top of other buildings, often without needing to reinforce the building below.

FIGURE 10.28 House in Malmö, Sweden.

FIGURE 10.29 Low energy house built in Stockholm.

FIGURE 10.30 Houseboat in Sweden.

FIGURE 10.31 Houseboat in Germany.

FIGURE 10.32 A plus energy house.

FIGURE 10.33 A hotel in Wenngarn castle in Sigtuna.

FIGURE 10.34 A low-energy house in Stockholm.

FIGURE 10.35 "A house without cord," plus energy house in Stockholm.

- Plus energy houses and off-grid houses.
- Living on the water.
- Temporary buildings that need fast assembly, such as emergency housing.
- Less conservative building industry.

Fig. 10.35 shows a plus energy house in Stockholm, which is called "A house without cord." Here are some of the advantages:

- Produces 17,000 kWh.
- In total, 68 solar modules: 48 on the roof and 20 on a fence next to the house are installed.
- Uses 8000 kWh, which means it can sell 9000 kWh to the grid or charge 2 Tesla cars.
- Heats warm water when overproducing solar electricity.
- Owns a drainage system.

What do the customer and environment gain from this technology? Let us look at the CO_2 emission produced from these panels and compare it with concrete:

Concrete CO_2 emissions 809 Kg CO_2/ton = 219 ton × 809= 177,171 Kg CO_2 emissions
Novasip CO_2 emissions 1.4 Kg CO_2 per Kg panel weight = 8580 kg × 1.4= 12,012 Kg CO_2 emissions

Difference in CO_2 emissions = 165,159 Kg CO_2

Freight from factory-Stockholm: For Novasip panels = 2300 USD
Freight from factory-Stockholm. For equivalent concrete= 27,600 USD
Difference in freight= 25,300 USD

And for the comparison between the weight of these panels with concrete with equivalent area:
Load in Novasip (Fig. 10.36):

10 pcs house foundation: Length 12,100 mm
3 pcs walls with the plastered outside: Length 12,100 mm
U-value: about 0.11
Weight: about 8580 Kg
Area: 410 m^2

Load in concrete

10 pcs house foundation: Weight= 157,300 Kg
3 pcs walls in concrete: Weight = 62,000 Kg
U-value: about 0.2
Same 410 m^2 area of panels from concrete: Total weight – 219,300 Kg

Weight difference between concrete and sandwich panels: 210,720 Kg = 210.72 ton.
A load in concrete needs 11 more trucks!
The modular manufacture and construction with sandwich panel is particularly suitable to tailor the material design for desirable performance of energy efficient buildings. Fig. 10.37 shows a bank building using well insulated sandwich panels to get the best possible U value on the facade while keeping a very sharp architectual view. The sandwich construction

FIGURE 10.36 Delivery from Novasip.

FIGURE 10.37 Bank building in Copenhagen. Well insulated sandwich panels to get the best possible U value on the facade.

FIGURE 10.38 House in Stockholm. House and garage were built in 2 days.

technology can minimize the field work by integrating it into the modular manufacture. The whole building envelope and floor have been prefabricated with standard quality control, and the building can be rapidly erected and completed. Fig. 10.38 shows that a large house and its garage are built in 2 days at Stockholm with the modular sandwich construction.

FIGURE 10.38, cont'd.

References

Aiche, 2020. https://www.aiche.org/resources/publications/cep/2020/may/catalyzing-commercialization-novel-concrete-wall-panels-lower-buildings-energy-footprint. (Accessed: 2021).

Al Dakheel, J., Del Pero, C., Aste, N., Leonforte, F., 2020. Smart buildings features and key performance indicators: A review. Sustain. Cities Soc. 61, 102328.

ASHRAE, 2009. IAQ guide: best practices for design, construction, and commissioning. American Society for Heating, Refrigeration, Air Conditioning Engineers, Peachtree Corners, Georgia.

Bellini, E., 2020. 25.17% efficient perovskite solar cell via new photoactive layer https://www.pv-magazine.com/2020/11/23/25-17-efficient-perovskite-solar-cell-via-new-photoactive-layer/. (Accessed: 2020).

Berawi, M.A., Miraj, P., Sayuti, M.S., Berawi, A.R.B., 2017, November. Improving building performance using smart building concept: Benefit cost ratio comparison. In AIP Conference Proceedings, Vol. 1903, No. 1, p. 030001. AIP Publishing LLC.

Blackout in cold weather, 2020. https://www.theguardian.com/us-news/2021/feb/17/texas-power-blackout-weather-cold. (Accessed: 2021).

Chen, Z., Jiang, C., Xie, L., 2018. Building occupancy estimation and detection: a review. Energy Build. 169, 260–270.

DallasPower Outage, 2021. https://www.wfaa.com/article/news/local/texas/dallas-texas-electrical-power-outage-ercot-failures/287-50797307-0afe-43eb-8175-b78e7e4fc13a. (Accessed: 2021).

Elkhateeb, A.M., Fikry, M.A., Mansour, A.A., 2018. Dynamic building and its impact on sustainable development. Alexandria Eng. J. 57 (4), 4145–4155.

Fabiani, C., et al., 2019. Assessing the potentiality of animal fat based-bio phase change materials (PCM) for building applications: an innovative multipurpose thermal investigation. Energies 12 (6), 1111.

Falkeis, A., 2015. Thinking out of the urban design toolbox. Arts, Research, Innovation and Society. Springer, Springer Nature, Switzerland, pp. 211–222.

Gallart-Sirvent, P., et al., 2017. Fatty acid eutectic mixtures and derivatives from non-edible animal fat as phase change materials. RSC Adv. 7 (39), 24133–24139.

Geissdoerfer, M., et al., 2017. The circular economy–a new sustainability paradigm? J. Cleaner Prod. 143, 757–768.

GSA. Turning marginal farmlands into a win for farmers and ecosystems, Geological Society of America (GSA). 2018; Available from: https://www.sciencedaily.com/releases/2018/11/181105160905.htm.

Gutierrez, L., Williams, E., 2016. Co-alignment of comfort and energy saving objectives for US office buildings and restaurants. Sustain. Cities Soc. 27, 32–41.

Heder, M.H.L., Sunflowers. http://harriesheder.com/project/sunflowers/. (Accessed: 2021).

Hu, Z., et al., 2017. Design, construction and performance testing of a PV blind-integrated Trombe wall module. Appl. Energy 203, 643–656.

Husain, A.A., et al., 2018. A review of transparent solar photovoltaic technologies. Renew. Sustain. Energy Rev. 94, 779–791.

Kelly, N.A., Gibson, T.L., 2009. Improved photovoltaic energy output for cloudy conditions with a solar tracking system. Sol. Energy 83 (11), 2092–2102.

Kumar, R.R., et al., 2020. Phase change materials and nano-enhanced phase change materials for thermal energy storage in photovoltaic thermal systems: a futuristic approach and its technical challenges. Renew. Sustain. Energy Rev. 133, 110341.

Li, Z., et al., 2016. Correlated perovskites as a new platform for super-broadband-tunable photonics. Adv. Mater. 28 (41), 9117–9125.

Lin, Q., et al., 2020. Design and experiment of a sun-powered smart building envelope with automatic control. Energy Build. 223, 110173.

Lunt, R.R., 2012. Theoretical limits for visibly transparent photovoltaics. Appl. Phys. Lett. 101 (4), 043902.

Lunt, R.R., Bulovic, V., 2011. Transparent, near-infrared organic photovoltaic solar cells for window and energy-scavenging applications. Appl. Phys. Lett. 98 (11), 61.

Mandal, J., et al., 2018. Hierarchically porous polymer coatings for highly efficient passive daytime radiative cooling. Science 362 (6412), 315–319.

Matiko, J.W., et al., 2013. Review of the application of energy harvesting in buildings. Meas. Sci. Technol. 25 (1), 012002.

McLellan, E.L., K.G.C., Eagle, A.J., Woodbury, P.B., Sela, S., Tonitto, C., Marjerison, R.D., van Es, H.M., The nitrogen balancing act: tracking the environmental performance of food production. Bioscience 2018, 68 (3): 194–203.

Mendell, M.J., et al., 2002. Improving the health of workers in indoor environments: priority research needs for a national occupational research agenda. Am. J. Public Health 92 (9), 1430–1440.

NPR, 2021. Biden moves to have U.S. rejoin climate accord. https://www.npr.org/sections/inauguration-day-live-updates/2021/01/20/958923821/biden-moves-to-have-u-s-rejoin-climate-accord. (Accessed: January 20, 2021).

PCM, 2015. PCM products.https://www.pcmproducts.net/files/Custom-made%20solutions-2015-1.pdf. (Accessed: 2021).

Pugsley, A., et al., 2020. BIPV/T facades–a new opportunity for integrated collector-storage solar water heaters? Part 1: state-of-the-art, theory and potential. Sol. Energy 207, 317–335.

Raza, M.Q., Khosravi, A., 2015. A review on artificial intelligence based load demand forecasting techniques for smart grid and buildings. Renew. Sustain. Energy Rev. 50, 1352–1372.

Recycle PV Solar LLC, 2021. Recycle PV Solar LLC.https://recyclepv.solar. (Accessed: 2021).

Research, G.V., 2017. Building-integrated photovoltaics market size, share & trends analysis report by technology (crystalline silicon, thin film), by application (roofs, glass), by end use (industrial, commercial), and segment forecasts, 2021–2028. https://www.grandviewresearch.com/industry-analysis/building-integrated-photovoltaics-bipv-market. (Accessed: 2021)..

Research, Z.M., 2016. Global smart home market size. Industry Analysis & Trends Up to 2022 n.d. https://www.zionmarketresearch.com/report/smart-home-market?utm_source=joseph/FREE/SM/jump.

Roslan, N., et al., 2018. Dye sensitized solar cell (DSSC) greenhouse shading: new insights for solar radiation manipulation. Renew. Sustain. Energy Rev. 92, 171–186.

SEIA, 2021a. SEIA National PV Recycling Program. https://www.seia.org/initiatives/seia-national-pv-recycling-program. (Accessed: 2021).

SEIA, 2019b. Lowering soft costs—a major opportunity for growth. https://www.seia.org/sites/default/files/2019-07/Solar-Soft-Costs-Factsheet.pdf. (Accessed: 2021).

SEIA, 2020. Solar as an economic engine https://www.seia.org/solar-industry-research-data#:~:text=Solar%20as%20an%20Economic%20Engine,investment%20in%20the%20American%20economy. (Accessed: 2020).

Sica, D., et al., 2018. Management of end-of-life photovoltaic panels as a step towards a circular economy. Renew. Sustain. Energy Rev. 82, 2934–2945.

Sutterlin, W.R., 2011. Phase change materials, a brief comparison of ice packs, salts, paraffins, and vegetable-derived phase change materials. Pharm. Outsourc..

Tabadkani, A., et al., 2021. Design approaches and typologies of adaptive facades: a review. Autom. Constr. 121, 103450.

Tsanakas, J.A., et al., 2020. Towards a circular supply chain for PV modules: Review of today's challenges in PV recycling, refurbishment and re-certification. Prog. Photovolt. Res. Appl. 28 (6), 454–464.

Tzortzakis, K., Papafotis, K., Sotiriadis, P.P., 2017, November. Wireless self powered environmental monitoring system for smart cities based on LoRa. In 2017 Panhellenic Conference on Electronics and Telecommunications (PACET), pp. 1–4. IEEE.

Wang, J., et al., 2020. Highly efficient all-inorganic perovskite solar cells with suppressed non-radiative recombination by a Lewis base. Nat. Commun. 11 (1), 1–9.

Wang, L., Lee, E.W., Yuen, R.K., 2018. Novel dynamic forecasting model for building cooling loads combining an artificial neural network and an ensemble approach. Appl. Energy 228, 1740–1753.

Werling, E., Building America Research-to-Market Plan. 2015, Office of Energy Efficiency and Renewable Energy, Washington, DC

Wikipedia, 2021. Texas power crisis. https://en.wikipedia.org/wiki/2021_Texas_power_crisis. (Accessed: 2021).

Yin, H., Pao, F., Zadshir, M., Lou, J., Wei, Z., 2021. Building integrated photovoltaic thermal systems. U.S. Patent Application No. 63/248,502.

Yu, G., et al., 2020. A review of designs and performance of façade-based building integrated photovoltaic-thermal (BIPVT) systems. Appl. Therm. Eng., 116081.

Yu, N., Capasso, F., 2014. Flat optics with designer metasurfaces. Nat. Mater. 13 (2), 139–150.

Index

Page numbers followed by "*f*" and "*t*" indicate, figures and tables respectively.

C

D

E

T

9780128210642